# Sensor Networks: Signals and Communication Technology

# Sensor Networks: Signals and Communication Technology

Editor: Sharon Garner

**NY** RESEARCH PRESS

New York

Published by NY Research Press
118-35 Queens Blvd., Suite 400,
Forest Hills, NY 11375, USA
www.nyresearchpress.com

Sensor Networks: Signals and Communication Technology
Edited by Sharon Garner

International Standard Book Number: 978-1-63238-640-3 (Hardback)

**Cataloging-in-Publication Data**

Sensor networks : signals and communication technology / edited by Sharon Garner
    p. cm.
Includes bibliographical references and index.
ISBN 978-1-63238-640-3
1. Sensor networks. 2. Wireless sensor networks. 3. Signal processing.
4. Communication and technology. I. Garner, Sharon.
TK7872.D48 S46 2019
681.2--dc23

# Contents

**Permissions**


**List of Contributors**


**Index**

# Preface

This book was inspired by the evolution of our times; to answer the curiosity of inquisitive minds. Many developments have occurred across the globe in the recent past which has transformed the progress in the field.

Sensor networks are wireless networks or systems of spatially scattered sensors that are used to record physical conditions or surroundings. They collate and organize data at a centralised location. Today, sensors are used across industries and fields for various purposes such as forest fire detection, machine health monitoring, air pollution monitoring, etc. The ever growing need of advanced technology is the reason that has fueled the research in the field of sensor networks in recent times. This book is a compilation of chapters that discuss the most vital concepts and emerging trends in this area. Some of the diverse topics covered herein have been contributed by international experts to address the varied branches that fall under this category. This book will prove to be immensely beneficial to students and researchers in this field.

This book was developed from a mere concept to drafts to chapters and finally compiled together as a complete text to benefit the readers across all nations. To ensure the quality of the content we instilled two significant steps in our procedure. The first was to appoint an editorial team that would verify the data and statistics provided in the book and also select the most appropriate and valuable contributions from the plentiful contributions we received from authors worldwide. The next step was to appoint an expert of the topic as the Editor-in-Chief, who would head the project and finally make the necessary amendments and modifications to make the text reader-friendly. I was then commissioned to examine all the material to present the topics in the most comprehensible and productive format.

I would like to take this opportunity to thank all the contributing authors who were supportive enough to contribute their time and knowledge to this project. I also wish to convey my regards to my family who have been extremely supportive during the entire project.

Editor

# Collaborative Signal and Information Processing for Target Detection with Heterogeneous Sensor Networks

**Minghui Li[1]\*, Yilong Lu[2] and Bo He[3]**

[1]School of Engineering, University of Glasgow, Glasgow, United Kingdom
[2]School of Electrical and Electronic Engineering, Nanyang Technological University, Singapore
[3]School of Information Science and Engineering, Ocean University of China, Qingdao, China

## Abstract

In this paper, an approach for target detection and acquisition with heterogeneous sensor networks through strategic resource allocation and coordination is presented. Based on sensor management and collaborative signal and information processing, low-capacity low-cost sensors are strategically deployed to guide and cue scarce high-performance sensors in the network to improve the data quality, with which the mission is eventually completed more efficiently with lower cost. We focus on the problem of designing such a network system in which issues of resource selection and allocation, system behavior and capacity, target behavior and patterns, the environment, and multiple constraints such as the cost must be addressed simultaneously. Simulation results offer significant insight into sensor selection and network operation, and demonstrate the great benefits introduced by guided search in an application of hunting down and capturing hostile vehicles on the battlefield.

**Keywords:** Sensor management; Collaborative signal and information processing; Heterogeneous sensor networks; Array signal processing

## Introduction

With the advancements in technologies, it is becoming increasingly feasible to conceive and deploy large-scale sensor networks for a wide variety of applications such as elderly assistance, traffic control, homeland security, military surveillance, and environmental monitoring [1]. Significant work has gone into the development of algorithms to perform a variety of tasks including detection, localization, classification, identification and tracking of one or more targets in the sensor field, and numerous approaches based on Collaborative Signal and Information Processing (CSIP) have been proposed in the literature, see, for example [2,3]. In the scenarios involving multiple targets, data association of measurements from multiple sensors is known to be NP-hard. Multiple hypothesis tracking (MHT) [4] and Markov chain Monte Carlo data association (MCMCDA) [5] are possible solutions to this tough problem at the cost of long latency and extensive computation.

Processing the signal from distributed but networked sensors collaboratively can be treated as an extension of multi-channel array signal processing, where the multiple sensors are collocated with a typical spacing of half a wavelength of the impinging waves and the centralized algorithms based on the relative phases and amplitudes of the wave across the sensors are utilized to obtain the optimal solutions. Array signal processing is essential in most sensor array based systems and has made significant success in a wide variety of applications including modern radar [6-12], underwater sonar [13-15], wireless communications [16-21], intelligent transportation systems [22-24], non-destructive evaluation [25-27], and ultrasound imaging [28]. However, the applications of these algorithms to wireless sensor networks are not straightforward, mainly due to the constraints intrinsic to most of the sensor nodes on energy, communications, computation and size, and the distributed, simple and efficient alternatives that can deliver sub-optimal solutions while meeting the practical constraints have received considerable attention.

We address the problem of target detection and acquisition with a heterogeneous sensor network in this paper. We take as a sample task the problem of hunting down and capturing hostile vehicles on the battlefield, with a sensor network consisting of Unattended Ground Sensors (UGS) and Unmanned Aerial Vehicles (UAV). A UGS may be one sensor node equipped with collocated multi-modality sensors or a small cluster of distributed sensors, but assume to be simple in function, cheap, static but rapidly deployable. The location of each UGS is assumed to be known in advance obtained from on-board GPS receivers or a self-localization algorithm. A UAV is very expensive, mobile and with high performance, equipped with cameras and the necessary image recognition software, and would be able to identify and localize the potential targets with high accuracy. The goal of the whole network system is to distinguish the hostile vehicles from the civilian vehicles, and then to locate and track the hostile vehicles, and eventually capture or destroy the hostile vehicles. There are many real-world applications with parallel environment and requirements such as search and rescue options, surveillance, tracking of moving parts, and search and capture missions. In some cases, the potential targets are intelligent and actively avoiding being captured as in the application addressed in this paper, whereas in other cases their motion is approximately random as in rescue options.

The problem is to study how to guide the UAVs effectively with a network of UGS nodes, and then capture the targets efficiently and accurately. A key element of this problem is the assumption that resources are limited, for example, constrained by the budget, and the choices are driven by trade-offs between the performance and the cost. The capability of the above-mentioned network to perform its tasks depends on several factors, including the amount of available UGS/

---

**\*Corresponding author:** Minghui Li, School of Engineering, University of Glasgow, Glasgow, United Kingdom, E-mail: minghui.li@ieee.org

UAV sensors, the sensory range and accuracy, the sensor deployment strategy and lifetime, environment characteristics, target behaviors, and etc. The question that we are investigating is that: with a finite budget, what is the best way to combine and allocate resources so as to maximize the performance level of the sensor network in terms of, for example, the target destruction rate and time?

In this paper, we propose an efficient algorithm to tackle this problem and show that it is effective in reducing the capturing time via simulations. With the data provided by the UGS network, the target vehicles' moving patterns are identified and they are further classified into wheeled and tracked categories of vehicles. The locations of the potential hostile vehicles are determined, predicted and tracked using different vehicle moving models, and the most efficient strategy for the deployment of the UAV fleet or other UGS clusters is decided, and eventually the UAV will confirm the targets and capture them. The rationale of our approach is that the low-cost low-capability sensors are strategically employed to cue and guide the high-performance but scare sensors to perform the task, through sensor management and strategic resource allocation and coordination placing the right sensors at the right locations on the right time.

The paper is organized as follows. Section 2 describes and formulates the problem at hand, and models this high dimensional design problem with a collection of autonomous decision-making entities, or agents. Section 3 discusses in details the collaborative signal and information processing algorithms for strategic resource allocation and coordination. Simulation results are given in Section 4 to evaluate the algorithms and demonstrate the benefits. The paper is then concluded.

## Problem Formulation and Modelling

### Problem description and formulation

To illustrate target detection and identification in complex environments with the aid of strategic resource allocation and coordination in heterogeneous sensor networks, we consider a specific problem where a sensor network consisting of UGS and UAVs is deployed on the battlefield, for the purpose of acquiring and eventually capturing the hostile vehicle targets. The vehicles are assumed to consist of three categories: hostile tracked vehicles, hostile wheeled vehicles, and civilian wheeled vehicles. In this scenario, the objective is to design rules for coordination and information sharing between UGS and UAVs to reduce the time required for clearing all the hostile vehicles with the constraint of total cost.

The system under consideration consists of three major components: a complex environment, vehicles, and sensors. The environment is a simplified two-dimensional urban grid of a pre-set, configurable size. The map is generated using a stochastic algorithm that ensures the resulting map is randomly created, but still follows a consistent pattern that resembles the layout of a city to a degree. Some buildings and forests are added into the map to complicate the terrain, which have impact on the sensors' accuracy such as the probability of detection and the probability of false alarm. Potential targets are road-bounded including hostile vehicles (either tracked or wheeled) and civilian vehicles (wheeled only). The vehicles traverse the environment by selecting starting points and destinations at random. They plan their routes by following the shortest path to their destinations, and are assumed to be confined in the environment if they are alive. The sensors consist of UGS and UAVs, which are imperfect, having a probability of reporting that a target is present when it isn't, as well as reporting that it

is not present when it is. It is assumed that the local data fusion within a UGS node is available, and no fusion within neighboring nodes is provided locally, but All UGS can transmit the detections back to the base station in the command and control (C2) centre. UGS has the ability to detect a vehicle in the vicinity, and classify it into tracked or wheeled vehicles using for example the spectrum based method [3], and cannot accurately localize or track the target. Each UAV is equipped with cameras, the necessary image recognition facility, and the laser range finder, and would be able to identify and localize the targets if they are within the range and field of view (FOV) of the camera. Each UAV maintains a direct communication link to the base station.

### Agent-Based Modelling (ABM)

We use a modeling approach known as agent-based modeling [29] to study and simulate the high dimensional design space problem at hand. In this approach, systems are represented as collections of autonomous decision-making entities, called agents. Each agent individually assesses its situation and makes decisions based upon a set of behavioral rules. At the simplest level, an agent-based model consists of a system of agents and the relationships between them. Even a simple agent-based model can exhibit complex behavioral patterns and provide valuable insight about the dynamics of the real-world applications that it emulates.

The benefits of ABM over other modelling techniques can be captured in three statements [30]: 1) ABM captures emergent phenomena; 2) ABM provides a natural description of the complex systems; 3) ABM is flexible. The ability of ABM to deal with emergent phenomena is the main driving force behind its success as a complex adaptive system modelling tool. The fact that it provides a natural framework to describe complex systems combined with its flexibility makes ABM the tool of choice for the problem at hand.

The parameters modelling the environment, sensors and vehicles, and the relationship and interaction between them are shown in Figure 1. The environment has impact on the vehicle behaviors, detection

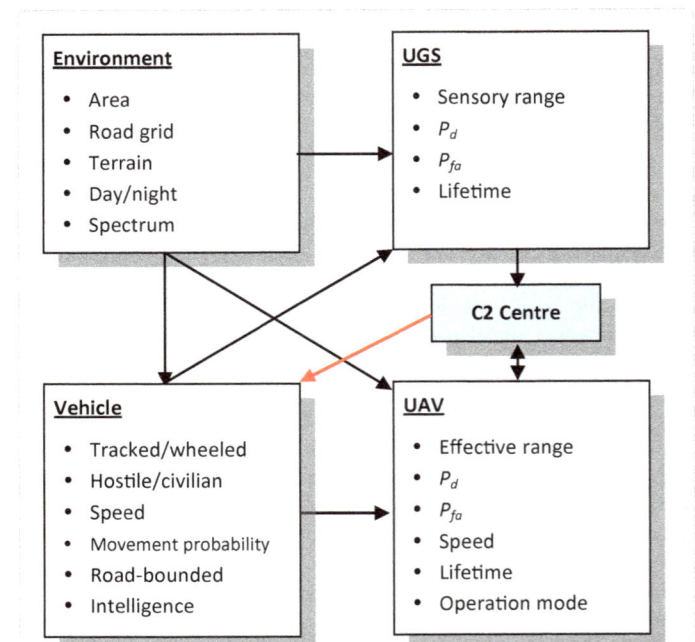

Figure 1: Agent-based system modelling.

probability $P_d$ and false alarm probability $P_{fa}$ of UGS and UAVs due to the terrain and day/night variation. The C2 centre collects the rough target information from UGS, makes optimal task allocation and path scheduling for UAVs, and deploys munitions to destroy targets that are confirmed and localized by UAVs. Both UGS and UAVs have limited lifetime due to battery and petrol consumption, respectively.

## Solution Approach

The algorithm for strategic resource allocation and coordination is shown below. Our approach specifies relatively simple local behavior to be adopted by each UGS and UAV, and the desirable global behavior arises from the interactions between them. N UGS are initially deployed at the junctions of roads to enhance the coverage efficiency because all the vehicles are assumed to be road bounded. If the vehicle behavior is known a priori, for example, the hostile vehicles are most likely coming from certain directions along certain roads, UGS can be placed strategically in those areas; otherwise, they are deployed randomly. At the beginning, M UAVs make autonomous search, where the term "autonomous" is compared to guided search cued by UGS. The UAVs may work in two ways: dividing the whole area into M zones and dispatching one UAV to each zone, or cooperatively searching for targets based on swarm intelligence [31]. Each UAV maintains a direct connection to the base station at the control centre, and all the UGS are networked and the detection information is sent to the base station either directly via the communication links or through multiple hops. The control centre maintains a two-dimensional map of the environment and the roads.

As illustrated in Figure 2, each UGS obtains its own sensor measurements, and makes a decision via the on-board data fusion. The results are contained in three categories: if no vehicles are detected in the vicinity, the result is 0; if tracked vehicles are detected, the result is 1; if only wheeled vehicles are detected, the result is 2. Assume UGS'

### The algorithm

Initial placement of N UGS
Autonomous search by M UAVs
**repeat**
    **for** each UGS **do**
        Get own sensor measurements
        Data fusion for target detection and classification
        Send the decisions to C2 centre
    **end for**
    C2 centre: task allocation and path scheduling for UAVs
             performing lower-priority tasks
    **for** each UAV **do**
        Receive commands from C2 center
        Search the vicinity of destination UGS or make
           autonomous search
        **if** any targets are confirmed
           Localize the target
           Track the target while C2 center prepares
             munitions to destroy it
           Perform battle damage assessment (BDA)
        **end if**
    **end for**
**until** termination of operation

**Figure 2:** Algorithm for UGS and UAV coordination.

function to be simple: it cannot count, accurately locate, or track the targets. This assumption helps to alleviate UGS' cost requirement. The results from the UGS are reported back to the control centre for data fusion which forms a preliminary search of the field. A probability method is then applied to the initial decisions to predict the vehicle locations to overcome the uncertainty caused by the delayed message, movement of the vehicles, and the inaccurate local information. A major challenge is how to associate the UGS detections with the correct targets, a method based on the principle of machine learning [32] is utilized in this work. After predicting the vehicle locations, the control centre designs the capture strategies. There are three options: 1) plan the most efficient pursuit trajectories of the UAV fleet based on the swarm intelligence; 2) rapidly deploy another set of UGS (with artillery or by operators) in the direction of the vehicle movement to track the vehicles; and 3) combined actions of decisions 1) and 2).

Assume $n_t$ UGS report tracked detection 1, $n_w$ UGS report wheeled detection 2, and the other $N-n_t-n_w$ UGS report null detection 0. If, $n_t+n_w \geq M$ all M UAVs are involved in the task allocation and coordination; otherwise only $n_t+n_w$ UAVs are involved and the remaining UAVs make autonomous search. Strategic deployment of multiple UAVs means task allocation and path planning for some destinations that must be visited and some risk sites that must be avoided. The flying time is a criterion used to measure the effectiveness of a schedule. A good schedule can guide the UAVs to all destination UGS as quickly as possible, before the targets move far away. When only one UAV is considered, the problem is similar to traveling salesman problem (TSP). In the scenarios involving multiple UAVs, the problem resembles Job-Shop Scheduling Problem (JSSP) [33]. Both TSP and JSSP are NP-hard and intensively studied in literature. In our system, the Particle Swarm Optimization (PSO) algorithm [34] is used to allocate tasks and design paths for each UAV. PSO is a recent addition to evolutionary algorithms first introduced by Eberhart and Kennedy in 1995. The foundation of PSO is based on the hypothesis that social sharing of information among conspecifics offers an evolutionary advantage. Partially inspired by animal social behaviors such as flocking of birds, PSO originally intends to graphically mimic the graceful way in which they find their food sources and save themselves from predators. It should be noted that if $n_t+n_w >> M$, it is not necessary to devise a complete schedule to visit all $n_t+n_w$ UGS, since after a long time, the targets have moved far away. The number of destination UGS considered in our algorithm is determined by the amount of UAVs, the sensory ranges of UGS and UAVs, the speed of the target vehicles and UAVs, target behavior, and etc. Another rule that should be pointed out is that, the task of searching for tracked vehicles has higher priority than that for wheeled vehicles, which has higher priority than autonomous search. Based on this rule, UAVs can be interrupted from current task and reallocated to perform higher priority tasks.

The algorithm will not terminate automatically. Even if all hostile vehicles have been destroyed (but this information has not been confirmed and notified to the control centre), civilian vehicles continue triggering UGS' detections and UAVs' guided search.

## Simulation Results

The design space for the problem of resource allocation and behavioral coordination is complicated and high-dimensional, involving the amount of sensors, detection range and accuracy of sensors, coordination rules between sensors, in addition to the environment characteristics and target behaviors. In this section, we demonstrate 1) the effectiveness of agent-based modelling to this type of problem, 2) sensor selection and coordination strategies at the constraint of limited budget, and 3) benefits introduced by strategic sensor management and coordination. For the sake of simplicity, only two key dimensions, the amount and capability of sensors are explored, but we stress that other dimensions can be similarly treated.

We consider an urban area of 40km by 40km including 2500 uniformly distributed junctions. The target vehicles consist of 40 hostile tracked vehicles, 40 hostile wheeled vehicles and 400 civilian wheeled vehicles. All the vehicles are road bounded with a movement probability of 50%, and at a speed between 40km/h and 80km/h. UAV has two operation modes: cruising mode at a speed of 200km/h, and searching mode at a speed of 100km/h. The results are calculated based on an average over 30 runs.

In the first experiment, the amount of required UGS and UAVs is studied. Assume UGS' sensory range r=0.5km, the probability of detection $P_d$=0.8, and the probability of false alarm $P_{fa}$=0.15; UAV's sensory range $r$=3km, $P_d$=0.8, $P_{fa}$=0.05. Figure 3 shows the effectiveness, in terms of the half target destruction time, versus the number of UAVs, with certain amount of UGS. It is evident that, in all cases we have a consistent reduction in the target destruction time with the increase of UAVs, which corresponds to a higher cost. It is interesting to note that when the number of UAVs is greater than 5, too many UGS (for example, when the number of UGS increases from 40 to 100) may cause drawbacks, since $P_{fa}$ is virtually increased due to large civilian vehicle detections which then triggers more vain UAV search. Another interesting point is the nonlinear change nature of the curves, which is common in the complex systems. For example, we can tell that when the UGS number is 40, increasing UAVs from 3 to 4 has a much greater impact than, say increasing UAVs from 9 to 10. The ability to find "tipping points" is a key driver of success in the analysis of complex systems of this type.

In the second experiment, the impact of UGS' detection range $r$, $P_d$ and $P_{fa}$ on the system effectiveness is studied. Figure 4 shows the half target destruction time versus r with certain $P_d$ value and $P_{fa}$=0.15. Although the curves are not strictly monotonic due to the average over small number of trials, the trends and impact of these parameters on the target destruction time are clearly demonstrated. In general,

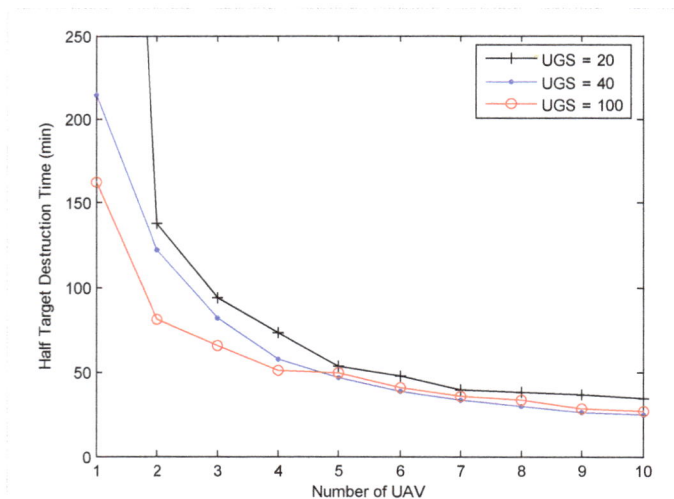

**Figure 4**: Impact of sensory range and accuracy on system performance.

larger detection range and higher $P_d$ of UGS, which corresponds to a higher cost, lead to faster target destruction. Figure 5 shows the target destruction time versus $P_d$ with certain $P_{fa}$ and $r$=1km. It is interesting to note that when $P_d$ is high (for example, above 0.8), the system is more sensitive to $P_{fa}$ by showing large fluctuations in the curves with $P_d$ changing; however, when $P_d$ is moderate (for example, between 0.6 and 0.8), the performance is more dominated by $P_d$, and lower $P_{fa}$ causes more fluctuations as $P_d$ varies. This phenomenon is of great practical interests to the system designers. Again, tipping points are identified in the system.

In the third experiment, the proposed method of UGS-guided search is compared with UAV's autonomous search. The UGS settings are $r$=1, $P_d$=0.8, $P_{fa}$=0.2. As shown in Figure 6, the solution involving 4 UAVs and 40 UGS is much more effective in terms of the half target destruction time, especially when the number of remaining targets is moderate or small. UAV is expensive high-end equipment, so the solution of 4 UAVs and a set of UGS is much cheaper than that involving 8 UAVs. For guided search, tracked vehicles are diminished faster than wheeled vehicles, since higher priority attention is given to tracked detections in our system based on our rules, and due to the assumption that all civilian vehicles are wheeled vehicles. For autonomous search, two types of vehicles are destroyed with a similar rate.

## Conclusions and Future Work

In this paper, an approach for target detection and acquisition with a heterogeneous sensor network through sensor management and strategic resource allocation and coordination is presented. We focus on the problem of designing a network system in which resource constraints (such as the limited budget) and system behavior and performance must be addressed simultaneously. Although our work is only embryonic in nature, the results already offer significant insight into the resource selection and network operation, and demonstrate the great benefits in a sample battlefield application.

As future work, we will explore the applications of evolutionary algorithms as techniques to search the high dimensional problem space for an optimal solution, in terms of the amount of each type of sensors, the sensor range and accuracy, the inter-node data fusion algorithms,

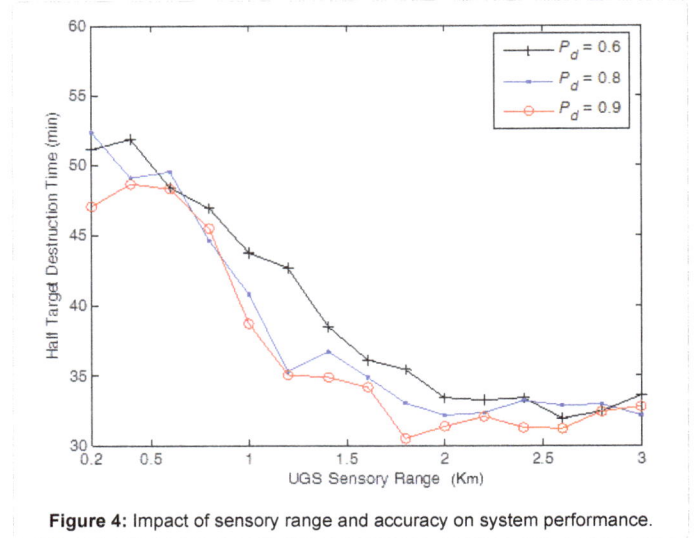

**Figure 3**: Impact of sensor number on system performance.

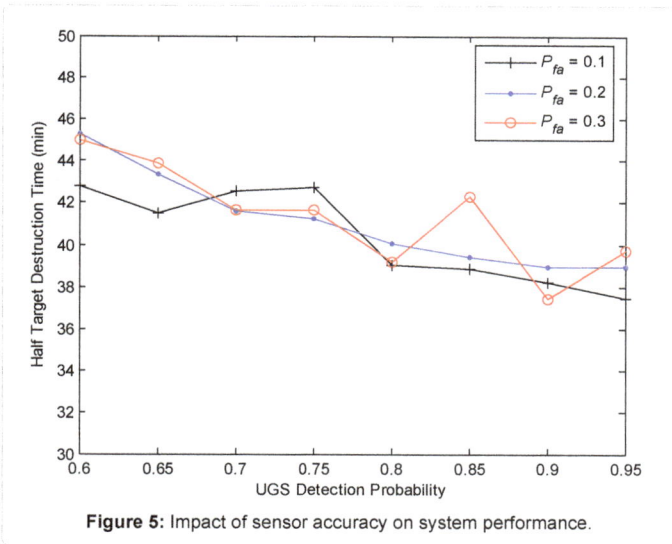

**Figure 5:** Impact of sensor accuracy on system performance.

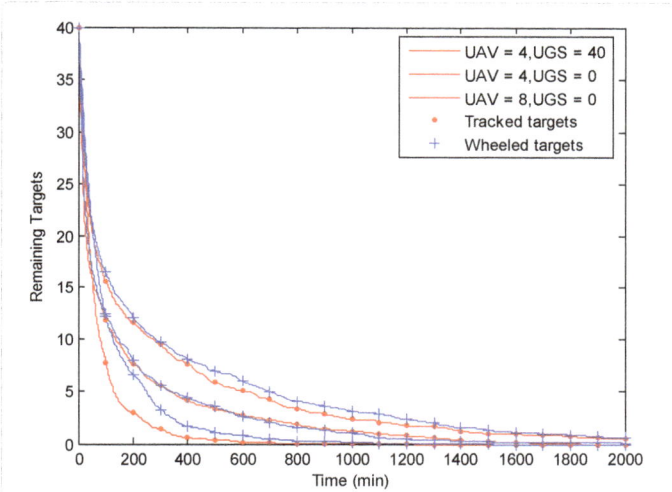

**Figure 6:** Comparison between UGS-guided search and autonomous search.

network level resource allocation rules, with multiple constraints such as budget, communication range and link failure, and etc.

### References

1. Culler D, Estrin D, Srivastava M (2004) Overview of sensor networks. IEEE Computer 37: 41-49.

2. Sheng X, Hu Y (2005) Maximum likelihood multiple-source localization using acoustic energy measurements with wireless sensor networks. IEEE Transactions on Signal Processing 53: 44-53.

3. Li D, Wong K, Hu Y, Sayeed AM (2002) Detection classification and tracking of targets. IEEE Signal Processing Magazine 19: 17-29.

4. Reid DB (1979) An algorithm for tracking multiple targets. IEEE Transactions on Automatic Control 24: 843-854.

5. Oh S, Russell S, Sastry S (2004) Markov chain Monte Carlo data association for general multiple-target tracking problems. Proceedings of the 43rd IEEE Conference on Decision and Control Bahamas.

6. Li MH, Lu YL (2004) Improving the performance of GA-ML DOA estimator with a resampling scheme. Signal Processing 84: 1813-1822.

7. Li MH, Lu YL (2006) Dimension reduction for array processing with robust interference cancellation. IEEE Transactions on Aerospace and Electronic Systems 42: 103-112.

8. Li MH, Lu YL (2005) Null-steering beam space transformation design for robust data reduction. Proceedings of 13th European Signal Processing Conference.

9. Li MH, Lu YL (2009) Source bearing and steering-vector estimation using partially calibrated arrays. IEEE Transactions on Aerospace and Electronic Systems 45: 1361-1372.

10. Li MH, Lu YL (2008) Maximum likelihood DOA estimation in unknown colored noise fields. IEEE Transactions on Aerospace and Electronic Systems 44: 1079-1090.

11. Li MH, Lu YL (2002) Genetic algorithm based maximum likelihood DOA estimation. Proceedings of Radar Conference 502-506.

12. Li K, Lu YL, Li MH (2005) Approximate formulas for lateral electromagnetic pulses from a horizontal electric dipole on the surface of one-dimensionally anisotropic medium. IEEE Transactions on Antennas and Propagation 53: 933-937.

13. He B, Liang Y, Feng X, Nian R, Yan T, et al (2012) AUV SLAM and experiments using a mechanical scanning forward-looking sonar. Sensors 12: 9386-9410.

14. Li MH, McGuire M, Ho KS, Hayward G (2010) Array element failure correction for robust ultrasound beamforming and imaging. Proceedings of 2010 IEEE International Ultrasonics Symposium.

15. Li MH, Ho KS, Hayward G (2009) Beamspace transformation for data reduction using genetic algorithms. Proceedings of IEEE International Ultrasonics Symposium.

16. Li MH, Lu YL (2007) A refined genetic algorithm for accurate and reliable DOA estimation with a sensor array. Wireless Personal Communications 43: 533-547.

17. Li MH, Lu YL (2008) Angle-of-arrival estimation for localization and communication in wireless networks. Proceedings of 16th European Signal Processing Conference.

18. Li MH, Lu YL, Chen HH, Wang B, Chen IM (2009) Angle of arrival (AOA) estimation in wireless networks. Invited chapter Wireless networks: Research technology and applications Nova Science Publishers Inc. New York. 135-164.

19. Wang B, Li MH, Lim HB, Ma D, Fu C (2009) Energy efficient information processing in wireless sensor networks. Guide to Wireless Sensor Networks London Springer. Chapter 1: 1-26.

20. Li MH, Lu YL, Wee L (2006) Target detection and identification with a heterogeneous sensor network by strategic resource allocation and coordination. Proceedings of 6th International Conference on Intelligent Transportation Systems Telecommunications 992-995.

21. Li MH, Ho KS, Hayward G (2010) Accurate angle-of-arrival measurement using particle swarm optimization. Wireless Sensor Network 2: 358-364.

22. Li MH, Wang B, Lu YL, Zhou MT, Chen IM (2010) Smart antenna in intelligent transportation systems. Wireless Technologies in Intelligent Transportation Systems Nova Science Publishers Inc. New York. Chapter 3: 51-84.

23. Li MH, Lu YL (2007) Optimal direction finding in unknown noise environments using antenna arrays in wireless sensor networks. Proceedings of 7th International Conference on Intelligent Transportation Systems Telecommunications 332-337.

24. Li MH, Lu YL (2007) Maximum likelihood processing for arrays with partially unknown sensor gains and phases. Proceedings of 7th International Conference on Intelligent Transportation Systems Telecommunications 185-190.

25. Gongzhang R, Li MH, Lardner T, Gachagan A, Hayward G (2012) Robust defect detection in ultrasonic nondestructive evaluation (NDE) of difficult materials. Proceedings of 2012 IEEE International Ultrasonics Symposium.

26. Lardner T, Li MH, Gongzhang R, Gachagan A (2012) A new speckle noise suppression technique using cross-correlation of array sub-apertures in ultrasonic NDE of coarse grain materials. Proceedings of Review of Progress in Quantitative Nondestructive Evaluation (QNDE).

27. Li MH, Hayward G, He B (2011) Adaptive array processing for ultrasonic non-destructive evaluation. Proceedings of 2011 IEEE International Ultrasonics Symposium.

28. Li MH, Hayward G (2012) Ultrasound nondestructive evaluation (NDE) imaging with transducer arrays and adaptive processing. Sensors 12: 42-54.

29. Bonabeau E (2002) Agent-based modeling: Methods and techniques for simulating human systems. Proceedings National Academy of Science 99: 7280-7287.

30. Parunak HV, Savit R, Riolo RL (1998) Agent-Based Modeling vs. Equation-Based Modeling: A case study and users' guide. Proceedings of Multi-agent systems and Agent-based Simulation (MABS'98): 10-25.

31. Kadrovach BA, Lamont GB (2002) A particle swarm model for swarm-based networked sensor systems. Proceedings of ACM symposium on Applied computing: 918-924.

32. Vercauteren T, Guo D, Wang X (2004) Joint multiple target tracking and classification in collaborative sensor networks. Proceedings of IEEE International Symposium on Information Theory (ISIT).

33. Fang HL, Ross P, Corne D (1993) A promising genetic algorithm approach to job-shop scheduling rescheduling and open-shop scheduling problems. Proceedings of the 5th International Conference on Genetic Algorithms: 375-382.

34. Eberhart RC, Kennedy J (1995) A new optimizer using particle swarm theory. Proceedings of the 6th Symposium on Micro Machine and Human Science, Nagoya, Japan:39-43.

# Impact of Frame Loss Aspects of Mobile Phone Networks on Forensic Voice Comparison

**Balamurali BT Nair[1,2]\*, Esam AS Alzqhoul[1,2] and Bernard J Guillemin[1,2]**

[1]Forensic and Biometrics Research Group (FaB), The University of Auckland, Auckland, New Zealand
[2]Department of Electrical and Computer Engineering, The University of Auckland, Auckland, New Zealand

## Abstract

The analysis of mobile phone speech recordings can play an important role in criminal trials. However it may be erroneously assumed that all mobile phone technologies, such as the Global System for Mobile Communications (GSM) and Code Division Multiple Access (CDMA), are similar in their potential impact on the speech signal. In fact these technologies differ significantly in their design and internal operation. This study investigates the impact of an important aspect of these networks, namely Frame Loss (FL), on the results of a forensic voice comparison undertaken using a Bayesian likelihood ratio framework. For both networks, whenever a frame is lost or irrecoverably corrupted, it is synthetically replaced at the receiving end using a history of past good speech frames. Sophisticated mechanisms have been put in place to minimize any resulting artefacts in the recovered speech. In terms of accuracy, FL with GSM-coded speech is shown to worsen same-speaker comparisons, but improve different-speaker comparisons. In terms of precision, FL negatively impacts both sets of comparisons. With CDMA-coded speech, FL is shown to negatively impact the accuracy of both same- and different-speaker comparisons. However, surprisingly, FL is shown to improve the precision of both sets.

**Keywords:** GSM; CDMA; Forensic voice comparison; Likelihood ratio; Frame loss; Frame error rate

**Abbreviations:** AMR: Adaptive Multi Rate; APE: Applied Probability of Error; BN: Background Noise at the Transmitting End; CDMA: Code Division Multiple Access; CELP: Code Excited Linear Prediction; CI: Credible Interval; $C_{llr}$: Log-Likelihood-Ratio Cost DRC: Dynamic Rate Coding; EVRC: Enhanced Variable Rate Codec; FER: Frame Error Rate; FL: Frame Loss Mechanism; FVC: Forensic Voice Comparison; GMM-UBM: Gaussian Mixture Model-Universal Background Model; GSM: Global System for Mobile Communications; LR: Likelihood Ratio; LLR: Log-Likelihood Ratio; MFCCs: Mel-Frequency Cepstral Coefficients; MOS: Mean Opinion Score; MVKD: Multivariate Kernel Density; OP: Anchor Operating Point; PCA: Principal Component Analysis; PCAKLR: Principle Component Analysis Kernel Likelihood Ratio; PESQ: Perceptual Evaluation of Speech Quality; PPP: Pitch Period Prototype

## Introduction

Mobile phone recordings are often used as evidence in courts of law. Analysis of such recordings using a range of forensic voice comparison (FVC) techniques can assist the court in establishing the guilt or innocence of a suspect. Forensic speech scientists when undertaking such analysis may erroneously assume that all mobile phone networks impact the speech signal in a similar manner. The most widely used mobile phone technologies in use today are Global System for Mobile Communications (GSM) and Code Division Multiple Access (CDMA). There are three key aspects of these networks which can directly impact the speech signal and thus the outcome of a FVC analysis: (i) dynamic rate coding (DRC), (ii) strategies for handling lost or corrupted frames (FL), and (iii) strategies for overcoming the effects of background noise at the transmitting end (BN). In [1] we examined the 1st of these. This paper directly follows on from that work and examines the 2nd factor, the impact of FL in these two networks.

In mobile phone networks speech is coded into 20 ms frames. The wireless channel associated with these networks can often be quite poor, necessitating the need for innovative techniques to try and ensure reliable transmission. Notwithstanding this, the following could happen to a transmitted frame: (i) it is lost, (ii) it is received, but in a corrupted state, or (iii) it is received without error. In the case of a corrupted frame, techniques such as convolutional coding [2,3] are used to try and correct for errors. If correction is not possible, the FL mechanism is initiated. For both networks this broadly involves replacing lost speech data with speech data from the past.

Much of the experimental methodology of this study is the same as that of our previous DRC study and the reader is referred to that paper for an in depth explanation and justification of our approach [1]. We again use the Bayesian likelihood ratio (LR) framework for the evaluation of speech forensic evidence. A number of methods have been proposed for evaluating speech evidence in the FVC arena, such as Gaussian mixture model universal background model (GMM-UBM) [4,5], multivariate kernel density (MVKD) [4,6] and principal component analysis kernel likelihood ratio (PCAKLR) [7]. Each of these computes a LR, which is a ratio of probabilities. The numerator of the LR is the probability of the evidence given the prosecution hypothesis; the denominator is the probability of the evidence given the defence hypothesis. GMM-UBM has been primarily designed for data-stream-based analysis scenarios, whereas MVKD and PCAKLR are primarily designed for token-based analysis scenarios [8]. The difference between MVKD and PCAKLR is principally in respect to the number of parameters that can be handled, this being 3-4 in the case of MVKD [6], and much larger than this in the case of PCAKLR [9]. Given

**\*Corresponding author:** Balamurali BT Nair, Forensic and Biometrics Research Group (FaB), The University of Auckland, Auckland, New Zealand
E-mail: bbah005@aucklanduni.ac.nz

that, as for our previous study, we use vowel tokens for the experiments reported here, these being represented by 23 Mel-Frequency Cepstral Coefficients (MFCCs), PCAKLR has been chosen for computing LRs.

To quantify the performance of a FVC experiment, we use here the same tools used in our DRC study, namely log-likelihood-ratio cost $(C_{llr})$, Tippett plots, applied probability of error (APE) plots, and credible interval (CI). The reader is again referred to our earlier paper for more details on these [1].

The remainder of this paper is structured as follows. The FL mechanisms for both the GSM and CDMA networks are discussed first in great detail, followed by our experimental methodology to study the impact of these on FVC. We then present our results and conclusions.

## FL Mechanisms in the GSM and CDMA Networks

With both the GSM and CDMA networks, lost or irrecoverably corrupted frames are replaced with synthetically generated frames using speech data derived from the past, a process which is implemented by the decoding section of the speech codec used in the network. The most widely used speech codecs in the GSM and CDMA networks are the adaptive multi rate (AMR) codec and enhanced variable rate codec (EVRC), respectively. For both networks, if successive frames are lost, the codec will continue replacing those, while at the same time gradually decreasing the output level until silence results, a process called muting [10]. A maximum of 16 successive frames (i.e., 320 ms) could be replaced in this manner before silence results [11,12].

From the perspective of a FVC, the automatic replacement of lost frames with synthetically generated frames is clearly of concern, unless their occurrence can be detected *a priory* and the synthetically generated sections excluded from an analysis. But the codecs have been designed with speech quality in mind and sophisticated strategies have been incorporated, such as smoothing out any abrupt amplitude transitions from one speech frame to another, to minimize or even eliminate any resulting perceptual artefacts. So effective are these strategies that subsequent detection of the FL process from the received speech signal is likely to be very difficult, if not impossible.

We believe it is important for forensic speech scientists to clearly appreciate that with mobile phone speech much of the decoded speech waveform could be artificially generated, and that this must necessarily impact upon the confidence they ascribe to any of their analysis findings. With the intention of convincingly making this point, the following discussion is deliberately detailed. However in reality it is not the specifics of the process that the forensic scientist needs to understand, but rather they need to have an overall appreciation of how much of the speech waveform, and in what respects, it might have been changed during transmission.

## GSM FL mechanism

The AMR codec processes speech frames using code excited linear prediction (CELP) into one of eight source coding bit rates: 4.75, 5.15, 5.90, 6.70, 7.40, 7.95, 10.20 and 12.20 kbps [13,14]. This multi-bit-rate capability is designed to allow the GSM network to use available transmission bandwidth as efficiently as possible in response to changing channel conditions [15].

The AMR FL mechanism is quite sophisticated [16]. The example of Figure 1 is intended to illustrate some of the key features of this process. Figure 1a shows a sequence of seven received speech frames. Four of these have been received without error and are therefore 'Good' (labelled with a superscript G), while the remaining three,

having been identified as containing irrecoverable errors, are therefore 'Bad' (labelled with a superscript B in the figure). Figure 1b shows the resulting speech frames that would be used to generate the decoded speech waveform. With this example, in order to convey the broad aspects of a process which in reality is quite complicated, we draw a distinction between data in a frame that could be classified as speech data (i.e., spectral shaping, voiced/voiceless classification, pitch, etc.) and data related to amplitude. We first consider how speech data gets impacted, then amplitude data.

The first two received frames, Frames 1 and 2, being 'Good', remain unchanged. The speech data of Frame 3, being 'Bad', is thrown away and replaced by speech data derived from the last 'Good' frame, namely Frame 2. The result is an artificially generated frame to replace the 'Bad' Frame 3. There is also an amplitude adjustment process associated with the generation of such frames, namely a gain reduction, as will be described below. In Figure 1b this new Frame 3 is labelled $3^R(2/)$, where the superscript R indicates a replaced frame, '2' indicates that its speech data has been derived from Frame 2, and '/' indicates that an amplitude adjustment has been applied. A similar process happens for the 'Bad' Frame 4, its speech data being derived from the synthetically-generated Frame 3, but with a further level of amplitude reduction. Thus in Figure 1b the new Frame 4 is labelled $4^R(3//)$ to indicate that its speech data has been derived from Frame 3, but now with two levels of amplitude adjustment. Frame 5 is 'Good', so its speech data is retained, but because it was preceded by a 'Bad' frame, its amplitude is also adjusted in an attempt to minimise amplitude discontinuities. Thus in Figure 1b it is labelled $5^R(5/)$. Frame 6 is 'Good', and given that it was preceded by a 'Good' frame, it is used without modification. Frame 7 is 'Bad', so its speech data is derived from Frame 6, but with an amplitude adjustment. It is therefore labelled as $7^R(6/)$.

In rather simplistic terms, the amplitude adjustment process associated with the AMR's FL mechanism works as follows. Each 20 ms frame is segmented into four 5 ms sub-frames, each with its own amplitude. Since Frames 1 and 2 are 'Good', the amplitudes of their sub-frames remain unchanged. In order to determine the amplitude, $\beta_{31}$, of the 1st sub-frame of the new Frame 3 (i.e., Frame labeled 3 (2/)), the median value $\beta_{median}$, of the amplitudes of the previous five sub-frames is determined. For illustration purposes, these amplitudes will be referred to as $\beta_{14}$, $\beta_{21}$, $\beta_{22}$, $\beta_{23}$, and $\beta_{24}$, where $\beta_{xy}$ is the amplitude of the $y^{th}$ sub-frame of Frame $x$. If $\beta_{24} \leq \beta_{median}$ then $\beta_{31} = \beta_{24} \times \alpha$, otherwise $\beta_{31} = \beta_{median} \times \alpha$, where $\alpha$ is some attenuation factor. (Note that the value of $\alpha$ is not constant, but changes dependent upon such factors as the sequence of 'Good' and 'Bad' frames received) [16]. An identical process

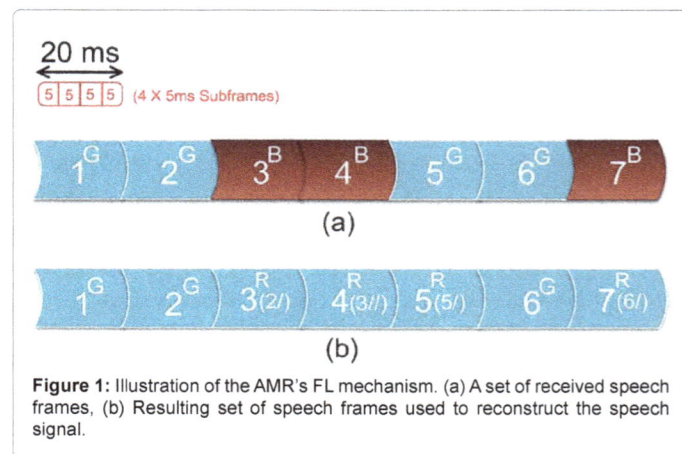

**Figure 1:** Illustration of the AMR's FL mechanism. (a) A set of received speech frames, (b) Resulting set of speech frames used to reconstruct the speech signal.

is used to determine the amplitudes of the remaining sub-frames of the new Frame 3, as well as of all of the sub-frames in the new Frame 4. So for instance, in respect to determining the amplitude, $\beta_{32}$, of the 2nd sub-frame of Frame 3, the $\beta_{median}$ value used is determined from $\beta_{21}$, $\beta_{22}$, $\beta_{23}$, $\beta_{24}$ and $\beta_{31}$, where $\beta_{31}$ has the same meaning as before, namely the amplitude of the 1st sub-frame of the new Frame 3 just determined.

The process in respect to deciding on the amplitudes of the sub-frames of the new Frame 5, namely $\beta_{51}$, $\beta_{52}$, $\beta_{53}$ and $\beta_{54}$, is different to the one just described for the sub-frames of the new Frames 3 and 4 because Frame 5 was a 'Good' frame preceded by a 'Bad' frame. So the only changes made to it are in respect to the amplitudes of its sub-frames, and this for the sole reason of minimising discontinuities.

The process here is based on a single comparison between the amplitude of each of the sub-frames of Frame 5 and the amplitude of the previous 'Good' sub-frame received (relative to it). So in respect to deciding on the amplitude of the 1st sub-frame of Frame 5, the previous 'Good' sub-frame received (again relative to it) is the 4th sub-frame of Frame 2. If $\beta_{51} \leq \beta_{24}$, then $\beta_{51}$ remains unchanged, else $\beta_{51}$: $\beta_{24}$. Similarly a value for $\beta_{52}$ is decided by comparing it to $\beta_{51}$ (i.e., the value just decided in the previous step). Specifically, if $\beta_{52} \leq \beta_{51}$, then $\beta_{52}$ remains unchanged, else $\beta_{52} = \beta_{51}$. This recursive process is repeated for the remaining two sub-frames of Frame 5. Frame 6 remains unchanged in the FL process, and this is in all respects including the amplitudes of its sub-frames. The amplitudes of the sub-frames of the final replaced frame, namely Frame 7, are determined in exactly the same manner as for Frames 3 and 4. So, for example $\beta_{median}$, the value used for determining $\beta_{71}$ will be based on $\beta_{54}$, $\beta_{61}$, $\beta_{62}$, $\beta_{63}$ and $\beta_{64}$ [16].

It can be seen from this example that though there were four 'Good' frames and three 'Bad' frames in the received seven-frame sequence of Figure 1a, only three of these 'Good' frames, along with four artificially generated replacement frames, have been used to produce the decoded speech waveform. But the bit rates for all frames, whether 'Good' and therefore unchanged, or artificially generated to replace "Bad" frames, remain the same. It is also evident from this example that a considerable degree of sophistication has been designed into the AMR's FL process to mask any resulting perceptual artefacts, making its subsequent detection from the decoded speech waveform, in all likelihood, impossible.

## CDMA FL mechanism

Before discussing the specifics of the FL strategy implemented by the EVRC decoder, it is necessary to give a brief overview of the codec itself. It operates in one of three modes, referred to as anchor operating points (OP), namely OP0, OP1, and OP2. The selection of a particular mode is made by the network according to the number of users accessing it. Once selected, the mode then defines the general behaviour of the codec as well as playing a role in determining the source coding bit rate for each speech frame. Upon selecting an OP, a speech frame is categorised as either voiced, voiceless, transient or silence. A source coding bit rate is then selected accordingly [17]. The codec produces output frames at one of four source coding bit rates, namely 8.55, 4, 2 and 0.8 kbps, with the latter being used to code silence frames. It also uses a number of coding techniques such as code excited linear prediction (CELP), pitch prototype period (PPP) and silence coder, these being selected for an individual frame on the basis of its speech category and the OP chosen.

The EVRC's FL mechanism also involves replacing 'Bad' frames with 'Good' frames using speech data from the past, but unlike with the AMR codec, an artificially created 'Good' frame is not necessarily at the

same bit rate as the 'Bad' frame it replaces. Usually it is set to the highest bit rate of 8.55 kbps. To illustrate the various aspects of the EVRC's FL strategy [17], we again use a similar example as for the AMR codec. We also again draw a distinction between speech data and amplitude data in a frame. But, unlike for the AMR codec, the EVRC's FL process is much simpler, so we discuss the handling of speech data and amplitude data at the same time.

Figure 2a shows a sequence of received frames, four 'Good' and three 'Bad'. The superscripts G and B associated with individual frames have the same meaning as before. Subscripts refer to the speech frame type such as silence (identified with S) or active speech (identified with an associated bit rate, namely 2, 4 or 8.55 kbps). Figure 2b shows the resulting speech frames that would be used to generate the decoded speech waveform. Again, the superscript R associated with an individual frame identifies it as a replacement frame and a subscript has the same meaning as in Figure 2a. Frames 1 and 2 are both 'Good' and therefore remain unchanged. Frame 1 is silence and Frame 2 is active speech at one of the three bit rates, namely 2, 4 or 8.55 kbps. Frame 3 is 'Bad' and is replaced by a synthetically-generated frame at a bit rate of 8.55 kbps. Essentially the speech data used in the new Frame 3 is the same as that in the last 'Good' speech frame, namely Frame 2, except for a possible modification needed to correct for any change in bit rate between the two frames. If the bit rate of Frame 2 was 8.55 kbps, then this will be used for the new Frame 3. If the bit rate for Frame 2 was either 2 or 4 kbps, a sophisticated bandwidth expansion of its speech data is performed to match the higher bit rate of the new Frame 3. As far as amplitude data for the new Frame 3 is concerned, this is made the same as for Frame 2. Thus the new Frame 3 in Figure 2b is identified as $3^R_{8.55}$ (2).

Frame 4 is also 'Bad' and its speech data would be replaced in an identical manner to Frame 3 (i.e., based on the speech data from the last 'Good' frame, namely Frame 2, but again with a possible bandwidth expansion). However, unlike for the new Frame 3, there would be an associated reduction in amplitude by a factor of 0.75 because Frame 4 is the second 'Bad' frame in a sequence. Thus the new Frame 4 in Figure 2b is identified as $4^R_{8.55}$ (2/). (Note: if a sequence of frames is 'Bad', the same process would be repeated, but with the amplitude of all subsequent replaced frames being reduced by a factor of $(0.75)^{N-1}$, where $N$ is the consecutive 'Bad' frame number ($N \geq 2$).) Frame 5 is 'Good', so remains essentially unchanged, except for its associated pitch parameter. Again with the goal of minimising discontinuities in the recovered speech signal, in this case in respect to pitch, the pitch information of Frame 5 would be altered to become essentially

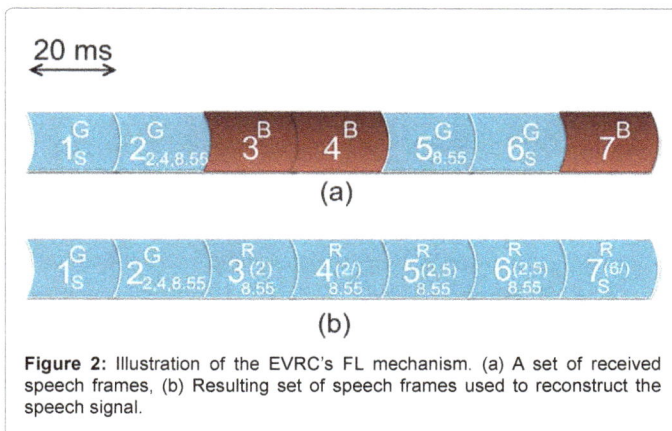

**Figure 2:** Illustration of the EVRC's FL mechanism. (a) A set of received speech frames, (b) Resulting set of speech frames used to reconstruct the speech signal.

an interpolation between the pitch of Frame 2 (and thus of the new Frames 3 and 4 which would have the same pitch as Frame 2) and that of Frame 5. Thus in Figure 2b the new Frame 5 is labelled as $5^R_{8.55}$ (2,5) to indicate that it has derived its speech data from Frames 2 and 5. Frame 6, which is a silence frame, is also 'Good'. However, one of the rules associated with the EVRC's FL mechanism is that a silence frame cannot be preceded by a replaced frame that is high quality (i.e., a frame with a bit rate of 8.55 kbps). So Frame 6 is discarded and is replaced by a copy of the previous frame, namely the new Frame 5. It is thus labelled as $6^R_{8.55}$ (2,5) in Figure 2b. Finally, Frame 7 is 'Bad' and so is replaced by essentially a copy of the 'Good' Frame 6 that was received, the only modification being in respect to its amplitude, this being recalculated slightly differently to other frames using procedures outlined in [17] because it was preceded by a silence frame. The new Frame 7 then becomes $7^R_S$ (6/) in Figure 2b.

It can be seen from this example that though there were four 'Good' frames and three 'Bad' frames in the received seven-frame sequence of Figure 2a, this has resulted in only two of these 'Good' frames, together with five synthetically-generated frames, being used to generate the decoded speech waveform. It is also clear from this example that, as with the AMR codec, a considerable degree of sophistication has been incorporated into the EVRC's FL mechanism, the underlying goal again being to conceal as far as possible, from a perceptual standpoint, that data has been lost or corrupted during transmission. The unfortunate consequence from the standpoint of a FVC analysis is that determining from the recovered speech signal when this process has occurred is likely to be very challenging, if not impossible.

## Experimental Methodology

### Speech database and speech parameters used

We used the same 130 male speakers from the XM2VTS database [18] as used in our DRC study [1], these being judged perceptually to have the same Southern British accent. The speakers were recorded on four different occasions separated by a one month interval. During each session each speaker read the following random digit sequences: 1. "zero one two three four five six seven eight nine", and 2. "five zero six nine two eight one three seven four". The speech files in the XM2VTS are sampled at 32 kHz with 16 bit digitization. We down-sampled these to 8 kHz to align with the input speech requirements of mobile codecs. Three recording sessions out of the four available have been used in our experiments. We focused on the three words, "nine", "eight" and "three" from these recordings and extracted their corresponding vowel segments /aɪ/, /eɪ/ and /i/ (i.e., two diphthongs and a monophthong) using a combination of auditory and acoustic procedures [19]. In summary, three non-contemporaneous sessions have been used with three vowels per session and four tokens per vowel.

As for our previous DRC study, the 130 speakers were divided into three groups: 44 speakers in the Background set, 43 speakers in the Development set and 43 speakers in the Testing set. (Note: the purpose of the Development set is to train the logistic regression-fusion system [20], the resulting weights of which are then used to combine LRs calculated from individual vowels for each comparison in the Testing set.) Two same-speaker comparison results were obtained for each speaker in the Testing set by comparing their Session 1 recording with their own recordings in Sessions 2 and 3. Similarly, three different-speaker comparisons were produced for each speaker by comparing their Session 1 recording with all other speakers' recordings from Sessions 1, 2 and 3 (refer to Table 2 of our DRC paper [1]). The

Background set remained the same for all comparisons and contained two recording sessions for all 44 speakers. 23 MFCCs were then computed for coded speech under various conditions of FL using the same MFCC extraction process as used in our DRC study. $C_{llr}$ values were calculated using the mean LRs for same- and different-speaker comparisons (note: two LRs were calculated for each same-speaker experiment and three LRs for each different-speaker experiment). CI was calculated by finding the variation in LR values (again, using two LRs for same-speaker comparisons and three LRs for different-speaker comparisons).

### Strategies to understand the impact of FL

Our goal with this study was to, as far as possible, study the impact of FL on FVC in isolation to the other two factors in a mobile phone network that can impact speech quality, namely DRC and BN. Clearly any approach involving the transmission of speech across an actual network would not make this possible. So in this study we have again chosen to pass speech through software implementations of the codecs under investigation and introduce FL in a controlled manner, while endeavouring to disable both DRC and BN. Disabling DRC and BN is straight forward for the AMR codec. For the EVRC codec, however, though disabling BN is also straightforward, this is not so for DRC because the bit rate for a frame depends partly upon its classification (i.e., voiced, voiceless or transition) [17] and partly upon the codec mode (i.e., OP0, OP1 or OP2). Obviously a frame's classification can't be changed, but the mode can be constrained to one of the three.

Figure 3 shows a block diagram of the processing stages used in our experiments. The speech files were processed by each codec under two scenarios: one assuming no lost or corrupted frames and the other with speech frames lost or irrecoverably corrupted between the coder and decoder stages of the codec in some controlled manner, as will be discussed in the following section.

### Simulating FL

The first aspect that needs to be considered when designing experiments of this kind is what level of frame loss, often referred to as frame error rate (FER), is typical of a real network. In mobile networks this parameter is constantly monitored during a call. When the FER exceeds in the region of 10 to 15% it is known that the overall voice quality degrades to a level where the mean opinion score (MOS) is less than about 2.9 [21]. Mobile network operators realise that such voice quality is unpleasant to the listener and they have therefore put procedures in place to automatically drop a call if this limit is reached.

In reality this monitoring of FER would be done over hundreds of frames corresponding to many seconds of speech. In our experiments, however, we have used vowel segments that are typically 12 to 15 frames

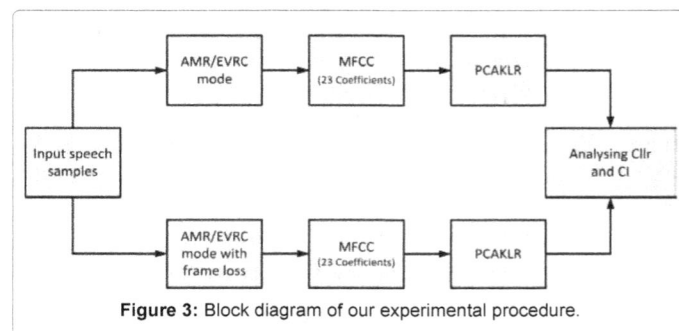

**Figure 3:** Block diagram of our experimental procedure.

in duration. In comparison to the duration of a vowel segment, the FER monitoring process described could be classified as a long-term statistical measure, and there would likely be short periods of time in which the actual FER was much higher. The question then arises as to whether this same upper value for FER of 10 to 15% is also appropriate for much shorter segments typical of vowels. To answer this question we conducted experiments where we examined the speech quality of vowels using PESQ [22] for a range of values of FER. In the interests of space we do not reproduce these experimental results here, but they showed that for vowel segments an FER in the region of 10 to 15% again translates into MOS values of the order of 2.9. So we used this same upper range for FER in our experiments as well. Given that the durations of our vowel segments were of the order of 12 to 15 frames, this FER rate translates into a maximum number of lost frames per vowel segment being typically one, or at most two. In the interests of investigating worst-case conditions, we have fixed the number of lost frames per vowel segment to two. For each vowel token, the locations of these lost frames has then been determined randomly according to a uniform distribution.

As shown in Figure 4, the speech files were coded at two different modes for each mobile network, these modes roughly translating into low and high quality speech coding. In the case of the GSM network, this was the 4.75 and 12.2 kbps modes, respectively, whereas in the CDMA network it was OP2 and OP0, respectively. For each mode, speech was coded twice, first without FL, then with it. The rationale behind conducting FVC experiments at two different speech qualities was to try and separate out the impact of speech coding quality from the impact of FL. It is important to note that the Background set used in these experiments contained coded speech at the specific mode being investigated, but without FL. This was done in an endeavour to minimise mismatch.

## Results

### Impact of FL on the decoded speech waveform

Before investigating the impact of FL on FVC performance, it is informative to examine how the temporal location of a lost frame, together with of course the associated FL corrective mechanism that it would have triggered, might impact upon the decoded speech waveform, both in terms of its temporal and spectral characteristics. To illustrate this, a set of time waveforms and spectrograms have been produced for a token of the vowel /aI/ coded with either the AMR or EVRC codec. A single lost frame has been introduced between the coded and decoded speech paths, but at three different temporal locations, namely at Frames 3, 4 and 5.

Figures 5 and 6 show the results for the AMR codec, with speech coded at 12.2 kbps. Figure 5a shows 180 ms of the time waveform of the vowel segment (i.e., 9 frames) without FL. Figures 5b-d show the resulting decoded speech waveform for Frames 3, 4 and 5 being lost, respectively. For the purpose of comparison, the amplitude of each of these time waveforms has been normalized to the maximum absolute value of the waveform in Figure 5a. Figure 6 shows the spectrograms corresponding to the time waveforms shown in Figure 5. Examination of both Figures 5 and 6 shows that the loss of a single frame can have quite an impact on all subsequent frames and that this impact depends very much on exactly which frame is lost.

The corresponding results for the EVRC are shown in Figures 7 and 8, with speech coded at OP0.

It is interesting to note that though it is exactly the same vowel segment that has been coded by both codecs, there are even differences between the resulting coded speech waveforms for the situation of no FL. As for the AMR codec, with the EVRC the loss of a single frame can have quite an impact on the subsequent decoded speech frames.

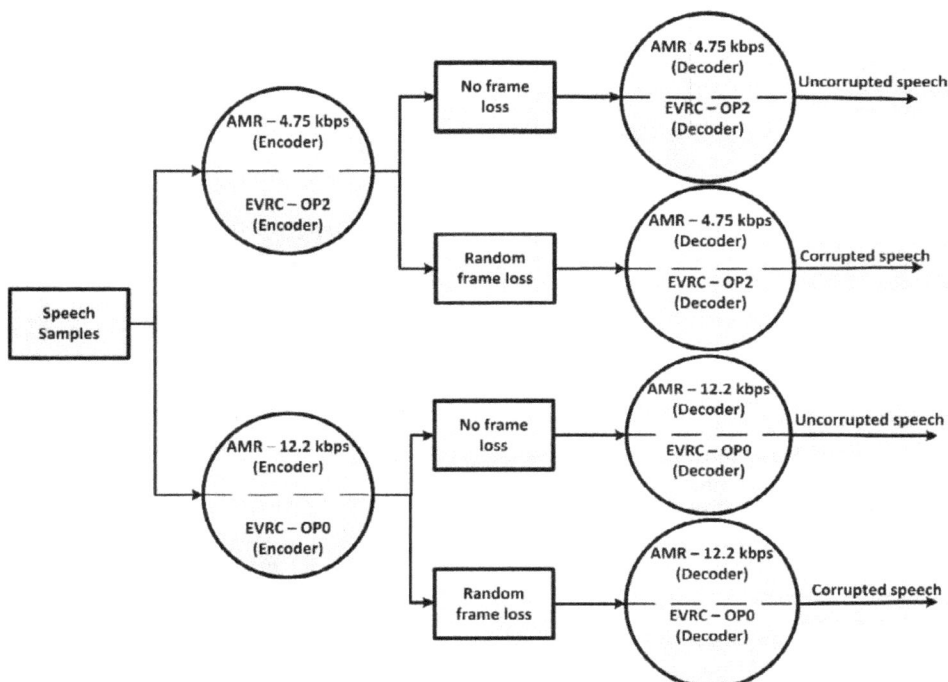

**Figure 4:** Processing of speech files using the AMR and EVRC codecs at low and high quality coding modes.

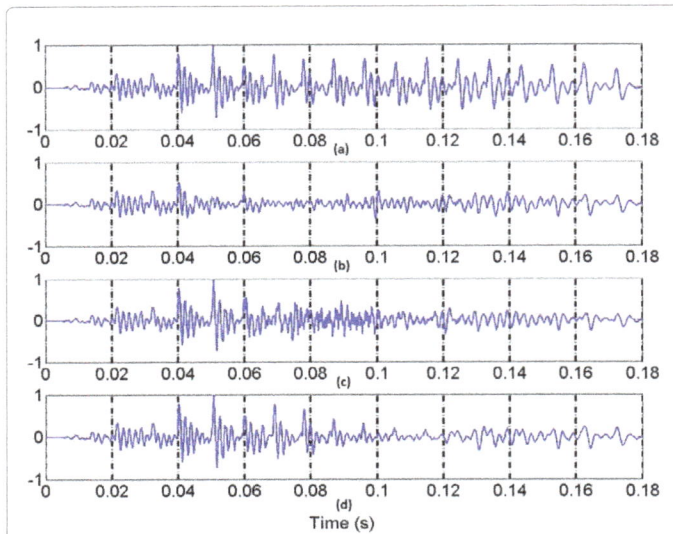

**Figure 5:** A set of time waveforms produced for /aI/ with the AMR codec and corrupted at different frame locations. (a) with no FL, (b) with FL at the 3rd frame, (c) with FL at the 4th frame, (d) with FL at the 5th frame. (Dashed lines show the frame boundaries).

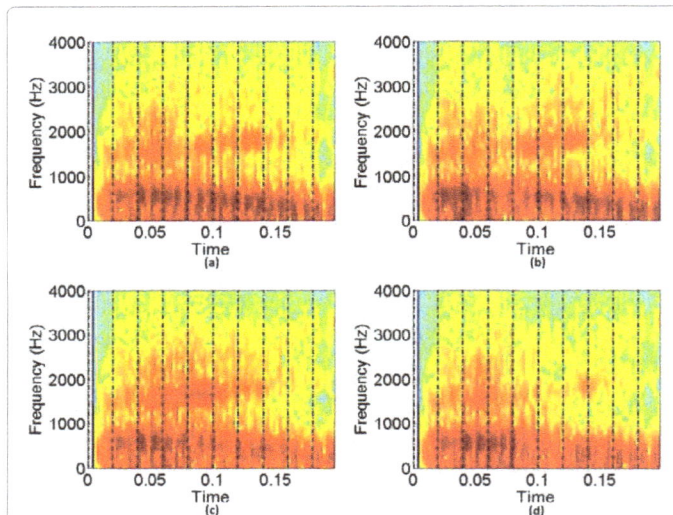

**Figure 6:** Spectrograms of the time waveforms shown in Figure 5 for the AMR codec. (a) with no FL, (b) with FL at the 3rd frame, (c) with FL at the 4th frame, (d) with FL at the 5th frame. (Dashed lines show the frame boundaries).

## Impact of FL on FVC performance

This section presents results showing the impact on FVC performance arising from FL for both the AMR and EVRC codecs. Exactly two lost frames have been introduced into each vowel segment, their temporal locations being randomly determined according to a uniform distribution. LR values have been computed separately for each of the vowels /aI/, /eI/ and /i/ and the results then fused using logistic-regression fusion. The resulting FVC performance is shown in terms of $C_{llr}$, CI, Tippett plots and APE plots.

**AMR codec:** Figure 9 examines the impact of FL on FVC performance in terms of CI and $C_{llr}$ for the AMR codec at 4.75 kbps and 12.2 kbps. Results are presented without and with FL for both cases. It is clear from these results that FL does have a negative impact upon

FVC performance, both in terms of accuracy ($C_{llr}$) and reliability (CI). Further, this impact is more severe for coded speech at the lower bit rate.

In order to investigate this latter aspect further, Figures 10 and 11 show Tippett plots for AMR-coded speech at 4.75 kbps, without and with FL, respectively. The corresponding results at 12.2 kbps are shown in Figures 12 and 13. The blue solid curve in these plots represents same-speaker comparison results and the red solid curve different-speaker comparison results. The dashed line on the either side of the blue and red curves represents the variation found in a particular LLR (i.e., LLR $^+$_ CI). Considering first the results at 4.75 kbps, it is clear that a major impact of FL is on same-speaker classifications. The strength of both same-speaker and different speaker comparisons has increased

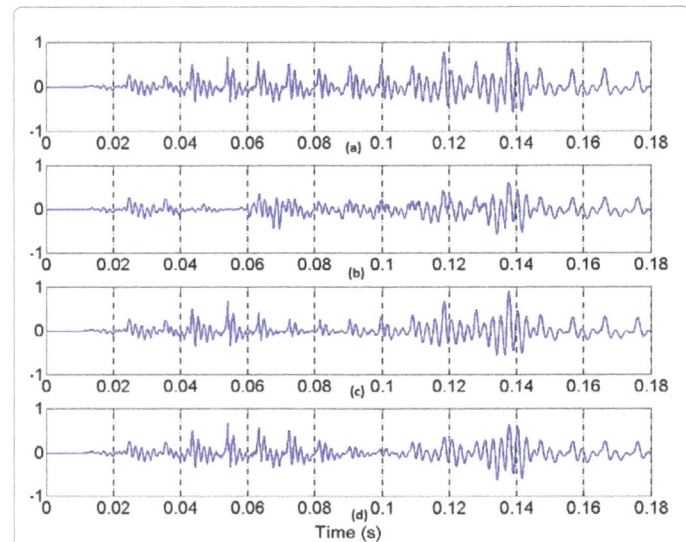

**Figure 7:** A set of time waveforms produced for /aI/ coded with the EVRC codec and corrupted at different frame locations. (a) with no FL, (b) with FL at the 3rd frame, (c) with FL at the 4th frame, (d) with FL at the 5th frame. (Dashed lines show the frame boundaries).

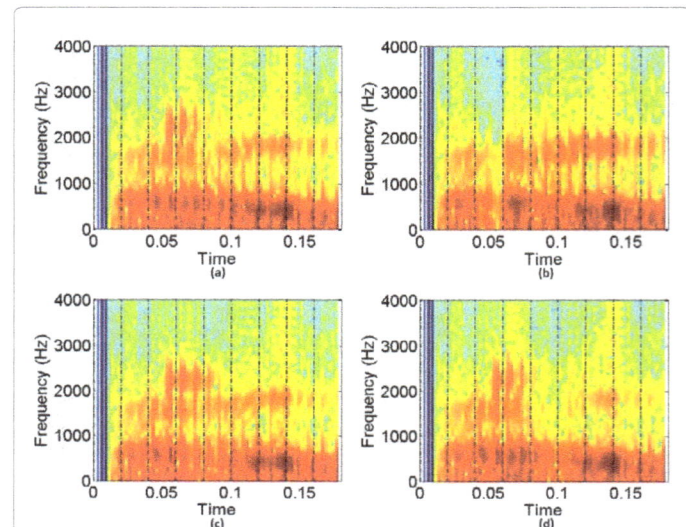

**Figure 8:** Spectrograms of the time waveforms shown in Figure 7 for the EVRC codec. (a) with no FL, (b) with FL at the 3rd frame, (c) with FL at the 4th frame, (d) with FL at the 5th frame. (Dashed lines show the frame boundaries).

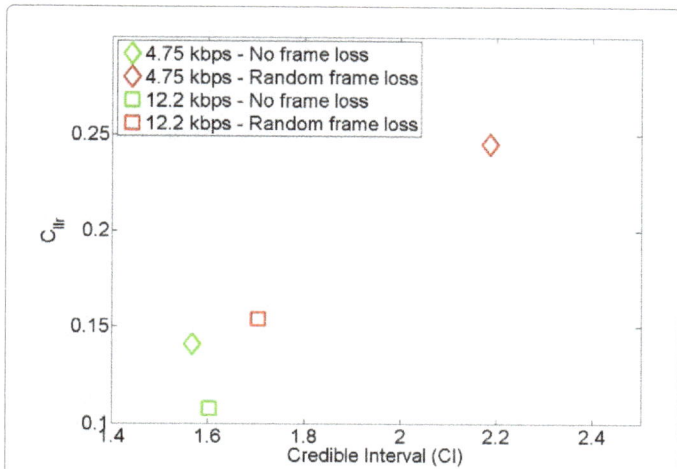

**Figure 9:** $C_{llr}$ v.s. CI for AMR-coded speech at 4.75 kbps and 12.2 kbps without and with FL.

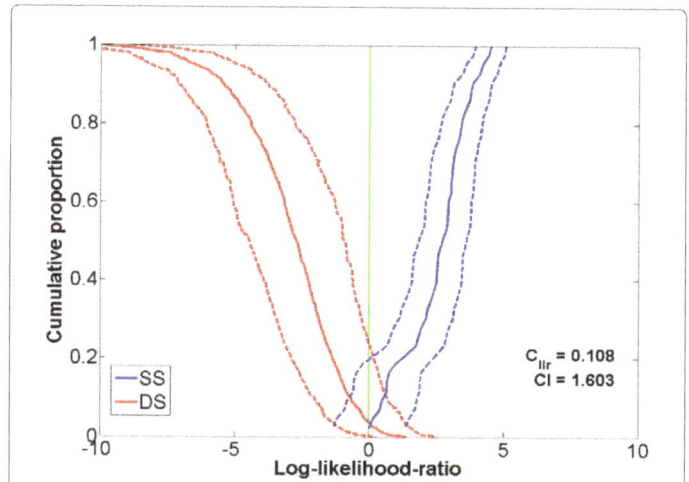

**Figure 12:** Tippett plot showing the performance of the AMR codec at 12.2 kbps without FL.

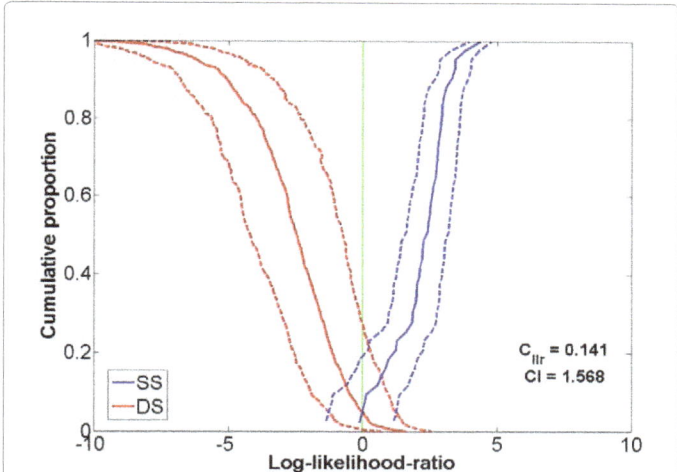

**Figure 10:** Tippett plot showing the performance of the AMR codec at 4.75 kbps without FL.

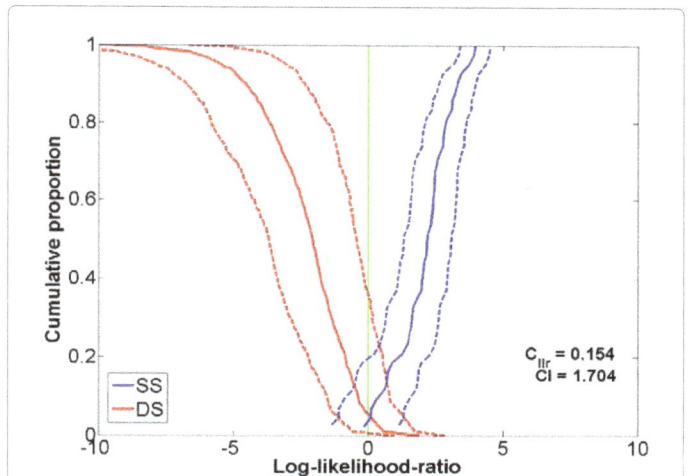

**Figure 13:** Tippett plot showing the performance of the AMR codec at 12.2 kbps with FL.

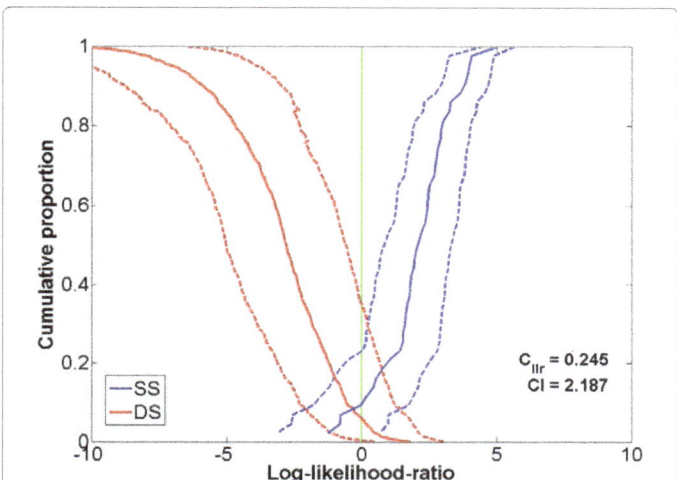

**Figure 11:** Tippett plot showing the performance of the AMR codec at 4.75 kbps with FL.

slightly, but importantly the number of same-speaker misclassifications has increased. Both of these findings are intuitively to be expected. In respect to different-speaker comparisons, the strength of these has improved slightly, which is again a finding one might expect. As far as reliability is concerned (i.e., CI), FL would appear to have a similar negative impact upon both same-speaker and different speaker comparisons. The results at 12.2 kbps (Figures 12 and 13) confirm that the impact of FL at the higher bit rate is fairly minimal, both in terms of same- and different-speaker comparisons.

To further understand what has contributed to the worsening of $C_{llr}$ values as a result of FL, Figures 14 and 15 show APE-plots for the two cases of 4.75 kbps and 12.2 kbps, respectively. Considering first Figure 14 for 4.75 kbps, it is clear that FL has resulted in a significant increase in discrimination loss of almost 95%. Calibration loss has also increased, but only by about 20%. In the case of high bit rate coding (Figure 15), the calibration and discrimination loss components have both increased by about 40%.

**EVRC Codec:** Figure 16 examines the impact of FL on FVC performance in terms of CI and $C_{llr}$ for the EVRC at OP2 (low quality

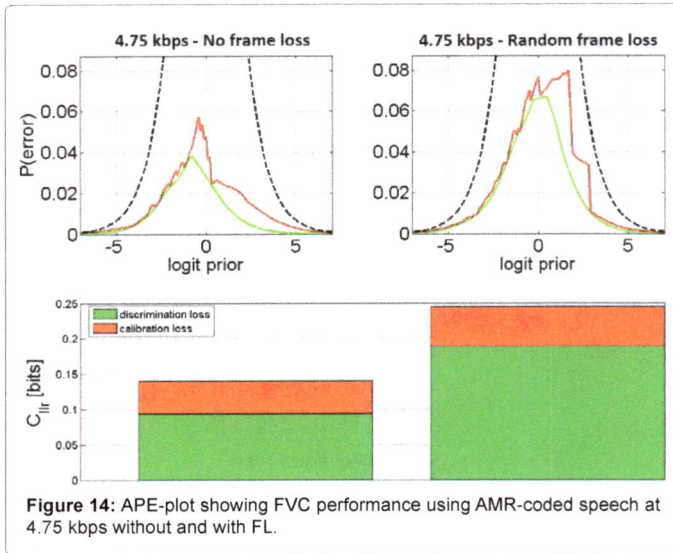

**Figure 14:** APE-plot showing FVC performance using AMR-coded speech at 4.75 kbps without and with FL.

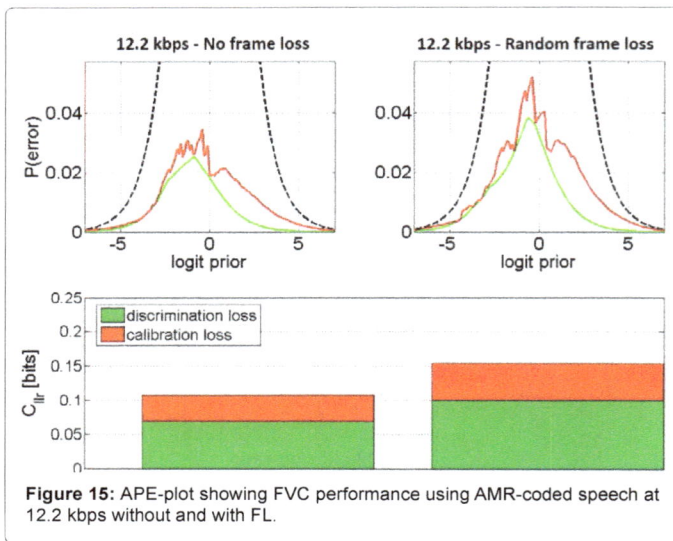

**Figure 15:** APE-plot showing FVC performance using AMR-coded speech at 12.2 kbps without and with FL.

coding) and OP0 (high quality coding). For purposes of comparison, results are presented without and with FL for both cases. In respect to $C_{llr}$, the results for the EVRC are very similar to those for the AMR codec, namely, FL negatively impacts FVC accuracy and this is worse for low quality speech coding. Unlike for the AMR codec, however, for the EVRC for both low and high quality speech coding the CI has improved as a result of FL. Why this might be so is not clear at this stage.

To further understand the degradation in $C_{llr}$ values, Tippett plots are shown for OP2 without and with FL (Figures 17 and 18, respectively) and OP0 without and with FL (Figures 19 and 20, respectively). The first observation from these figures is that FL has negatively impacted both same- and different-speaker classifications, but this is less at the higher quality coding. Secondly, it has caused both same-and different-speaker misclassifications to increase, though for high quality coding this increase is minimal. As far as CI is concerned, Figures 17-20 confirm the previous finding for the AMR codec, namely FL has caused this aspect to improve.

Figures 21 and 22 show APE-plots for OP2 (low quality coding)

and OP0 (high quality coding), respectively. As was the case for the AMR codec, FL in low quality coded speech causes the discrimination loss to increase significantly, in this case by about 110%. There is also a small increase in calibration loss of about 30%. The situation for high quality speech is somewhat different. Here the major impact of FL is to increase the calibration loss by about 65%, with discrimination loss increasing by only about 15%.

## Conclusions

In this paper we have presented the impact of FL on FVC for speech transmitted through two major mobile phone networks: GSM and CDMA. We have noted that it is quite incorrect to assume that there is such a thing as 'generic' mobile phone speech. The GSM and CDMA mobile phone networks are fundamentally different in their design and implementation and this necessarily translates into differences in the characteristics of the speech they produce and in the subsequent impact of these differences on FVC. We have described in considerable detail the FL processes implemented by the AMR (GSM network) and EVRC (CDMA network) codecs. Our reason for describing these

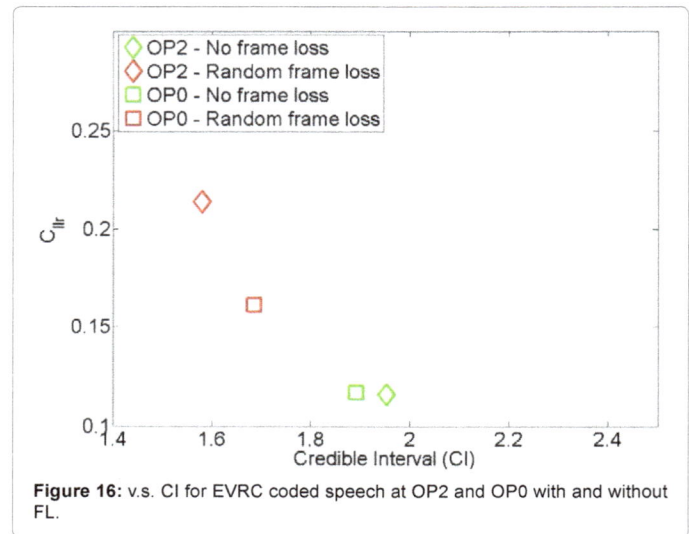

**Figure 16:** v.s. CI for EVRC coded speech at OP2 and OP0 with and without FL.

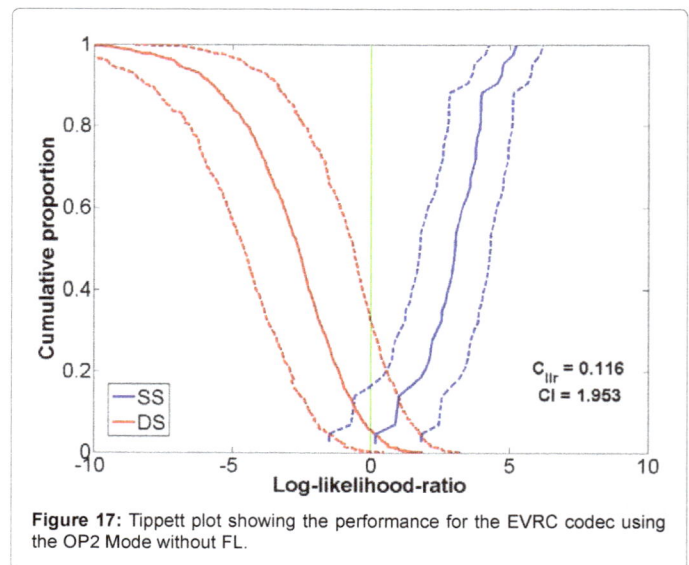

**Figure 17:** Tippett plot showing the performance for the EVRC codec using the OP2 Mode without FL.

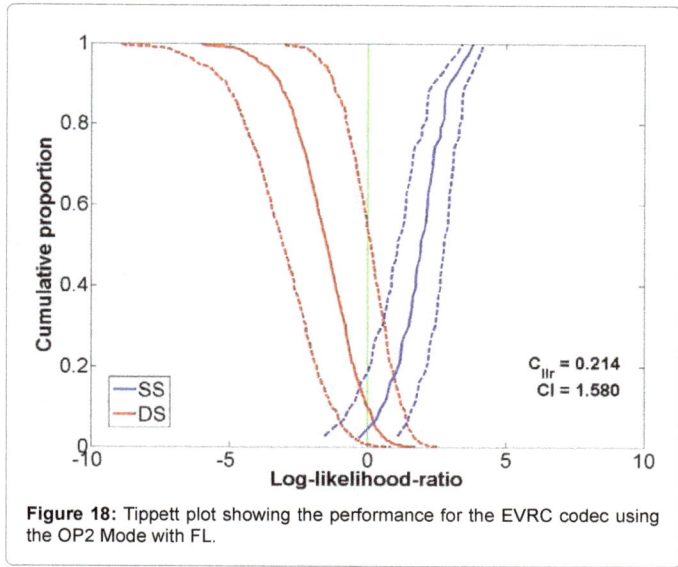

**Figure 18:** Tippett plot showing the performance for the EVRC codec using the OP2 Mode with FL.

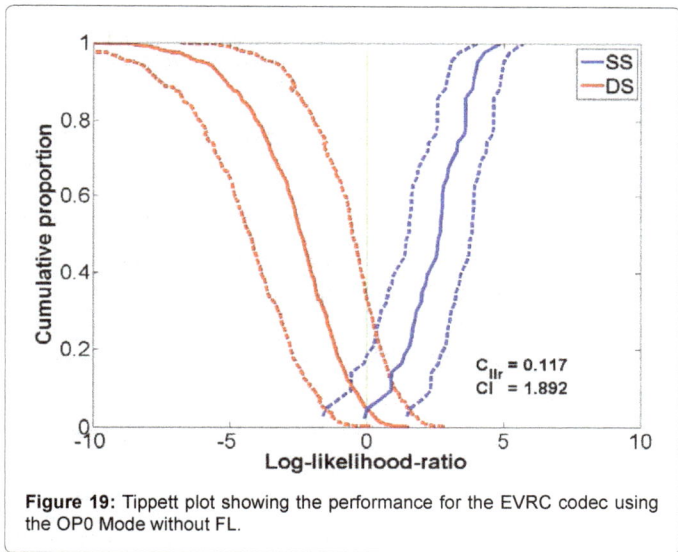

**Figure 19:** Tippett plot showing the performance for the EVRC codec using the OP0 Mode without FL.

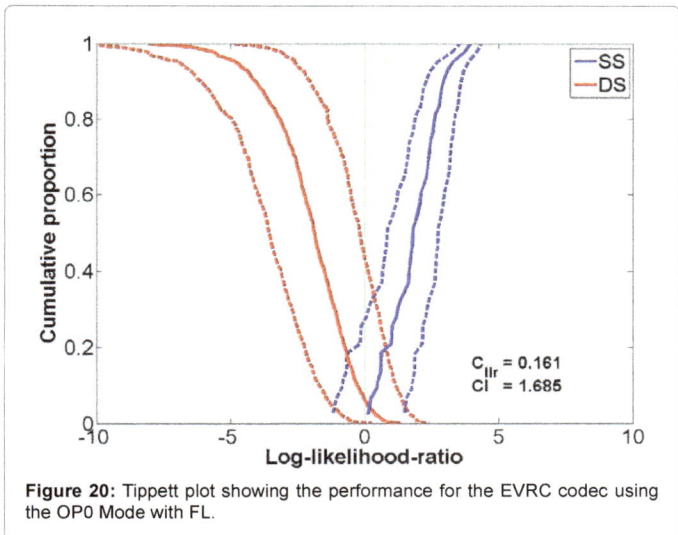

**Figure 20:** Tippett plot showing the performance for the EVRC codec using the OP0 Mode with FL.

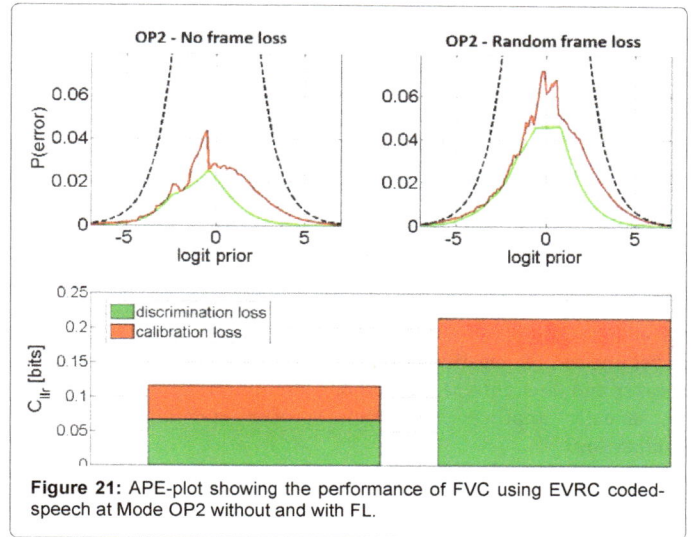

**Figure 21:** APE-plot showing the performance of FVC using EVRC coded-speech at Mode OP2 without and with FL.

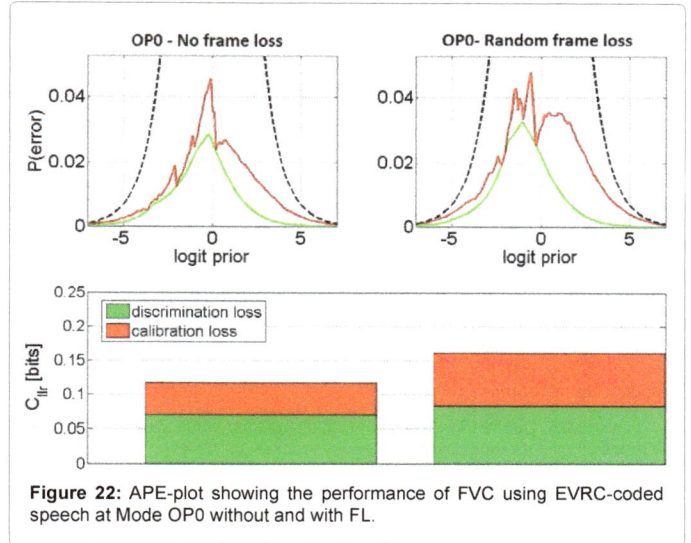

**Figure 22:** APE-plot showing the performance of FVC using EVRC-coded speech at Mode OP0 without and with FL.

processes in such detail is because we believe it essential for forensic speech scientists to have a clear appreciation of the nature and extent to which speech acquired from a mobile phone network could contain artificially generated sections. An important conclusion from this presentation is that these processes embody a considerable degree of sophistication designed specifically to mask, as far as possible, any resulting perceptual artefacts. Whether the occurrence of these processes is nonetheless still detectable from the recovered speech signal is clearly a matter for further research, but at this stage we are quite sceptical of this possibility.

We have noted that the operators of mobile phone networks permit a call to continue even if the percentage of lost frames is in the region of 10 to 15%. Given that a single lost frame will also impact upon a number of the subsequent 'Good' frames that follow it, the amount of artificially generated material in a mobile phone speech recording could well be higher than 10 to 15%. Our experiments have focused on vowel segments of typically 12 to 15 frames in duration. In the interests of considering worst-case conditions, we have introduced two lost frames into these segments, the temporal locations of which have been determined randomly according to a uniform distribution.

We have shown that with AMR-coded speech, FL causes a worsening of same-speaker comparisons in terms of accuracy, and noted this is more problematic for low quality coded speech than high quality. Perhaps not surprisingly, our experimental results also suggest that FL with AMR-coded speech can improve the accuracy of different-speaker comparisons. As far as reliability is concerned, FL negatively impacts upon both same- and different-speaker comparisons in a similar manner.

With the EVRC, though a number of our experimental results are similar to those of the AMR codec, there are also some important differences. One such difference is in respect to the impact of FL on the accuracy of different-speaker comparisons. For reasons which are as yet unclear, FL negatively impacts upon the accuracy of both same-speaker and different-speaker comparisons, but in terms of reliability, FL actually improves both same-speaker and different-speaker comparisons.

Though much more research needs to be done on this aspect of the impact of FL on FVC undertaken using mobile phone speech, it is clear from the results presented in this paper that it can be significant, a fact that must necessarily impact on the confidence a forensic scientist ascribes to their analysis results.

## References

1. Alzqhoul EAS, Nair BB, Guillemin BJ (2015) Impact of dynamic rate coding aspects of mobile phone networks on forensic voice comparison. Science & Justice 55: 363-374.

2. Kuhn V (1997) Applying list output Viterbi algorithms to a GSM-based mobile cellular radio system. In: Universal Personal Communications Record, Conference Record, IEEE 6th International Conference on IEEE pp: 878-882.

3. 3GPP, TS 45.003; 3rd Generation Partnership Project; Technical Specification Group GSM/EDGE Radio Access Network; Channel coding.

4. Morrison GS (2011) A comparison of procedures for the calculation of forensic likelihood ratios from acoustic–phonetic data: Multivariate kernel density (MVKD) versus Gaussian mixture model–universal background model (GMM–UBM). Speech Comm. 53: 242-256.

5. Reynolds DA, Quatieri TF, Dunn RB (2000) Speaker verification using adapted Gaussian mixture models. Digital Signal Process 10: 19-41.

6. Aitken CG, Lucy D (2004) Evaluation of trace evidence in the form of multivariate data. Journal of the Royal Statistical Society: Series C (Applied Statistics) 53: 109-122.

7. Nair BB, Alzqhoul E, Guillemin BJ (2014) Determination of likelihood ratios for forensic voice comparison using Principal Component Analysis. International Journal of Speech Language and the Law 21: 83-112.

8. Jessen M (2014) Comparing MVKD and GMM-UMB applied to a corpus of formant-measured segmented vowels in German. In: International Association for Forensic Phonetics and Acoustics Annual Conference (IAFPA 2014), Zurich, Switzerland.

9. Alzqhoul EA, Nair BB, Guillemin BJ (2014) An Alternative Approach for Investigating the Impact of Mobile Phone Technology on Speech. In: Proceedings of the World Congress on Engineering and Computer Science.

10. Bruhn S, Error concealment in relation to decoding of encoded acoustic signals. US. patent No. 6,665,637.

11. ETSI, Substitution and Muting of Lost Frames for Full Rate Speech Channels. Retrieved on 2 June 2013, last retrieved from http://www.3gpp.org/.

12. Alzqhoul EA, Nair BB, Guillemin BJ (2012) Speech Handling Mechanisms of Mobile Phone Networks and Their Potential Impact on Forensic Voice Analysis. In: SST 2012, Sydney, Australia.

13. 3GPP, TS 26.071 V11.0 3rd Generation Partnership Project; Technical Specification Group Services and System Aspects; Mandatory speech CODEC speech processing functions; AMR speech CODEC; General description.

14. 3GPP, TS 26.101 V11.0.0 3rd Generation Partnership Project; Technical Specification Group Services and System Aspects; Mandatory speech codec speech processing functions; Adaptive Multi-Rate (AMR) speech codec frame structure.

15. 3GPP, TS 45.009 3rd Generation Partnership Project; Technical Specification Group GSM/EDGE Radio Access Network;Link adaptation.

16. 3GPP, TS 26.091 3rd Generation Partnership Project; Technical Specification Group Services and System Aspects; Mandatory Speech Codec speech processing functions; Adaptive Multi-Rate (AMR) speech codec; Error concealment of lost frames.

17. 3GPP2, S0018-D, Minimum Performance Specification for the Enhanced Variable Rate Codec, Speech Service Options 3, 68, 70, and 73 for Wideband Spread Spectrum Digital Systems.

18. Messer K, Matas J, Kittler J, Luettin J, Maitre G (1999) XM2VTSDB: The extended M2VTS database. In: Second international conference on audio and video-based biometric person authentication, Citeseer pp: 965-966.

19. Rose P (2004) Forensic speaker identification, CRC Press.

20. Ramos-Castro D, Gonzalez-Rodriguez J, Ortega-Garcia J (2006) Likelihood ratio calibration in a transparent and testable forensic speaker recognition framework. In: Speaker and Language Recognition Workshop IEEE Odyssey, The IEEE pp: 1-8.

21. D. Networks, Voice Quality Solutions for Wireless Networks. Retrieved on 21 June 2013, last retrieved from http://www.ditechnetworks.com/.

22. Rix AW, Beerends JG, Hollier MP, Hekstra AP (2001) Perceptual evaluation of speech quality (PESQ)- a new method for speech quality assessment of telephone networks and codecs. In: Acoustics, Speech, and Signal Processing, Proceedings (ICASSP'01). 2001 IEEE International Conference on IEEE pp: 749-752.

# 3

# RSSI-based Indoor Localization Using RSSI-with-Angle-based Localization Estimation Algorithm

**Ambassa Joel Yves\* and Peng Hao**

*School of Electronics and Information, Jiangsu University of Science and Technology, 2 Mengxi Road Jingkou Zhenjiang Jiangsu 212003, PR China*

### Abstract

For the scenarios of indoors localization and tracking, the solutions generally need complex infrastructure because they would require either a grid of antennas, each having a well-known position (proximity based approach), or a sophisticated algorithm that uses scene fingerprint to estimate the location or the zone of an object by matching the online measurement with the closest offline measurement. Those techniques may not be available in unknown zones, which will make it difficult to locate a lost node. In this paper, with no additional hardware costs, we propose a new RSSI-based approach in order to find a lost node using a known node. By rotating the known node at the same spot we can collect different RSSI for different polar angles. Two pairs of angles with the strongest RSSI will indicate the main lobes of the radiation pattern, namely, zone of the unknown node. Experimental results illustrate a very close estimation of the unknown node zone, reducing up to 84% of the zone uncertainty.

**Keywords:** RSSI; Node; Polar angles; Rotation; Localization

## Introduction

The Received Signal Strength Indication (RSSI) is the measurement of the power present in a received radio signal. It has commonly been used to estimate the distance between nodes. Techniques have been discussed, using distance estimation from multiple reference points to determine the position of a node. The distance estimation is processed through the radio signal velocity and the time spent by the signal to reach its target. This method uses the concept of Time of Arrival (TOA) [1] and Time Difference of Arrival (TDOA) for estimating distance. Two main issues are encountered in TOA: first transmitters and receivers in the system must be perfectly synchronized in order to get a meaningful estimation, then a timestamp must be embedded in the transmitting signal in order for the measuring unit to more accurately estimate the corresponding distance. These two issues may lead to errors or even meaningless estimation. RSSI is widely adapted in wireless communication protocols such as Bluetooth, ZigBee [2,3] and other wireless techniques, because it indicates the strength of the signal transmitted. Furthermore, this signal strength can be measured without connecting or setting up devices. And this makes things easier since neither synchronization nor timestamp is required.

We mostly use antennas for signals transmission. The zone where we can possibly get a signal is a radiation pattern. The main lobe or main beam is the zone containing the maximum power of the signal. This is the lobe that exhibits the greatest strength. The radiation pattern of most antennas shows a pattern of "lobes" at various angles, directions where the radiated signal strength reaches a maximum, separated by "nulls" [1-5], angles at which the radiation falls up to zero. For dipole antennas, we have two main lobes, usually opposite. Because of the canonical form of the pattern radiation, we will assume that we may get at least 90° between two main lobes [4].

In this paper a new approach for unknown node location is proposed. It consists of determining the zone of an unknown node using the angles associated to the strongest RSSI transmitted by that node. This paper is arranged as follow: In Section 2 we will review the related work, in Section 3 we will propose the RALE approach, in Section 4 we will show the experiment results and their analysis and finally in Section 5 we will give a conclusion.

## Related Work

Various studies have proposed different localization algorithms which vary on whether the algorithm depends on known nodes or not. The anchor-based algorithms use some nodes with known position. The higher number of known nodes guarantees a better location accuracy of the unknown node that has an unknown position. Those algorithms may need at least 3 anchor nodes, with the TOA and the Angle of Arrival (AOA) of the signal [1] respectively, in order to estimate the distance between known nodes and the unknown node. This technique requires a very large and complex hardware set up and many reference points. In addition, a big number of nodes can affect noisily the location of the unknown node [5].

The proximity approach requires in getting location using antennas [6]. The AOA and the signal strength of those antennas will help determine the zone of the unknown node. Sometimes the blind node can calculate its position because the AOA and signal strength information are transmitted in the signal received by the blind node.

Several methods have been used for indoor positioning, such as trilateration [1,5], triangulation [1], proximity approach [7], radio map [8], and fingerprinting [9]. The results reported in ref. [5] shows the trilateration has an accuracy of 50% and it needs extra hardware to send two types of data simultaneously; further confirmations of the accuracy of trilateration in ref. [10] shows that it can attain a better accuracy

Some literature, tells us that signal strength and its fluctuation may differ according to antenna angle variation [11]. Particularly, it is also observed that, human movement from slower to faster paces increases

**\*Corresponding author:** Ambassa Joel Yves, School of Electronics and Information, Jiangsu University of Science and Technology, 2 Mengxi Road Jingkou Zhenjiang Jiangsu 212003, PR China, E-mail: ambassajy@yahoo.fr

the fluctuations of the RSSI. However the height between the receiver and the transmitter has less effect on the RSSI values [12].

A technique using radio map's data, accelerometer and magnetic sensor has been discussed. It enables somebody whose initial position is known to be located on a radio map through the compass and the signal strength read in his mobile phone. The algorithm then consists of comparing his current data with the one attached with the nearest radio map's data [8] saved in a server. The compass measures the polar angle which is one of the coordinate components defining the user location, the other component of the user coordinate is the radius which corresponds to the distance between the initial point and the current point, and this is where the accelerometer will be useful. This distance is compared with the distance calculated using the maximum distance estimation made through the RSSI values. Indeed there are ranges of RSSI values that correspond to maximum distance estimations. Factors such as diffraction, reflection and scattering can deeply affect the RSSI.

Every node use an antenna to transmit its signal; but the signal radiation depends on the type of antenna it has and this can deeply affect the value of the RSSI [13]. Mobile phones for the most part have a broad bandwidth and circular polarized antennas that make them very reliable and efficient when it comes to receive signals [14]. Most of the time the issue encounter in signal reception is to know the type of antenna the transmitter used and its signal radiation mode: methods of radiation calculation based on directional function calculation are also found in the literature [4]. These methods reveals that transmitters antennas build radiation lobes which have different, and sometimes almost opposite directions.

## RALE Approach

This section describes the RSSI-with-Angle-based Localization Estimation (RALE) approach in order to determine one unknown node. RSSI value can increase as we get close to the unknown node; in order to avoid fluctuant RSSI due to angle variation [11], we assumed that both nodes are stable and on the same horizontal plane.

We collected the pair (RSSI, angle) for every random direction during the rotation, with a short pause in many directions as shown in Figure 1; we averaged the RSSI values for each direction in order to get more meaningful RSSI values.

The meaningful RSSI ($RSSI_i$, $i=1, 2,..., n$) and its corresponding angle ($\theta_i$) will be considered as weighted point for our approach. In the database, we searched the two maximum RSSI values, $RSSI_{m1}$ and $RSSI_{m2}$ with $RSSI_{m1} > RSSI_{m2}$ and we read the corresponding angles $\theta_1$ and $\theta_2$.

We required two pair belonging to different lobes of the antenna radiation thus we proceeded to a test in order to separate measurement from the same lobe. Because of the canonical form of the antenna's pattern radiation we defined that $|\theta_1 - \theta_2| > 90$ ⊠. When that difference doesn't match the condition we consider that those two angle values are associated to RSSI belonging to the same lobe. We then have to change $RSSI_{m2}$ with the next strongest RSSI saved in the database ($RSSI_{m3}$) and proceed the same condition. Then the fourth pair will then be regarded as $RSSI_{m3}$ .we can repeat this process till the conditions are respected as illustrated in Figure 2.

From this we get two directions that divide the plane in four zones. Our target is to determine the zone containing the unknown device and for that we have to make a choice. The third greatest RSSI ($RSSI_{m3}$) will be used to identify the zone and its associated angle $\theta_3$ should be

greater or equal to $\theta_1 \pm 90°$. The third strong RSSI ($RSSI_{m3}, \theta_3$) will help us decide in which zone the node might be located because now we get four possible zones of interest: $[\theta_1; \theta_2]$ $[\theta_1; \theta_2 -\pi]$ $[\theta_1 -\pi; \theta_2]$ $[\theta_1 -\pi; \theta_2 -\pi]$ and thus $\theta_3$ should belong to one of these 4 zones. We called this process zone determination

RALE has four main advantages:

• It doesn't require a survey of the scene (like in the fingerprint method), and can be implemented in every indoor environment: this is one of the common requirement in indoor positioning location, for new scene and less time it' an indoor localization.

• It doesn't require heavy or sophisticated infrastructure for his implementation like in trilateration or triangulation, no need TOA or DOA. Indeed in TOA or DOA techniques we need to be able to measure the angle of arrival and the time of arrival of the signal. This requires special tools that most of time can be difficult to get.

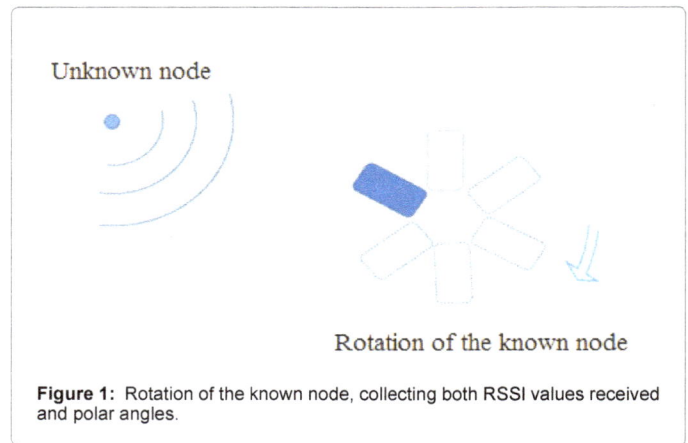

**Figure 1:** Rotation of the known node, collecting both RSSI values received and polar angles.

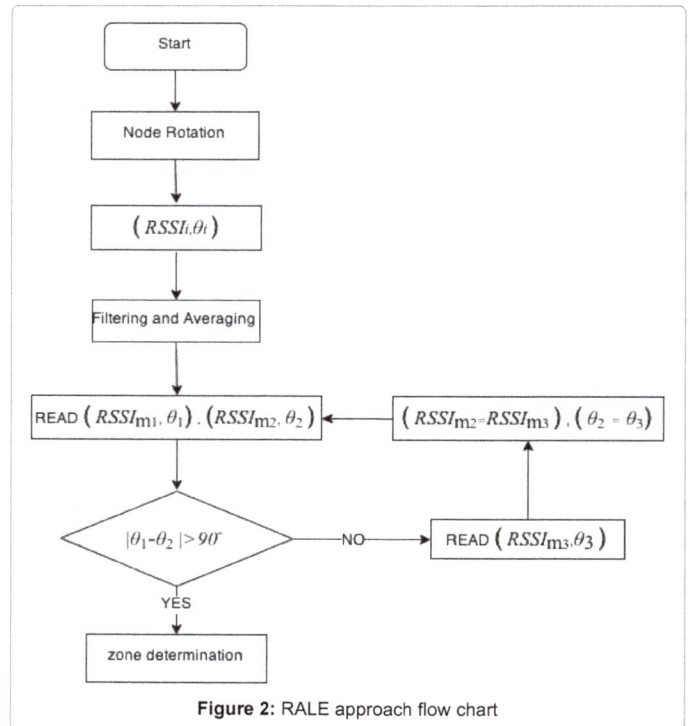

**Figure 2:** RALE approach flow chart

- It is low cost, we just need 2 nodes and the algorithm is simple to execute. Basically a mobile phone and a BLE device can do the job; those are not expensive and are accessible to everybody. Beside an app can execute the algorithm in the phone.

- We just need the values of RSSI during the rotation. A key is to record RSSI during the rotation we don't need other measurement to estimate the location of the node.

RALE analytically interpreted the data stored during the rotation of the node. Its efficiency was guaranteed by a huge amount of data. In the next section we will describe how experimentally we were able to implement this approach.

## Experiments

The rotation of a Bluetooth sensor (known node) has been performed in 4 different positions. We put, during the whole experiment the Bluetooth sensor and a Bluetooth Low Energy (BLE) device (unknown node) on the same flat ground as shown in Figure 3.

The BLE device is made on the CC2541 chip from Texas Instrument, working on 2.4-GHz applications. During the rotation of our sensor we have collected and saved in a database the polar angles and RSSI values, and we could see the repartition of the pair point (angle, RSSI) in Figures 4-7: we noticed that the aligned points represent the data during the pause of the smartphone and we can see that for a fixed

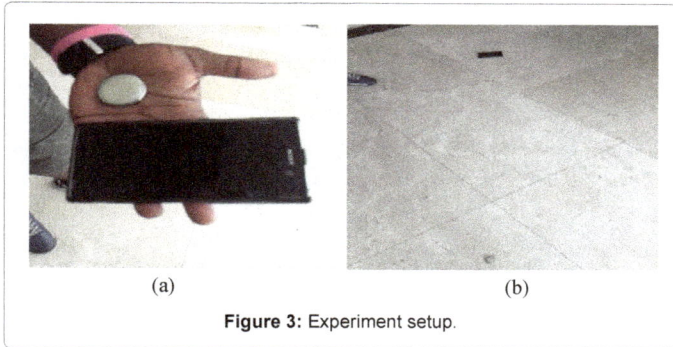

**Figure 3:** Experiment setup.

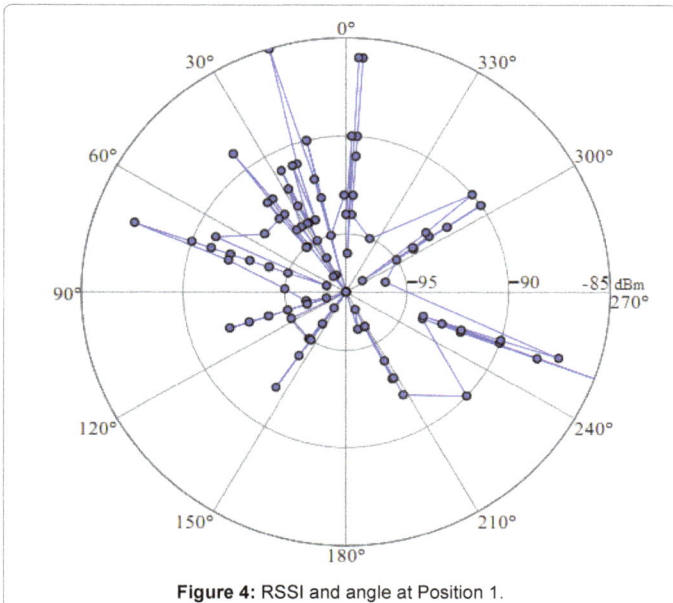

**Figure 4:** RSSI and angle at Position 1.

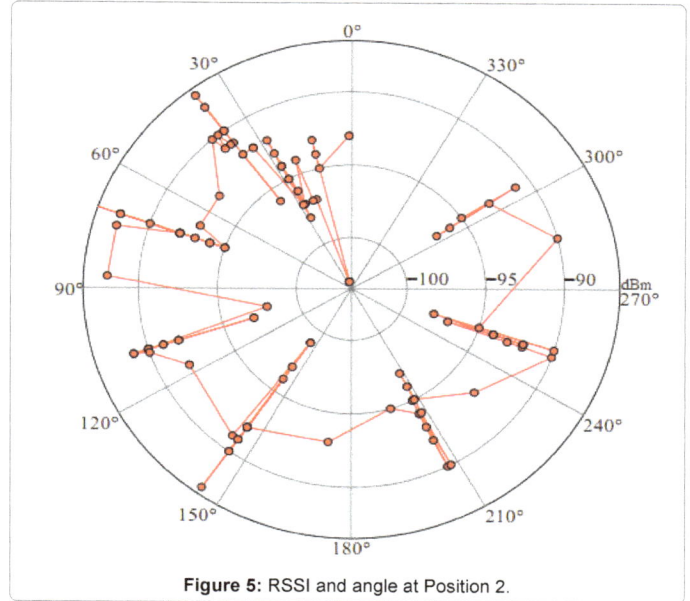

**Figure 5:** RSSI and angle at Position 2.

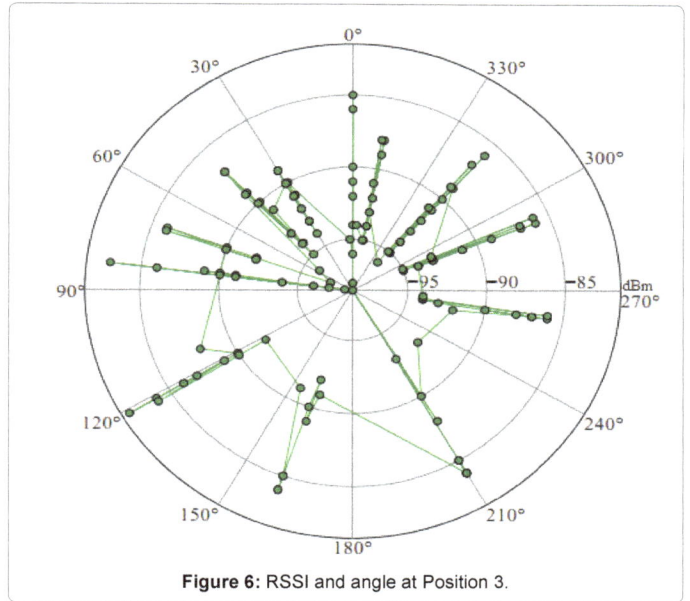

**Figure 6:** RSSI and angle at Position 3.

direction the RSSI fluctuates a lot, that's why we averaged those values and filtered the unaligned points in order to get weighted pair point as shown in Figure 8-11. The weighted points clearly showed the maximum RSSI, their angles and we observed for example Strong RSSI in Figure 9 belonging to the same lobe radiation (around 120° and 180°).

The angular velocity (Table 1) is determined by the angles covered by the rotation over the time used to cover them. The time is measured by the number of data times 0.5 s which is the period set up for every data collection.

In Table 2 we display the zone calculated by our algorithm and the angle measured with the Bluetooth sensor pointing the direction of the BLE devise. We come to the conclusion that sometimes the zone might not always contain the direction or angle measured, but can be very close to it. It's the case at Position 2; this can be explained by the

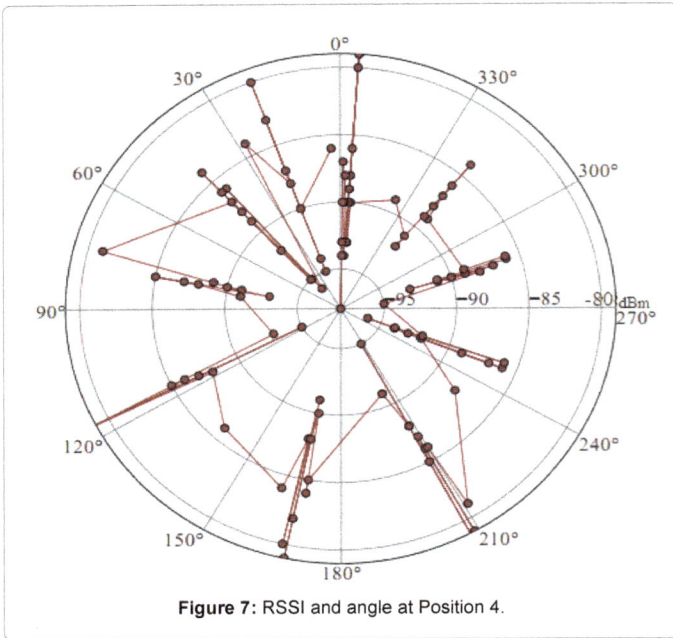

**Figure 7:** RSSI and angle at Position 4.

It is calculated as follows Accuracy $\frac{(360 - \partial Zone)}{360} \times 100, \partial$ zone is the angle difference between the zone intervals. When compared to our method, which has an accuracy of up to 84%, it can be shown that our method performs considerably well using limited resources. Also, our method implies that you can use only one device to obtain high accuracy.

Again, the results of triangulation, as reported in ref. [1] or fingerprint in ref. [9], or even proximity approach in ref. [7] show that it can have a pretty good accuracy but requires very expensive and complex infrastructure.

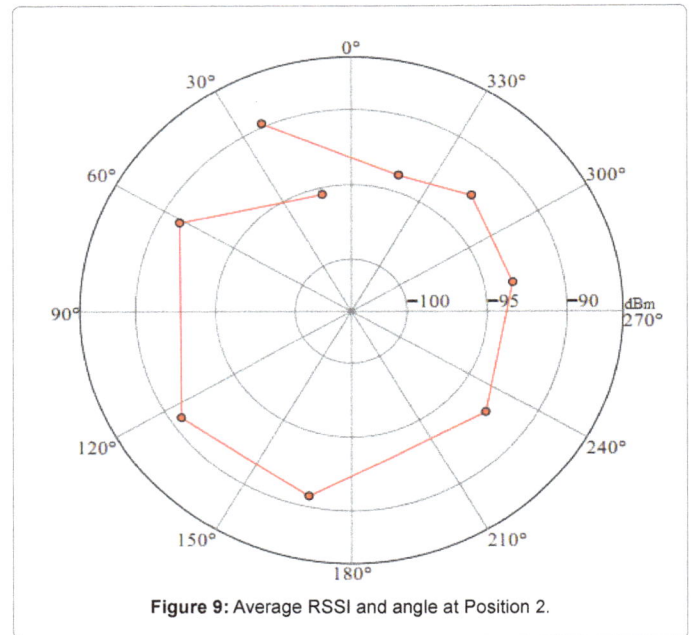

**Figure 9:** Average RSSI and angle at Position 2.

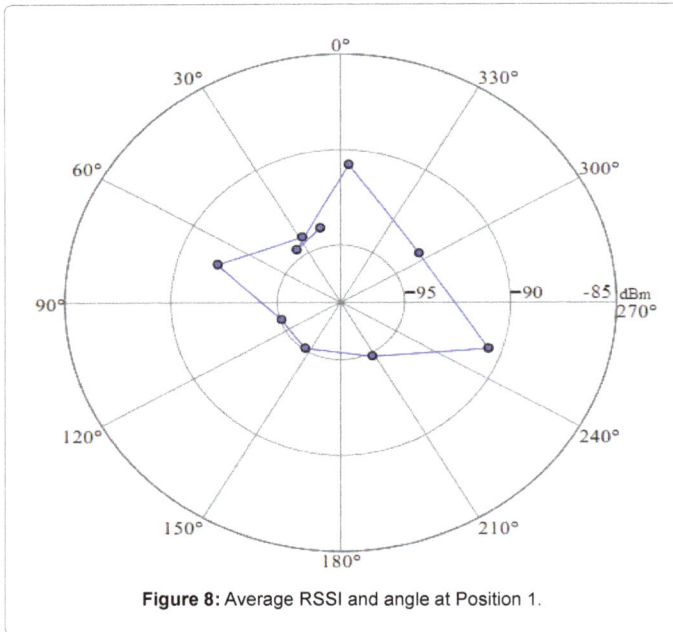

**Figure 8:** Average RSSI and angle at Position 1.

number of pause during the rotation of the smartphone. Indeed in Position 2 we counted 7 pauses during the rotation, when for the other positions we had at least 8 pauses.

As for techniques like trilateration, or triangulation the location accuracy is improved by increasing the number of reference points. So does our approach, but instead of adding reference points we increase the number of pause directions as we noticed at Position4 where we made 9 pauses direction.

The zone represents the area of certainty, where we may actually find the unknown node. The accuracy here represents the percentage of the remaining area over 360°.

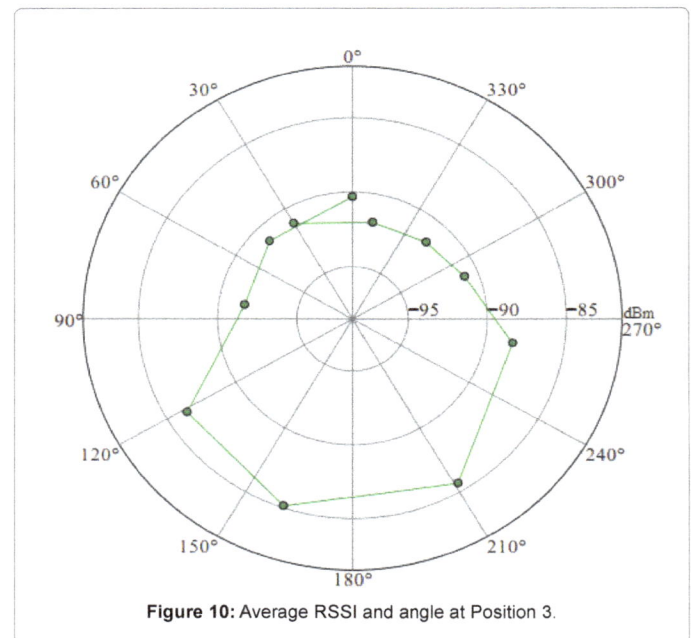

**Figure 10:** Average RSSI and angle at Position 3.

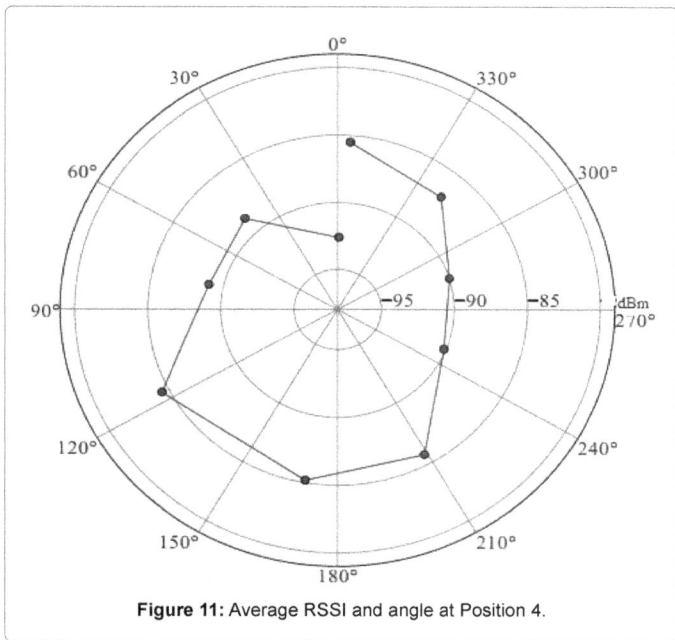

**Figure 11:** Average RSSI and angle at Position 4.

| Position | Position 1 | Position 2 | Position 3 | Position 4 |
|---|---|---|---|---|
| Angular velocity | 5.16°/s | 5.42°/s | 4.9°/s | 6.3°/s |

**Table 1:** Angular velocity of the Smartphone rotation.

| Angles | $\theta_1$ | $\theta_2$ | $\theta_3$ | Zone interval | Direction measured | Accuracy |
|---|---|---|---|---|---|---|
| Position 1 | 357° | 251° | 71° | [357°;71°] | 10° | 79.45% |
| Position 2 | 24° | 124° | 61° | [24°;124°] | 13° | 72.23% |
| Position 3 | 161° | 261° | 0° | [-19°;81°] | 0° | 72.23% |
| Position 4 | 117° | 356° | 320° | [297°;356°] | 356° | 83.68% |

**Table 2:** Example of Zone determination based on θ1, θ2, θ3.

## Conclusion

In this paper we have implemented an algorithm that determined the location of an unknown device by using RALE approach. The experiment results give the zone where the unknown device is located. RALE just requires RSSI and angles, regardless of the technology used (Bluetooth, Wi-Fi…). It doesn't require also prior survey of scene or special map, sophisticated infrastructure or resources, not even complex algorithm. The accuracy of this method in locating the node depends on the number of pauses during the rotation of the node and also the type of antenna used. Further research may look in a way to get the node position dynamically at different heights and environment.

### References

1. Wang Y, Cuthbert L (2013) Bluetooth positioning using RSSI and triangulation methods. IEEE 10th Consum Commun Netw Conf, pp: 837-842.

2. Subaashini K, Dhivya G, Pitchiah R (2012) Zigbee RF Signal strength for indoor location sensing – Experiments and results. IEEE Conf, pp: 12-17.

3. Galván-Tejada CE, Carrasco-Jiménez JC and Brena RF (2013) Bluetooth-Wi-Fi based Combined Positioning Algorithm, Implementation and Experimental Evaluation. Procedia Technol 7:37-45.

4. Vestenicky M, Vaculik M, Vestenicky P, Mravec T (2014) Pattern Calculation in MA TLAB Environment 1:118-121.

5. Jung J, Bae C (2013.) M2M Distance Estimation in Indoor Wireless Network. IEEE Conf, pp: 287 – 290.

6. Pirzada N, Nayan MY, Subhan F, Hassan MF and Khan MA (2013) Comparative Analysis of Active and Passive Indoor Localization Systems. AASRI Procedia 5:92-97.

7. Abdellatif MM, Oliveira JM, Ricardo M (2014) Neighbors and relative location identification using RSSI in a dense wireless sensor network. 2014 13th Annu Mediterr Ad Hoc Netw. Work, pp: 140-145.

8. Wu Y, Hu J, Chen Z (2007) Radio Map Filter for Sensor Network Indoor Localization Systems. 5th IEEE Int Conf Ind Informatics, pp: 63-68.

9. Khan (2012) Bluetooth Indoor Positioning.

10. Khan R, Khan SU, Khan S, Khan MUA (2014) Localization Performance Evaluation of Extended Kalman Filter in Wireless Sensors Network. Procedia Comput. Sci 32:117-124.

11. Ahmed SH, Bouk SH, Javaid N, Sasase I (2012) Combined Human Antenna Orientation in Elevation Direction and Ground Effect on RSSI in Wireless Sensor Networks. 10th Int Conf Front Inf Technol, pp: 46-49.

12. Min BC, Lewis J, Schrader DK, Matson ET and Smith A (2012) Self-orientation of directional antennas, assisted by mobile robots, for receiving the best wireless signal strength. 2012 IEEE Sensors Appl Symp, pp: 261-266.

13. Kong Y, Yang Q (2013) An Improved Location Algorithm Based on CC2431. IEEE Int Conf Green Comput Commun and IEEE Internet Things IEEE Cyber Phys Soc Comput, pp: 646-652.

14. Liang Z, Li Y, Long Y, Member S (2014) Multiband Monopole Mobile Phone Antenna with Circular Polarization for GNSS Application. IEEE 62: 1910-1917.

# 4

# Estimating Node Density for Redundant Sensors in Wireless Sensor Network

**Dhruv Sharma[1]\*, Kayiram Kavitha[2] and R Gururaj[3]**

[1]Software Engineer, PayPal, Chennai, India
[2]Researcher, Pune, India
[3]Department of CS and IS, BITS Pilani, Hyderabad Campus, Hyderabad, India

## Abstract

The lifetime of a sensor network depends on the judicious utilization of the resource-constrained nodes. Practices like data aggregation, sleep scheduling play a major role in conserving the node's energy. But in most cases, we observe a disparity in energy consumption rates among different sensors. This disparity results from higher utilization of a small set of deployed sensors in the field leaving these sensors drained out of power. To overcome this problem, it is often required to deploy redundant sensors to act as replacements for a faulty node. Secondly, the sensor network technology, being an application oriented technology, experiences variance in network parameters from application to application because of the various dynamics in nature. It is not often viable to go for a theoretically determined sensor distribution technique. Thus, it is often required to place sensors by studying the geographical constraints. These have proven to be highly valuable in designing energy efficient routing schemes and network topologies for sensor networks. In this paper we propose a scheme to decide how the distribution of available redundant sensor nodes should take place around sensor nodes. The scheme gives the flexibility to determine sensor positions based on application and geographical constraints. We propose to use the probability estimates of the utilization of a sensor in a given deployment to achieve desired network lifetimes. We also show how in some cases we can leverage the relative position from source(s) and sink be used for the same.

**Keywords:** Wireless sensor network; Deployment of nodes; Energy-efficiency; Probabilistic analysis

## Introduction

A Wireless Sensor Network (WSN) is a collaboration of a large number of sensor nodes. The sensor node operates on small batteries with limited lifetime. Some sensors, acting as source nodes, transmit their sensed data to a sink node via multi-hop communication. Often nodes nearer to the sink expend more energy than the nodes farther from the sink. Therefore, it is highly desirable to have an efficient sensor node distribution. The sensor nodes are often deployed in remote geographical locations or hazardous environment, where the replacement of batteries is very difficult and expensive. Therefore the prime consideration in WSN is to prolong the battery lifetime by efficient utilization. It is observed that the energy spent in routing data is about 80% of the total energy in the network while the remaining energy is used in sensing and other operations. Hence, various schemes were proposed in the past for energy efficient data routing, data aggregation, query processing etc. in a WSN.

Network topology is also one of the fundamental issues in wireless sensor networks (WSNs) that affects not only routing but also energy efficiency in a WSN. An efficient topology can reduce the number of hops in the network aiding to energy efficiency. The sensors are spread across the region of interest for satisfactory coverage. A deployment strategy determines the number of sensor nodes required in the Region of Interest (ROI). The deployment strategy impacts the routing schemes. At present, the focus is on developing power-efficient topologies [1] and routing schemes [2] that ensure not only the neighbors are at appropriate distances but the next hop for transmission is chosen based on the energy availability of neighboring nodes.

In WSNs, network deployments can be broadly categorized into: random and deterministic deployments. The geographic locations like volcano, seismic zone etc., are physically inaccessible. Here, the sensor nodes are deployed by means of a helicopter or any other means and

termed as random deployment. As the sensor node location cannot be determined, the deployment strategy is termed as random deployment. This often leads to randomly distributed node densities in various portions of the network.

In contrast, deterministic deployments are preferred in scenarios when the deployment area is physically accessible. Deterministic techniques focus on coverage, network longevity, improving connectivity and improving data reliability. Not just that, deterministic schemes have more control over placement of the nodes and also provide a lower bound on the number of nodes needed to cover the area which proves helpful in achieving pre-determined performance.

A considerable research has been done in formulating algorithms [3] to determine the optimal node locations to achieve maximum coverage, connectivity and network lifetime. But, it does not consider application specific factors like inaccessible terrains; environmental obstructions etc. Therefore, deterministic deployments after careful sampling of environment specific factors are becoming increasingly popular. The City Sense [4] network for urban monitoring, the Soil Monitoring [5] etc., are some of the cases where sensors are placed by monitoring physical conditions of the application. Significant work has been done in estimating the node density and number of redundant nodes required in a WSN application. The density of redundant nodes

**\*Corresponding author:** Dhruv Sharma, Software Engineer, Paypal, L404 Lancor Appt, Central Park South, Sholinganallur, Chennai, Tamil Nadu 600119, India
E-mail: dhruvshubham@gmail.com

is estimated based on the distance of a node from the sink while dealing with connectivity and coverage.

In general, the sensors closer to the sink tend to consume more energy than those farther from the sink. The reason for this disparity is that the sensors closer to the sink are sending data, sensed by them, to the sink as well as routing the transmissions originating at the nodes further from the sink. Due to the higher rates of energy consumption, the sensors closer to the sink die early leaving the sink disconnected with a portion of the network. This disparity has a serious impact on network lifetime and connectivity. The work in [6] proposed that the sensor density should increase from source to sink. However, it was assumed that each node is equally likely to serve as the source for forwarding its data to the sink. This scenario is not true in all cases. In some cases it is possible that only outermost nodes or nodes unevenly distanced from the sink serve as sources. In such a case, the observation that the node density should increase from source to sink would not hold true and we have to look for an alternative approach to determine the critically affected regions.

Deploying unequal number of redundant sensors around each sensor is also an effective measure to resolve the discrepancies in energy consumption. A node whose area of coverage overlaps with another node is often regarded as a redundant node. The idea is to first cover the ROI with a minimal number of sensor nodes using any distribution scheme and topology, as the same is subject to application conditions, and then deploy the remaining nodes as redundant nodes around each sensor. In this paper we propose a simple concept exclaiming that the number of redundant nodes around a particular sensor should be proportional to the estimate of the energy consumption by the sensor. We first prove that in certain scenarios it's not always the best solution to increase density of redundant nodes from source to sink (Section 5). Then, for $n$-neighbor topologies, we show the hop-distance from the sink and source can be used to derive an estimate for the number of redundant nodes.

The rest of the paper is organized as follows. The literature survey is discussed in section 4. In section 5 we show, by comparing a geometrically increasing and uniform distribution model, that an increasing distribution model cannot prove to be the best solution for all scenarios. In Section 6, 7 and 8 we present the proposed scheme with description, simulation and experimental results. Finally, we conclude the work in Section 9.

## Literature Survey

In this section we discuss the various approaches for achieving desired network lifetime challenged by disparity in energy consumptions in different regions of the network. There are a lot of deployment schemes [6-9] to determine node densities in different regions of the network suggested for the achieving elevated network lifetime in a scenario where sensors have different rates of energy consumptions. Often deploying redundant sensors around each sensor based on its distance from a sink [10] is considered to be an effective measure to solve the problem. The scheme proposed in [10] is most related to our work as we determine an estimate number for these redundant sensors to be deployed in different regions of the network based on its estimated utilization in the course of the network's lifetime.

In certain scenarios, it is observed that the nodes closer to the sink tend to experience more traffic than other nodes. As a result, their energy consumption rates tend to be higher than those nodes that are distant from the sink. The nodes closer to the sink tend to die early

leaving a hole near the sink and therefore, disconnecting the sink from some nodes in the network. This phenomenon is common in WSNs where the sensor nodes are homogeneous and report events generated at a constant rate to the sink and is known as the energy-hole problem [10,11].

One of the solutions to the energy-hole problem is proposed in [6]. In this scheme, the first task is to divide the entire Region of Interest (ROI) into concentric coronas or rings around the sink. Then, one of the sensors in each ring is chosen to forward data from outmost rings to the sink at equal intervals. A message transmitted from the outermost ring (say) $C_i$ is forwarded by sensor nodes in $C_{i-1}$, $C_{i-2}$ to the sink (centre of the circle) as shown in Figure 1. With a goal to achieve an equal energy dissipation rate in all rings except the outermost one, the authors proposed that the number of nodes should increase geometrically from the outermost ring towards the sink. This is also observed in the deployment scheme proposed in [8] where nearly balanced energy depletion in the network is achievable by deploying a geometrically increasing node density from the outer coronas to the inner ones. The work in [12] proposed a technique to estimate the number of nodes to be deployed to achieve a predetermined lifetime. The ROI is divided into equal sized strips. Here the density of sensors deployed increases as the distance between a strip and the sink decreases. Hence the distribution is effective.

In [10] the author proposes a sensor deployment strategy for tree based data forwarding observing the sleep-scheduling scheme used for sensor nodes. Here, the root of the tree acts as a source, forwarding data to a sink, which may be one of the leaf nodes. It estimates number of nodes in the rooted sub-tree of an intermediate node in the data forwarding tree. Based on the estimation, the number of redundant nodes required to be placed is calculated. The model assumes each region is $w$-covered by sensor nodes, out of which one node is in active state and others are redundant. An intermediate node in the data forwarding tree should have at least $z \times W$ number of redundant nodes where $z$ increases from the source towards the sink.

We propose to increase the network longevity by estimating only the number of redundant nodes to be deployed around each sensor given an upper limit on the total number of sensor nodes that can be deployed in the network. However, the fundamental assumption in the above works is that all sensors are likely to act as source which leads to the observation that the number of redundant nodes or the node density should increase as we move from source/root towards sink. In section 5 we show that the above observation (z, the number of redundant nodes, should increase from root to sink) is not true in some models.

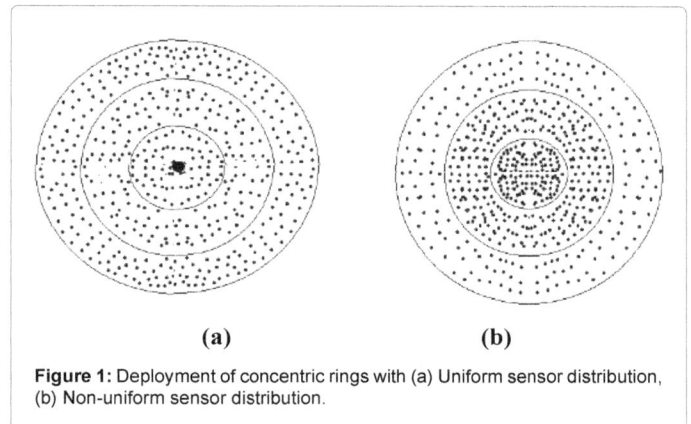

**Figure 1:** Deployment of concentric rings with (a) Uniform sensor distribution, (b) Non-uniform sensor distribution.

## Limitations in Existing Approach

As discussed in section 4, a geometrically increasing node distribution is favorable solution for the energy-hole problem. The underlying assumption in these schemes was that each sensor transmits its data to a central static sink. This model is known as a Single Static sink Centre Placement (SSCP) deployment model. Secondly, in applications such as volcano monitoring [13], where the sensor nodes are placed only in safe areas near gas plumes, it cannot be assumed that each sensor would act as a source sensor at all times. There may be nodes that act only as relays to constitute a multi-hop communication between the source sensors and the sink. Thus, it is important to study whether the same observation (a geometrically increasing node distribution) holds true for such scenarios. In this section we try to prove that "an increasing distribution model cannot prove to be the best solution for all scenarios".

In this section, we assume two popular node distribution models, mainly an increasing node distribution for redundant nodes from source to sink and a uniform distribution of redundant nodes and run simulations to understand how the network lifetime varies for both of them under different scenarios. The simulation environment is explained in Section 8. Further, we simulate our experiments over different scenarios based on the following factors:

1) The sensors send data to the sink acting as source nodes, i.e., the set of sensors acting as source during the simulation.

a. Each node acts as a source, forwarding data to the sink at equal intervals,

b. Only the end nodes or the nodes in the outermost rings act as the source with each one sending data to the sink at equal intervals, and

c. A generalized scenario where there are multiple source nodes, each at a different hop-distance from the sink and are transmitting data to the sink in unequal time intervals.

2) The position of the sink i.e.,

a. Single Static sink cOrner Placement (SSOP) [14] model, where the sink is placed at one corner of the mesh. Figure 2(a) describes a SSOP model with two sources forward data to a single sink placed in the corner.

b. Single Static sink Centre Placement (SSCP) [14] model, where the sink is placed at the centre of the mesh. Figure 2(b) describes a SSCP model with two sources forward data to a single sink placed in the centre.

3) The topology of the network. We have considered two different topologies explained as follows:

a. The 4-neighbour topology, where each sensor is surrounded by four neighbours as shown in Figure 3(a).

b. The 3-neighbour topology, where each sensor is surrounded by three neighbours as shown in Figure 3(b).

Using this experiment we try to prove that the observation of selecting a distribution where the sensor node densities increase geometrically from source to sink is not the best solution in obtaining increased network lifetime in some scenarios.

We compare the network lifetime of the sensor networks (Y-Axis) over total number of sensors (X-axis) in the network simulated using simulation parameters defined in Section 8 over these six different scenarios for two popular node distribution models, i.e., an increasing node distribution for redundant nodes from source to sink and a uniform distribution of redundant nodes:

1) A 4-neighbour, SSOP model where each node acts as a source with equal probability (results in Figure 4(a)),

2) A 3-neighbour, SSCP model where each node acts as a source with equal probability (results in Figure 5(a)),

3) A 4-neighbour, SSOP model where only the outermost nodes act as source nodes (results in Figure 4(b)),

4) A 3-neighbour, SSCP model where only the outermost nodes act as source nodes (results in Figure 5(b)),

5) A 4-neighbour, SSOP model where there are m source nodes, each at a different hop-distance from the sink and are transmitting data to the sink in unequal time intervals (results in Figure 4(c)),

6) A 3-neighbour, SSCP model where there are m source nodes, each at a different hop-distance from the sink and are transmitting data to the sink in unequal time intervals (results in Figure 5(c))

With the help of these experiments we try to figure out which factors mentioned above tend to perform better and offers an increased network lifetime for the increasing node distribution and the uniform distribution of redundant nodes.

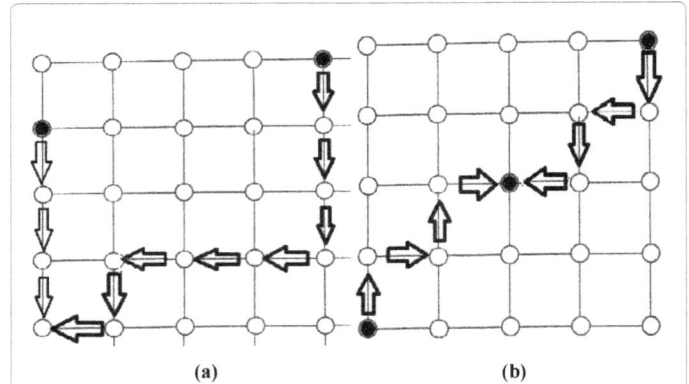

**Figure 2:** Different models based on the position of the sink (a) Single Static sink cOrner Placement (SSOP) (b) Single Static sink Centre Placement (SSCP).

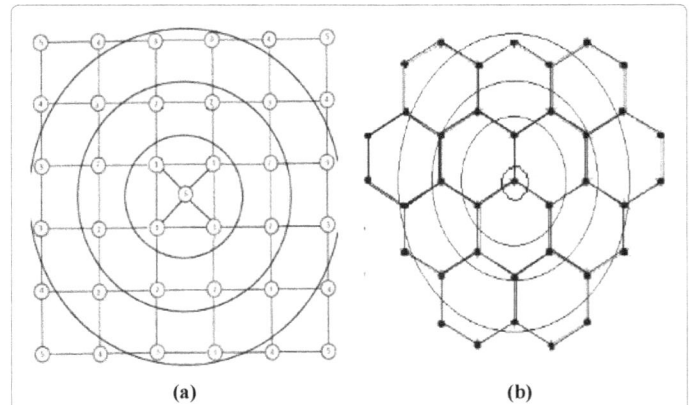

**Figure 3:** *n*-neighbour topologies (a) 4-Neighbour mesh (b) 3-Neighbour mesh.

assumption aligns with the works mentioned in Section 4 and we also see a similar result in favour of the increasing node distribution.

However, in Figure 4(b) we see the uniform distribution of redundant nodes out-performs the increasing node distribution for a SSOP model where only the outermost sensors act as source nodes and matches the results (Figure 5(b)) posed by the increasing distribution for an SSCP model. This proves our claim that there exist scenarios where the increasing node distribution cannot be assumed as the best choice for distributing redundant sensors. Intuitively, we

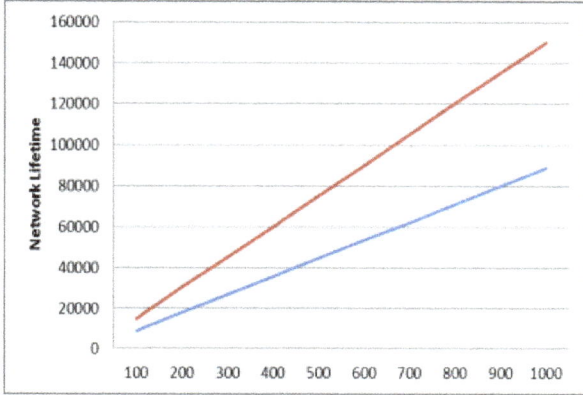

4(a)

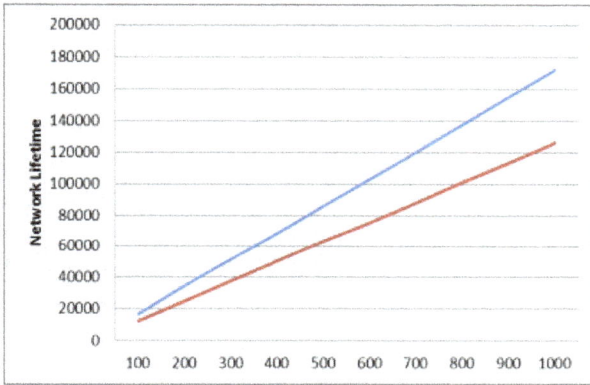

4(b)

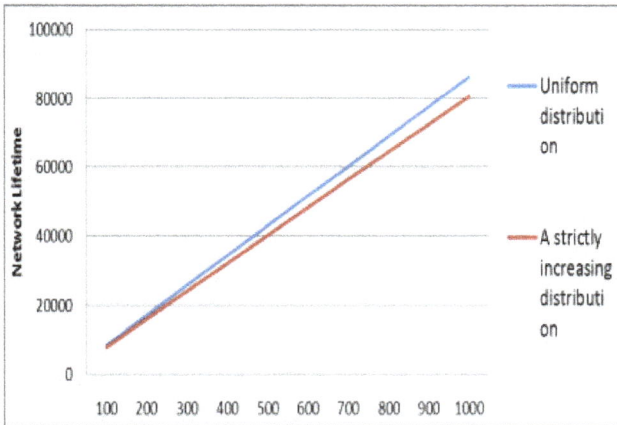

4(c)

**Figure 4:** Network Lifetime of sensor network for Uniform and Geometrically Increasing node density for 4-Neighbour SSOP topology when (a) Each active sensor acts as a source, (b) Sensors in the outermost ring acts as a source, (c) m sources each at different hop distance from the sink act as source.

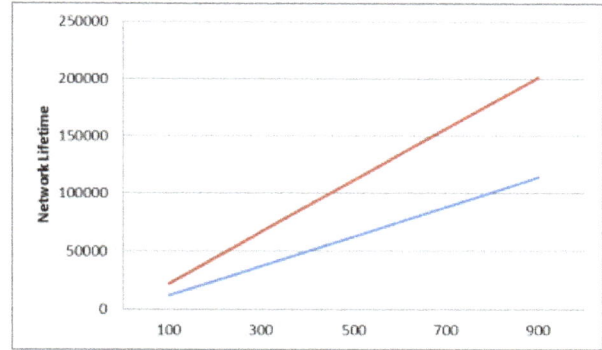

5(a)

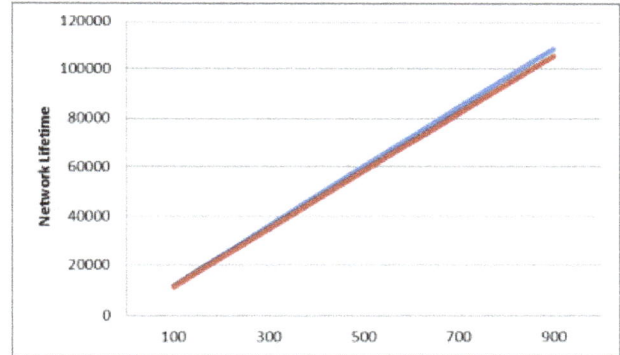

5(b)

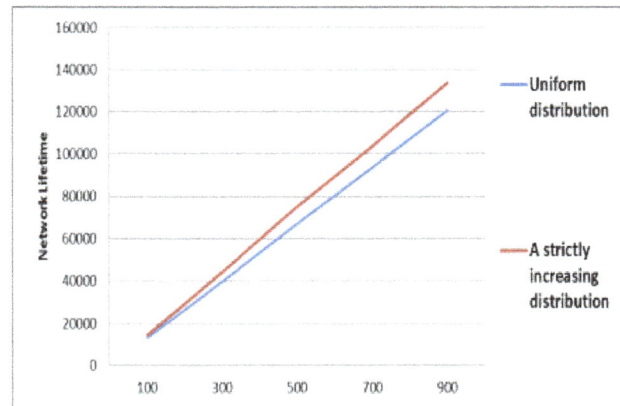

5(c)

**Figure 5:** Network Lifetime of sensor network for Uniform and Geometrically Increasing node density for 3-Neighbour SSCP topology when (a) Each active sensor acts as a source, (b) Sensors in the outermost ring acts as a source, (c) m sources each at different hop distance from the sink act as source.

In Figures 4(a) and 5(a), we observe that the geometrically increasing node distribution out-performs the uniform distribution for both SSOP model and the SSCP model. Here, we assume that each sensor acts as a source in sending data regularly to the sink. This can imagine as only the outermost sensors are required to forward data to the sink, the remaining nodes act as only relay nodes, thus expending approximately same energy for transmission/reception (data communication/message transmission).

We generalize the results by assuming a scenario where $m$ source nodes, each at a different hop-distance from the sink, are transmitting data to the sink in unequal time interval. We observe that for an SSOP model the uniform distribution slightly out-performs the increasing distribution model and in contrast the increasing distribution model slightly out-performs the uniform distribution model for an SSCP model. Since the difference in network lifetime is marginal, though we cannot generalize that an SSOP model always favours a uniform distribution and an SSCP model favours an increasing node distribution model.

However, our claim that "an increasing distribution model cannot prove to be the best solution for all scenarios" is verified.

## Proposed Scheme

In Section 4, we have seen different approaches like determining node densities, adding redundant nodes etc. to solve the energy-hole problem. These approaches have distributed nodes in geometrically increasing number from source to sink. But, as discussed in Section 5, the observation cannot be chosen as a universal solution for different network models. We have seen how in few cases, even a uniform distribution has out-performed the increasing distribution model. In this section, we propose a generalized scheme which can fit the trends for both models and provides a generic solution for any network model which could be considered a hybrid of both.

We consider an application with a known regular topology, a power-efficient routing algorithm to estimate the number of redundant nodes that can be installed around the sensors to increase network lifetime. The scheme provides a proportionality equation and shows how it can be used to calculate the desired numbers.

Formally: Given the total number of extra sensor nodes available to be used for redundancy and a pre-determined deployment, it determines how the distribution of these redundant sensors should take place to maximize network lifetime.

We assume that each source forwards a fixed size of data to the sink. We also assume the use of a regular topology to ensure that the distance between 2 sensors is not a variable factor in determining the energy required to transmit a packet and the use of power efficient algorithm to ensure, (though cannot completely) that a data forwarding path takes a form: source → 1-hop →2-hop → 3- hop→ n hop to reach the sink. We assume that each source sends a packet of same size to the sink.

We propose that the number of redundant nodes around a sensor should be directly in proportion with the estimate of energy consumed by the sensor.

$$n \propto \bar{e} \tag{1}$$

We consider $\bar{e}$ as the estimate energy used for communication. The communication energy $\bar{e}$ spent during one run of data forwarding from a source to sink is taken to be the linear combination of the product of probability of utilization and the energy spent in doing so. Since a node can act as both a source and relay, $\bar{e} = \bar{e}_{source} + \bar{e}_{relay}$. Substituting the probabilities:

$$\bar{e} = (\bar{e}_T \ast P_S) + ((\bar{e}_T + \bar{e}_R) \ast P_R) \tag{2}$$

Where $\bar{e}_T$ and $\bar{e}_R$ are the estimated energy used to transmit and receive the packets respectively and $P_S$ and $P_R$ are the probability of acting as a source and relay respectively. The energy estimates used here are corresponding to the energy spent during that action in one round of packet is forwarded from source to sink. We consider the scenario where we have a single sink and m sources where 0<m< total number of sensors T. Let the source nodes be $S_1, S_2, S_3 \dots S_m$ and let the probability of the source $S_i$ to act as the source in a query plan be $p_i$. Thus, for a sensor j, the probability of acting as a relay is

$$P_R(j) = \sum_{i=1}^{i=m} p_i \times u_{ij} \tag{3}$$

Here $u_{ij}$ is the probability of sensor j to act as a relay node for a transmission originating from the $i^{th}$ sensor as the source. $u_{ij}$ is a probability estimate and its estimation especially in an irregular topology is dependent on various physical and deployment factors. The accuracy of the estimate dictates the efficacy of the proposal. The scheme does not lay any rule to calculate the probabilities but an experimental procedure of determining the probabilities by running a simulation of the experiment should be best suited. The values can also be estimated by using heuristics.

Substituting equation (2) and equation (3) into equation (1) we get the number of redundant nodes around $j^{th}$ sensor,

$$n_j \propto (\bar{e}_T \times P_S(j) + (\bar{e}_T + \bar{e}_R) \times (\sum_{i=1}^{i=m} p_i \times u_{ij}) \tag{4}$$

Since, this equation only utilizes the energy spent during data forwarding it can be generalized even further by adding the term $e_{others}$, which would be estimate of energy spent during other tasks like sensing, aggregation etc.

$$n_j \propto e_{others} + (\bar{e}_T \times P_S(j) + (\bar{e}_T + \bar{e}_R) \times (\sum_{i=1}^{i=m} p_i \times u_{ij}) \tag{5}$$

To calculate actual numbers we need to decide upon a constant condition. We assume that the total number of redundant sensors available has an upper limit N. Hence,

$$N = \Sigma n_j$$

Also, the energy dissipated at each level should be normalized. Hence, for sensors $S_i$ and $S_j$:

$$n_j / n_i = e_j / e_i$$

The two equations are sufficient to solve for all n by assigning any one of the n values.

Now we try to elaborate the equation for two special cases where:

1) Each node acts as a source, forwarding data to the sink at equal intervals,

2) Only the end nodes or the nodes in the outermost rings act as the source

We try to use the number of sensors at a hop-distance to estimate the values for various variables declared above.

Let $t_1, t_2, t_3 \dots t_n$ be the total number of sensors at 1, 2, 3 and n hop distances from the sink. The participation of a sensor at k-hop distance in a query as a relay node is only practical when it acts as a relay node for queries originating from sensors at hop-distances >k from the sink.

### Scenario 1: *Every active sensor acts as a source*

Assume that the furthest node from the sink is at n-hop distance from the sink.

Here, $P_S = 1/\Sigma t_i = 1/T$, as the probability of a sensor acting as a source is 1/total number of sensors (T). Now, as each sensor acts as a source with equal probability.

From equation (3)

$$P_R(j) = \sum_{i=1}^{i=m} p_i \times u_{ij}$$

Here,

$$P_i = \frac{1}{\Sigma t_i} = \frac{1}{T} \qquad (6)$$

Also, $u_{ij} = 0$, if source sensor $S_i$ is at hop distance <k as the sensors at the $k^{th}$ hop distance would not be utilized to act as a relay node. The value of $u_{ij} = 1/t_k$ if source sensor Si is at hop distance >k as the participation of a sensor $S_k$ to act as a relay node is shared by all the $t_k$ sensors at k-hop distance from the sink.

For a sensor at k-hop distance from the sink, to calculate $P_R(j)$ we would require to summate over the sources at a hop-distance >k. The total number of such source nodes would be a summation of all nodes at hop-distance >k. Thus, the number of source nodes becomes $\sum_{i=k+1}^{i=n} t_i$.

Substituting $p_i$ in eq (3),

$$P_R(j) = \sum_{i=1}^{i=m} \frac{1}{T} \times u_{ij} \qquad (7)$$

Substituting the total number of effective source nodes in equation (7),

$$P_R(j) = \frac{1}{T} \times \frac{\sum_{i=k+1}^{i=n} t_i}{t_k} \qquad (8)$$

Substituting equation (8) in equation (5),

$$n_j \propto \left(\overline{e_T} \times \frac{1}{T} + (\overline{e_T} + \overline{e_R}) \times (\frac{1}{T} \times \frac{\sum_{i=k+1}^{i=n} t_i}{t_k})\right) \qquad (9)$$

**Scenario 2:** *A sensor at n-hop distance from the sink acts as a source*

Here, $P_s = 0$ for relay nodes, as only the sensors at n-hop distance from the sink act as source nodes.

From eq(3)

$$P_R(j) = \sum_{i=1}^{i=m} p_i \times u_{ij}$$

Here,

$$p_i = 1/t_n \qquad (10)$$

Also, $u_{ij} = 1/t_k \qquad (11)$

As the participation of a sensor $S_k$ to act as a relay node is shared by all the $t_k$ sensors at k-hop distance from the sink.

For a sensor at k-hop distance from the sink, to calculate $P_R(j)$ we would require to summate over all the sources. The total number of source nodes would be $t_n$ as only the nodes at n-hop distance act as source nodes.

Substituting $p_i$ in equation (3),

$$P_R(j) = \sum_{i=1}^{i=t_n} \frac{1}{t_n} \times u_{ij} \qquad (12)$$

Substituting $u_{ij}$ in equation (12),

$$P_R(j) = \sum_{i=1}^{i=t_n} \frac{1}{t_n} \times \frac{1}{t_k} \qquad (13)$$

Substituting equation (13) in equation (5),

$$n_j \propto \overline{e_T} \times 0 + (\overline{e_T} + \overline{e_R}) \times (\sum_{i=1}^{i=t_n} \frac{1}{t_n} \times \frac{1}{t_k}) \qquad (14)$$

In the next section, we compare the network lifetime obtained by using the proposed scheme using equation (4), equation (9) and equation (14) against two popular node distribution models i.e., an increasing node distribution for redundant nodes from source to sink and a uniform distribution of redundant nodes.

## Experiment and Results

In this section, we compare the network lifetime of the proposed scheme against two popular node distribution models i.e., an increasing node distribution for redundant nodes from source to sink and a uniform distribution of redundant nodes. We compare the network lifetime of the sensor networks (Y-axis) over total budget of redundant sensors (X-axis) in the network simulated using simulation parameters defined in section 8 over these six different scenarios:

I. A 4-neighbour, SSOP model where each node acts as a source with equal probability (results in Figure 6(a)),

II. A 3-neighbour, SSCP model where each node acts as a source with equal probability (results in Figure 7(a)),

III. A 4-neighbour, SSOP model where only the outermost nodes act as source nodes (results in Figure 6(b)),

IV. A 3-neighbour, SSCP model where only the outermost nodes act as source nodes (results in Figure 7(b)),

V. A 4-neighbour, SSOP model where there are m source nodes, each at a different hop-distance from the sink and are transmitting data to the sink in unequal time intervals (results in Figure 6(c)),

VI. A 3-neighbour, SSCP model where there are m source nodes, each at a different hop-distance from the sink and are transmitting data to the sink in unequal time intervals (results in Figure 7(c))

In Figures 4(a) and 5(a) we have seen a case where when each sensor node acts as source node, the increasing node distribution for redundant nodes out-performs the uniform distribution and contrastingly through Figures 4(b) and 5(b) (Section 5), the uniform distribution out-performs the former when only the outermost sensors act as source nodes. We also observed how the SSCP model slightly favours the increasing distribution model and the SSOP model favours the uniform distribution model. In Figures 6(a) and 7(a) we extended our experiment to compare these results against the numbers obtained from the proposed scheme. In Figures 6(a) and 7(a) we can see how the proposed scheme out-performs both the distribution schemes for the SSOP model and nearly matches the numbers for increasing distribution for the SSCP model. Also, through Figures 6(b) and 7(b) we can observe a similar trend where the proposed model out-performs both the distribution schemes for an SSCP model and nearly matches the numbers obtained for uniform distribution for the SSOP model. Thus, as proposed, the scheme is able to generalize the trends mentioned above. We obtained the numbers for a generalized scenario where *m* different sensors act as source nodes forwarding data at unequal intervals for both SSOP and SSCP models. This is done to observe the three distribution schemes performance for a generalized model and in both cases; the proposed scheme is able to out-perform the other distribution schemes.

## Simulation Environment

For the purpose of simulation, we have used AlgoSenSim to simulate

6(a)

6(b)

6(c)

**Figure 6:** Comparing network lifetime of sensor network between proposed schemes, uniform and geometrically increasing node density for 4-Neighbour SSOP topology when (a) Every active sensor acts as a source, (b) outermost nodes act as source nodes, (c) m sources each at different hop distance from the sink act as source.

Now, we give a brief description of assumptions made for simulation. Each packet will be of fixed size i.e., 1000 bytes, originating from a source to a single static sink. For battery usage, we assume that only the energy spent in message transmission in the network, contribute to depletion in energy. We also assume that at any point of time only 1 sensor is active in the cluster and all other nodes (acting as redundant sensors) are in a Sleep State.

For each simulation, the total number of nodes in the budget was varied from 100 to 1000 and the results are plotted to see the effect of number of total nodes on the experiment.

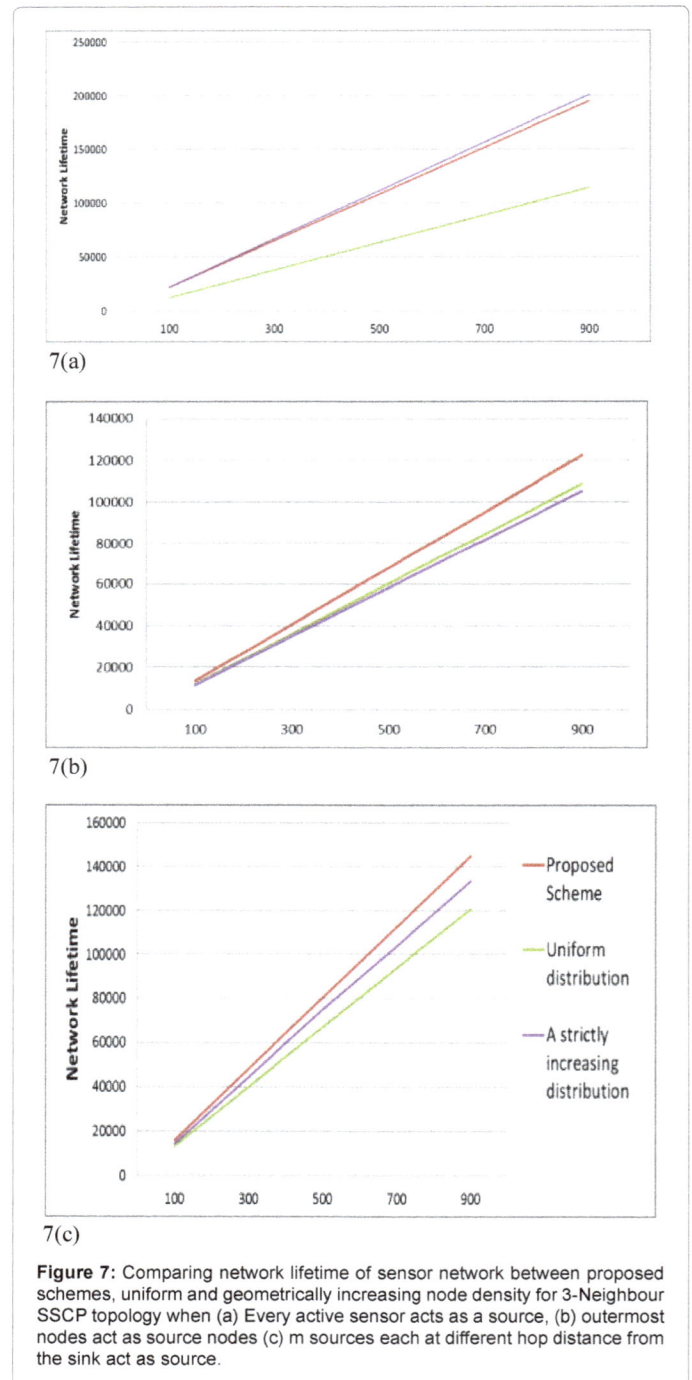

7(a)

7(b)

7(c)

**Figure 7:** Comparing network lifetime of sensor network between proposed schemes, uniform and geometrically increasing node density for 3-Neighbour SSCP topology when (a) Every active sensor acts as a source, (b) outermost nodes act as source nodes (c) m sources each at different hop distance from the sink act as source.

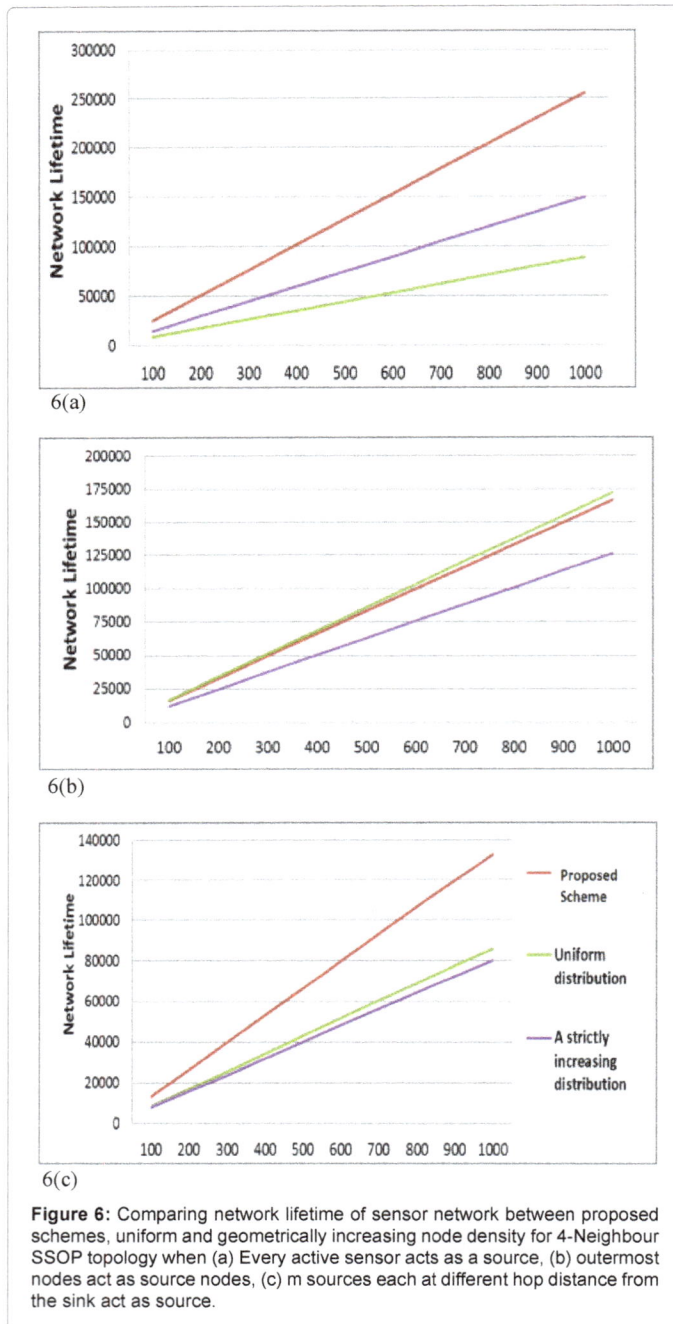

the WSN environment for various cases. AlgoSenSim performs the simulation in iterations called generations. Each generation is basically an instance in time in which it evaluates all the tasks that have to be performed across all nodes in the network. A task it performs at each node is called an Algorithm. We used the Power-DSAP (Directional Source-Aware Protocol) [2] scheme, a power efficient greedy routing algorithm, as the routing algorithm for the purpose of simulation. Power-DSAP has the advantage of being able to be used as a power-efficient routing scheme and the flexibility of using it with 3-neighbour, 4-neighbour generalized to $n$-neighbour mesh topologies. The metrics on which the evaluation is done is network lifetime. So, for the purpose of the tests, the number of generations in AlgoSenSim it takes for the first cluster to fail is taken as the estimate of network lifetime.

## Conclusion

In this paper we contemplated about the energy-hole problem persisting in wireless sensor network and how placing redundant nodes around sensors receiving more traffic can substantially increase the lifetime of the sensor network. We first discussed that how the previous studies were able to resolve the same for a model where each sensor can act as a sink and how increasing network density from source towards sink can solve that problem. In general, abiding by the concept of application specific network design, we assume, given that initial position of nodes are decided based on application parameters like terrain, areas to monitor, region accessibility etc., there should be a way to decide that in a budget of total $N$ sensors, how can we distribute the redundant nodes around each sensor to increase the network lifetime. We showed that a single model for density distribution is not favourable for multiple cases. Thus, we generalize the trends using a simple equation. We see how it is effective in resolving the problem for different cases and can be generalized using experimentally or heuristically computed estimations. The future work is to generalize for irregular topologies and a larger variance of routing schemes. This enables us to decide how effectively we can distribute the power consumption among the limited resource in hand, in turn increasing network utilization and lifetime.

## References

1. Seung yun K, Guzide O, Cook S (2009) Towards an Optimal Network Topology in Wireless Sensor Networks: A Hybrid Approach. International Conference on Sensor Networks and Application pp: 13-18.

2. Salhieh A, Weinmann J, Kochhal M, Schwiebert L (2001) Power efficient topologies for wireless sensor networks. International Conference on Parallel Processing pp: 156-163.

3. Cheng P, Chuah CN, Liu X (2004) Energy-aware Node Placement in Wireless Sensor Networks. in Global Telecommunications Conference GLOBECOM '04 IEEE 5: 3210-3214

4. Murty R, Mainland G, Rose I, Roy Chowdhury A, Gosain A, et al. (2008) CitySense: An urban-scale wireless sensor network and testbed. In: Proc. IEEE Int. Conf. Technol. Homeland Secur pp: 583-588.

5. Ramanathan N, Schoelhammer T, Kohler E, Whitehouse K, Harmon T, et al. (2009) Suelo: Human-assisted sensing for exploratory soil monitoring studies. In: Proc. ACM SenSys pp: 197-210.

6. Olariu S, Stojmenovic I (2006) Design guidelines for maximizing lifetime and avoiding energy holes in sensor networks with uniform distribution and uniform reporting. IEEE INFOCOM.

7. Liao WH, Lin MS (2011) An energy-efficient sensor deployment scheme for wireless sensor networks. IEEE ICVES, pp: 76-81.

8. Wu X, Chen G, Das SK (2008) Avoiding Energy Holes in Wireless Sensor Networks with Nonuniform Node Distribution. IEEE transactions on Parallel and Distributed systems 19: 710-720.

9. Liu X, Mohapatra P (2005) On the Deployment of Wireless Sensor Nodes. Proc. Sen Metrics pp: 78-85.

10. Chakraborty S, Chakraborty S, Nandi S, Karmakar S (2013) Exploring Gradient in Sensor Deployment Pattern for Data Gathering with Sleep based Energy Saving. Wireless Communications and Mobile Computing Conference (IWCMC) pp: 1394-1399.

11. Li J, Mohapatra P (2005) An analytical model for the energy hole problem in many-to-one sensor networks. Proc. of Vehicular Technology Conference pp: 2721-2725

12. Seetharam A, Bhattacharyya A, Naskar MK, Mukherjee A (2008) Estimation of node density for an energy efficient deployment scheme in wireless sensor network. Proc. of the 3rd COMSWARE pp: 95-98.

13. Zhang Y (2005) Wireless sensor network for volcano monitoring. Master's Thesis, KTH, School of Information and Communication Technology (ICT), Microelectronics and Information Technology, IMIT p. 49.

14. Lian J, Naik K, Agnew GB (2006) Data Capacity Improvement of Wireless Sensor Networks Using Non-Uniform Sensor Distribution. International Journal of Distributed Sensor Networks 2: 121-145.

# 5

# Surface Micro Defect Detection of Tapered Rollers Based on Laser Diffraction

**Li Cao[1], Shuncong Zhong[1,2]*, Qiukun Zhang[1] and Xinbin Fu[3]**

[1]Laboratory of Optics, Terahertz and Non-destructive Testing & Evaluation, School of Mechanical Engineering and Automation, Fuzhou University, Fuzhou 350108, China
[2]Fujian Key Laboratory of Medical Instrument and Pharmaceutical Technology, Fuzhou 350108, P.R. China
[3]Xiamen Special Equipment Inspection Institute, Xiamen 361000, P. R. China

## Abstract

A method based on laser diffraction was reported to improve the defect recognition accuracy of surface micro defect detection of tapered rollers. According to Fraunhofer diffraction theory, the fluctuations of the width of a tiny slit can be transformed to obvious changes of diffraction fringes, which can be employed to measure the micro surface defect of tapered rollers. These optical diffraction fringes were captured by a CCD camera, and subsequently were transmitted to a developed image processing system. The system includes image de-noising based on anisotropic diffusion, automatic extraction of fringe center lines by the derivative-sign binary image method, and analysis of the extracted fringe center lines for automatic defect detection. The experimental results demonstrated that the proposed method could magnify defect effect and therefore improve the accuracy of defect detection, making it attractive for industrial applications on tapered roller defect detection.

**Keywords**: Laser diffraction; Surface micro defect detection; Tapered roller; Image processing

## Introduction

Recently, due to high precision, high quality, and high speed demand of mechanical equipment becomes higher and higher, the bearing plays an increasingly prominent role in industries. Rolling element (known as roller), the most important one of the rolling bearing components, is crucial to the rotating accuracy, kinematic performance, and working life of the bearing. For a tapered roller bearing, the roller is even more important since it acts as bearing and transmitting role, which makes great influence on working performance [1]. According to the experiments, surface roughness of the bearing roller takes fifty percent influence of the every factor: the lifetime, working noise, and rotating accuracy [2]. Thus, it can be seen that any surface micro defect of the bearing may seriously affect the working performance. Therefore, strict quality testing to the surface of the roller is necessary. Improving this detection precision is a problem demanding prompt solution for bearing industries.

At present, most bearing manufacturers adopt artificial visual inspection method. However, it is inefficiency and works on low precision. This method is also unable to overcome the error due to human fatigue. It no longer meets the requirement of modern industry. Magnetic particle flaw detection [3] is a common way for nondestructive testing. It can defect small defects on the surface as well as near the surface of the work piece, but it is not suitable to detect shallow and wide faults. Eddy current testing technique [4] is good for surface crack detection, but it is not sensitive to other cosmetic defects, and the anti-jamming performance of the system is poor. Thus, missing detection and false detection are unavoidable if eddy current technique is used. Acoustic vibration [5,6] is mainly used to test the bearing roller surface crack; however, it is only suitable for some defect detection. Detection based on CCD system for testing the roller surface defect has high precision, fast processing speed, high resolution, and strong anti-interference ability, etc. That makes it become the major trend of the current research work. However, the method lacks comprehensive knowledge of defect generation mechanism and presentation; also, there is no general digital signal processing platform and pattern recognition algorithm suitable for the detection [7]. With the development of computer technology, neural theory and pattern recognition technology has been being mature. Machine vision detection technology will make due contributions to the detection of the surface defects of bearing rollers. Therefore, the on-line automatic detection system based on machine vision technology to detect bearing roller surface defect is particularly important in the field. However, the ability of machine vision detection technology to test surface tiny defect of high precision bearing roller is limited. In the present work, laser diffraction measurement technique [8] was employed for surface micro defect detection of tapered rollers and it improved the accuracy of defect detection by magnifying defect effects.

## Defect Detection Methodology for Tapered Rollers

### Theory of laser diffraction measurement technique

Laser diffraction technique, based on Fraunhofer diffraction [9], is non-contact detection method with good stability and high precision. The measurement system employed in the present work is shown in Figure 1. The light beam from a laser source passes through the collimating lens and exposures to the slit formatted by the reference object (a high-precision cylinder) and a tapered roller sample. The generated diffraction field is received by a CCD camera located in the focal plane of lens2. The computer collects and records the information of the diffraction stripes; subsequently the image is processed by a developed image processing system. The surface defects of the tapered

*Corresponding author: Shuncong Zhong, Laboratory of Optics, Terahertz and Non-destructive Testing & Evaluation, School of Mechanical Engineering and Automation, Fuzhou University, Fuzhou 350108, China
E-mail: Chinazhongshuncong@hotmail.com

**Figure 1:** Detection system: the light beam from a laser source passes through the collimating lens 1and exposures to the slit formatted by the reference object (a high-precision cylinder) and a tapered roller sample. The generated diffraction field is received by a CCD camera located in the focal plane of lens2. $f$ is the focal length of the objective lens2. $\theta$ is the diffraction angle. Cylinder 1 and cylinder 2 are used to support the sample. Cylinder 1 is rotating during the inspection.

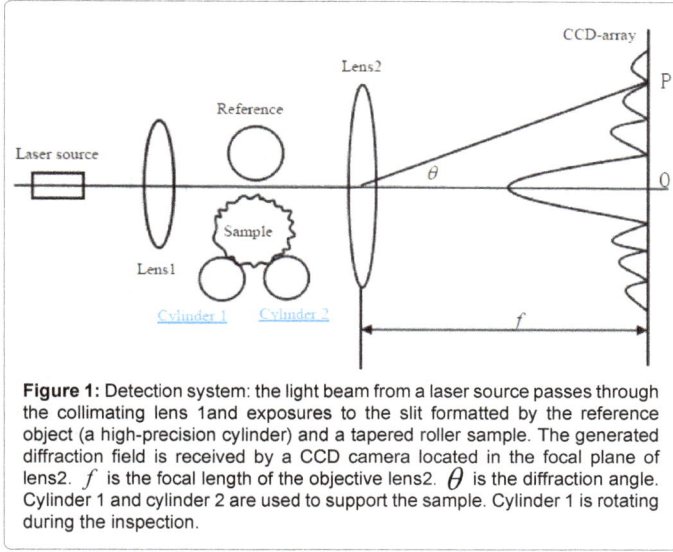

roller can be identified by analyzing the spacing between the $k$-th dark stripe and 0-th stripe.

The experimental setup for laser diffraction measurement system is shown in Figure 2. The sample is supported by cylinder 1 and cylinder 2. Cylinder 1 controlled by the driving motor is rotating when the detection system works. The sample will rotate during the inspection by the frictional force between cylinder1 and the sample. Noted that the rotation is driven by the motor by gear transmission. However, the limitation of this system is that the ends of the tapered roller sample cannot be tested because they are used to contact the reference to generate the slit.

The light intensity received by a CCD camera and the coordinate position of $k$-th diffraction dark stripe can be calculated approximately by the following formulas:

$$I = I_0 (\sin^2 \beta) / \beta^2 \qquad (1)$$

$$X_k = \frac{kf\lambda}{d} \qquad (2)$$

Where $\lambda$ is the wavelength of the laser source; $f$ is the focal length of lens2; $\beta = (\pi d / \lambda) \sin \Theta$, and $I_0$ is the light intensity when the diffraction angle $\Theta=0$. The width of the slit $d$ can be calculated as

$$d = \frac{kf\lambda}{X_k} \qquad (3)$$

If there exists a defect on the surface of a tapered roller (sample), as shown in Figure 3a, the diffraction fringes will be different in different locations when the slit is formatted by the contour lines of the sample and the reference. Due to the slit widths are different at defect and no defect locations, the coordinates of the $k$-th stripe will have enlarged gap in the corresponding locations. Figure 3b shows a real tapered roller with a defect (dented surface), which is used in the experiment. Detection of other defects will be considered as our future work. Figure 4a and 4b show the diffraction fringes captured from the locations where a defect and no defect exists. There is large fluctuation in the fringe marked by a red rectangle box, which results from the defect located on the surface of the tapered roller.

From Equation (2), the amplification factor of diffraction $A_f$ can be deduced as

$$A_f = \frac{\delta X_k}{\delta d} = \frac{kf\lambda}{d^2} \qquad (4)$$

In our experiment, 3-th dark stripe was used, i.e., $k=3$. A lens with focus length of $f=25$ mm and a light with wavelength of$=0.63$ $\mu m$ were employed. For the width of the slit, the suggested value of $d$ is between 0.01 and 0.5 mm [10] . Since a smaller slit width will result in more clear diffraction and higher amplification factor, we set the width of the slit to 0.02 mm. So, the amplification factor of diffraction $A_f$ is about 118, i.e., the slit width could be amplified by 118.2 times. For the measurement system, the resolution in fact depends on the pixel size of CCD and also the precision of the image processing algorithm used. The CCD camera we used is DC140M (ShenZhen D-image Technology Co.,Ltd., China) whose pixel size is $4.65 \times 4.65 \mu m$. In our current setup, we could obtain the defect information from at least 4 pixels of the CCD. Therefore, the resolution of $X_k$ is 18.6 $\mu m$ and subsequently the

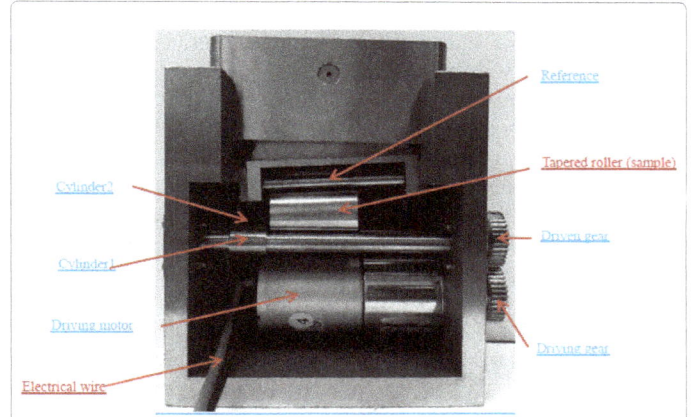

**Figure 2:** Experimental setup of the laser diffraction measurement system: The tapered roller sample is supported by cylinder 1 and cylinder 2. Cylinder 1 controlled by a driving motor is rotating during inspection. The sample will rotate simultaneously by the frictional force between cylinder1 and the sample.

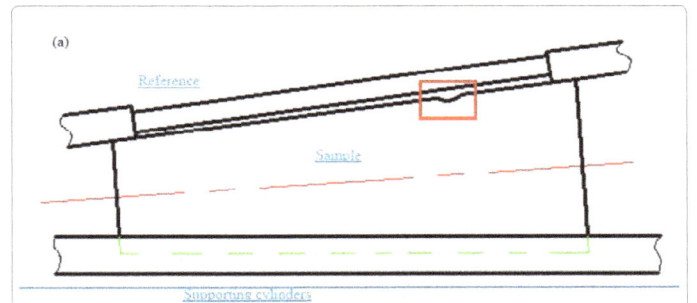

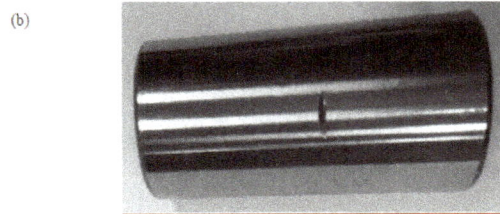

**Figure 3:** (a) The widths of the slit formed by the sample (tapered roller) and the reference are different at different locations with and without a defect. The defect is marked by a red rectangle box. (b) A real tapered roller with a defect.

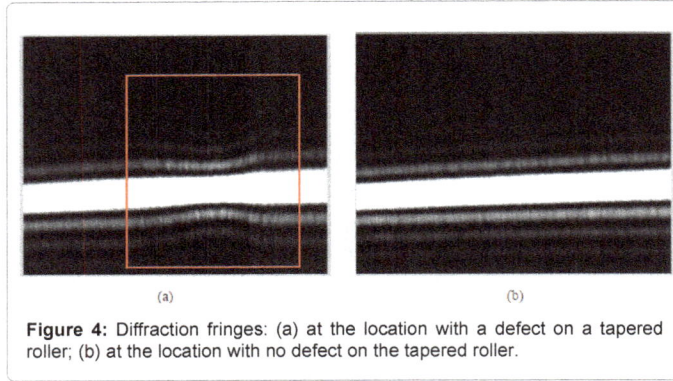

**Figure 4:** Diffraction fringes: (a) at the location with a defect on a tapered roller; (b) at the location with no defect on the tapered roller.

minimum recognizable defect depth of our system is about 0.176 $\mu m$ which can be calculated using Equation (4).

### Derivative-sign binary image method

From Figures 4a and 4b, the defect can be identified automatically by extracting fringe center lines and analyzing its characteristic. At the present work, derivative-sign binary image (DSBI) [11,12] method was employed to extract the fringe center lines. From Equation (1), we know that the derivatives in the fringe normal direction on both sides of a fringe center line have opposite signs whilst the signs of the normal derivatives in the area between adjacent dark fringes and bright fringes center lines are the same. This feature is always true for both dark and bright fringes, no matter how dark and how bright the fringe and no matter what the fringe density. The feature of diffraction fringes in Figure 4 is the same as the feature of interference fringe patterns described in ref [11]. Hence, this feature is also our criterion for constructing a binary image. In the binary image, we set the points equal to the dark value if the derivative-sign is negative, or to equal the bright value, as shown in Figure 5.

The boundaries of the binary fringes are the center lines of the original fringes without deviation. The points of the derivative-sign binary-fringe image can be presented as

$$G_{ij} = \begin{cases} B & \partial G / \partial r \big|_{ij} \geq 0 \\ 0 & \partial G / \partial r \big|_{ij} < 0 \end{cases} \tag{5}$$

where $G_{ij}$ is the gray level at point $ij$ for the binary image; $G$ is the gray level of the original image; $r$ is the position vector in the direction perpendicular to the local fringe tangent, with the positive direction defined in Figure 5; $B$ is the bright value of the binary image. We take the derivative signs as the judgment for that they are more sensitive to the gray-level variation of fringes. The derivatives of equation (5) are derived by a nominal derivative algorithm as

$$\frac{\partial G}{\partial r}\bigg|_{ij} = \sum_{r=1}^{n} \left( G_{j+r} - G_{j-r} \right) \tag{6}$$

where $G_{ij}$ is the gray level of a current point and $G_{ij+r}$ is the gray level of the $r$-th point counted from the current point in the local fringe normal direction. The $n$ normally takes a value of 2 or 3, depending on the fringe density.

In order to present the result of DSBI, a simulation was done according to Fraunhofer diffraction of single slit. Figure 6a is the diffraction fringe pattern, and Figure 6b is the corresponding derivative-sign binary-fringe image. The Figures demonstrate that the

boundaries of the binary fringes are just the center lines of the original fringes without any deviation.

### Extraction of binary-fringe boundaries

A new method is proposed in the present work for extracting binary-fringe boundaries. Because the fringes of images are not closed and are along the horizontal direction, the edge pixel points of the derivative-sign binary-fringe image have the following property

$$G_{ij} = B \,\, \& \,\, \left| G_{i+1j} - G_{i-1j} \right| = B \tag{7}$$

This feature can be the criterion for extracting binary-fringe boundaries. We set the points, in the binary images obtained by DSBI, meeting the property expressed by equation (7) to equal bright value, or to equal the dark value. There is a problem that the noise pixel points from the binary image may meet the requirement. However, the problem can be avoided by filtering the binary image before extracting binary-fringe boundaries. In this work, we adopted morphological filtering method [13].

### Image de-noising based on anisotropic diffusion

Many filtering methods treated the high frequency component as noise, which sometimes will result in losing the edge information of image. Researchers hoped to find methods that could distinguish the high frequency characteristics of image from noise, and then nonlinear diffusion methods arose. Perona and Malik first proposed nonlinear

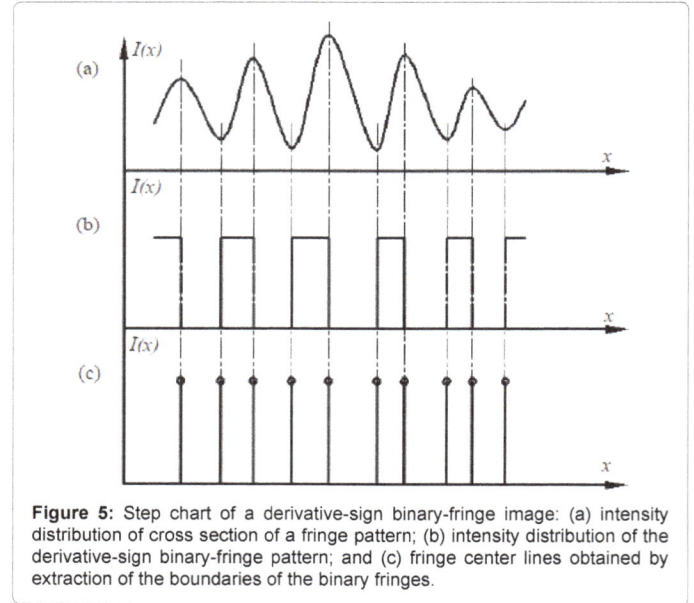

**Figure 5:** Step chart of a derivative-sign binary-fringe image: (a) intensity distribution of cross section of a fringe pattern; (b) intensity distribution of the derivative-sign binary-fringe pattern; and (c) fringe center lines obtained by extraction of the boundaries of the binary fringes.

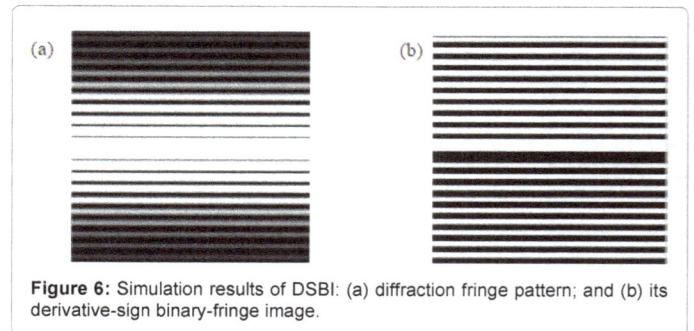

**Figure 6:** Simulation results of DSBI: (a) diffraction fringe pattern; and (b) its derivative-sign binary-fringe image.

diffusion equation: P-M equation [14]. It is mathematically formulated as a diffusion process, and it encourages intra-region smoothing in preference to smoothing across the boundaries. The estimation about local image structure is guided by knowledge about the statistics of the noise degradation and the edge strengths.

The following is the theory of the anisotropic diffusion proposed by Perona and Malik.

We assumed that $I_0$ represents a piece of gray image, $I_0(x,y)$ is the gray level at point $(x,y)$ of the image. Introducing the time factor $t$, the anisotropic diffusion equation used to modify the noisy image can be expressed as

$$\begin{cases} \dfrac{\partial I}{\partial t} = div(c(|\nabla I|) \cdot \nabla I) \\ I(x,y,0) = I_0(x,y) \end{cases} \qquad (8)$$

The function $c(|\nabla I|)$ is a monotonically decreasing function called diffusion coefficient and is anisotropic [14]. To achieve the above properties, $c(|\nabla I|)$ needs to satisfy the following conditions:

$$\begin{cases} c \longrightarrow 0 \quad \text{for} \quad |\nabla I| \longrightarrow \infty \\ c \longrightarrow 0 \quad \text{for} \quad |\nabla I| \longrightarrow \infty \end{cases}$$

Perona and Malik proposed two functions for $c(|\nabla I|)$ expressed by

$$c(|\nabla I|) = \exp(-(\frac{\nabla I}{k})^2) \qquad (9)$$

$$c(|\nabla I|) = \frac{1}{1+\left(\dfrac{|\nabla I|}{k}\right)^2} \qquad (10)$$

where $k$ is the edge magnitude parameter also is called smoothing parameter. Equation (9) is called 'option 1' and equation (10) is called 'option 2' when the anisotropic diffusion is implemented. The diffusion coefficient in equation (9) favors high-contrast edges over low-contrast ones; the diffusion coefficient in equation (10) favors wide regions over smaller ones [14].

Thus, anisotropic diffusion is an efficient nonlinear technique for simultaneously performing contrast enhancement and noise reduction. It preserves the edges by allowing no diffusion over the edges but allowing diffusion parallel to the edges. This is an iterative equation on the initial condition of image $I_0$. The solution $I(x,y,t)$ is the new image obtained after $t$ times iteration, and stop the iterating process when getting the satisfied image.

## Results and Discussions

The image processing of diffraction fringes is implemented through the following five steps: 1) Image de-noising based on anisotropic diffusion; 2) Making the DSBI and filtering the DSBI by morphology filtering method; 3) Extracting binary-fringe boundaries; 4) Extracting the 0-th and $k$-th fringe center lines; 5) Analyzing the result obtained from step 4).

Figures 7a and 7b show two captured diffraction fringes. Figure 7a is different from Figure 7b since it has a fluctuation, which reflects there is defect on the surface of the tapered roller. Five-step signal processing method is employed to process the images.

The de-noising results based on anisotropic diffusion is shown in

Figure 8. The derivative-sign binary images of Figure 8a and 8b after filtering by morphology method are shown in Figure 9. In order to illustrate the advantage of this de-noising method based on anisotropic diffusion, the result of block-matching 3D filtering (BM3D) [15], one of the state-of-the-art de-noising methods, is shown in Figure 10. The result shown in Figure 10 is poor than the one shown in Figure 9. The

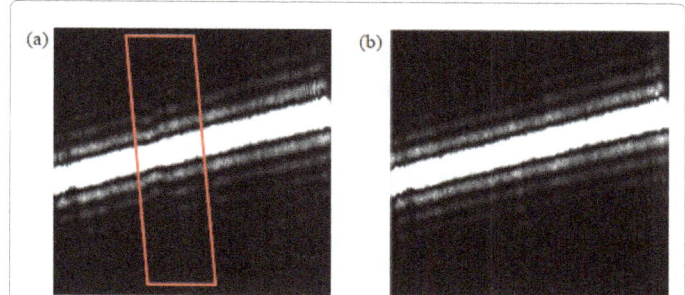

Figure 7: Original images: (a) diffraction fringes at the location with a defect on a tapered roller; (b) diffraction fringes at the location with no defect on the tapered roller.

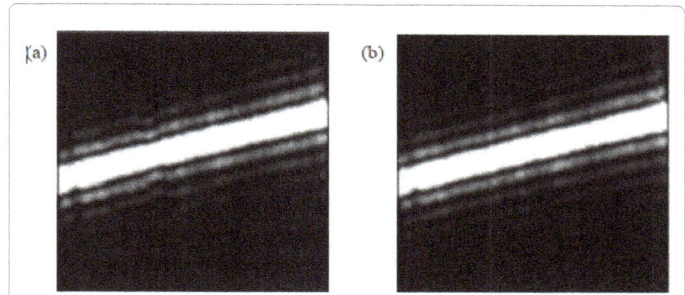

Figure 8: De-noising result based on anisotropic diffusion: (a) a defect on a tapered roller; (b) no defect on the tapered roller.

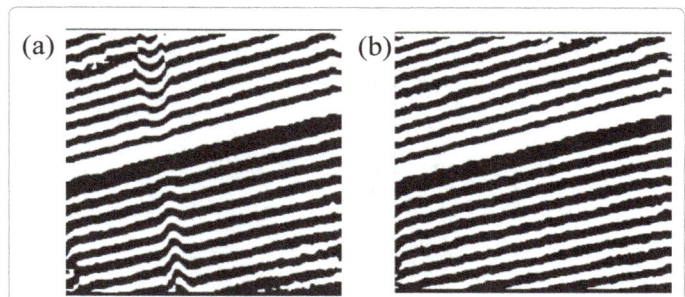

Figure 9: DSBI after morphology filtering of: (a) Figure 9 (a); (b) Figure 9 (b).

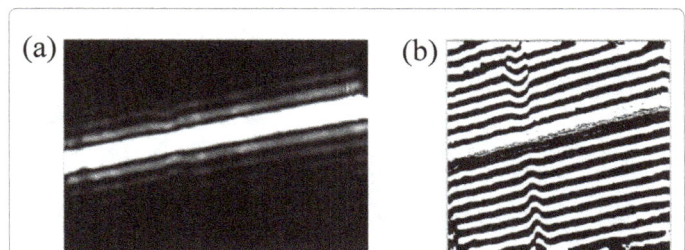

Figure 10: Result of block-matching 3D filtering: (a) the filtered image of Figure 8 (a) by BM3D; (b) DSBI after morphology filtering of (a).

DSBI of the original images are shown in Figure 11. It can be seen that there is much noise in Figureures if no filtering is not applied before making DSBI. Comparing Figures.9 (a), 10 (b), and 11 (a), we can make a conclusion that the de-noising method based on anisotropic diffusion is more consistent with our filtering purpose than the BM3D since it not only preserves the useful information but also have better filtering performance.

After making the DSBI and filtering the DSBI by morphology method, the binary-fringe boundaries are extracted by the method proposed in section 2.3, as shown in Figure 12. In order to analyze the fluctuation characteristic of the fringes, the center lines of 0-th stripes and 3-th dark stripes are extracted, as shown in Figure 13.

To recognize the defect automatically, the analyzing process could be described in two steps as following: 1) Extraction of the gray value and the horizontal coordinates of Figure 8 at the location of center lines of 0-th and 3-th dark stripes. In this step, we discarded the beginning and ending data of the center lines for that they were contaminated by the background. 2) Correction of the obtained curves by making them not slanted and by subtracting the dc component. Subsequently, the distance of the two corrected curves is calculated. The result is shown in Figure 14. It can be found that there is a peak in the distance curve as shown in Figure 14 (a). This indicates a defect exists on the tapered roller. If a proper threshold is set, defect detection of tapered roller could be achieved.

## Conclusions

In the present work, we have demonstrated a methodology for detect detection of the surface micro defect on tapered roller. Experimental results show that the method can detect the defect accurately. The whole implementation process is automatic and the image processing method is very simple and robust. According to Fraunhofer diffraction

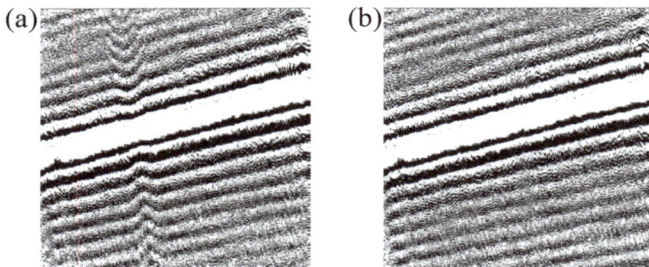

**Figure 11:** (a) DSBI of the original image shown in Figure.7 (a); (b) DSBI of the original image shown in Figure 7 (b).

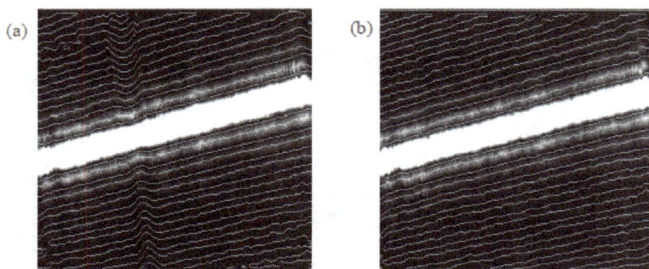

**Figure 12:** Extracted fringe skeletons of Figure 9. The skeletons are shown on the original images.

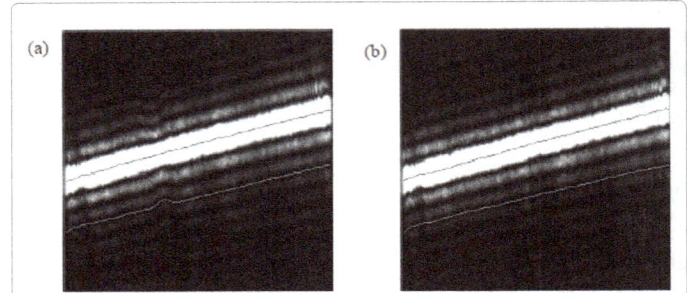

**Figure 13:** Center lines of 0-th and 3-th dark stripes. The lines are shown on the original images.

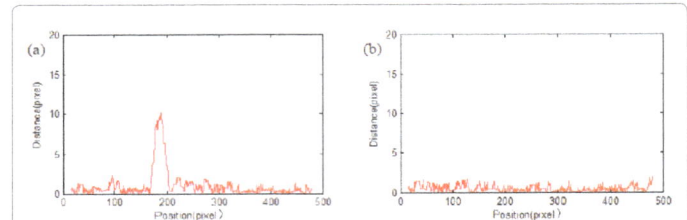

**Figure 14:** Distance of the corrected curves of center lines extracted from 0-th and 3-th dark stripes using (a) the center lines obtained in Figure.13 (a); and using (b) the center lines obtained in Figure 13(b).

theory, the fluctuations of the width of the tiny slit can be transformed to obvious changes of the coordinates of the fringes, which can be employed to measure the micro surface defect of the rollers. These optical diffraction fringes were captured by CCD-array, and then were transmitted to a developed image processing system, which includes image de-noising based on anisotropic diffusion, automatic extraction of fringe center lines by the derivative-sign binary image method, and analysis of the extracted fringe center lines for automatic defect detection on tapered roller. Our approach has great potential in real application in surface micro defect detection of taper roller and other cylindrical components.

**Acknowledgements**

We gratefully acknowledge support from the National Natural Science Foundation of China (51005077), the Fujian Provincial Excellent Young Scientist Fund (2014J07007, 2011J06020), the Specialised Research Fund for the Doctoral Program of Higher Education, the Ministry of Education, P. R. China (20133514110008), the Ministry of Health, P.R. China (WKJ-FJ-27) and Fujian Provincial Natural Science Foundation (2015J01234).

**References**

1. Wenlong W (2013) Automatic detection system for detecting the surface of tapered roller. Changchun University of Science and Technology.

2. Wen H, Yan W, Jianying Lv (2005) Analysis of factors influenced on grinding surface roughness of tapered roller. J. Bearing 12: 15-16.

3. Stadthaus M (1991) Investigations on magnetic hand-yokes for the magnetic particle inspection. J. Materials Testing: 88-91.

4. Chuanbin P, Huifen G (2002) Research on surface defects detection for railway vehicle bearing roller. J. Sci-Tech Information Development & Economy 12: 169-171.

5. Jingliang Z, Zhixi L, Yaozhi H (2006) Nondestructive detection of roller bearing with acoustic vibration. J. Mechanical & Electrical Technology 29: 36-37.

6. Jingliang Z, Zhixi L, Yaozhi H (2007) Data acquisition and processing of nondestructive testing for bearing roller. Journal of Fujian University of Technology 5: 79-82.

7. Fangqing J (2007) Research on Surface Defects Detection for Steel Strips Based on Machine Vision [D]. 2007: 68-69.

8. Suhong D, Hui F, Huisheng W (2005) Study of the Radial Beat Measurement System of High Speed Drilling Machine Using CCD Laser Diffraction. J Measurement & Control Technology 24: 71-72.

9. Xiaoai G, Xiangning L, Jiabi C (2002) System of thin cylinder diameter measurement by Fraunhofer diffraction with digital camera. M. College Physics 21: 36-39.

10. Changku S, Mingxia H, Peng W (2008) Laser measurement technology. M Tianjin: Tianjin University press pp: 84-85.

11. Yu Q, Andresen K (1994) Fringe orientation maps and 2D derivative-sign binary image methods for extraction of fringe skeletons. J. Appl. Opt 33: 6873-6878.

12. Yu Q, Liu X, Sun X (1998) Generalized spin filtering and improved derivative-sign binary image method for extraction of fringe skeletons. J. Appl. Opt. 37: 4504-4509.

13. Cheng F, Venetsanopoulos AN (2000) Adaptive morphological operators, fast algorithms and their applications. Pattern Recognit 33: 917-933.

14. Perona, Malik J (1990) Scale-space and edge detection using anisotropic diffusion. J. IEEE Trans on PAMI 12: 629-639.

15. Dabov K, Foi A, Katkovnik V, Egiazarian K (2006) Image de-noising with block-matching and 3D filtering. J Proc. SPIE Electronic Imaging: Algorithms and Systems 1: 6064A-30.

# 6

# Pedestrian Indoor Navigation System Using Inertial Measurement Unit

**Maryam Banitalebi Dehkordi\*, Antonio Frisoli, Edoardo Sotgiu, Claudio Loconsole**

*Perceptual Robotics laboratory (PERCRO), Scuola Superiore Sant' Anna, Pisa, Italy*

## Abstract

This paper presents a method for an indoor pedestrian localization, based on the data that solely are measured by a foot-mounted Inertial Measurement Unit (IMU). To locate the user accurately, a comprehensive Extended Kalman Filter (EKF) with five states is developed. Five different error reduction methods are employed to estimate the errors of all five states. These error reduction methods feed EKF independently, at stance phases or different time intervals of swing phases. The navigation system is developed using the accelerometer and gyroscope measurements and without magnetometer, thus it is insensitive to the presence of metal and magnetic fields, and it is able to estimate the user's tracked trajectory with the same accuracy in both indoor and outdoor environments. The system does not rely on the measurement from external infrastructure (e.g., RFID). To evaluate the accuracy of the system, several experimental tests are carried out over the known trajectories. Results demonstrate that the error of the estimated tracked trajectory is less than 1% of the total traveled distance.

**Keywords:** Pedestrian tracking; Dead reckoning; Inertial navigation system; Extended Kalman filter; Inertial measurement unit; Localization

## Introduction

The large potential applications of a pedestrian navigation system have attracted the attention of many researchers worldwide. Improving the performance of such systems as an important issue is still an open research challenge.

Navigation systems can be considered as indoor and outdoor. Outdoor navigation systems are mostly developed using Global Positioning System (GPS), however GPS is degraded in urban areas, tunnels, thick forest and indoor environments. Indoor navigation systems can be categorized as beacon-based (infrastructure-based) and beacon-free (infrastructure-free) approaches [1]. Beacon-based system relies on the infrastructure such as Wireless Fidelity (WIFI), Ultra-Wideband (UWB), Radio Frequency Identification (RFID) and infrared ultrasound [2-4]. Although these systems estimate the user's position with tracking error rate of less than 1% of the total traveled distance, their main drawbacks are the expenses and difficulty of infrastructure deployment. Beacon-free system relies on Inertial Measurement Unit (IMU). Utilizing the latest advancements in Micro Electro Mechanical Systems (MEMS) technology, several types of off-the-shelf IMUs are produced. IMU incorporates an assortment of inertial sensors, including orthogonal gyroscopes, accelerometers, and sometimes magnetometers and barometric pressure sensor [5,6]. However, magnetometer is sensitive to the properties of environments, e.g., presence of metals might influence the magnetometer performance, thus in general it is not reliable enough to be used in an indoor navigation system.

The signal generated by gyroscope and accelerometer can be employed to estimate the pedestrian orientation and position. Pedestrian Dead Reckoning (PDR) and Inertial Navigation System (INS) are two conventional methods that provide such estimation, taking advantage of the sequential nature of human bipedal locomotion.

In PDR methodology, position of the pedestrian is obtained based on the step length and heading angle at each detected step. The step length is time-varying, and can be estimated online using its linear relationship with step frequency [7], while the heading is obtained from the data measured by gyroscope or magnetometer. The accuracy of navigation system based on PDR alone can be as low as 2%-10%,

where acceleration can be measured on the torso, waist, foot and head [8-11]; however the error may grow quickly due to the inherent instrumental error, disturbances from the environment and unsteady walking, in addition to error of step length calculation and increment of the heading error as a non-observable state [12,13].

On the other hand, inertial navigation technology can be employed for pedestrian tracking [1,8,14]. In an INS by integrating the angular velocity signals measured by gyroscope, the IMU orientation is obtained. In addition these signals are used to transform the measured accelerations from the local coordinates system to the global coordinates system. Knowing the starting point, the pedestrian position in turn is estimated by double integrating the acceleration after subtracting the gravitational acceleration. However, the data from IMU is susceptible to drift error. Integration and double integration of biases exist on the signals measured by accelerometers and gyroscopes, together with the incorrect projection of gravitational acceleration on the horizontal axes, will result in the large tracking error. This error grows cubically in time and may exceed one meter in a few seconds [12-15].

A human gait consists of stance and swing phases [16,17]. In order to limit the INS error growth, many researchers mounted IMU on the user's shoe (as shown in Figure 1) and used different techniques to detect the stance phase, based on IMU signals [12,18-21]. Among them Foxlin [22] was one of the initiative researchers who used Extended Kalman Filter (EKF) to estimate the error and correct the states in every stance phase using Zero velocity UPdaTe (ZUPT). Zero Angular Rate Update (ZARU) is another common technique that estimates the bias of the gyroscope in every so called *still* phase [19,23]. Based on the fact that most of the paths in indoor environments are straight, Borenstein et al. [24] proposed a technique called Heuristic Drift Reduction (HDR) to

---

**\*Corresponding author:** Dr. Maryam Banitalebi Dehkordi, Perceptual Robotics laboratory (PERCRO), Scuola Superiore Sant' Anna, Pisa, Italy
E-mail: m.banitalebidehkordi@sssup.it

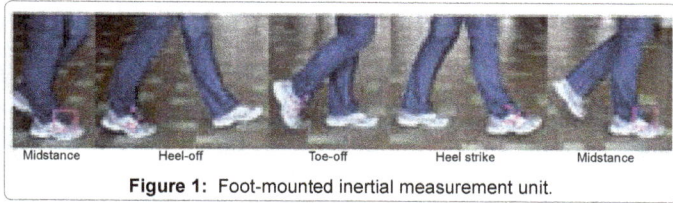

**Figure 1:** Foot-mounted inertial measurement unit.

reduce the heading error. Jimenez et al. [1,19] presented the IMU-alone EKF-based INS algorithm for indoor navigation with tracking error rate of 0.8%-1.8% of the total traveled distance. However the accuracy of these systems needs to be improved.

This paper presents an infrastructure-free navigation system which accurately locates the pedestrian indoor. The accelerometers and gyroscopes measurements are used to calculate the user's position and orientation, and the 15-element EKF is implemented to correct the estimations. Five different error reduction methods are used to measure the errors of five states, which feed EKF independently. This system is developed as a part of the blind navigation aid project [25]. Taking into account the fact that blind in indoor environments usually follows straight paths and walks parallel to the building's walls, the proper error reduction methods are implemented. In addition, while magnetometer is not used, the system is insensitive to the environment equipment and supplies. Compare to the state-of-the-art, in this work more error reduction methods are employed that feed EKF independently and at both stance and different time intervals of swing phases. It causes to obtain the higher accuracy from an EKF-based INS algorithm. The performance of the navigation system is evaluated in several tests and results show the tracking error is less than 1% in terms of the percentage of the total traveled distance.

The remainder of the paper is organized as follows. Section II presents the IMU sensor used in this work. Section III describes the EKF-based INS method. Section IV details the error reduction methods. Section V discusses an evaluation of several indoor tracking tests, and finally, conclusion and future research directions are summarized in the last section.

## Sensor Module Selection

Considering many factors including size, dynamic range, sampling rate and bias, we select the commercially available IMU, NavChipISNC01 from InterSense Inc. [5]. This miniature IMU is a high precision MEMS six-axis inertial measurement unit. Its size with castellations and frame is 92×58×22 mm (length×width×height), and it weighs 51 g. IMU is configured to provide inertial data at 200 Hz. The IMU outputs are a(t), dθ(t) and m(t) that corresponds to the output of the three-axis accelerometer, gyroscope and magnetometer, respectively.

In this work we mount the IMU on the user's foot, to take advantage of the sequential nature of the pedestrian motion and to employ error reduction methods at foot stances. Figure 2 shows how sensor can be fixed on the right foot of the user, using shoe laces. Assuming the user is walking along global y-axis, the sensor local (body) coordinates and the global coordinates are shown. Our algorithm is developed irrespective of the exact position and orientation of IMU on the user's shoe.

At every sample time, IMU measures accelerations and angular velocities. Figure 3 shows the typical signals measured by IMU during five steps of walking. As can be seen, the sequential nature of human

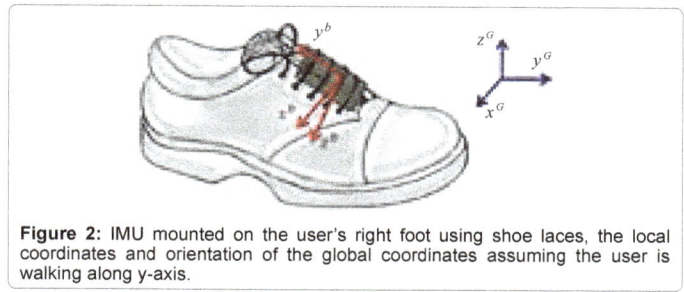

**Figure 2:** IMU mounted on the user's right foot using shoe laces, the local coordinates and orientation of the global coordinates assuming the user is walking along y-axis.

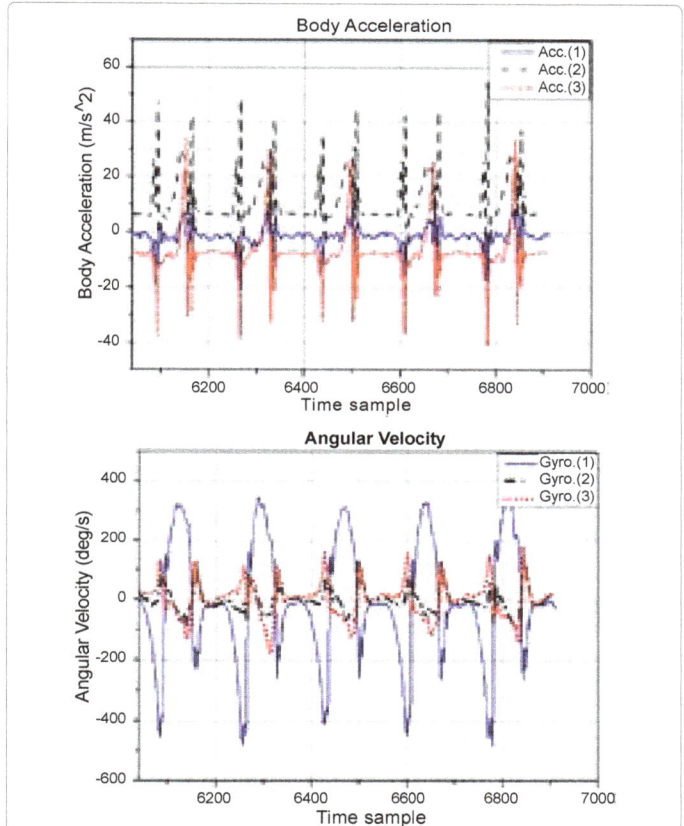

**Figure 3:** The typical signals measured by IMU over five steps of human walking, in the local coordinates. (a) Acceleration (b) Angular velocity.

bipedal walking yields to obtain the sequential similar signals.

## Inertial Tracking Methods

This section details our proposed pedestrian tracking algorithm. In comparison to the state-of-the-art on EKF-based navigation systems, in this work: (i) we update EKF with the error measurements from all five states and not only one (υ in [22]) or three states (υ, ψ and ω in [19]); (ii) the correction is applied not only at stance phases (alike previous works) but also at different time intervals of swing phases; (iii) different from the previous works, the error reduction methods update EKF independently, which improves the tracking accuracy, significantly; and (iv) the system is developed without using magnetometer, thus it can work properly in the existence of metal.

Figure 4 illustrates the framework of the proposed navigation system. As can be seen, IMU measures the user's foot acceleration **a** and angular velocity **ω**. These data are used to detect stance phases, and

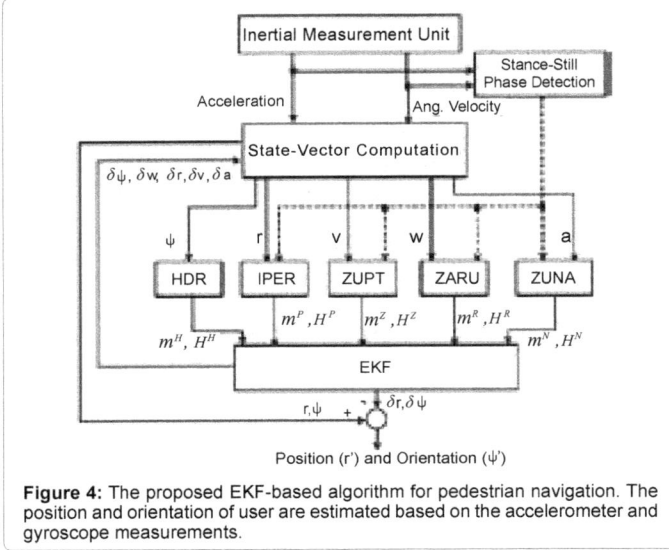

**Figure 4:** The proposed EKF-based algorithm for pedestrian navigation. The position and orientation of user are estimated based on the accelerometer and gyroscope measurements.

to calculate position $\mathbf{r}$, velocity $\mathbf{v}$ and heading $\Psi$. Five error reduction methods include Heuristic Drift Reduction (HDR), Zero Angular Rate Update (ZARU), Investigative Position Error Reduction (IPER), Zero velocity UPdaTe (ZUPT) and Zero UNrefined Acceleration update (ZUNA) are used to compensate gravity, sensor's bias and errors of EKF states. Error reduction methods feed EKF independently, with error measurement matrices $m^{(.)}$ and $H^{(.)}$ which are detailed in section IV. The final output of the system is the estimated position and orientation of the user in the global coordinates, which are parallel to the building's main walls (North-East, North-West, Up). The rotation matrix that transforms data from the local coordinates to the global coordinates is obtained employing gyroscopes signals [8]. The details of the proposed framework are as follows.

**A. Stance detection**

In order to detect the stance phase, some conditions are defined. Once all of these conditions are satisfied, the stance is detected. These conditions are as follows

- The norm of the gravity-free acceleration becomes less than the defined threshold $thr_{sa}$.

- The norm of the gravity-free acceleration variance remains under the defined threshold $thr_{sv}$.

- The norm of the angular velocity becomes less than the defined threshold $thr_{gyro}$.

If all of these conditions are satisfied for the time more than the defined threshold $thr_{time}$, the foot is detected to be at the *still* phase.

**B. Inertial navigation system**

The conventional INS algorithm estimates the user's position and orientation, based on acceleration (superscript $b$ refers to the body coordinates) and angular velocity $\omega_k^b$ which are measured by IMU's accelerometers and gyroscopes, respectively. The main difficulty of using IMU data is the accumulation of sensors's biases, which leads to a huge error on the estimated position and orientation over a short period of time. In order to correct this error EKF can be used. In this work the error state is a 15-element (five states) vector and is defined as

$$\delta\mathbf{x} = \left[\delta\Psi^T, \delta\omega^{b^T}, \delta\mathbf{r}^T, \delta\mathbf{v}^T, \delta\mathbf{a}^{b^T}\right]^T \tag{1}$$

where all five components are 3×1 matrices and, $\delta\Psi$, $\delta\mathbf{r}$ and $\delta\mathbf{v}$ are the estimated error in heading, position and velocity in the global coordinates, and $\delta\omega^b$ and $\delta\mathbf{a}^b$ are the estimated biases for gyroscopes and accelerometers in local body coordinates, respectively.

To compensate the biases exist in the IMU data, the raw acceleration $\mathbf{a}_k^b$ and angular velocity $\omega_k^b$ are subtracted by the biases estimated by EKF over the previous sample time. In addition, the error estimated by EKF is used to correct the obtained orientation $\Psi_k^{'G}$, position $r_k^{'G}$ and velocity $v_k^{'G}$ in the global coordinates (superscript G refers to the global coordinates). Thus, the procedure of states biases compensation and error correction can be written as

$$\mathbf{x}'_k = \begin{bmatrix} \Psi_k^{'G} \\ \omega_k^{'b} \\ \mathbf{r}_k^{'G} \\ \mathbf{v}_k^{'G} \\ \mathbf{a}_k^{'b} \end{bmatrix} = \begin{bmatrix} \Psi_k^G \\ \omega_k^b \\ \mathbf{r}_k^G \\ \mathbf{v}_k^G \\ \mathbf{a}_k^b \end{bmatrix} - \begin{bmatrix} \delta\mathbf{x}_k(1:3) \\ \delta\mathbf{x}_{k-1}(4:6) \\ \delta\mathbf{x}_k(7:9) \\ \delta\mathbf{x}_k(10:12) \\ \delta\mathbf{x}_{k-1}(13:15) \end{bmatrix} = \begin{bmatrix} \Psi_k^G - \delta\Psi_k^G \\ \omega_k^b - \delta\omega_{k-1}^b \\ \mathbf{r}_k^G - \delta\mathbf{r}_k^G \\ \mathbf{v}_k^G - \delta\mathbf{v}_k^G \\ \mathbf{a}_k^b - \delta\mathbf{a}_{k-1}^b \end{bmatrix} \tag{2}$$

**C. Dynamic orientation tracking**

The dynamic orientation is obtained by integrating the angular velocity measured by gyroscope, over the time. Assume $\omega_k^b = [\omega_x^b, \omega_y^b, \omega_z^b]^T$ be the angular velocity vector at sample time k, in a short sampling time, the axes will be rotated by small rotated angle vector of $\delta\Psi$ which at sample time k can be written as

$$\delta\Psi_k = \left[\delta\psi_k, \delta\theta_k, \delta\phi_k\right]^T = \omega_k^b \delta t \tag{3}$$

Due to this fact that $\delta t$ is short, $\delta\psi$, $\delta\theta$ and $\delta\phi$ are small. Therefore using approximation [8] the orientation update equation at sample time k can be obtained as

$$\mathbf{R}_{b_k}^G = \mathbf{R}_{b_{k-1}}^{'G} \cdot \left(\mathbf{I} + \frac{sin(\bar{\omega}_k \delta t)}{\bar{\omega}_k}\Omega_k + \frac{1 - cos(\bar{\omega}_k \delta t)}{\bar{\omega}_k^2}\Omega_k^2\right) \tag{4}$$

where $\bar{\omega} = \|\omega_x, \omega_y, \omega_z\|$ and $\Omega_k$ is the skew symmetric matrix at sample time k, and can be written as

$$\Omega_k = \begin{bmatrix} 0 & -\omega_{zk}^{'b} & \omega_{yk}^{'b} \\ \omega_{zk}^{'b} & 0 & -\omega_{xk}^{'b} \\ -\omega_{yk}^{'b} & \omega_{xk}^{'b} & 0 \end{bmatrix} \tag{5}$$

Obtaining the rotation matrix, acceleration in the global coordinates at time k [8] can be estimated as

$$\mathbf{a}_k^G = \mathbf{R}_{(b_k)}^G \cdot \mathbf{a}_k^{'b} \tag{6}$$

Following this notation, in order to obtain velocity and position, the gravity compensated acceleration is integrated and double integrated over the time, which can be written using trapeze integration as

$$\mathbf{v}_k^G = \mathbf{v}_{k-1}^{'G} + \left(\frac{\mathbf{a}_k^G + \mathbf{a}_{k-1}^G}{2} - g^G\right).\delta t \tag{7}$$

$$\mathbf{r}_k^G = \mathbf{r}_{k-1}^{'G} + \frac{\mathbf{v}_k^G + \mathbf{v}_{k-1}^G}{2}.\delta t \tag{8}$$

where $\mathbf{v}^G$ and $\mathbf{r}^G$ are the velocity and position of the user in the global coordinates. Using the error estimated by EKF for velocity and acceleration at time k, the corrected position $\mathbf{r}_k^{'G}$ and velocity $v_k^{'G}$, in the global coordinates, can be obtained from Equation 2. Finally, the rotation matrix is corrected using angles' errors estimated by EKF. Assuming the angles' errors are small, using Pad'e approximation [19], the corrected rotation matrix can be updated as

$$\mathbf{R}_{b_k}^{'G} = \frac{2\mathbf{I}_{3\times3} + \Xi_k}{2\mathbf{I}_{3\times3} - \Xi_k}.\mathbf{R}_{b_k}^G \tag{9}$$

where $\Xi_k$ is the skew symmetric matrix at time t, and can be written as

$$\Xi_k = -\begin{bmatrix} 0 & -\delta\Psi_k(3) & \delta\Psi_k(2) \\ \delta\Psi_k(3) & 0 & -\delta\Psi_k(1) \\ -\delta\Psi_k(2) & \delta\Psi_k(1) & 0 \end{bmatrix} \tag{10}$$

## D. Extended Kalman filter

We follow the EKF model addressed in [19,22] and proposed our EKF-based navigation system with dynamic error measurement matrices. Assume $\delta\mathbf{x}_k'$ is the error state vector at time k, which is a nonlinear function of the states, the linearized state transition at time k can be written as

$$\delta\mathbf{x}_k = \mathbf{\Phi}_k\delta\mathbf{x}'_{k-1} + w_{k-1} \tag{11}$$

where $\delta\mathbf{x}_k$ is the estimated error state at time k, $\delta\mathbf{x}'_{k-1}$ is the corrected error state at time k-1 (can be obtained from Eq. 14), $w_{k-1}$ is the process noise, and $\mathbf{\Phi}_k$ is a state transition matrix of size 15×15, which can be obtained as

$$\mathbf{\Phi}_k = \begin{bmatrix} \mathbf{I} & \Delta t.R_{b_k}^G & 0 & 0 & 0 \\ 0 & \mathbf{I} & 0 & 0 & 0 \\ 0 & 0 & \mathbf{I} & \Delta t.\mathbf{I} & 0 \\ -\Delta t.\mathbf{S}(a_k'^G) & 0 & 0 & \mathbf{I} & \Delta t.R_{b_k}^G \\ 0 & 0 & 0 & 0 & \mathbf{I} \end{bmatrix} \tag{12}$$

where $\mathbf{S}(a_k'^G)$ is the skew symmetric matrix of global acceleration and is used to estimate the IMU's orientation variations [19]. The measurement model [22] at time k can be written as

$$\mathbf{z}_k = \mathbf{H}\delta\mathbf{x}'_k + v_k \tag{13}$$

where $\mathbf{z}_k$ is the measurement error, $\mathbf{H}$ is the measurement matrix, and $v_k$ is an additive white zero-mean Gaussian noise, with covariance $\mathbf{R}_k = E\left(v_k v_k^T\right)$.

Once the measurement at k is obtained, using the Kalman update equation, the filtered error state at time k can be estimated as

$$\delta\mathbf{x}'_k = \delta\mathbf{x}_k + \mathbf{K}_k.[\mathbf{m}_k - \mathbf{H}\delta\mathbf{x}_k] \tag{14}$$

where $\mathbf{K}_k$ is the Kalman gain, $\mathbf{m}_k$ is the actual error measurement, and $\delta\mathbf{x}_k$ is the estimated error state [19].

The Kalman gain can be obtained as

$$\mathbf{K}_k = \mathbf{P}_k\mathbf{H}^T(\mathbf{H}\mathbf{P}_k\mathbf{H}^T + \mathbf{R}_k)^{-1} \tag{15}$$

where $\mathbf{P}_k$ is the covariance matrix of the estimated error. The process noise $w_k$ is due to the sensor noise and calibration residual [22] and its covariance matrix at time k can be obtained as $\mathbf{Q}_k = E(w_k w_k^T)$. At each sample time, the covariance can be propagated forward employing

$$\mathbf{P}_k = \mathbf{\Phi}_{k-1}\mathbf{P}'_{k-1}\mathbf{\Phi}_{k-1}^T + \mathbf{Q}_{k-1} \tag{16}$$

where $\mathbf{P}'_k$ is the estimation error covariance matrix at time k, based on the measurement received at the last sample time k-1, and in turn is obtained as

$$\mathbf{P}'_k = (\mathbf{I} - \mathbf{K}_k\mathbf{H})\mathbf{P}_k(\mathbf{I} - \mathbf{K}_k\mathbf{H})^T + \mathbf{K}_k\mathbf{R}_k\mathbf{K}_k^T \tag{17}$$

where $\mathbf{I}$ is a 15×15 identity matrix.

Following this notation we propose a comprehensive EKF which corrects the estimations based on the five states' error measurements. In our method the estimated biases of gyroscope and accelerometer are compensated in the EKF, at every sample time, while the non-bias estimated error term is obtained and compensated at sample times that the relevant conditions are satisfied. We define the dynamic measurement matrix $\mathbf{H}$ and the dynamic error measurement matrix

$\mathbf{m}$. Both $\mathbf{H}$ and $\mathbf{m}$ matrices are equal to 15×15 and 15×1 zero matrices at initialization, respectively. We employed five methods to reduce the error, each of them feeds EKF independently. We split the $\mathbf{H}$ and $\mathbf{m}$ matrices to five submatrices, each submatrix can be updated based on its relevant condition and method, and separated from the other submatrices. If the relevant condition of submatrix is not satisfied the submatrix will be remained at zero, and then will be neglected. It means that at time k, EKF would be fed based on all, non or some of the five methods. Matrices $\mathbf{H}$ and $\mathbf{m}$ can be written as

$$\mathbf{H} = [\mathbf{H}^{Z^T}_{3\times15}, \mathbf{H}^{H^T}_{3\times15}, \mathbf{H}^{R^T}_{3\times15}, \mathbf{H}^{P^T}_{3\times15}, \mathbf{H}^{N^T}_{3\times15}]^T \tag{18}$$

and

$$\mathbf{m} = [\mathbf{m}^{Z^T}_{3\times1}, \mathbf{m}^{H^T}_{3\times1}, \mathbf{m}^{R^T}_{3\times1}, \mathbf{m}^{P^T}_{3\times1}, \mathbf{m}^{N^T}_{3\times1}]^T \tag{19}$$

where $\mathbf{H}^{(\cdot)}$s are 3×15 and $\mathbf{m}^{(\cdot)}$s are 3×1 matrices, and the superscripts Z, H, R, P and N stand for ZUPT, HDR, ZARU, IPER and ZUNA, respectively.

## Error Reduction Methods

In this work in the absence of the external infrastructures, EKF is fed with the combination of five error reduction methods to correct the estimations. These methods are used to decrease the accumulated errors in orientation, velocity and position, and to compensate the sensors' biases exist on the measured angular velocity and acceleration. This section details these employed error reduction methods.

### A. Zero Velocity Update (ZUPT)

The Zero velocity UPdaTe or ZUPT method, is based on this fact that when the user's foot is in the stance phase, its velocity is zero. In practical works, this velocity is small but not equal to zero, this difference can be used as an error measurement in velocity to feed EKF [18]. Therefore the actual error measurement submatrix $\mathbf{m}^Z$ at time k [1], can be obtained as

$$\mathbf{m}_k^Z = \mathbf{v}_k^G - [0,0,0] \tag{20}$$

The measurement submatrix $\mathbf{H}^z$ is defined in the way that selects the velocity error components $\delta\mathbf{v}$ of error state matrix $\delta\mathbf{x}$ (terms 10 to 12), which can be written as

$$\mathbf{H}^Z = [0_{3\times3} \quad 0_{3\times3} \quad 0_{3\times3} \quad \mathbf{I}_{3\times3} \quad 0_{3\times3}] \tag{21}$$

### B. Zero Angular Rate Update (ZARU)

The term ZARU stands for Zero Angular Rate Update, as addressed in [19]. It feeds EKF, when the foot is in the so called still phase. The idea is based on this fact that when the foot is motionless on the floor, the measurements from three orthogonal gyroscopes should be equal to zero; thus the data from gyroscope in this case, is in fact the measured error of the angular velocity. Following this notation the angular velocity error measurement submatrix $\mathbf{m}^R$ can be written as

$$m_k^R = \omega_k^b - [0,0,0] \tag{22}$$

where $\omega_k^b$ is an angular velocity in the local coordinates, at the $k^{th}$ sample time. The measurement submatrix $\mathbf{H}^R$ is defined to select the angular velocity error components $\delta\omega^b$ of $\delta\mathbf{x}$ (terms 4 to 6), which can be written as

$$\mathbf{H}^R = [0_{2\times3} \quad \mathbf{I}_{3\times3} \quad 0_{3\times3} \quad 0_{3\times3} \quad 0_{3\times3}] \tag{23}$$

### C. Heuristic Drift Reduction (HDR)

The term HDR stands for Heuristic Drift Reduction, as originally proposed in [24]. The idea is based on the fact, that for the indoor applications, most of the corridors and paths are straight. Employing HDR by detecting the moments that user walks straight, the error in the heading is obtained. In this work, we follow the same idea as [19,24], but we perform it in the different way. We define the sliding window of $\lambda$ samples and we compare heading at time $k$ to heading at time $k$ - $\lambda$. If the difference is less than the threshold, it is detected as an error in heading estimation, and EKF will be fed by this detected error, otherwise it is detected as a real variation of heading due to the user's turning. The advantage of using sliding window is to apply HDR at every time interval (during both swing and stance phases) and not only at stance phase, which yields to improve the accuracy.

Assume $Er_\psi$ is the difference between the sliding window's last data $\psi_k^G$ (current heading) and the first data $\psi_{k-\lambda}^G$ (heading at $\lambda^{th}$ preceding sample time), one can obtain

$$Er_{\psi_k} = \psi_k^G - \psi_{k-\lambda}^G \tag{24}$$

Following this notation, $m_k^H$ can be obtained as

$$\mathbf{m}_k^H = \begin{cases} Er_{\psi_k} & \left|Er_{\psi_k}\right| \le thr_{Er_\psi} \\ 0 & Otherwise \end{cases} \tag{25}$$

where $thr_{Er_\psi}$ is the threshold and we empirically chose it as 0.15 radians. To select the heading components of $\delta\mathbf{x}$, the measurement submatrix $\mathbf{H}^H$, is defined as

$$\mathbf{H}^H = \begin{bmatrix} 0_{2\times3} & 0_{2\times3} & 0_{2\times3} & 0_{2\times3} & 0_{2\times3} \\ [0,0,1] & 0_{1\times3} & 0_{1\times3} & 0_{1\times3} & 0_{1\times3} \end{bmatrix} \tag{26}$$

The first two rows of $\mathbf{H}^H$ are defined as zero, to reserve the further development of yaw and pitch errors estimations.

**D. Investigative Position Error Reduction (IPER)**

IPER is used to obtain the error in calculating the estimated displacement. This method obtains the error of estimated displacement along x and y in every time interval (fixed windowing) based on the idea of the generality of straight paths in indoor applications, and obtains the error of displacement along z-axis at stance phases. The corrections for displacements along x, y and z-axis can feed EKF, independently. Assume the user walks straight and the estimated tracked trajectory is deviated from the expected straight line by angle $\rho$. Comparing $\rho$ to $thr_x$ and $thr_y$ the direction of movement is detected to be along x, y or a free walk neither along x nor along y, where $thr_x$ and $thr_y$ are thresholds of movements along x and y axes and can be chosen based on the actual location of straight paths in the global coordinates (e.g. to detect if user walks along the corridor or parallel to building walls). If the movement is detected to be along x (y) axis, in this case the real movement of pedestrian along y (x) axis is actually equal to zero, therefore the variations in the estimated displacement along y (x) axis is an error, which is used to feed EKF to estimate and correct the position error. Therefore $\mathbf{m}_k^P$ is obtained as

$$\mathbf{m}_k^P = \begin{cases} \left[0, r_{yl} - r_{yf}, m_{z_k}^P\right] & if \left|\rho - thr_x\right| \le thr_\rho \\ \left[r_{xl} - r_{xf}, 0, m_{z_k}^P\right] & if \left|\rho - thr_y\right| \le thr_\rho \\ \left[0, 0, m_{z_k}^P\right] & Otherwise \end{cases} \tag{27}$$

where $thr_\rho$ is the threshold of straight movements detection, and $thr_x$, $thr_y$ and $thr_\rho$ are equal to 0, $\pi/2$ and 0.2 radians, respectively. To correct the displacement along z-axis, at every stance phase of walking the displacement along z-axis is compared to the position of the IMU at the start point. Without loss of generality it is assumed that the position of IMU along z-axis at the start point is equal to zero, while walking the

estimated displacement along z-axis in the stance phase should be equal to the one at the start point and thus should be zero. This difference is used to feed EKF and correct the displacement along z-axis. Thus $m_{z_k}^P$ can be written as

$$m_{z_k}^P = r_{zk} - 0 = r_{zk} \tag{28}$$

where $r_{zk}$ is the displacement along z-axis at the current sample time. The zero terms of $\mathbf{m}^P$ do not imply that the error is zero, while determining the proper matrix of $\mathbf{H}^P$ only the non-zero terms of $thr_\rho$ feed EKF, and the zero terms are neglected and thus they are ineffective. Following this notation, $\mathbf{H}^P$ can be obtained as

$$\mathbf{H}^P = [0_{3\times3} \quad 0_{3\times3} \quad I_{k_{3\times3}}^P \quad 0_{3\times3} \quad 0_{3\times3}] \tag{29}$$

where $I_k^P$ is a 3×3 matrix and can be written as

$$\begin{array}{c} P \\ k \end{array} \begin{bmatrix} 0 & 0 \\ 0 & P \\ & y \\ 0 & 0 \end{bmatrix} \tag{30}$$

The diagonal components of $I_k^P$ can be determined independently. $I_x^P = 1$ only if the movement is along y-axis, $I_y^P = 1$ only if the movement is along x-axis and $I_z^P = 1$ only at stance phases, and they are equal to zero, otherwise.

**E. Zero UNrefined Acceleration update (ZUNA)**

At the start point and during system initialization, when the user's foot is still motionless on the floor, the acceleration is measured over three seconds and its mean value is calculated ($\bar{a}^b$). During the midstance phase of walking (when the entire foot is stationary on the floor), the measured acceleration is expected to be equal to the mean value at the start point and the difference is used to feed EKF to correct the measured acceleration. Following this notation $m_k^N$ at midstance phase can be obtained as

$$\mathbf{m}_k^N = \mathbf{a}_k^b - \bar{\mathbf{a}}^b \tag{31}$$

The midstance phase is detected in the same way as stance phase but by the smaller thresholds on acceleration and the variance of acceleration and larger threshold of time. The measurement submatrix $\mathbf{H}^N$, during midstance phase can be written as

$$\mathbf{H}^N = [0_{3\times3} \quad 0_{3\times3} \quad 0_{3\times3} \quad 0_{3\times3} \quad I_{3\times3}] \tag{32}$$

## Localization Testing and Results

This section presents the implementation of the navigation algorithm, and analyzes the system performance in different tests. The positioning error can be considered as a percentage of the total traveled distance. From the state-of-the-art the acceptable error of self-contained (IMU-alone) navigation system is about 2% of the total traveled distance [8]. We carried out several indoor tests including repetitive and closed-form trajectories. During the tests IMU was fixed on the user's right foot using shoe laces, as shown in Figure 5. In addition, IMU was connected to the netbook computer through USB port, and the estimated tracked path was displayed in the netbook screen online, while the data were recorded to replicate the test for more processing. The user walked normally at a steady speed of approximately 0.5 - 1 $m/s$ in the forward direction.

**A. Rectangular path**

In the first test the user tracked along a known-length rectangular path, clockwise. To evaluate the accuracy of orientation estimation, at each rectangle's vertex, the user stopped and then turned 90° clockwise, but the system also works well for continuous tracking. User stood at the start point for three seconds, and after tracking the path, stopped at the same point. The error of walking along the desired path and standing at the exact same stop/start point by user was a few centimeters. Figure 6 shows the original acceleration and angular velocity in the local coordinates, measured by IMU during this test. As can be seen during the stance phase the angular velocity is not zero due to the sensor's biases. In addition, the norm of the accelerations in the stance phase at the beginning of the experiment was 1.0129$g$, due to the gravitational acceleration and sensor's biases. The value of angular velocity along x-axis is larger than other components, because user's foot had more rotation about this axis during walking, compare to other axes.

However, at vertices and during turning the value of angular velocity along z-axis becomes significant because the rotation was on the leveled floor and thus about z-axis. Employing the discussed algorithm, the gravity and biases were removed from the acceleration. The upper part of Figure 7 shows the gravity and bias-free acceleration in the global coordinates. From Figure 7 it is possible to estimate the direction of walking, while in the first and third parts of the graph, the acceleration has the greater value along the y-axis and in the second and last parts the value along x-axis is greater. The norm of the global acceleration at the stance phases has the average value of 0.0017$g$, which demonstrates the gravity and sensor biases, were compensated properly. The global velocity was obtained by integrating the global acceleration over the

time. The second part of the Figure 7 shows the global velocity. As can be seen, the value of the velocity is greater along either y or x axes in the same order as acceleration.

Figure 8 shows the displacements along x and y axes, and the orientation of the user during test, which is indeed the yaw angle respect to the global coordinates. The global displacement was obtained by double integrating the global acceleration over the time. In addition, Figure 8 illustrates the user walked and turned clockwise. The reference heading shows the ideal yaw angle, where in a normal walking of human the turning of direction cannot be sharp.

Figure 9 illustrates the tracked trajectory in the x-y plane. The reference path was the rectangle with a perimeter of 25.8 m, and its sides were along x and y axes of global coordinates. The estimated path was alike the real tracked trajectory, in the sense of shape and closed-form. The difference between the start and estimated end points was 0.05 m and therefore the tracking error was 0.19% of total traveled distance. Figure 10 shows the estimated tracked path in the three dimensions, as can be seen the estimated path is corrected along all three axes successfully. At the rectangle vertices where the user's foot was on the floor for a longer time, the displacement along z approached zero.

## B. Comparison of error reduction methods

To compare the EKF-based navigation system with five error reduction methods to the state-of-the-art (systems with up to three

Figure 5: IMU mounted on the user's shoe using the laces.

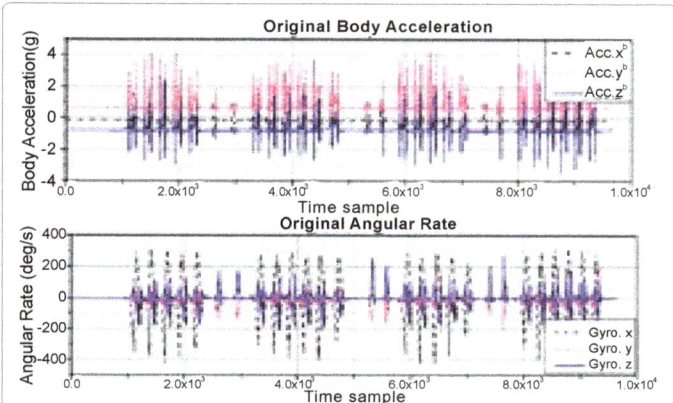

Figure 6: Accelerations and angular velocities in the local coordinates measured by IMU relevant to a human walking along a rectangular path.

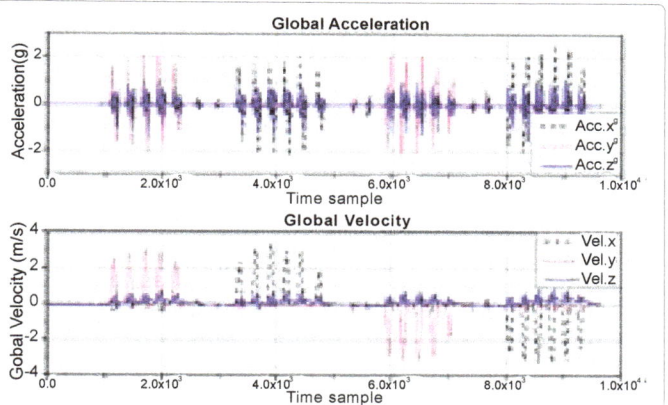

Figure 7: Acceleration and velocity in the global coordinates, relevant to the human walking along a rectangular path.

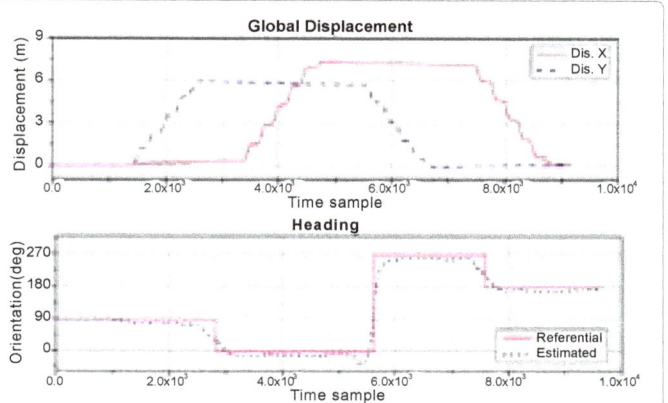

Figure 8: Estimated global displacement and orientation of a human walking along a rectangular path.

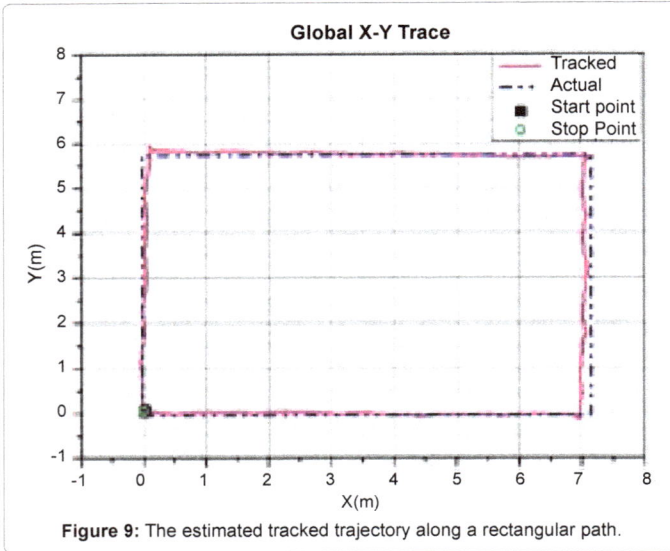

**Figure 9:** The estimated tracked trajectory along a rectangular path.

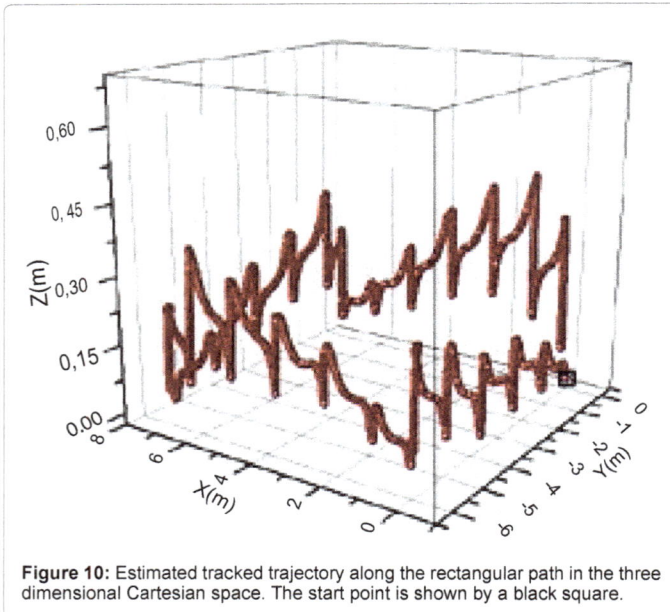

**Figure 10:** Estimated tracked trajectory along the rectangular path in the three dimensional Cartesian space. The start point is shown by a black square.

error reduction methods), and to evaluate the effect of applying correction at different time intervals and not only at stance phases, a rectangular path was tracked, and the recorded signals were used to replicate the estimation employing EKF with different error reduction methods. Figure 11 shows the different trajectories obtained from EKF in the existence of three, four or five error reduction methods. Figure 12 shows the heading obtained from EKF using different methods. The third part of heading graph tends to approach -270° or 90° which in fact gives the same direction. The error between the start point and estimated stop point is reported in Table 1 in terms of distance and the percentage of Total Traveled Distance (TTD%).

## C. Tracking on other indoor paths

To evaluate the performance of the proposed EKF-based INS algorithm for indoor navigation, we carried out some indoor tests over longer trajectories. To consider the repeatability of the algorithm, the rectangular path is tracked six times continuously clockwise in the

forward direction. The results demonstrate the difference between the start and estimated end point was $0.38\ m$ thus the tracking error was $0.25\%$ in terms of the percentage of total traveled distance. The point to point comparison shows the maximum error occurred at the upper right vertex of the triangle where the error is about $0.52\ m$, despite of estimation error, this error is due to this fact that at the turning point during normal walking, each time the user passed about the same point and not necessarily over the exact point (as start/end point). Figure 13 shows the estimated tracked trajectory. We performed another test in our main building. During the test the user walked forward and stopped if the person was in the path. User started to walk from a point inside the building, and after walking indoor, exited the building and walked along the building's wall and finally entered the building from a different door and stopped at the same point as start point.

Figure 14 shows the estimated trajectory, where the real trajectory length was $78\ m$. The estimated tracked trajectory, was alike the real path in sense of shape and closed-form, and it was shorter than the real trajectory. The difference between the start and estimated end point was $0.6382\ m$ and the tracking error in terms of the percentage of total traveled distance were $0.82\%$.

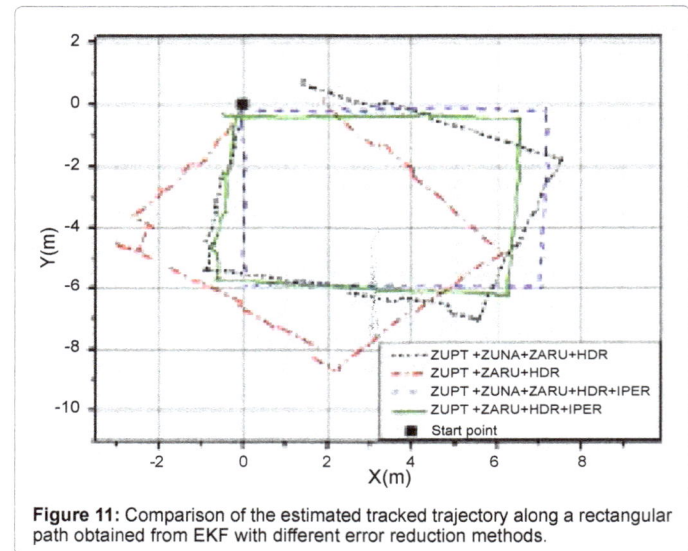

**Figure 11:** Comparison of the estimated tracked trajectory along a rectangular path obtained from EKF with different error reduction methods.

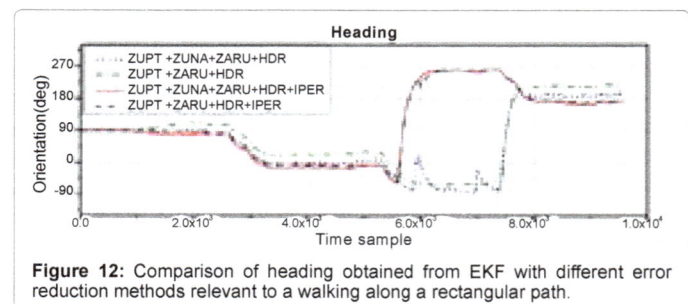

**Figure 12:** Comparison of heading obtained from EKF with different error reduction methods relevant to a walking along a rectangular path.

| Error Reduction Methods | Error (TTD %) | Error (m) |
|---|---|---|
| ZUPT + ZARU + HDR + ZUNA + IPER | 0.2% | 0.052 m |
| ZUPT + ZARU + HDR + ZUNA | 6.6% | 1.72 m |
| ZUPT + ZARU + HDR + IPER | 2.3% | 0.60 m |
| ZUPT + ZARU + HDR | 7.9% | 2.05 m |

**Table 1:** Tracking error of EKF-based navigation system using different error reduction methods relevant to a human walking along a rectangular path.

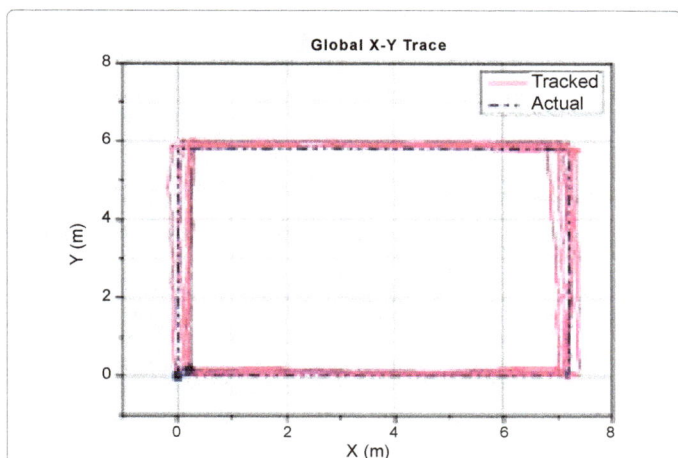

**Figure 13:** Estimated tracked trajectory of the repetitive test along the rectangular path. The path is tracked six times to obtain the total traveled distance of 154.8 m, the tracking error was 0.25% in terms of the percentage of total traveled distance.

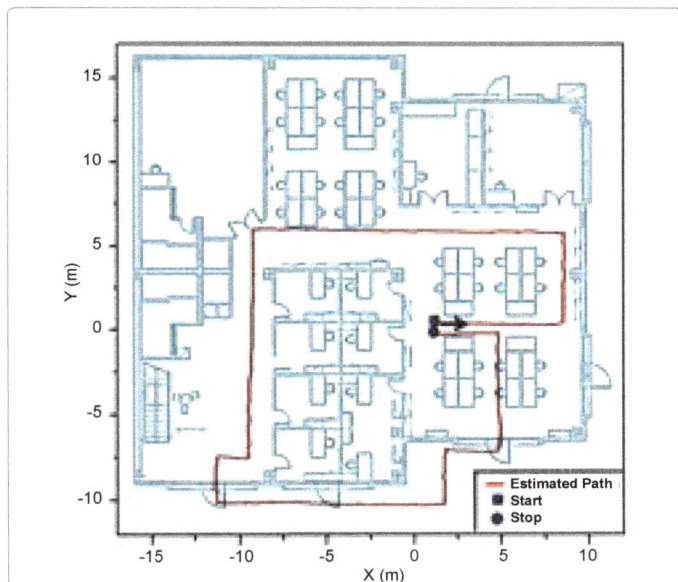

**Figure 14:** Estimated tracked trajectory over 78 m trajectory with tracking accuracy of 0.82% of total traveled distance. (a) On the building map. (b) On the building satellite view.

## Conclusion

This paper presented an accurate infrastructure-free EKF based navigation system for pedestrian indoor navigation, employing only the measurements of accelerometers and gyroscopes of IMU. The contributions of this work in comparison to the conventional systems were as follow: (i) in this work the dynamic error measurement matrix ($H$) was used, thus different error reduction methods fed EKF independently and at different times, which yielded to obtain the higher accuracy; (ii) the error correction was applied not only at the stance phase but also at different time intervals during the swing phase of human walking; (iii) the algorithm was developed without using magnetometer, thus environment's equipment and supplies have not affected the system performance; therefore it is reliable for indoor applications; and (iv) in our system five different error reduction methods were used, which fed EKF with estimated error of all five states; thus as a result the tracking error were obtained less than 1% (0.19%-0.82%) in terms of the percentage of the total traveled distance, which means the proposed system accuracy is higher than the conventional systems. As a future work the proposed navigation system will be implemented on the user's smartphone.

### Acknowledgment

This work was partially supported from DOC project funded within the "Industria 2015" Research Program, and it is dedicated to the memory of Eng. Franco Vitucci, who threw light on the importance of developing a navigation aid for blinds and visually impaired. He inspired and supported this work with his long-term vision.

### References

1. Jimenez Ruiz A, Seco Granja F, Prieto Honorato J, Guevara Rosas J (2012) Accurate pedestrian indoor navigation by tightly coupling foot-mounted imu and rfid measurements. IEEE Trans Instrum Meas, 61: 178–189.

2. Evennou F, Marx F (2006) Advanced integration of wifi and inertial navigation systems for indoor mobile positioning. EURASIP J Adv Signal Process, 2006:164–164.

3. Pittet S, Renaudin V, Merminod B, Kasser M (2008) Uwb and mems based indoor navigation. J NAVIGATION, 61: 369–384.

4. House S, Connell S, Milligan I, Austin D, Hayes TL, Chiang P (2011) Indoor localization using pedestrian dead reckoning updated with rfid-based fiducials. Annual International Conference of the IEEE, Engineering in Medicine and Biology Society, EMBC Boston, MA, 7598–7601.

5. NavChip (2013) I. InterSense.

6. VectorNav T (2013) Inertial Measurement Unit.

7. Levi RW, Judd T (1996) Dead reckoning navigational system using accelerometer to measure foot impacts US Patent 5,583,776.

8. Huang C, Liao Z, Zhao L (2010) Synergism of ins and pdr in selfcontained pedestrian tracking with a miniature sensor module. IEEE Sens J, 10: 1349–1359.

9. Wang Q, Chen Y, Chen X, Zhang X, Chen R, Chen W (2011) A novel pedestrian dead reckoning solution using motion recognition algorithm with wearable emg sensors. Journal of Global Positioning Systems, 10: 39–49.

10. Ojeda L, Borenstein J (2007) Non-gps navigation for security personnel and first responders. J NAVIGATION, 60:391.

11. Alvarez JC, Lo´pez AM, Gonzalez RC, A´ lvarez D (2012) Pedestrian dead reckoning with waist-worn inertial sensors. Instrumentation and Measurement Technology Conference (I2MTC), 2012 IEEE International. IEEE, Graz, 24–27.

12. Zampella F, Khider M, Robertson P, Jim´enez A (2012) Unscented kalman filter and magnetic angular rate update (maru) for an improved pedestrian dead-reckoning. Position Location and Navigation Symposium (PLANS), 2012 IEEE/ION, 129–139.

13. Jirawimut R, Ptasinski P, Garaj V, Cecelja F, Balachandran W (2003) A method for dead reckoning parameter correction in pedestrian navigation system. IEEE Trans Instrum Meas, 52:209–215.

14. Chatfield AB (1997) Fundamentals of high accuracy inertial navigation. AIAA (American Institute of Aeronautics & Astronautics), 174.

15. Woodman OJ (2007) An introduction to inertial navigation. University of Cambridge, Computer Laboratory, Tech. Rep. UCAMCL-TR-696.

16. Trew M, Everett T (1997) Human movement: an introductory text. Churchill Livingstone Elsevier, China, 10.

17. Bebek O, Suster MA, Rajgopal S, Fu MJ, Huang X, et al. (2010) Personal navigation via high-resolution gait-corrected inertial measurement units. IEEE Trans Instrum Meas, 59:3018–3027.

18. Str̈omb̈ack P, Rantakokko J, Wirkander S, Alexandersson M, Fors I, et al. (2010) Foot-mounted inertial navigation and cooperative sensor fusion for indoor positioning. Proc. ION.

19. Jim´enez A, Seco Granja F, Prieto J, Guevara J (2010) Indoor pedestrian navigation using an ins/ekf framework for yaw drift reduction and a foot-mounted imu. 7th Workshop on Positioning Navigation and Communication (WPNC), 135–143, 2010.

20. Sabatini AM, Martelloni C, Scapellato S, Cavallo F (2005) Assessment of walking features from foot inertial sensing. IEEE Trans Biomed Eng, 52:486–494.

21. Cavallo F, Sabatini AM, Genovese V (2005) A step toward gps/ins personal navigation systems: real-time assessment of gait by foot inertial sensing in Intelligent Robots and Systems. 2005 IEEE/RSJ International Conference, IEEE, 1187–1191.

22. Foxlin E (2005) Pedestrian tracking with shoe-mounted inertial sensors. IEEE Comput Graph Appl, 25:38–46.

23. Rajagopal S (2008) Personal dead reckoning system with shoe mounted inertial sensors. Masters Degree Project, Stockholm, Sweden,013.

24. Borenstein J, Ojeda L, Kwanmuang S (2009) Heuristic reduction of gyro drift in imu-based personnel tracking systems. SPIE Defense, Security, and Sensing Conference. International Society for Optics and Photonics, 73 061H–73 061H.

25. Banitalebi Dehkordi M (2014) Design and Experimental Evaluation of Advanced Human-Robot and Human-Computer Interfaces for Assistive Applications. PhD dissertation, Pisa, Italy, 275.

# 7

# Delivery Delay Analysis of Selective Repeat ARQ in Underwater Acoustic Communications

**Mingsheng Gao\*, Jian Li, Wei Li and Ning Xu**

*College of IoT Engineering, Hohai University, P.R. China*

## Abstract

Despite being the most efficient automatic repeat request (ARQ) protocol, the selective-repeat ARQ (SR-ARQ) is previously thought to be infeasible in underwater acoustic communications owing to the half-duplex property of typical underwater acoustic modems. However, with the help of the juggling-like stop-and-wait (JSW) transmission scheme, it has now become feasible. In this paper, we evaluate the delivery delay (consisting of transmission and resequencing delays) of the SR-ARQ operating over the JSW scheme under the static assumption that the relative radial velocity between the transmitter and the receiver is zero, aiming to provide system designers with a valuable reference for delay evaluation under more general scenarios. We model the underwater acoustic channel as a two-state Discrete Time Markov Channel, and derive the closed-form expression for the delivery delay under heavy traffic situation. Unlike most analytical approaches for the delay analysis of SR-ARQ in terrestrial communications whose computational complexities grow exponentially with round-trip delay, our proposed analytical approach is immune to the round-trip delay. This also makes our approach suitable for terrestrial communications. To highlight the accuracy of our approach, we also provide comparisons between analytical and simulation results.

**Keywords:** Underwater acoustic channels; SR-ARQ; Transmission scheme; Radial velocity; Delay analysis

## Introduction

Underwater acoustic communication is deemed as the technique that will be widely used for oceanography, data collection, pollution monitoring, offshore exploration and tactical surveillance applications [1]. However, the characteristics of underwater acoustic channels, such as high bit error rate (BER), large round-trip delay, multipath coherence, as well as the half-duplex property of typical underwater acoustic modems, make the underwater acoustic link very poor [2].

In communication systems, the main error control mechanisms used on a per-hop basis to preserve data integrity include the use of forward-error-correction (FEC) schemes at the physical layer, and the use of automatic-repeat-request (ARQ) techniques at the data link layer. For ARQ protocols, the performance measures typically used are throughput [3,4], delay [5,6] and energy efficiency [7]. Delay analyses for ARQ protocols, especially for the most efficient SR-ARQ, have been extensively performed for terrestrial communications over the past few decades [8,9]. In underwater acoustic communications, however, little work has been done in analyzing the delay of ARQ protocols so far; this is because the least efficient stop-and-wait ARQ (SW-ARQ) and its variants are potentially viewed as the only ones that can be used in *half-duplex* underwater acoustic links, which make their delay analyses simple and straightforward.

In [10], we have proposed a transmission scheme, known as the juggling-like stop-and-wait (JSW) scheme [10], enabling the use of continuous ARQ protocols over half-duplex underwater acoustic links. Recently, the JSW scheme has been received much attention in the underwater community. For example, Chitre et al. validated the effectiveness of the JSW scheme in [11], and put forward a modification of this scheme when operated in small file scenarios. In [12], a variant of the JSW scheme was proposed, and applied in underwater networks. Also, we proposed a hybrid protocol for underwater acoustic networks by making use of the JSW scheme [13]. It is of importance to analyze the performance of the JSW scheme.

In this paper, we study the SR-ARQ when it is operated over our JSW scheme, focusing on analyzing its delay performance with the aim of providing some useful insights. As reported in [14], the total delay of an ARQ protocol can be attributed to three components, namely, queueing delay, transmission delay, as well as resequencing delay. The queueing delay is the duration from the time a packet arrives at the transmitter until its first transmission attempt, which correlates to the channel behavior and the packet arrival process. The transmission delay is the time from a packet's first transmission until its successful arrival at the receiver (including all retransmission delays). This component is only related to the channel behavior. The last term is the resequencing delay, which is the most complicated, and defined as the waiting time of the packet in the receiver resequencing buffer. In this case, packets with higher identifiers must wait in the receiver resequencing buffer until all the packets with lower identifiers have been correctly received. Note that "delivery delay" consists of transmission delay and resequencing delay only. In this paper, we study the delivery delay performance of the SR-ARQ by considering both time-varying channel and finite round-trip time. Also, the effect of bursty channel errors is taken into account. For this purpose, we model the underwater acoustic channel by means of the two-state Markov chain that has been extensively used in wireless networks. Some assumptions are made to simplify the formal description; however, they do not affect the generality of the results, as they can be relaxed if needed. We have extended previous studies in terrestrial communications, and derived an exact closed-form expression for the delivery delay in the static case where the relative radial velocity between the transmitter and the receiver is zero.

\*Corresponding author: Mingsheng Gao, College of IoT Engineering, Hohai University, P.R. China, E-mail: gaoms@hhu.edu.cn

To our best knowledge, it is the first time an attempt has been made to analyze the delay performance of *continuous* ARQ protocols in *half-duplex* underwater acoustic links. The objec-tives of this paper are two-fold: 1) to provide system designers with a valuable reference for delay evaluation under more general scenarios in underwater acoustic communications; 2) to provide an alternative effective approach that can also be applied in terrestrial communications, as our proposed analytical approach is insusceptible to round-trip delay, whereas the computa-tional complexities of most existing analytical approaches for the delay analysis of SR-ARQ in terrestrial communications tend to grow exponentially with round-trip delay.

The rest of this paper is organized as follows. In Section 4, we review some related work on the delay analysis of the SR-ARQ. In Section 5, we first introduce the JSW transmission scheme which allows the SR-ARQ to operate in half-duplex underwater acoustic links. We then develop an analytical model and derive a closed-form expression for the delivery delay of the SR-ARQ under the static assumption. To verify the analytical results from the derivations, we compare them with simulation results in Section 5. Finally, Section 7 concludes the paper.

## Related Work

The SW-ARQ and its variants are generally perceived as the only class of ARQ protocols that are suitable for half-duplex underwater acoustic links. As a result, there is no attempt to study the delay performance of *continuous* ARQ protocols (e.g., go-back-N ARQ (GBN-ARQ) and SR-ARQ) in underwater acoustic communications so far. Nevertheless, there has been extensive research on the delay performance of the SR-ARQ in terrestrial communications, as it is the most efficient among the continuous ARQ protocols. For this reason, we give a brief survey on the existing delay performance analysis of the SR-ARQ protocol in terrestrial communications.

In [5], Konheim assumed a renewal traffic source, and derived the probability generating functions (PGF) for the transport delays and queue lengths of both the GBN-ARQ and SR-ARQ. In [6], Anagnostou and Protonotarios developed an exact analytic approach to analyze the delay performance, and also proposed another approach that is based on the *ideal SR-ARQ* approximation (i.e., the queueing process is assumed to be dependent on the history of the transmission process). However, the results obtained in [5] and [6] are both based on the assumption of a static channel model; furthermore, the computational complexities of these approaches increase exponentially with round-trip delay. Rosberg and Shacham [8] derived the distributions of the buffer occupancy and the resequencing delay at the receiver under heavy traffic assumption. In [9], Rosberg and Sidi analyzed the joint distribution of buffer occupancy at the transmitter and the receiver, deriving the mean transmission and resequencing delays under the assumption of a renewal arrival process. By using the flow graph analysis method, the authors in [3] addressed the effect of forward/backward channel memory on ARQ error strategies, and compared GBN-ARQ with SR-ARQ in terms of their throughput efficiency. In [15], Shacham and Towsley investigated the buffer occupancy and resequencing delay in a wireless environment in which a single transmitter and multiple receivers communicate, assuming heavy-traffic conditions and a static channel. Again, the results obtained in [3,8,9] and [15] were still under the assumption of a static channel. Based on the assumption of a nonstationary channel model, Fantacci [16] considered the channel's time-varying feature, deriving the mean packet delay and the mean queue length of the SR-ARQ for both the zero and the finite round-trip

delay cases. However, the author has made a simplifying assumption that the arrival process is Bernoulli.

In [17], Lu and Chang considered both the kth-order Markov model and the gap error model, and investigated how different error statistics would affect ARQ performance with the help of signal flow graphs. In [14], Kim and Krunz considered a time-varying channel with finite round-trip delay and a Markovian traffic source. A mean analysis was developed accordingly for all the ARQ delay contributions. Similarly, Rossi et al. [18] took time-varying channel, finite round-trip delay, and the effect of bursty channel errors into consideration together in their investigation of the SR-ARQ's delay performance. A closed-form expression for delivery delay was subsequently derived. However, the computational complexities of the analytical approaches presented in [17] and [14] show an exponential increase with round-trip delay.

From the discussion above, we can observe that the main drawbacks of existing analytical approaches include: 1) overly simplified channel model assumptions, 2) very simple arrival process assumption, and/or 3) extremely high computational complexity. More specifically, some approaches are based on the static channel assumption, which results in significant underestima-tion of the delay performance. As is shown in [14], using a time-varying channel model can lead to a remarkable increase in the mean transport delay. The exponential growth in the computational complexity of these approaches with the round-trip delay is also a tiresome issue that makes their usage very difficult in a practical system. Hence, an intuitive and simple expression for the delay calculation is very much needed.

## The Exact Analysis of Delivery Delay

In this section, we briefly describe the JSW transmission scheme, and then characterize the underwater acoustic channel by means of an embedded Markov model. Some assumptions are accordingly made to simplify our analysis. Note that these assumptions can be extended to a more general situation if necessary. We thereafter derive the closed-form expression for the delivery delay of the SR-ARQ when it operates over the JSW scheme.

### The JSW transmission scheme

Consider a point-to-point, half-duplex underwater acoustic communication system with a transmitter and a receiver. We assume that the transmitter has already acquired the control of the channel successfully, and has a series of packets to send to the receiver. Although the channel is half-duplex, a pair of nodes can still send packets that cross each other in the medium and yet receive each others' packets correctly, so long as each node has finished its transmission and has switched to listening mode by the time the packet arrives. Based on this property, the transmitter alternates between transmitting a data packet to the receiver, and listening for an earlier packet's ACK/NAK, while one or more other packets are still in transit. For the receiver, it transmits the ACK/NAK immediately after receiving each data packet. Apart from the time it spends on transmitting the ACK/NAK, the receiver listens for data packet at other times. In order for the scheme to work correctly, an appropriate window size (denoted by s), as well as the appropriate inter-packet spacings (denoted by $t_0$) to transmit the first s packets, must be chosen properly. The flow of the JSW transmission scheme is illustrated in Figure 1 when s=3.

Our focus in this paper is on the static case where the relative radial velocity between the transmitter and the receiver is zero. Thus, the propagation delays in both directions are constant, which we denote by $d_0$. The value of $t_0$ is also chosen in such a way that the inter-packet

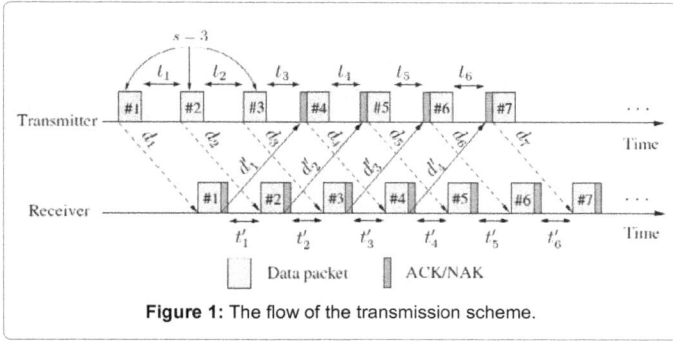

**Figure 1:** The flow of the transmission scheme.

spacing between any two consecutive data packets will be equal to $t_0$ throughout the entire session (not just for the first s packets). From Figure 1, we see that the maximum window size s that can be chosen for the static case is given by

$$s = \left\lceil \frac{2d_0}{\delta + \gamma} \right\rceil + 1, \tag{1}$$

where $\delta$ is the transmission time of a data packet, and $\gamma$ is the transmission time of an ACK/NAK packet, respectively. The value of $t_0$ is in turn given by

$$t_0 = \frac{2d_0 - (s-1)\delta + \gamma}{s} \tag{2}$$

As can be seen in Figure 1, the transmitter undergoes a cyclical behavior of alternating between sending a data packet, and listening for an ACK/NAK. We refer to this cyclical behavior as a "Transmit-ACK cycle (TAC)", whose cycle time is defined to be from the instant when the transmitter starts sending the first bit of the data packet, till the instant when it finishes receiving the last bit of the following ACK/NAK. All TACs thus have the same cycle time of $\delta + t_0$. For more details on how the JSW transmission scheme operates, especially for the dynamic case where the relative radial velocity is non-zero, interested readers are encouraged to refer to [10].

## The channel model

The underwater acoustic channel is generally characterized by poor quality physical link, due to time-varying multipath propagation and motion-induced Doppler distortion. As a result, the BER of the underwater acoustic link can vary with time as the propagation conditions change. Errors in the received bit stream are thus inevitable. Here, we model the underwater acoustic channel as a two-state Markov chain (as shown in Figure 2) based on the following assumptions:

• Time is divided into slots, each of which equals to the cycle time of a TAC (i.e., $\delta + t_0$).

• The transmitter and the receiver are synchronized with the timeslots (It can be easily implemented using existing underwater synchronization schemes [19,20]); the transmission of a packet occurs at the beginning of a slot.

• The ACK/NAK messages are always received correctly. This can be justified by considering the use of a powerful error-correcting code to guarantee successful receptions with a high probability.

• A packet is released from the transmitter buffer only when an ACK message has been received for its last transmission attempt.

• There is an infinite data source at the transmitter, which implies a heavy-traffic situation.

Two possible channel states are defined, namely, state 0 (quiet, with BER $e_0$) and state 1 (noisy, with BER $e_1$). State 0 represents a channel propagation situation in which the transmitter can correctly receive data packets with a very high probability, while state 1 represents a channel propagation situation in which it is extremely difficult to correctly receive the data packets. The values of $e_0$ and $e_1$ are dependent on the characteristics of the propagation environment, the transmission modulation scheme, and the detection technique implemented at the receiving end. Of note is the fact that errors in state 1 usually occur in random-length bursts. We also assume that channel state transitions could only occur at the end of the slots with probabilities $p_{01}$ and $p_{10}$, respectively (see Figure 2). In this way, we do not take into account the fact that in actual non-stationary channels, the state transitions may occur anywhere in time. In particular, our assumption leads to a slight error in the estimation of the probability that a data packet would be received erroneously. As stated in [4], however, this may have a negligible impact on the accuracy of the analysis. With the channel model above, we can derive the following relationships: 1) the average burst length, i.e., the average number of data packets transmitted in state 1, is $N_b = \frac{1}{p_{10}}$; 2) The steady state channel probability in state 0 is $p_0 = \frac{p_{10}}{p_{01} + p_{10}}$; 3) The steady-state channel probability in state 1 is $p_1 = \frac{p_{01}}{p_{01} + p_{10}}$; 4) the average channel error rate is $\varepsilon = \frac{p_{10}e_0 + p_{01}e_1}{p_{01} + p_{10}}$.

Denote by $p_{ij}^{(k)}$ the transition probability from state $i$ ($i=0$, 1) to state $j$ ($j=0$, 1) after k slots. The four possible k-step transition probabilities are then given by [4]:

$$p_{00}^{(k)} = \frac{p_{10}}{p_{01} + p_{10}} + \frac{p_{01}}{p_{01} + p_{10}}[1 - (p_{01} + p_{10})]^k, \tag{3}$$

$$p_{01}^{(k)} = \frac{p_{01}}{p_{01} + p_{10}} - \frac{p_{01}}{p_{01} + p_{10}}[1 - (p_{01} + p_{10})]^k, \tag{4}$$

$$p_{10}^{(k)} = \frac{p_{10}}{p_{01} + p_{10}} - \frac{p_{10}}{p_{01} + p_{10}}[1 - (p_{01} + p_{10})]^k, \tag{5}$$

and

$$p_{11}^{(k)} = \frac{p_{01}}{p_{01} + p_{10}} + \frac{p_{10}}{p_{01} + p_{10}}[1 - (p_{01} + p_{10})]^k. \tag{6}$$

From the above, it follows that parameters $p_{01}$, $p_{10}$ and $\varepsilon$ completely define the two-state Markov channel model. Parameters $N_b$ and $p_1$ characterize the burstiness of the channel with $p_{01}$ and $p_{10}$ describing the time variation of the channel behavior. The special case $p_1 = p_{01} = 1 - p_{10}$, corresponds to the two-state block interference (BI) channel model. Evidently, the BI channel model is entirely determined by parameters $\varepsilon$ and $p_1$. This paper is based on the assumption that all transmissions in state 0 are error free, while all those in state 1 are erroneous. It is a

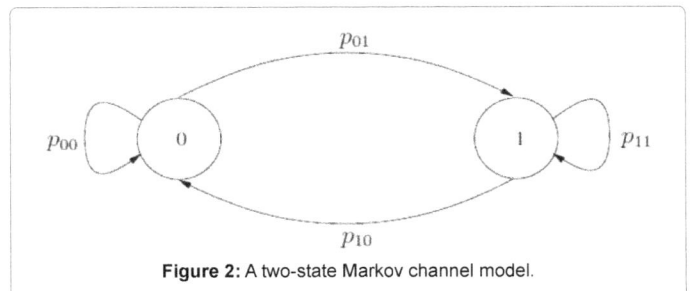

**Figure 2:** A two-state Markov channel model.

reasonable assumption which has also been used in [21]. In this case, $\varepsilon$ degrades to $p_1$. Note, however, that: 1) the two-state Markov model can be extended to a more common scenario characterized by $N_b$, $p_1$ and $\varepsilon$; 2) it can also be extended to account for a higher order Markov chain, which might yield slight improvements in the results, but at the expense of more complicated and tedious computations.

## Calculation of delivery delay

Consider a packet of interest (referred to as *tagged* packet) upon its first transmission. Denote by X $(t)=(X_1(t), X_2(t), \ldots, X_s(t))$ the set of identifiers of those packets that are transmitted during window t. The process $(X_1(t), X_2(t), \ldots, X_s(t))$ governs the evolution of the resequencing buffer occupancy. We refer to the position k $(1 \leq k \leq s)$ in each window as column k. Note that, one column corresponds to one slot. Any packet that was unsuccessfully transmitted on column k in window t is assumed to be retransmitted at the same column in window t + 1. Without loss of generality, we assume the *tagged* packet is transmitted for the first time at slot (denoted by $\tau$ ) s in a window, which implies the channel state at $\tau=0$ is 0 (otherwise a retransmission will occur at $\tau=s$). Note that the packets that block the *tagged* packet in the receiver buffer at the time of its successful reception are those that were transmitted during the same window where the *tagged* packet was first transmitted and have not been correctly received. Denote by $\Gamma_p$ the period from the instant when the *tagged* packet is transmitted for the first time on column s to the instant when the *tagged* packet, as well as all those with lower identifiers in the same window, is correctly received. To find the delivery delay, we define the last blocking packet, satisfying: 1) it is on column j $(1 \leq j \leq s)$, and has been transmitted m times until success during $\Gamma_p$; 2) $\forall \alpha \in [1, j-1]$, there is at least one successful transmission on column $\alpha$ during $\Gamma_p$; 3) $\forall \beta \in [j+1, s]$, there is at least one successful transmission on column $\beta$ within the prior m − 1 transmissions of $\Gamma_p$. Note that, j=s implies the case where the *tagged* packet itself, as well as all those with lower identifiers in the same window, will be correctly received at the receiver for the *tagged* packet's first transmission.

Let $A_j^k$ denote the event {the packet on column *j* has succeeded in the $k^{th}$ transmission}, $\overline{A}_j^k$ denote the event {the packet on column *j* has failed in the $k^{th}$ transmission) , and $\overline{A}_j^1\overline{A}_j^2....\overline{A}_j^m \mid G$ denote the event {the packet on column *j* has not been successfully transmitted for m consecutive times, given that the channel state at time $\tau=0$ is 0 (equivalently termed G)}. The probability of the event $\overline{A}_j^1\overline{A}_j^2....\overline{A}_j^m \mid G$ is given by

$$\Pr\{\overline{A}_j^1\overline{A}_j^2....\overline{A}_j^m \mid G\} = \sum_{s_1=0}^{1}\sum_{s_2=0}^{1}....\sum_{s_m=0}^{1} p_{0s_1}^{(j)}p_{s_1s_2}^{(s)}......p_{s_{m-1}s_m}^{(s)} . \Pr\{\overline{A}_j^1\overline{A}_j^2....\overline{A}_j^m \mid s_1, s_2,...,s_m\} \quad (7)$$

Where

$$\Pr\{\overline{A}_j^1\overline{A}_j^2....\overline{A}_j^m \mid s_1, s_2,...,s_m\} = \Pr\{R_d\mid s_1\}.\Pr\{R_d\mid s_2\}....\Pr\{R_d\mid s_m\}. \quad (8)$$

Note that, for $1 \leq k \leq m$, $Pr\{R_d\mid s_k\}$ denotes the probability of the event {the data packet has been successfully transmitted, given that the channel state for the transmission is $s_k$ ($s_k=0$ or 1)}, defined as

$$Pr\{R_d \mid s_k\} = \begin{cases} 0, & if \ s_k = 0 \\ 1, & if \ s_k = 1 \end{cases} \quad (9)$$

Hence, (7) can be rewritten as

$$\Pr\{\overline{A}_j^1\overline{A}_j^2....\overline{A}_j^m \mid G\} = \sum_{s_1=0}^{1}\sum_{s_2=0}^{1}....\sum_{s_m=0}^{1} p_{0s_1}^{(j)}p_{s_1s_2}^{(s)}......p_{s_{m-1}s_m}^{(s)}$$
$$.\Pr\{R_d\mid s_1\}.\Pr\{R_d\mid s_2\}....\Pr\{R_d\mid s_m\} \quad (10)$$

To simplify the above expression, we define

$$U \stackrel{\Delta}{=} [\ 1 1\ ], \quad (11)$$

And

$$G \stackrel{\Delta}{=} \begin{bmatrix} p_{00}^{(s)}\Pr\{Rd\mid 0\} & p_{10}^{(s)}\Pr\{Rd\mid 0\} \\ p_{01}^{(s)}\Pr\{Rd\mid 1\} & p_{11}^{(s)}\Pr\{Rd\mid 1\} \end{bmatrix} \quad (12)$$

$$V_j \stackrel{\Delta}{=} \begin{bmatrix} p_{00}^{(j)}\Pr\{Rd\mid 0\} \\ \\ p_{01}^{(j)}\Pr\{Rd\mid 1\} \end{bmatrix} \quad (13)$$

Inserting into (10), we have

$$\Pr\{\overline{A}_j^1\overline{A}_j^2....\overline{A}_j^m \mid G\} = UG^{m-1}V_j \quad (14)$$

Further, denote by m|j, G the event {the last blocking packet on column j has been transmitted m times until success during $\Gamma_p$, given that the channel state at time $\tau=0$ is G}. For $1 \leq \alpha \leq j-1$, we let $\alpha$|m, j, G denote the event {there is at least one successful transmission on column $\alpha$ during $\Gamma_p$, given that the channel state at time $\tau=0$ is G}. Also, for $j+1 \leq \beta \leq s$, we let $\beta$|m, j, G denote the event {there is at least one successful transmission on column $\beta$ within the prior m −1 transmissions of $\Gamma_p$, given that the channel state at time $\tau=0$ is G}. The probability of the event m|j,G is given by

$$\Pr\{m \mid j, G\} = \Pr\{\overline{A}_j^1\overline{A}_j^2....\overline{A}_j^{m-1}A_j^m \mid G\}$$
$$= \Pr\{\overline{A}_j^1\overline{A}_j^2....\overline{A}_j^{m-1} \mid G\} - \Pr\{\overline{A}_j^1\overline{A}_j^2....\overline{A}_j^m \mid G\}$$
$$= \begin{cases} 1 - UV_j, & if \ m=1 \\ UG^{m-2}(I_{2\times 2} - G)V_j & if \ m \geq 2, \end{cases} \quad (15)$$

where $I_{2\times 2}$ is a two-by-two identity matrix.

For $1 \leq \alpha \leq j-1$, it is easy to see that

$$Pr\{\alpha \mid m, j, G\} = 1 - UG^{m-1}V_{\alpha}, \quad (16)$$

and for $j+1 \leq \beta \leq s$,

$$Pr\{\beta \mid m, j, G\} = 1 - UG^{m-2}V_{\beta}. \quad (17)$$

It can be easily inferred that, if the event m|j, G takes place, then the events $\alpha$|m, j, G and $\beta$|m, j,G take place concurrently, and vice versa. Given (m, j, G), the delivery delay of the tagged packet, T, can then be expressed as

$$T = (m-1)s + F, \quad (18)$$

where F is the column j where the last blocking packet is transmitted, which is uniformly distributed over the set $\{1, 2, \ldots, s\}$, and hence Pr $\{F=j\} = \dfrac{1}{s}$

From (15)-(17), we have

$$\Pr\{T = (m-1)s + j \mid m, j, G = \Pr\{m \mid j, G\}\prod_{\alpha=1}^{j-1}\Pr\{\alpha \mid m, j, G\}\prod_{\beta=j+1}^{s}\Pr\{\beta \mid m, j, G\} \quad (19)$$

By averaging (19) over all possible sets with regard to m, j and G, we can obtain

$$E\{T\} = \sum_{m=1}^{\infty}\sum_{j=1}^{s}\Pr\{T = (m-1)s + j \mid m, j, G\} .\Pr\{F = j\}.\Pr\{G\}.[(m-1)s + j], \quad (20)$$

where Pr $\{G\} = p_0$.

Note that the representation in (20) involves an infinite summation which is somewhat incon-venient. By performing some algebraic manipulations, nevertheless, a simpler formula for E{T } is provided in Appendix II.

## Analytical and Simulations

In this section, we present the numerical results of the delivery delay obtained using the above analytical results. To validate the accuracy of our analysis, we have simulated the transmission of packets based on the SR-ARQ protocol operating over the JSW transmission scheme, which we have presented in Section 5.1.

In our simulations, the data rate of the underwater acoustic channel is assumed to be 8 kbps, along with $d_0$=0.6667 s, $\delta$=0.064 s, and $\gamma$=0.005 s. When varying the value of one parameter, the other parameters are kept at their default values above, unless specified otherwise. We also assume that there is an infinite data source at the transmitter. The total simulation time is set to 1,000,000 s, and all the results presented are averaged over 20 simulation runs. Note that all the obtained delivery delays in this section have been normalized to the transmission time of a data packet, which is different from that of a TAC.

In Figure 3, the mean delivery delay T is shown as a function of the mean channel error rate $\varepsilon$ for various values of the average error burst length: $N_b$=3, 5, 10, 15. Here, $\varepsilon$ is assumed to range from 0.01 to 0.3. Note that T does not include the queueing time, i.e., the duration from the time a packet arrives at the transmitter until its first transmission attempt. Also, we vary $N_b$ by varying the transition probability $p_{10}$ (from state 1 to state 0). A good agreement is observed between simulation and analysis. It can be seen that T increases almost linearly with regard to the mean channel error rate $\varepsilon$. This can be explained as follows. With an ARQ protocol, the delivery delay is mainly affected by the number of retransmissions and the round-trip delay. However, for a given $\varepsilon$, the average number of retransmissions is almost constant. Therefore, the delivery delay is dominated by the round-trip delay, which is independent of the mean channel error rate. Moreover, for a given $\varepsilon$, we can see that, the greater the average error burst length $N_b$, the smaller the mean delivery delay T . This is due to the fact that the probability of encountering a long sequence of slots without errors increases as $N_b$ increases.

The impact of the average error burst length $N_b$ on the mean delivery delay T is illustrated in Figure 4. Here, we assume that $N_b$ varies from 1 to 50, and set the other parameters to their default values. We vary $N_b$ by tuning the transition probability $p_{10}$. Again, the simulation results show good agreement with the analytical results. For each $\varepsilon$, the mean delivery delay T decreases as $N_b$ increases, for the same reason as explained above.

Figure 5 demonstrates the significance of the propagation delay $d_0$ by contrasting the mean delivery delay for different $d_0$ (0.6667 s and 0.3333 s, respectively). As was done previously, $\varepsilon$ is varied by changing the parameter $p_{10}$ under a fixed $p_{01}$. From this figure, one can make the following remarks. Firstly, in both cases, it can be seen that the simulation results match with the numerical results quite well. Secondly, the relationship between the mean delivery delay and the channel error rate exhibits similar trends as what Figure 3 has shown previously. Finally, it can be seen that a smaller $d_0$ (which corresponds to a shorter distance between the transmitter and the receiver) results in a lower mean delivery delay. For example, when $N_b$=3, the mean delivery delay for $d_0$=0.3333 s is much lower than that for $d_0$=0.6667 s. The same conclusion about the mean delivery delay still holds when $N_b$=10. This can be attributed to the fact that, a smaller $d_0$ corresponds to a smaller round-trip delay s, hence resulting in a lower mean delivery delay under the condition that both $N_b$ and $\varepsilon$ are constant.

The mean delivery delay T as a function of the average error burst length is shown in Figure 6 for both $d_0$=0.3333 s and $d_0$=0.6667 s. These results reveal high variation of the mean delivery delay with the mean channel error rate for different $d_0$. As expected, the mean delivery delay for $d_0$=0.3333 s is lower than that for $d_0$=0.6667 s, which is also similar to what we have previously observed in Figure 5. Again, it suggests that delay largely depends on the distance between the transmitter and the receiver in underwater. For a given $\varepsilon$, it can be seen by comparing Figure 6 with Figure 4 that, the varying trend between the mean delivery delay and the average error burst length remains the same.

## Conclusions

Accurate delay analysis plays an important role in evaluating the performance of underwater acoustic communication systems. In previous studies, continuous ARQ protocols are perceived to be infeasible in underwater environments due to the half-duplex property of underwater acoustic modems, and thus the existing ARQ approaches have centered around the use of the classic SW-ARQ protocol and its variants, whose delay estimates are simple and straightforward.

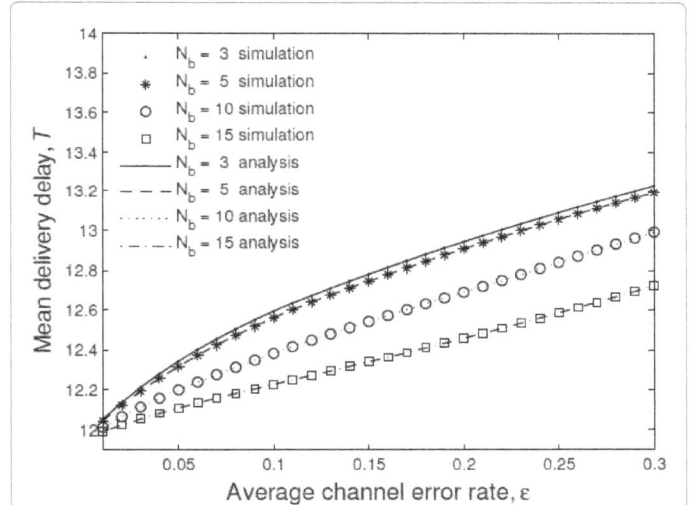

**Figure 3:** Mean delivery delay versus mean channel error rate $\varepsilon$ for d0=0.6667.

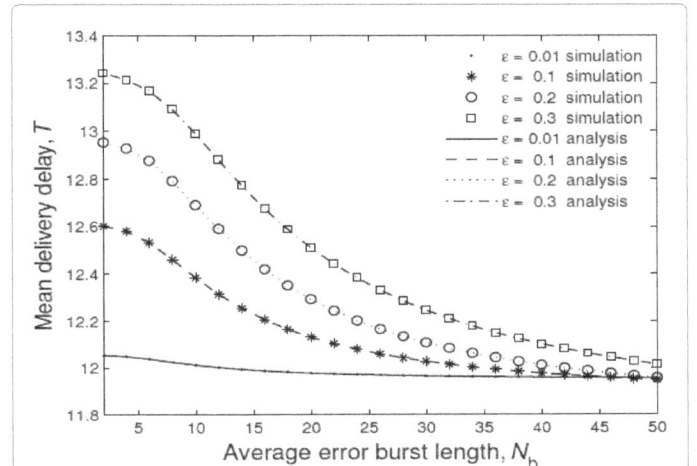

**Figure 4:** Mean delivery delay versus average error burst length Nb for d0=0.6667.

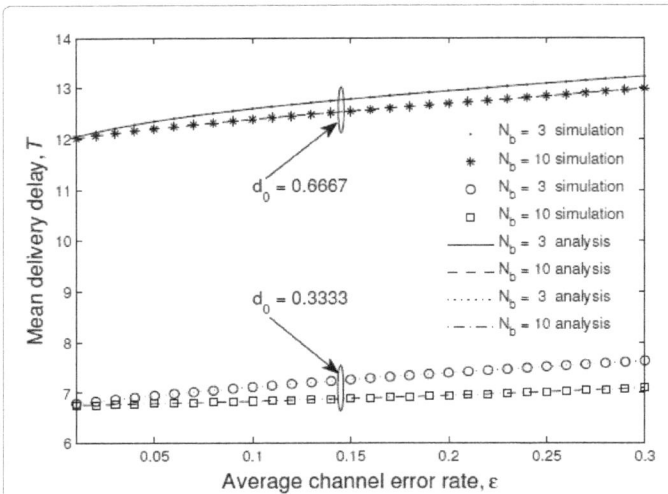

**Figure 5:** Mean delivery delay as a function of mean channel error rate ε for d0=0.6667 and d0=0.3333.

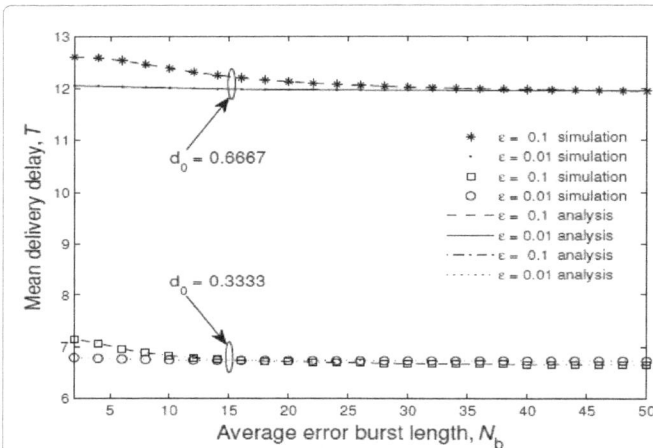

**Figure 6:** Mean delivery delay as a function of average error burst length Nb for d0=0.6667 and d0=0.3333.

In this paper, we operate the SR-ARQ protocol in half-duplex underwater acoustic channel for the first time based on the JSW transmission scheme that was tailored for underwater acoustic communications. Based on the static assumption that the relative radial velocity between the transmitter and the receiver is zero, we first characterize the non-stationary underwater acoustic channel by a two-state Markov channel model, which captures the time-varying and correlated nature of channel errors. Then, we investigate the delivery delay and derive a closed-form expression for the mean delivery delay of the SR-ARQ protocol. Finally, we validate the accuracy of the expressions obtained through simulations. The contributions of the work are: 1) to present an exact delay analysis for the static case, which serves as a valuable reference for delay estimation under dynamic scenarios for underwater acoustic communications; 2) to present a simple and straightforward approach to delay analysis, which can be effectively applied in terrestrial communications as well, because its computational complexity is immune to the round-trip delay.

Our future research will be focused on the following aspects: 1) to evaluate the end-to-end delay performance in underwater network settings; 2) to investigate the energy efficiency when applying the SR-ARQ protocol in underwater acoustic networks.

## References

1. Akyildiz IF, Pompili D, Melodia T (2006) State-of-the-art in protocol research for underwater acoustic sensor networks. In: WUWNet'06, Los Angeles, California, USA.

2. Proakis J, Sozer E, Rice J, Stojanovic M (2001) Shallow water acoustic networks. IEEE Commun Mag 39: 114-119.

3. Cho YJ, Un CK (1994) Performance analysis of ARQ error controls under markovian block error pattern. IEEE Trans Commun 42: 2051-2061.

4. Kallel S (1992) Analysis of memory and incremental redundancy ARQ schemes over a nonstationary channel. IEEE Trans Commun 40: 1474-1480.

5. Konheim AG (1980) A queueing analysis of two ARQ protocols. IEEE Trans Commun 28: 1004-1014.

6. Anagnostou ME, Protonotarios EN (1986) Performance analysis of the selective repeat ARQ protocol. IEEE Trans Commun 34: 127-135.

7. Chockalingam A, Zorzi M (2008) Adaptive ARQ with energy efficient backoff on Markov fading links. IEEE Trans Wireless Commun 7: 1445-1449.

8. Rosberg Z, Shacham N (1989) Resequencing delay and buffer occupancy under the Selective Repeat ARQ. IEEE Trans Inf Theory 35: 166-173.

9. Rosberg Z, Shacham N (1989) Resequencing delay and buffer occupancy under the Selective Repeat ARQ. IEEE Trans Inf Theory 35: 166-173.

10. Gao M, Soh WS, Tao M (2009) A transmission scheme for continuous ARQ protocols over underwater acoustic channels. IEEE, ICC, Dresden, Germany.

11. Chitre M, Soh WS (2015) Reliable point-to-point underwater acoustic data transfer: To juggle or not to juggle? " IEEE J Ocean Eng 40: 93-103.

12. Azad S, Casari P, Zorzi M (2013) The underwater selective repeat error control protocol for multiuser acoustic networksdesign and parameter optimization. IEEE Trans Wireless Commun 12: 4866-4877.

13. Yang M, Gao M, Foh CH (2015) Hybrid Collision-Free Medium Access (HCFMA) Protocol for Underwater Acoustic Networks: Design and Performance Evaluation. IEEE J Ocean Eng 40: 1-11.

14. Kim JG, Krunz MM (2000) Delay analysis of selective repeat ARQ for a Markovian source over a wireless channel. IEEE Trans Veh Technol 29: 1968-1981.

15. Shacham N, Towsley D (1991) Resequencing delay and buffer occupancy in selective repeat ARQ with multiple receivers. IEEE Trans Commun 39: 928-936.

16. Fantacci R (1996) Queuing analysis of the selective repeat automatic repeat request protocol wireless packet networks. IEEE Trans Veh Technol 45: 258-264.

17. Lu DL, Chang JF (1993) Performance of ARQ protocols in nonindependent channel errors. IEEE Trans Commun 41: 721-730.

18. Rossi M, Badia L, Zorzi M (2003) Exact statistics of ARQ packet delivery delay over Markov channels with finite round-trip delay. In: Proc IEEE Globecom pp. 3356-3360.

19. Chirdchoo N, Soh WS, Chua KC (2008) Mu-Sync: A Time Synchronization Protocol for Underwater Mobile Networks. In: Proc ACM WUWNet 2008, San Francisco, CA.

20. Liu J, Zhou Z, Peng Z, Cui JH, Zuba M, et al. (2013) Mobi-Sync: Efficient Time Synchronization for Mobile Underwater Sensor Networks. IEEE Trans Par Dist Sys 24: 406-416.

21. Zorzi M, Rao RR, Milstein L (1998) Error statistics in data transmission over fading channels. IEEE Trans Commun 46: 1468-1477.

# 8

# Study of Traps in Special Doped Optical Fiber Radiation Sensors via Glow Curve Analysis

**Mostafa Ghomeishi[1,4]\*, G Amouzad Mahdiraji[1], Peyman Jahanshahi[1], FR Mahamd Adikan[1] and David A Bradley[2,3]**

[1]Integrated Lightwave Research Group, Faculty of Engineering, University of Malaya, 50603 Kuala Lumpur, Malaysia
[2]Department of Physics, University of Malaya, 50603 Kuala Lumpur, Malaysia
[3]Department of Physics, University of Surrey, Guildford, GU2 7XH, UK
[4]Faculty of Science, Science and Research branch, Islamic Azad University, Tehran, 14778, Iran

## Abstract

Thermoluminescence dosimeters (TLDs) are widely used, serving the needs of various radiation applications. In recent times optical fibers have been introduced as alternatives to more conventional phosphor-based TLD systems, with many efforts being carried out to improve their thermoluminescence (TL) yield. While there have been extensive studies of many of the various TLD characteristics of optical fibers, including TL response, linearity, reproducibility, repeatability, sensitivity and fading, far more limited studies have concerned dependence on the type of TL activator used in optical fibers, promoting the TL mechanism. Present study focuses on TLD glow curves analysis for five different doped optical fibers that have been subjected to photon and electron irradiation. Trap parameters such as activation energy and frequency factors have been obtained from second order kinetics analysis, based on computerized glow curve deconvolution. An interesting observation is that co-doped fibers typically leads to enhanced TL characteristics, pointing to a need for optimization of the choice and levels in use of co-dopants.

**Keywords:** Glow curve; Fiber dosimeter; Defects; Thermoluminescence; CGCD

## Introduction

New perspectives are evolving in respect of the wide range of potential thermoluminescence dosimetry (TLD) applications of various forms of optical fiber [1,2]. As an instance, optical fiber TLDs has been shown to be effective for a range of photon sources, from visible and UV, through to x- and gamma-rays of different energies [3-5]. Further studies have concerned silica fibers of various shape and doped-core dimension, flat optical fibers being shown to lead to improved TL response [6-10], it also being demonstrated that the TL response of fibers will vary depending upon core dopant concentration as well as core size [11-13].

Although commercial TLDs have typically developed, following up upon the favourable outcome of studies of various constituents, for optical fibers the study of TL yield have been much more limited, most typically with germanium, aluminum, oxygen, phosphorus and nanomaterial cluster dopants [4,14-16]. Present study seeks to help address this lack, expanding investigations to include additional rare-earth dopants, examining TL response and associated kinetics analysis of such optical fibers.

The characterization of the defect centers forms a crucial step in understanding the mechanism of TL [17]. In this context, analysis of glow curves offers a sensitive and suitable technique for such study. Here, in present study, new materials are tested to examine their effect, both on TL response and kinetic parameters.

## Method and Materials

### Optical fiber fabrication and preparation

The samples are standard circular cross-section optical fiber of 125 µm outer diameter, fabricated using the facilities of the University of Malaya fiber pulling lab. The 9 µm central core that forms the dopant channel has been confirmed using the energy-dispersive x-ray (EDX) facility of an SEM, doped-core and silica cladding mappings being obtained for all but one of the samples (Figure 1). For Al-doped fiber,

EDX mapping of the core represents a severe challenge, due to the neighbouring atomic numbers of Al (Z=13) and Si (Z=14), providing for limited differential X-ray fluorescence production, further confounded by the associated low energy emissions (Figure 1b). The doped elements are compared with $SiO_2$, BaF, GaP, $Al_2O_3$, $TmF_3$ as a reference for EDX identification. The characteristic peaks through which the dopant materials are detected are $K_{a1}$ for O, Si, Al, Y, Ba and Ga and $L_{\alpha,\beta}$ for Tm. Table 1 shows the concentration of each element present in the different samples.

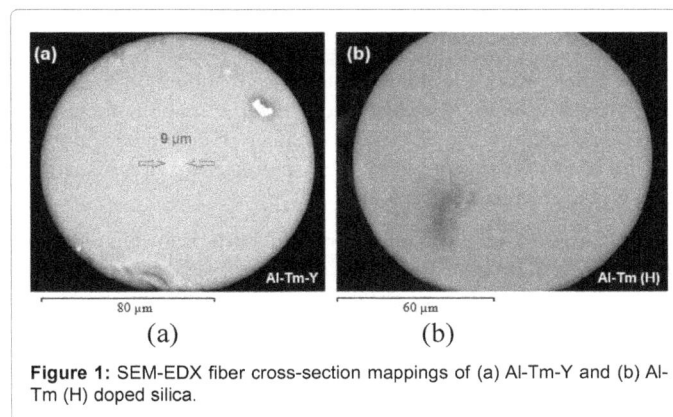

(a)　　　　　(b)

**Figure 1:** SEM-EDX fiber cross-section mappings of (a) Al-Tm-Y and (b) Al-Tm (H) doped silica.

**\*Corresponding author:** Mostafa Ghomeishi, Integrated Lightwave Research Group, Faculty of Engineering, University of Malaya, 50603 Kuala Lumpur, Malaysia, E-mail: mostafa.ghomeishi@gmail.com

| Element | Al-Y-Tm | Ba | Al-Tm (H) | Al-Tm (L) | Ga |
|---------|---------|-------|-----------|-----------|------|
| O | 52.66 | 51.13 | 54.52 | 55.09 | 52.79 |
| Si | 46.42 | 43.27 | 40.11 | 41.02 | 41.81 |
| Ba | - | 5.6 | - | - | - |
| Ga | - | - | - | - | 5.4 |
| Tm | 0.25 | - | 0.24 | 0.18 | - |
| Al | 0.54 | - | 5.13 | 3.71 | - |
| Y | 0.13 | - | - | - | - |

**Table 1:** EDX results for the different doped fiber samples. In this table the weight percentage of each material is shown.

The outer polymer cover to the fibers has been carefully removed using a chloroform solution, soaking the fibers in the solution for 30 seconds. The samples are subsequently cleaned with acetone and carefully cut to 5 mm lengths using a fiber cleaver. Two sets of five samples each were chosen from each type of fiber, the uncertainly in fiber length being ± 0.5 mm. Since the fiber samples have closely identical mass density, also being carefully selected for uniform length, mass normalization of the results has not been required in carrying out analysis and comparisons.

### Irradiation and readout

Prior to any irradiation run, in order to remove any prior radiation history, all samples are annealed at 400°C for one hour, subsequently cooled to room temperature. The samples have then been exposed to a dose of 8 Gy dose delivered either by 6 MeV electrons or 6 MV photons, use being made of a Varian 2100 C linear accelerator. The samples were positioned on the surface of a *solid-water*$^{TM}$ phantom. The field size and surface to source distance were set at $20 \times 20$ cm$^2$ and 100 cm respectively. A square applicator with $20 \times 20$ cm$^2$ aperture size was used for electron irradiation.

The TL response of radiated samples was obtained using a Harshaw 3500 TLD reader, the irradiated samples being thermally stimulated from 50°C to 400°C, provided at a heating rate of 20°C/s. The generated glow curves have been analyzed using a second order kinetics model for TL deconvolution.

## Results and Discussion

The differential TL responses of the five sets of optical fibers with respect to a dose of 8 Gy are illustrated in Figure 2 for both photon and electron irradiations. The TL yields are similar for both the photon and electron irradiations, with a deviation between them of less than 4% based on the mean values.

In Figures 3-7 the main glow curves for each of the fiber samples have been deconvoluted, with a figure of merit (FOM) for fitting of about 3.4-5.2% and a mean deviation of 0.8%. The extracted traps information from the deconvolution of glow curves is presented in Tables 2-6. In the tables, $E_a$ is the activation energy of each trap in units of eV, $s'$ is the first derivative of the frequency factor (s$^{-2}$), $n_0$ is the initial concentration of trapped electrons (cm$^{-3}$). The Peak-$I$ and Peak-$T$ values are the glow curves maximum intensity (μC) and associated relevant temperature (°C) respectively. The full-width-half-maximum (FWHM) value is calculated on the absolute temperature scale and the TL emission wavelength is in units of nm.

Tables 2-6 show the first peak (referred to as Trap 1) to have an activation energy of 0.9-1.1 eV, a peak intensity of 2-2.9 μC, and an emission wavelength of between 1100 and 1400 nm pointing to independence from the dopant material, instead being more related to

the substrate or the silica preform itself. The activation energy suggests association between $O_2$ and $O^{2-}$ defects in amorphous and fused silica [18].

Examining Figures 2-4, the numerous traps in Al-Y-Tm-doped and Ba-doped fiber provide for the greater response of these two forms over that of other samples, each of which are composed of lesser numbers of traps (Figures 5-7). This supports the underpinning basis of greater numbers of traps producing greater absorption. According to Figure 3 the combination of Al, Y and Tm as fiber co-dopants results in an increased number of trap centres and, as a result, enhanced absorption. Figure 4 for Ba-doped silica, provides for lower intensity glow peaks compared to that for Al-Y-Tm-doped fiber, albeit with similar energy traps formed in the two types of fiber.

In accord with Tables 2 and 3, the major difference between Al-Y-Tm-doped fiber and Ba-doped fiber is the intensity of the individual

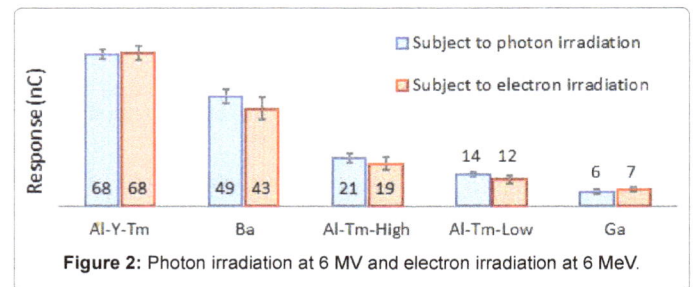

**Figure 2:** Photon irradiation at 6 MV and electron irradiation at 6 MeV.

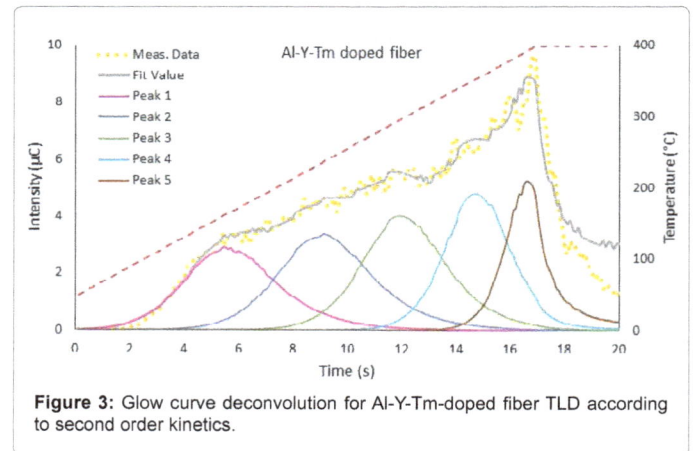

**Figure 3:** Glow curve deconvolution for Al-Y-Tm-doped fiber TLD according to second order kinetics.

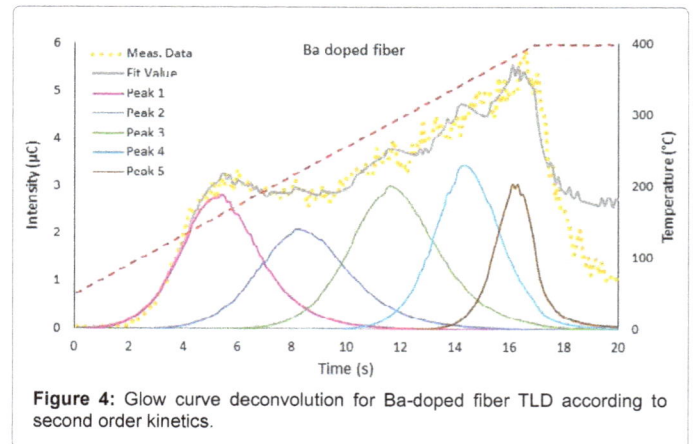

**Figure 4:** Glow curve deconvolution for Ba-doped fiber TLD according to second order kinetics.

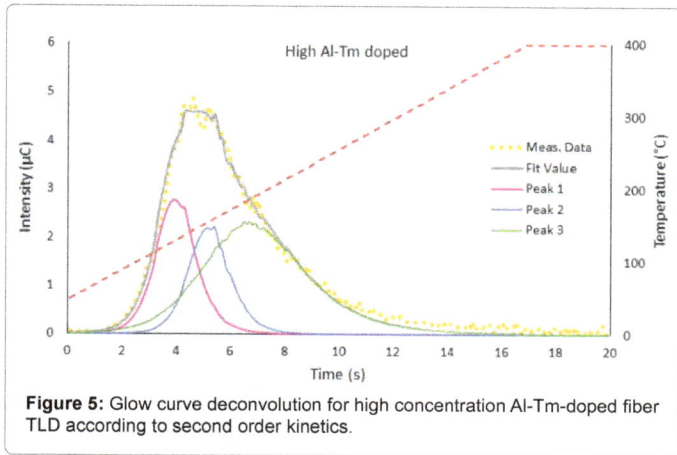

**Figure 5:** Glow curve deconvolution for high concentration Al-Tm-doped fiber TLD according to second order kinetics.

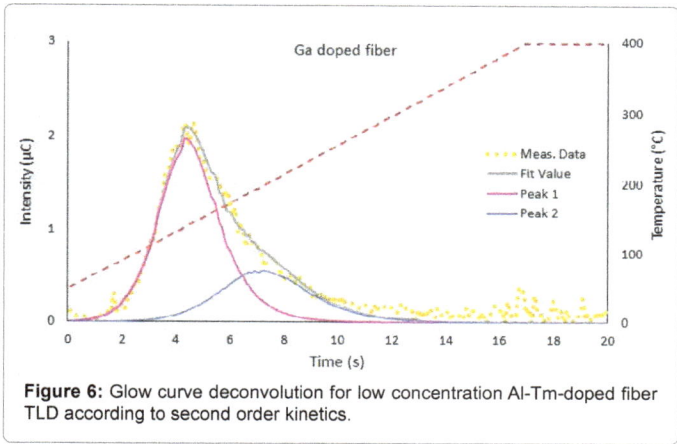

**Figure 6:** Glow curve deconvolution for low concentration Al-Tm-doped fiber TLD according to second order kinetics.

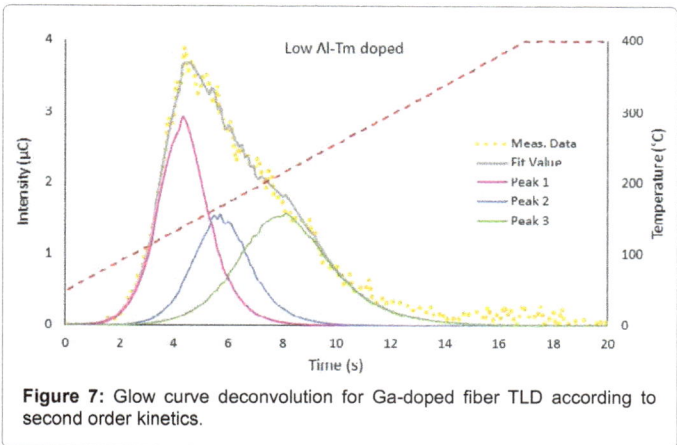

**Figure 7:** Glow curve deconvolution for Ga-doped fiber TLD according to second order kinetics.

peaks. In the first case, the more energetic traps have the greater capacity for storing memory of the irradiation. The complex shape of the Ba-doped fiber glow curve is posited to be due to deep interactions of this element with the host silica preform.

Figures 5 and 6 show deconvolutions for optical fibers containing the same dopant (Al-Tm), one with a greater concentration of Al inside the fiber core than the other. Tables 4 and 5 show that in increasing this dopant concentration, there is no practical change in the peak positions (peak $T \approx 160\text{-}166$ and $193\text{-}206$ for traps 2 and 3 respectively), but the intensity is increased, the exception being for trap 1 which is

independent of dopant as discussed earlier (peak $I_1$=1.5 and 1.6 → peak $I_2$=2.2 & 2.3 for traps 2 and 3 respectively).

The simple glow curve of Ga-doped fiber is shown in Figure 7. The

| Al-Y-Tm doped | | | | | |
|---|---|---|---|---|---|
| | **Trap: 1** | **Trap: 2** | **Trap: 3** | **Trap: 4** | **Trap: 5** |
| $E_a$ | 1.1 | 0.9 | 1.2 | 2.0 | 3.0 |
| $s'$ | 1.3e3 | 1.6e4 | 5.7e6 | 5.3e11 | 9.0e19 |
| $n_0$ | 1.4e4 | 1.6e4 | 1.6e4 | 1.5e4 | 4.1e3 |
| Peak $I$ | 2.9 | 3.4 | 4.0 | 4.8 | 5.2 |
| Peak $T$ | 162 | 237 | 295 | 353 | 393 |
| FWHM | 79 | 81 | 73 | 54 | 29 |
| Emission | 1138 | 1438 | 993 | 628 | 408 |

**Table 2:** Trap parameters for Al-Y-Tm-doped fiber.

| Ba doped fiber | | | | | |
|---|---|---|---|---|---|
| | **Trap: 1** | **Trap: 2** | **Trap: 3** | **Trap: 4** | **Trap: 5** |
| $E_a$ | 1.0 | 0.8 | 1.2 | 1.9 | 3.0 |
| $s'$ | 6.7e4 | 3.7e4 | 2.3e6 | 4.7e11 | 1.5e21 |
| $n_0$ | 1.1e4 | 9.2e3 | 1.2e4 | 1.1e4 | 1.0e3 |
| Peak $I$ | 2.8 | 2.1 | 3.0 | 3.4 | 3.1 |
| Peak $T$ | 166 | 241 | 297 | 345 | 403 |
| FWHM | 65 | 79 | 73 | 54 | 31 |
| Emission | 1238 | 1463 | 1055 | 644 | 408 |

**Table 3:** Trap parameters for Ba-doped fiber.

| High Al-Tm doped | | | |
|---|---|---|---|
| | **Trap: 1** | **Trap: 2** | **Trap: 3** |
| $E_a$ | 1.1 | 1.5 | 0.6 |
| $s'$ | 1.9e13 | 1.6e14 | 7.2e2 |
| $n_0$ | 5.8e3 | 4.5e3 | 1.1e4 |
| Peak $I$ | 2.8 | 2.2 | 2.3 |
| Peak $T$ | 128 | 160 | 193 |
| FWHM | 33 | 33 | 86 |
| Emission | 1145 | 829 | 1916 |

**Table 4:** Trap parameters for high concentration of Al-Tm-doped fiber.

| Low Al-Tm doped | | | |
|---|---|---|---|
| | **Trap: 1** | **Trap: 2** | **Trap: 3** |
| $E_a$ | 1.1 | 1.1 | 0.8 |
| $s'$ | 7.6e9 | 2.0e9 | 4.3e4 |
| $n_0$ | 6.8e3 | 4.3e3 | 6.7e3 |
| Peak I | 2.9 | 1.5 | 1.6 |
| Peak T | 137 | 166 | 206 |
| FWHM | 42 | 48 | 77 |
| Emission | 1128 | 1111 | 1503 |

**Table 5:** Trap parameters for low concentration of Al-Tm-doped fiber.

| Ga doped fiber | | |
|---|---|---|
| | **Trap: 1** | **Trap: 2** |
| $E_a$ | 0.9 | 0.8 |
| $s'$ | 1.3e7 | 6.9e4 |
| $n_0$ | 6.0e3 | 2.4e3 |
| Peak I | 2.0 | 0.6 |
| Peak T | 139 | 197 |
| FWHM | 50 | 77 |
| Emission | 1398 | 1600 |

**Table 6:** Trap parameters for Ga-doped fiber.

main TL is generated from the first glow peak which is from the silica preform, while lesser TL is contributed from peak 2, from Ga traps.

## Conclusion

The similarity in activation energy of first glow peak for all fiber samples indicates the TL response of the substrate, regardless of the dopant in the fiber. It is concluded that different concentration of similar dopant will only affect the intensity of relevant glow peaks while other factors will remain the same. An interesting observation from this study is that co-doping of the fiber is associated with an increase in the TL yield, pointing to a need for optimization of the choice and levels in use of co-dopants.

### Acknowledgment

The authors would like to acknowledge MMU/TM R&D research group for fabricating specialty-doped material optical fiber preforms; University of Malaya fiber pulling group for drawing the optical fibers; and UM-MOHE High Impact Research grant number A000007-50001 that financially supported this project.

### References

1. Yusoff AL, Hugtenburg RP, Bradley DA (2005) Review of development of a silica-based thermoluminescence dosimeter. Radiation Physics and Chemistry 74: 459-481.

2. Bradley DA, Hugtenburg RP, Nisbet A, Abdul Rahman AT, Issa F, et al. (2012) Review of doped silica glass optical fibre: their TL properties and potential applications in radiation therapy dosimetry. Appl Radiat Isot, 71: 2-11.

3. Abdul Rahman T, Abu Bakar NK, Chandra Paul M, Bradley DA (2014) Ultraviolet radiation (UVR) dosimetry system and the use of Ge-doped silica optical fibres. Radiation Physics and Chemistry.

4. Girard S, Ouerdane Y, Marcandella C, Boukenter A, Quenard S, et al. (2011) Feasibility of radiation dosimetry with phosphorus-doped optical fibers in the ultraviolet and visible domain. Journal of Non-Crystalline Solids 357: 1871-1874.

5. Issa F, Latip NAA, Bradley DA, Nisbet A (2011) Ge-doped optical fibres as thermoluminescence dosimeters for kilovoltage X-ray therapy irradiation. Nuclear Instruments and Methods in Physics Research Section A: Accelerators, Spectrometers, Detectors and Associated Equipment 652: 834-837.

6. Bradley DA, Abdul Sani SF, Alalawi AI, Jafari SM, Noor M, et al. (2014) Development of tailor-made silica fibres for TL dosimetry. Radiation Physics and Chemistry 104: 3-9.

7. Bauk AS, Abdul-Rashid HA, Gieszczyk W, Hashim, Mahdiraji GA, et al. (2014) Potential application of pure silica optical flat fibers for radiation therapy dosimetry. Radiation Physics and Chemistry 106: 73-76.

8. Ghomeishi M, Mahdiraji GA, Adikan FM, Ung N, Bradley (2015) Sensitive Fibre-Based Thermoluminescence Detectors for High Resolution In-Vivo Dosimetry. Scientific reports 5.

9. Amouzad Mahdiraji G, Ghomeishi M, Dermosesian E, Hashim S, Ung NM, et al. (2015) Optical fiber based dosimeter sensor: beyond TLD-100 limits. Sensors and Actuators A: Physical 222: 48-57.

10. Ghomeishi M, Mahdiraji GA, Adikan FRM, Hashim S (2013) The thermoluminescence response of undoped silica PCF for dosimetry application. In: Lasers and Electro-Optics Pacific Rim (CLEO-PR), 2013 Conference: 1-2.

11. Zahaimi NA, Ooi Abdullah MHR, Zin H, Abdul Rahman AL, Hashim S, et al. (2014) Dopant concentration and thermoluminescence (TL) properties of tailor-made Ge-doped SiO₂ fibres. Radiation Physics and Chemistry.

12. Ghomeishi M, Jahanshahi P, Dermosesian E, Adikan FR (2014) Analysis of optical fibre defects using thermoluminescence glow curve method. In: Photonics (ICP), 2014 IEEE 5th International Conference: 141-143.

13. Bradley DA, Mahdiraji G, Ghomeishi M, Dermosesian E, Adikan, et al. (2014) Enhancing the radiation dose detection sensitivity of optical fibres. Applied Radiation and Isotopes 100: 43-49.

14. Abdulla YA, Amin Y, Bradley DA (2001) The thermoluminescence response of Ge-doped optical fibre subjected to photon irradiation. Radiation Physics and Chemistry 61: 409-410.

15. Wagiran H, Hossain I, Bradley D, Yaakob ANH, Ramli (2012) Thermoluminescence Responses of Photon and Electron Irradiated Ge- and Al-Doped SiO₂ Optical Fibres. Chinese Physics Letters 29: 027802.

16. Salah N (2011) Nanocrystalline materials for the dosimetry of heavy charged particles: A review. Radiation Physics and Chemistry 80: 1-10.

17. Rasheedy MS (2005) A modification of the kinetic equations used for describing the thermoluminescence phenomenon. J Fluoresc 15: 485-491.

18. Salh R (2011) Defect related luminescence in silicon dioxide network: a review. In: Crystalline Silicon - Properties and Uses 135-172.

# 9

# Energy-Aware Cross Layer Framework for Multimedia Transmission over Wireless Sensor Networks

**Mohammed Ezz El Dien\*, Aliaa AA Youssif and Atef Zaki Ghalwash**

*Department of Computer Science, University of Helwan, Helwan, Egypt*

## Abstract

Wireless Multimedia Sensor Networks (WMSNs), is a network of sensors which are limited in terms of computational, memory, bandwidth, and battery capability. Multimedia transmission over WSN requires certain Qos guarantees such as huge amount of bandwidth, strict delay and lower loss ratio, which makes transmitting multimedia content over it, is a challenging problem. Recently adopting cross-layer design in WMSNs proved to be a promising approach, which improves quality of service of WSN under various operational conditions. In this work, an energy aware framework for transmitting multimedia content over WSN (ECWMSN) is presented, where packet and path scheduling were introduced, so that It adaptably selects optimum video encoding parameters at application layer according to current wireless channel state, and schedules packets according to its type to drop less important packets in case of network congestion. Finally, path scheduling is introduced so that different packets types/priority is routed through suitable path with suitable Qos taking into consideration the network lifetime. Simulation results show that ECWMSN optimizes video quality and prolongs network lifetime.

**Keywords:** Wireless multimedia sensor networks; Cross layer design; WMSNs; GOP structure; Packet scheduling; Queue scheduling; Path scheduling

## Introduction

WMSNs is an ad-hoc arrangement of wirelessly interconnected multifunctional sensor nodes that allow retrieving video and audio streams, still images, and scalar sensor data. Sensor nodes can be densely distributed over large even remote areas, and will continue to collaborate their efforts to the benefit of the network; even if a number of nodes malfunction, the network will continue to function.

It is expected that there will be different multimedia applications [1] with different Qos metrics will exist, such as family monitors [2], traffic routing [3], industrial process control [4], surveillance sensor networks [3], and healthcare delivery [5], but there are many challenges for transmitting multimedia over such type of networks [6] due to constrained nodes whether in terms of limited battery lifetime, limited memory and limited throughput. In addition, all of such wireless networks commonly uses wireless channel for communication, which suffers from interference problems, multipath fading, shadowing and high signal attenuation effects, that results in a high bit error rate and fluctuated bandwidth due to link failures and congested packets.

In addition to multimedia Qos guarantees, Network lifetime management and balanced usage of battery power add additional challenges, due to sensors operating using batteries and progress in battery development is limited and replacing such battery is costly or impossible.

Routing protocols plays an important role for increasing throughput and decreasing end-to-end delay. However single path routing protocols are not efficient for Qos requirements for multimedia transmission such as delay and loss. Multipath routing [7] is an alternative approach to handle multimedia transmission, as it increases bandwidth efficiency, reliability, decreases end-to-end delay, and evenly spread power consumption in the network.

Cross-layer design [8-12] shows that it is a promising approach to handle multimedia based transmission over WSN. Traditional OSI layered architecture's networking services are organized and provided by specific layers without sharing or communication between them.

Cross-layer approach optimizes the performance of the network through violating the traditional approach in many different ways such as: allowing sharing information across layers, direct communication between layers, creating new intermediate interfaces between layers or merging adjacent layers.

In this work, an adaptive multimedia transmission framework called Energy Aware Framework for multimedia Transmission over Wireless Sensor Networks (ECWMSN) is presented. It is an extension of previously introduced [13], so that current framework includes three main characteristics: adaptive MPEG-4 video encoder, packet and path scheduling.

It is a cross-layer framework that communicates current wireless channel state to application layer, so that application layer can apply proper video encoding parameters according to the current state of the wireless channel. Each video packet has different priority and effect on video transmitted, so for packet effect on video transmitted, Sink node analyzes suitable video encoding parameters and sends it back to source node to use it, while for packet priority it schedules packets into different queues and at time of congestion, it drops less important packets without affecting video quality. Finally framework extended to schedule different packet types on different routes according to packet type and matched path cost considering network lifetime.

So by combining cross layer design with multipath routing, packet and path scheduling in one single framework to transmit multimedia content over WSN shows a promising framework that can optimize bandwidth, multimedia quality and network lifetime.

---

**\*Corresponding author:** Mohammed Ezz El Dien, Department of Computer Science, University of Helwan, Helwan, Egypt
E-mail: mohamed.ezz011@gmail.com

The remainder of the paper is organized as follows: section two presents a survey on cross layer protocols for WMSN. In section three, the newly introduced ECWMSN framework is presented. In section four, an evaluation of the newly introduced framework ECWMSN using extensive simulation scenarios is presented. In section five the paper is concluded.

## Related Work

Path scheduling finds suitable path between source and destination node not only using optimal hop count as used by traditional routing protocols but also using other user defined QoS metrics such as delay, bandwidth, loss and energy requirements, which depends on the application nature of the WSN.

Path scheduling in a constrained networks such as WSN depends on various routing metrics [14] for example; in this work [15] it selects path based on packet type, so that delay sensitive packets to be routed through fastest path, while error sensitive packets are routed through the most reliable links, and non-constrained packets through least energy routes. While In this work [16] paths ranked considering interference, congestion and hop count through an integrated routing metric (bottleneck of node congestion level, hop count, bottleneck of node leisure level and the number of congested nodes), in addition videos frames are assigned to single or multipath depending on congestion level of each path.

In this work [17] a video transmission scheme which schedules different video packet types over different paths, so that critical packets are transmitted through highly rated paths. Source nodes send control messages periodically to sink node with updated state each path, and Sink node receive path status and score each path using (energy level, buffer status, hop count and path reliability) and sends back to source node the new path cost, so that source nodes later schedules packets according to its type through suitable path, While in this work [18] Ants-based multi-Qos routing algorithm that schedules paths using (loss ratio, available memory, queuing delay and remaining energy), Other work [19] uses AI technique for scoring paths, it uses link expiration time, probabilistic link reliable time, link packet error rate, link received signal strength and residual battery power to calculate index of each path using fuzzy logic. While this work [20] uses signal strength, remaining energy and available memory to score each path, while in [21] uses drain rate and delay for scoring path, other cross layer protocol [22] uses energy level and free buffer space.

In [23] CLAR's NWK layer chooses optimal route based on Ad-hoc routing's DSR protocol which selects a route that has an optimal value of PHY layer parameter CQI (channel quality indicator) that asses link reliability and stability. It is estimated from SNIR (signal to noise interference ratio) of the received signal and maintained for each neighbor node by [24] DRMACSN MAC layer protocol. Moreover; CLAR checks network status ( number of ongoing transmissions parameter which maintained for each neighbor node by DRMACSN MAC layer protocol ) before exchanging routing control packets to further minimize energy consumption in case channel is bad or current simultaneous transmissions are above a threshold value.

In this work [25] a new architecture is presented for video transmission over WSN, which is called energy efficient and high quality video transmission architecture (EQV-Architecture). This architecture affects three layers of communication protocol stack. It considers wireless video sensor nodes constraints like limited process and energy resources in addition to preserving video quality at the sink. Application, transport, and network layers are the layers in which the compression

protocol (Modified MPEG), transport protocol, routing and dropping scheme are proposed respectively.

The dropping scheme presented in this work decides to discard packets based on energy level of each node and priority information that had been provided by video compression layer inside the received packets.

In this work [26] the author believes that it is resource efficient to have single scheme which aggregates common protocol layers into a single cross layer protocol XLM for resource constrained sensor nodes. XLM objectives are to provide, high reliable communication with minimal energy consumption, adaptive communication and local congestion control. XLM is based on an initiative determination concept, which allows each node to decide to participate in a communication based on four conditions that must be satisfied before a node decides to start communication. First condition ensures reliable links for communication, while second condition limits the traffic a node can relay to prevent congestion, the third condition ensures no buffer overflow could occur at this node to prevent congestion and the last condition guarantees even distribution of energy consumption by keeping remaining energy level at this node above minimum value.

While previous work focus on multimedia Qos, there are several work [27,28] aim to maximize network lifetime of each node and use battery wisely to prolong network lifetime before network get partitioned, it transmits data at the minimum power level to maintain links or adaptly choose transmit range of each node to minimize energy consumption. While other work [29,30] distribute energy usage of all mobile nodes by selecting under-utilized route instead of shortest path. Other work [31] chooses inactive communication to minimize energy consumption where some nodes are put to sleep to keep minimum number of nodes waken for transmission while others get reduce energy consumption energy consumed while nodes is inactive.

In this work [32] a secure routing protocol is presented which is based on cross layer design and energy-harvesting mechanism. It uses a distributed cluster-based security mechanism. It is a Duty-Cycled WSN, where nodes are initially in active state, where nodes actively participate in WSN operations. However, as long as the energy value of sensor node decreases, it switches to semi-active state. In semi-active state, nodes are in wake and sleep conditions. In wake position, nodes take part in network operations, while in sleep position, nodes only harvest environmental energy. In idle state, nodes only harvest energy until it switches back either to active or semi-active states. Routing in this work is based on energy level, which is exchanged between PHY and NWK layer to ensure efficient use of energy. Energy harvesting system which is based on photovoltaic cells are used to convert sun light energy into electric energy to extract and store energy, which is used to take decisions for the node state and thus for the routing issues.

## Proposed ECWMSN Energy Aware Cross Layer Framework for Multimedia Transmission over WSN

It consists of four components (Adaptive Video Encoder, Packet Scheduling, Queue Scheduling, Path Scheduling) as will be explained in next subsections, to solve multimedia transmission problems like limited bandwidth, wireless link failures, congested packets and limited battery resources.

It uses cross-layer communication between Application, Network, Data link and Physical layer as shown in Figure 1. At application layer MPEG-4 encoder adjusts encoding parameters according to current wireless channel status, which is communicated from physical layer and recommendation from Sink node [13]. Moreover, it apply path scheduling to routes packets through different paths according to packet type and path's Qos guarantees; finally it apply queue scheduling so that it can drop less important packets in case of network congestion without affecting video quality.

There is continuous feedback messages sent from Sink node to source nodes, to propagate back both recommended video encoding parameters and path cost. Received video will be analyzed at sink node [13] and could be assessed using PLSR (Partial Least Squares Regression) [33] or other video quality assessment techniques (such as VQM, MS-SSIM,...) [34,35], so that, the optimum MPEG-4 encoding parameters suitable for current wireless channel state is sent back to source nodes. Source node will configure MPEG-4 encoder using new video encoding parameters sent from sink, in addition source node will select suitable path at time of sending new video packet according to current packet type and communicated path cost from sink node.

### Packet scheduling component

MPEG (Moving Pictures Experts Group) employed today by most of the video compression systems; it is inter-frame video compression algorithm which exploits temporal correlation between frames to achieve high level of compression by independently coding reference frames. It is found [36] that the number of distorted frames due to loss of I, P or B depends on the type and position of the lost frame. I-Frame will cause distortion to $(N + M-1)$ frames; while P-Frame will cause distortion to $(2M + N-2) / 2$ and finally B-Frame contains temporal information and is not used as a reference, their loss only causes motion artifacts and it does not spread errors. So each packet type will need to be scheduled differently according to its type [37], where I-Frames will need more protection than P-Frame and B-Frames to reduce its loss effect of video quality. Queue and path scheduling will uses packet type information communicated from application layer so that higher priority packets such as I-Frame will be queued and routed in way better than lower priority packets as will be explained in next sub sections.

### Adaptive video encoder component

Group of Picture (GOP) structure affects video transmission over lossy wireless sensor network [38,39], it is determined by parameters N and M, which defines the sequence of I, P and B frames. Parameter N specifies the I-frame interval and parameter M determines I or P frame interval. The higher frequency of I-Frame [36,40] will reduce error propagation and causes better video quality but on the other hand it will reduce compression ratio of the video which will produces large sized video file. Based on results found in [13], the sink node periodically sends recommended video encoding parameters GOP total length $(G_L)$ and number of B-Frames $(B_f)$ to source node after analyzing video received during previous period. It recommends optimum parameters according to current wireless channel status (Packet Loss Ratio, BER ...).

### Queue scheduling component

ECWMSN framework differentiates between two types of packets, data and video packets, data packets are less important than video packets. Video packets are further prioritized according to its frame type, so that I-Frame packets are the highest priority followed by P-Frame and B-Frame and finally data packets. After identifying priority of packets, each intermediate node can schedule incoming packets as shown in Figure 2 as follows, buffer space is divided into high and low priority queues, and the priority of each packet determines whether such packet will be queued or dropped according to priority of packet and current status of the buffer.

In case of the incoming packet has high priority it will be queued in high priority queue if there is a room for it. Otherwise it will search low priority queue for lowest priority packet to replace it. Low priority packets is queued into low priority queue in case there is a room for it; otherwise it can replace lower priority packet in lower priority queue or to be dropped, and at time of dequeue, it will dequeue higher priority packet then lower priority.

### Path scheduling component

ECWMSN is a cross layer framework where different layers interact with each other, so after recognizing packet type at Application layer, it communicates it with Network layer so that path is scheduled according to each packet type, where higher priority packet will be scheduled through path with higher Qos measures, while lower priority packets will be routed through lower Qos guarantee path.

It is based on AOMDV [41] multipath routing protocol which route packets using optimal number of hops for routing packets; while ECWMSN routing protocol which is a modified version of [36] that uses path cost function to rank each path based on network energy status, available buffer, number of hops, lost packet. the new framework preserve power consumed in network by choosing paths with less power consumption until network reach energy threshold value, then it favor paths with better energy reserve regardless power consumed.

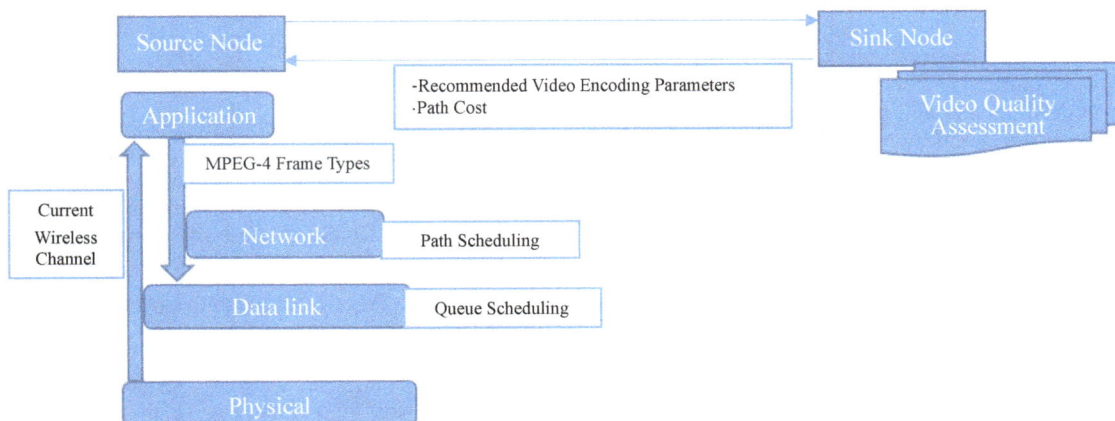

**Figure 1:** Energy-aware adaptive cross layer framework for video transmission over WSN.

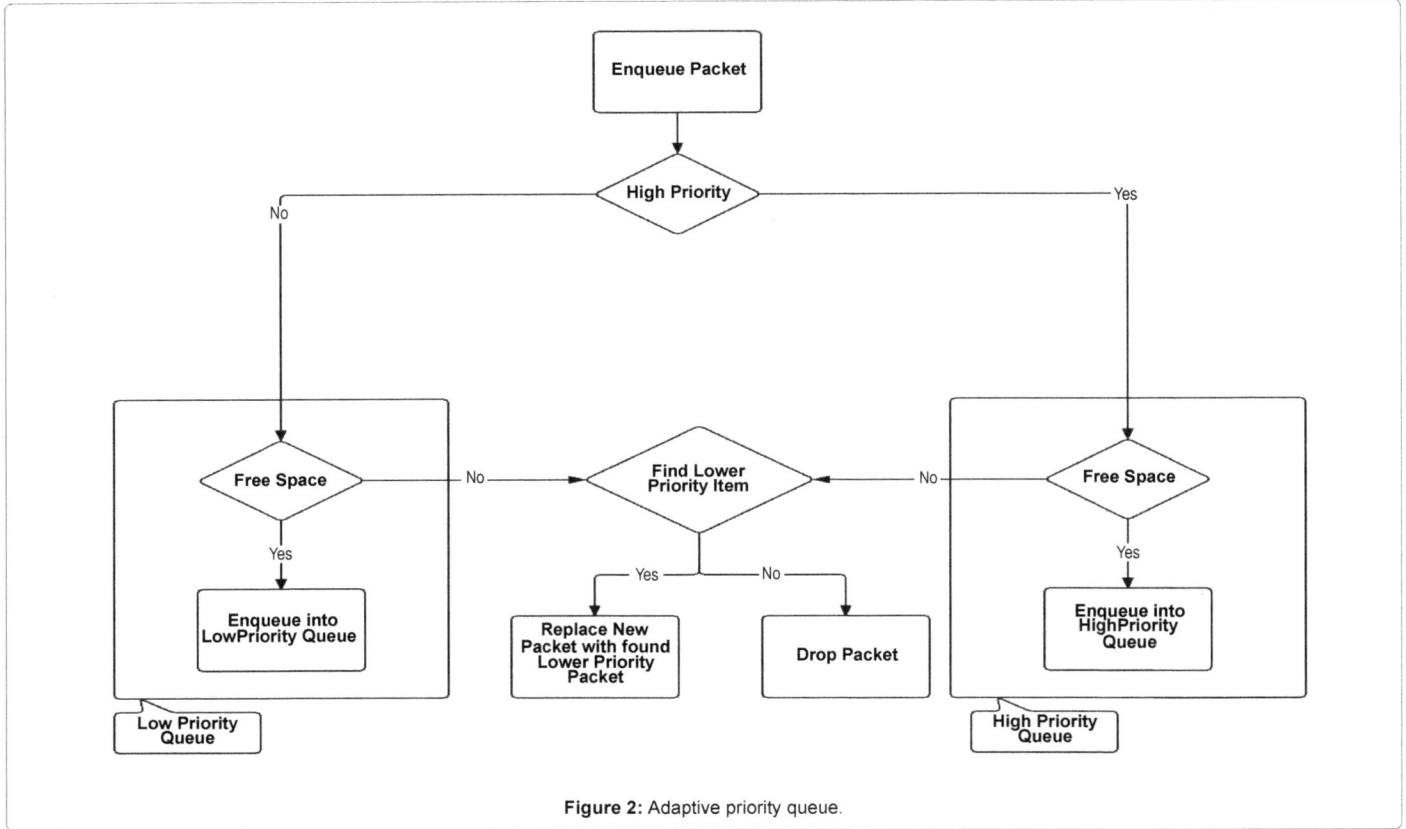

**Figure 2:** Adaptive priority queue.

ECWMSN introduces new messages "Metric-Update" to AOMDV protocol, which sent periodically from source node to sink node as "Forward-Metric-Update" message, which collects status of each node along different paths toward sink node. Sink node will evaluate each received message from different paths and generates "Backward-Metric-Update" message to be sent back through path which it came from, with calculated cost of such path.

Upon receiving "Forward-Metric-Update" message at intermediate nodes along path, it will update such message with its current energy level if it is less than minimum energy stored within message, otherwise no updates as shown in Equation1, where $re$ is remaining energy of node $S$ along path $P$.

$$\text{Min. Energy (p)} = \min_{S \in P} re(s) \qquad (1)$$

Intermediate nodes which received "Forward-Metric-Update" will updates message with its current free buffer count if it is less than minimum free buffer count within message, otherwise no updates as shown in Equation 2, where, $bf$ is free buffer at node $S$ along path $P$.

$$\text{Min. Buffer (p)} = \min_{S \in P} bf(s) \qquad (2)$$

Upon receiving "Forward-Metric-Update" message at intermediate nodes along path, it updates message with $Pw$ total average power consumed per each node along path whether at time of sending or receiving as shown in Equation 3. Where $Tx$ is average power consumed at sending time at node $S$ along path $P$, and $Tr$ is average power consumed at receiving time at node $S$ along path $P$.

$$Pw(p) = \sum_{s \in p} Tx(s) + \sum_{s \in p} Tr(s) \qquad (3)$$

Each message received along different paths at Sink node will be evaluated to calculate cost of each path as shown in Equation 4, where

$\alpha + \beta + \gamma + \delta = 1$ are weight factors for each term, so that it is configured by user according to current application of the network.

$$Cost(p) = \omega.\alpha + \min.buffer(p).\beta + \left(\frac{1 + max\ HC\text{-}HC(p)}{max\ HC}\right).\gamma + 1 - \frac{no.DelayedPkts}{totalPkets\,\text{Re}\,cv}\delta \quad (4)$$

The first component $\omega$ is the network lifetime term, which is evaluated as shown in Equation 4.1, where $\Omega$ is energy threshold, so that if minimum energy value along path p, is greater than energy threshold value $\Omega$, it uses power consumption term $pw(p)$ (Equation 3) which is total average power consumed along path p, otherwise it uses Min. Energy(p) (Equation 1) term which is the minimum remaining energy found along path p.

$$\omega = \begin{cases} \dfrac{min.energy(p)}{inital\ energy} & min.energy(p) < \Omega \\[3mm] \dfrac{1}{pw(p)} & min.enrgy\ (p) \geq \Omega \end{cases} \qquad (4.1)$$

For the second component it represents minimum buffer found along path p, while third component it measures hop count for this path where Max HC is maximum hop count in the network, while HC(p) is hop count of current evaluated path. Finally fourth component it measures path reliability as it measures ratio of packets delivered to sink node without delay to total packets received.

Sink node will evaluate each path's cost after receiving "Forward-Metric-Update" message using Equation 4, then it sends back along such path a new corresponding message "Backward-Metric-Update", which it came from, so that each intermediate node update its routing table with new cost for such evaluated path.

During data and video sending phase, source node will select path toward sink node according to type of packet and suitable path cost; so that higher priority packets will be routed through paths with higher cost value. Each packet is modified to carry path id that identify hops of the selected path, so that intermediate node will selects next hop according to path selected previously by source node.

## Simulation and Results

In this section ECWMSN framework will be compared to non-adaptive QPS, QPS+ [17], which uses static encoding parameters $G_L = 7$ and $B_f = 2$ regardless wireless channel state, while ECWMSN framework adaptively uses suitable $G_L$ and $B_f$ parameters [13] depending on wireless channel condition; in addition it apply packet, queue and path scheduling as explained in previous section. QPS-Scheme [17] applies packet, path and queue scheduling techniques, where queue priority scheduling is used at data link layer that uses 4 different queues to priotorize packets in round robin way. Finally it uses path scheduling that route packets according to its type but without considering fair usage of battery through network. QPS+ is QPS scheme but it uses Equation 4.1, which considers energy at scheduling time similar to ECWMSN.

The results in this work obtained using NS2 network simulator [42] to simulate packet transmission over the wireless network. The three schemes use similar settings as shown in Table 1. There are 150 nodes 2 of them are video nodes which sends video packets to sink node and every other node sends data packets of 255 bps to sink node. All nodes work as intermediate routing nodes, which are uniformly distributed in rectangular field of dimension 1000m x 1000m.

Evalvid framework [43-46] used to generate MPEG-4 video traffic using ffmpeg [47] to transmit Paris video with 1065 frame. Our used performance metrics are PSNR, End-to-End Delay, Loss, Network Lifetime and energy usage among nodes.

## PSNR

Figure 3 shows that ECWMSN gives better video quality of 30.40 dB than other schemes as it depends on communicated packet type that allow higher priority frames such as I-Frame to be kept without dropping in case of congestion and routed through more reliable paths in addition adaptively uses suitable encoding parameters according to current wireless channel. While other Schemes QPS, QPS+ show similar PSNR of 28 dB, where their queue scheduling does not keep high priority packets in case of congestion, in addition ECWMSN reduces energy usage of the network so that more packets can be transmitted.

## End - to - end delay

Figure 4 show that ECWMSN scheme recorded an average of 24 milliseconds while other schemes recorded average 27 milliseconds. E-ACWSN, QPS and QPS+ schemes use path scheduling that depends on an overhead of Metric-Update messages to propagate network status and path cost to nodes, such overhead causes delay or loss of packets flow in the network.

## Lost frames

Figure 5 shows that ECWMSN scheme recorded average loss ratio of 5.5, as ECWMSN in congestion time keeps higher priority packets by replacing lower priority packets in adaptive priority queue; while QPS schemes just drop packets even they are high priority in case of overflow.

In addition ECWMSN uses energy wisely among nodes to keep network lifetime long so that more packets can be transmitted without loss.

## Network energy

ECWMSN not only maximized network lifetime by choosing paths with higher remaining energy, but also used energy fairly through the network by choosing paths with lower energy consumption. It is shown in Figure 6 that ECWMSN and QPS+ schemes recorded better energy usage value 24.02 and 23.77 joules through simulation time, while QPS scheme is 23.72 joules as it only seeks to select paths which have better reserve of energy. while ECWMSN and QPS+ select paths with low cost of power consumption as explained earlier, in addition selecting paths with better reserve of energy in case network reach critical threshold energy value.

| Parameters | Value |
|---|---|
| Total Nodes | 150 |
| Queue length | 50 |
| Network Dimension | 1000 m × 1000 m |
| Routing Protocol | Modified version of AOMDV |
| Video | Paris |
| Encoding | MPEG-4 |
| Frame rate | 30 Hz |
| Format | QCIF, 176 × 144 |
| Bit rate | 64000 bps |
| No. video frames | 1065 |
| Initial energy | 50 joule |
| Traffic Model | Two video nodes send video to sink while every other node sends 255 bps data traffic to sink |

**Table 1:** Parameters of simulation scenario.

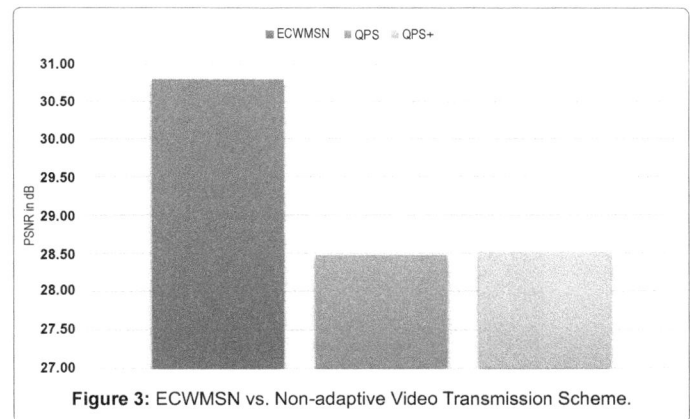

**Figure 3:** ECWMSN vs. Non-adaptive Video Transmission Scheme.

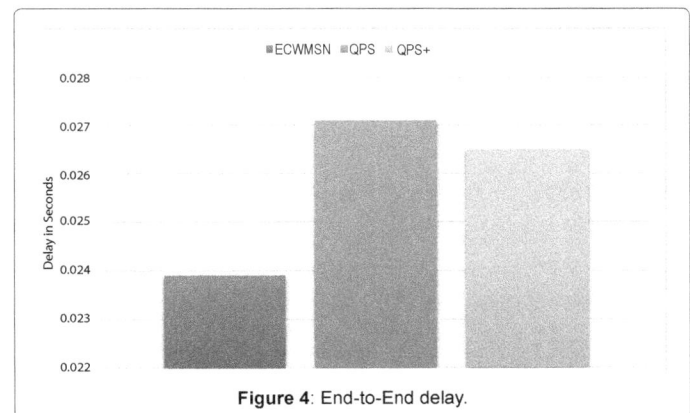

**Figure 4:** End-to-End delay.

In Figure 7 shows minimum energy recorded through simulations, both ECWMSN and QPS+ tries to save energy usage by using minimum energy cost of routing and that causes to lower their energy till network reach its energy threshold then it starts to switch back to avoid paths with nodes have lower energy values.

## Conclusion

Transmitting video over WSN is a challenging problem due to huge amount of bandwidth it requires, so by combining different protocols into a single framework is a promising approach to handle such challenge. In this work ECWMSN is an multimedia transmission framework that uses optimal GOP structure / size suitable to current wireless

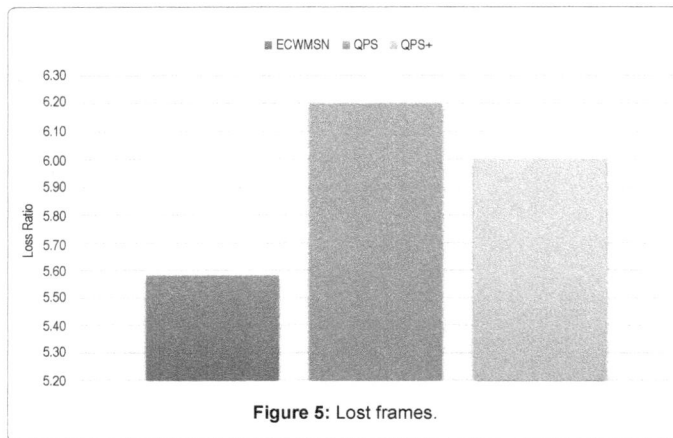

**Figure 5:** Lost frames.

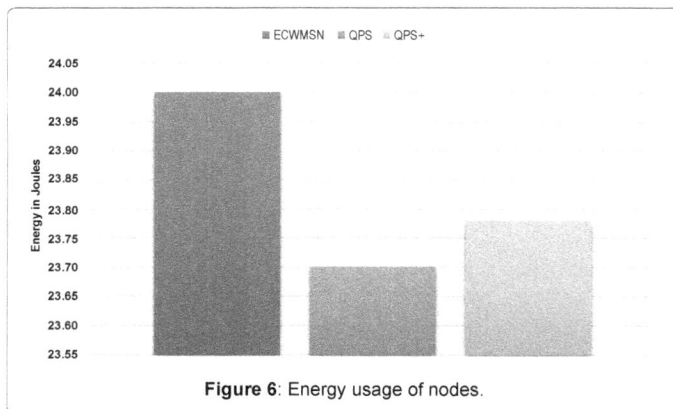

**Figure 6:** Energy usage of nodes.

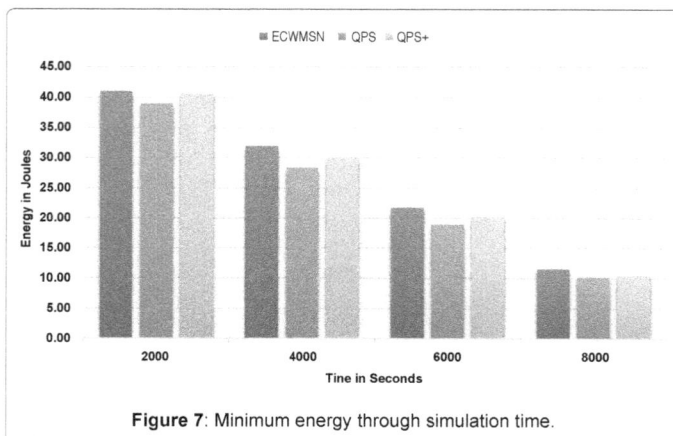

**Figure 7:** Minimum energy through simulation time.

channel state, an adaptive queue that schedules incoming packets to drop less important packets with affecting overall video quality. Finally, it modified AOMDV multi-path routing protocol to schedule paths according to packet type and balance traffic and energy over multiple paths instead of single path transmission. By comparing ECWMSN to QPS schemes QPS+, it gave better video quality with average PSNR value 30.40 dB. In addition ECWMSN show great energy saving and balanced energy consumption scheme over other schemes, it recorded 24.02 joule as average energy consumed by all nodes through simulation while QPS recorded 23.72 joule.

### References

1. Shen Z, Luo J, Zimmermann R, Vasilakos AV (2011) Peer-to-Peer Media Streaming: Insights and New Developments. Proceedings of the IEEE 99: 2089-2109.

2. Eeves AA (2005) Remote Monitoring of patients suffering from early symptoms of Dementia. In International Workshop on Wearable and Implantable Body Sensor Networks, London, UK.

3. Campbell J, Gibbons PB, Nath S, Pillai P, Seshan S, et al. (2005) IrisNet: an Internet-scale architecture for multimedia sensors. In Proceedings of the 13th ACM International conference on Multimedia, Singapore.

4. Guha A, Pavan A, Liu JCL, Roberts BA (1995) Controlling the Process with Distributed Multimedia. IEEE Multimedia 2: 20-29.

5. Hu F, Kumar S (2003) Multimedia query with QoS considerations for wireless sensor networks in telemedicine. Society of Photo-Optical Instrumentation Engineers-International Conference on Internet Multimedia Management Systems, Orlando, Florida, USA.

6. Xiong Z, Fan X, Liu S, Zhong Z (2010) Distributed image coding in wireless multimedia sensor networks: A survey. Third International Workshop on Advanced Computational Intelligence, Jiangsu.

7. Rosário D, Costa R, Santos A, T Braun, Cerqueira E (2013) QoE-aware Multiple Path Video Transmission for Wireless Multimedia Sensor Networks. Simpósio Brasileiro de Redes de Computadores e Sistemas Distribuídos, Brasil.

8. Farooq MO, Kunz T, St-Hilaire M (2011) Cross layer architecture for supporting multiple applications in Wireless Multimedia Sensor Networks. 7th International Wireless Communications and Mobile Computing Conference, Turkey.

9. Alikhani S, Kunz T, St-Hilaire M, Richard F (2010) A central-networked cross-layer design framework for wireless sensor networks. Proceedings of the 6th International Wireless Communications and Mobile Computing Conference, China.

10. Mendes L, Rodrigues J (2011) A survey on cross-layer solutions for wireless sensor networks. Journal of Network and Computer Applications 34: 523-534.

11. Costa D, Guedes L (2011) A Survey on Multimedia-Based Cross-Layer Optimization in Visual Sensor, Sensors 11: 5439-5468.

12. Gunasekaran R, Qi H (2008) XLRP: Cross Layer Routing Protocol for Wireless Sensor Networks. WCNC. IEEE Wireless Communications Networking Conference, USA.

13. Youssif A, Ghalwash A, Kader MA (2015) ACWSN: an adaptive cross layer framework for video transmission over wireless sensor networks. Wireless Networks 21: 2693-2710.

14. Zara H, Hussain F (2014) QoS in Wireless Multimedia Sensor Networks: A Layered and Cross-Layered Approach. Wireless Personal communication 75: 729-757.

15. Boukerche A, Araujo RB, Villas L, Novel A (2007) QoS Based Routing Protocol for Wireless Actor and Sensor Networks. IEEE Global Telecomm Conference, Washington, DC.

16. Wan Z, Xiong N, Yang LT (2015) Cross-layer video transmission over IEEE 802.11e multihop networks. Multimedia Tools and Applications 74: 5-23.

17. Karimi E, Akbari B (2013) Priority Scheduling for Multipath Video Transmission in WMSNS. International Journal of Computer Networks & communications 5: 167-180.

18. Cobo L, Quintero A, Pierre S (2010) Ant-based routing for wireless multimedia sensor networks using multiple QoS metrics. Computer Networks 54: 2991-3010.

19. Palaniappan S, Chellan K (2015) Energy-efficient stable routing using QoS monitoring agents in MANET. EURSIP Journal on Wireless Communications and Networking 2015: 1-11.

20. Pillai M, Sebastian M, Madhukumar S (2013) Dynamic multipath routing for MANETs -A QoS adaptive approach. 3rd International Conference on Innovative Computing Technology, London, UK.

21. Begum A, Iqbal F, Mohammad A (2015) Quality of service routing for multipath manets. International Conference on Signal Processing and Communication Engineering Systems, Guntur.

22. Chen J, Li Z, Liu J, Kuo Y (2011) QoS Multipath Routing Protocol Based on Cross Layer Design for Ad hoc Networks. In Internet Computing and Information Services, Hong Kong.

23. Chabalala SC, Muddenahalli TN, Takawira F (2011) Cross-layer adaptive routing protocol for wireless sensor networks. AFRICON, IEEE, Livingstone.

24. Muddenahalli T, Takawira F (2009) DRMACSN: New mac protocol for wireless sensor networks. SATNAC Conference on Network Planning & General Topics, RSS, Swaziland.

25. Aghdasi H, Abbaspour M, Moghadam M, Samei Y (2008) An Energy-Efficient and High-Quality Video Transmission Architecture in Wireless Video-Based Sensor Networks. Sensors 8: 4529-4559.

26. Vuran M, Akyildiz I (2010) XLP: A Cross-Layer Protocol for Efficient Communication in Wireless Sensor Networks. IEEE Transactions on Mobile Computing 11: 1578-1591.

27. Toh C (2011) Maximum battery life routing to support ubiquitous mobile computing in wireless ad hoc networks. In Communications Magazine IEEE 39: 38-147.

28. Prakash S, Saini J, Gupta S (2010) A review of Energy Efficient Routing Protocols for Mobile Ad Hoc Wireless Networks. International Journal of Computer Information Systems 4: 37-46.

29. Buddhdev B (2013) Energy Enhancement in AOMDV. International Journal of Advanced Research in Computer Engineering and Technology 2: 1924-1929.

30. Gálvez J, Ruiz P, Skarmeta A (2010) A feedback-based adaptive online algorithm for multi-gateway load-balancing in wireless mesh networks. In World of Wireless Mobile and Multimedia Networks IEEE, Canada.

31. Chen B, Jamieson K, Morris R, Balakrishnan H (2001) Span: an energy-efficient coordination algorithm for topology maintenance in ad hoc wireless networks. Proceedings of International Conference on Mobile Computing and Networking, Italy.

32. Alrajeh N, Khan S, Lloret J, Loo J (2013) Secure Routing Protocol Using Cross-Layer Design and Energy Harvesting in Wireless Sensor Networks . International Journal of Distributed Sensor Networks 2013: 1-11.

33. Wang Z, Wang W, Xia Y, Wan Z, Lin W (2014) No-reference hybrid video quality assessment based on partial least squares regression. Multimedia Tools and Applications 74: 10277-10290.

34. Chikkerur S, Sundaram V, Reisslein M, Karam LJ (2011) Objective video quality assessment methods: A classification, review, and performance comparison. IEEE Transactions on Broadcasting 57: 165-182.

35. Wang Z, Wang W, Xia Y, Wan Z, Wang J, et al. (2014) Visual Quality Assessment after Network Transmission Incorporating NS2 and Evalvid. The Scientific World Journal 2014: 1-7.

36. Huszak A, Imre S (2010) Analysing GOP Structure and Packet Loss Effects on Error Propagation in MPEG-4 Video Streams. 4th International Symposium on Communications, Control and Signal Processing, Cyprus.

37. Wan Z, Xiong N, Ghani N, Vasilakos AV, Zhou L (2014) Adaptive Unequal Protection for Wireless Video Transmission over IEEE 802.11e Networks. Multimedia Tools and Applications 72: 541-571.

38. Zulpratita S (2013) GOP Length Effect Analysis on H.264/AVC Video Streaming Transmission Quality over LTE Network. 3rd International Conference on Computer Science and Information Technology, Indonesia.

39. Fang T, Chau L (2005) An Error-Resilient GOP Structure for Robust Video Transmission. IEEE Ttransactions On Multimedia 7: 1131-1138.

40. Sousa R, Mota E, Silva EN, Paixão KSP, Faria B, et al. (2010) GOP size influence in high resolution video streaming over wireless mesh network. Computers and Communications, IEEE Symposium, Italy.

41. Marina, Mahesh K, Das SR (2006) Ad hoc on-demand multipath distance vector routing. Wireless Communications and Mobile Computing 6: 969-988.

42. Information Science Institute (2003) NS-2 network simulator, Software Package.

43. Klaue J, Shieh CK, Hwang WS, Ziviani A (2008) An Evaluation Framework for More Realistic Simulations of MPEG Video Transmission. Journal of Information Science and Engineering 24: 425-440.

44. Klaue J, Rathke B, Wolisz A (2003) EvalVid-A Framework for Video Transmission and Quality Evaluation. 13th International Conference on Modelling Techniques and Tools for Computer Performance Evaluation, Urbana, Illinois, USA.

45. NCKU (2015) Integrating EvalVid with NS2.

46. Wang H, Wang W, Wu S, Hua K (2010) A Survey on the Cross-Layer Design for Wireless Multimedia Sensor Networks. 3rd International Conference Mobile Wireless Middleware, operating systems and applications, USA.

47. FFMPEG (2015) MPEG-4 encoder.

# 10

# Data Encryption and Transmission Based on Personal ECG Signals

**Ching-Kun Chen[1], Chun-Liang Lin[1]\*, Shyan-Lung Lin[2] and Cheng-Tang Chiang[3]**

[1]Department of Electrical Engineering, National Chung Hsing University, Taichung, Taiwan
[2]Department of Automatic Control Engineering, Feng Chia University, Taichung, Taiwan
[3]Boson Technology Co., LTD, Taichung, Taiwan

## Abstract

ECG signal vary from person to person, making it difficult to be imitated and duplicated. Biometric identification based on ECG is therefore a useful application based on this feature. Synchronization of chaotic systems provides a rich mechanism which is noise-like and virtually impossible to guess or predict. This study intends to combine our previously proposed information encryption/decryption system with chaotic synchronization circuits to create private key masking. To implement the proposed secure communication system, a pair of Lorenz-based synchronized circuits is developed by using operational amplifiers, resistors, capacitors and multipliers. The verification presented involves numerical simulation and hardware implementation to demonstrate feasibility of the proposed method. High quality randomness in ECG signals results in a widely expanded key space, making it an ideal key generator for personalized data encryption. The experiments demonstrate the use of this approach in encrypting texts and images via secure communications.

**Keywords:** Communication security; Encryption; Chaos synchronization, Electrocardiogram

## Introduction

Digital information is increasingly applied in real-world applications as multimedia and network technologies continue to develop. A specific encryption system is therefore required to protect the information during transmission [1-3]. Cryptography is a basic information security measure that encodes messages to make them non-readable. However, conventional block cipher algorithms such as data encryption standard (DES), triple data encryption standard (Triple-DES), and international data encryption algorithm (IDEA) are unsuitable for image encryption because of the special storage characteristics of images [4,5]. Conventional image encryption algorithms are primarily based on the position permutation, such as Arnold transform, magic square matrix, and fractal curve scan [6]. In addition, permutation only algorithms are weak against known text attacks because they cannot change the grayscale of the pixel. Recently, the close relationship between chaos and cryptography has played an active role in data encryption [7-13] because of its significant features, including sensitivity to initial conditions, non-periodicity, and randomness. These features make the chaotic system an ideal tool for communication security [14].

There are no models accounting for all cardiac electrical activities because the human heart is an extremely complex biological system, which makes ECG signals vary from person to person. Compared with common biometric-based systems, the biometric feature of ECG signals is extremely difficult to duplicate. Therefore, an ECG signal could be a biometric tool for individual identification [15-22]. The theory of chaotic dynamical systems has used several features, including the correlation dimension, Lyapunov exponents, and approximate entropy, to describe system dynamics. These key features can explain ECG behavior for diagnostic purposes [23-25].

In ordinary telecommunication system, a specific frequency sine wave carrier is modulated and transmitted with certain message. A receiver system must be tuned to the particular frequency of the carrier sine wave to recover the message. Synchronization of chaotic systems provides a rich mechanism forming another application to personalize secure communications, which are noise-like and impossible to be guessed or predicted. The application of chaotic synchronization to secret communication was previously suggested by Pecora and Carroll [26,27]. There were many control techniques to synchronized chaotic systems, such as fuzzy control [28,29], delayed neural networks [30,31], impulsive control [32], and nonlinear error feedback control [33]. The chaotic signals can also be used to mask information or serve as the modulating waveforms [34-37].

In this research, we use Lyapunov exponent's spectrum to extract the features of human ECG and use them as a secret key to encrypt images and text messages for secure data transmission. The proposed approach uses a chaotic cryptosystem based on the private feature of ECG signals and chaotic functions for information encryption. We combine the previously developed information encryption/decryption system [7-9] with chaotic synchronization circuits to facilitate private key masking. The chaotic synchronization system consists of a driver circuit and a response circuit. This concept of the private key transmission is based on chaotic signal masking and recovery. The transmitter adds a noise-like masking signal to the private key and the receiver removes it by using two synchronization circuits. This configuration forms an indecipherable scheme that is useful for personalized data transmission, in which extreme security is of primary concern.

To implement the proposed the secure communication system, a pair of Lorenz-based synchronized circuits are realized by operational amplifiers, resistors, capacitors and multipliers. The experimental results contain numerical and hardware verification which demonstrates applicability of the proposed design method.

**\*Corresponding author:** Chun-Liang Lin, Department of Electrical Engineering, National Chung Hsing University, Taichung, Taiwan 402, R.O.C
E-mail: chunlin@dragon.nchu.edu.tw

## Description of Method

### Phase-space reconstruction

A phase space or diagram is a space in which each point describes two or more states of a system variable. The number of states that can be displayed in the phase space is called the phase space dimension or reconstruction dimension. The phase space in d dimensions displays a number of points $\{\vec{Z}(n)\}$ of the system, where each point is given by

$$\vec{Z}(n) = [z(n), z(n+n_T), \cdots, z(n+(d-1)n_T)] \qquad (1)$$

Where n is the moment in time of the state variables, $n_T = T/\Delta$ with $\Delta$ denoting the sampling period and T is the period between two consecutive measurements for constructing the phase plot. The trajectory in d dimensional space is a set of k consecutive points and $n = n_0, n_0 + n_T$, L, n0+ $(k-1)n_T$, where $n_0$ is the starting time (in terms of the number of sampling period) of observation.

Phase space reconstruction shows the state trajectories of $z(n)$ and $z(n+n_T)$ at the same time scale. Figure 1 shows the ECG signal from encryption person and the phase plot.

### Calculation of the lyapunov exponents

The Lyapunov exponent is an important feature of chaotic systems which quantifies sensitivity of the system to the initial conditions. Sensitivity to initial conditions means that a small change in the state of a system will grow at an exponential rate and eventually dominate the overall system behavior. Lyapunov exponents are defined as the long-term average exponential rates of divergence of the nearby states. If a system has at least one positive Lyapunov exponent, the system is chaotic. The larger the positive exponent, the more chaotic the system becomes. The exponents are generally arranged such that $\lambda_1 \geq \lambda_2 \geq \dots \geq \lambda_n$, where $\lambda_1$ and $\lambda_n$ correspond to the most rapidly expanding and contracting principal axes, respectively. Therefore, $\lambda_1$ may be regarded as an estimator of the dominant chaotic behavior of the system. This study uses the largest Lyapunov exponent $\lambda_1$ as a measure of the ECG signal using the Wolf algorithm [38].

### Logistic map

The logistic map is a polynomial mapping of the second order. Its chaotic behavior for different parameters was unveiled in [39]. The logistic map equation is given by the following equation and can be illustrated as in Figures 2 and 3:

$$L_{n+1} = A L_n (1 - L_n) \qquad (2)$$

where $n=0,1,2,\dots, 0 \leq L \leq 1, 0 \leq A \leq 4$, A is a (positive) bifurcation parameter. Figure 2 shows the bifurcation diagram of the logistic map in the range $1 \leq A \leq 4$. When the vertical slice A=3.4, the iteration sequence splits into two periodic oscillations, which continues until A is slightly larger than 3.45. This is called periodic-doubling bifurcation in chaos theory.

Successive doublings of the period quickly occur in the range of 3.45<A<3.6. When A increases to 3.6, the periodicity becomes chaotic in the dark area. Many new periodic orbits emerge as A continuously grows from 3.45 to 4. Figure 3 shows the property of the logistic map with different parameter A. The results converge on the same value after several iterations without any chaotic behavior when $A \in (0, 3)$, as shown in Figure 3(a). The system appears periodicity when $A \in [3, 3.6)$, as illustrated in Figure 3(b). The chaotic random-like behavior when $A \in [3.6, 4)$ is shown in Figure 3(c).

### Henon map

The Henon map is a 2-D iterated map with chaotic solutions proposed by M. Henon [40]. The Henon equation can be written as follows

$$\begin{cases} X_{n+1} = 1 - aX_n^2 + bY_n, \\ Y_{n+1} = X_n, \end{cases} \quad n = 1, 2, \dots \qquad (3)$$

where a and b are (positive) bifurcation parameters with b being a measure of the rate of area contraction. The Henon map is the most general 2-D quadratic map possessing the property that the contraction is independent of X and Y. Bounded solutions exist for the Henon map over the ranges of a and b, and some yield chaotic solutions. The demonstrative map was plotted for $a = 1.4$ and $b=0.3$. From which one can observe the chaotic behavior. Figure 4 shows the attractor of the Henon map on the X-Y space.

### Lorenz system

The Lorenz system was originally developed as a simplified mathematical model of atmospheric by Edward Lorenz in 1963 [41], which is a 3-dimensional dynamical system described by

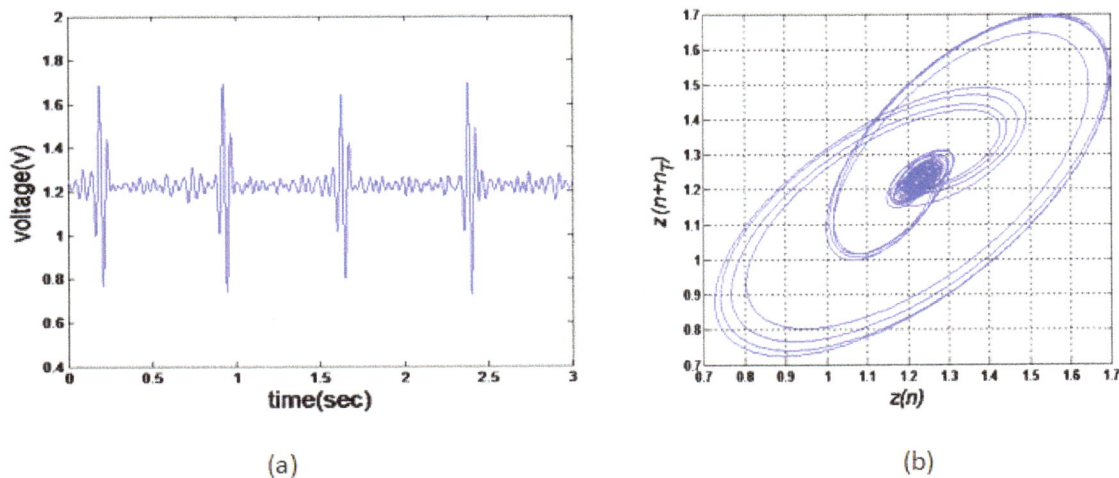

**Figure 1:** (a) ECG signal (b) phase plot taken from the encryption person.

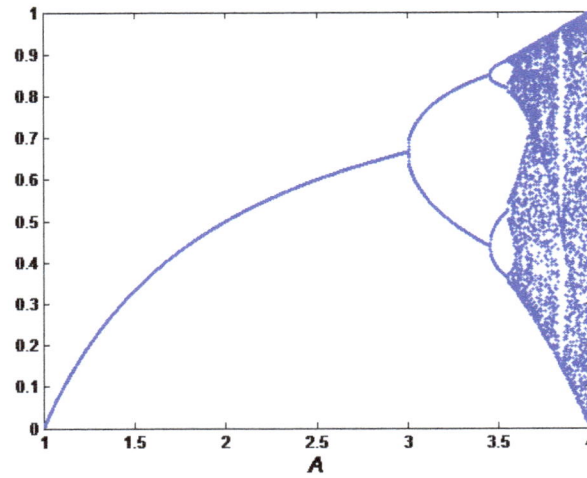

**Figure 2:** Bifurcation diagram of the logistic map.

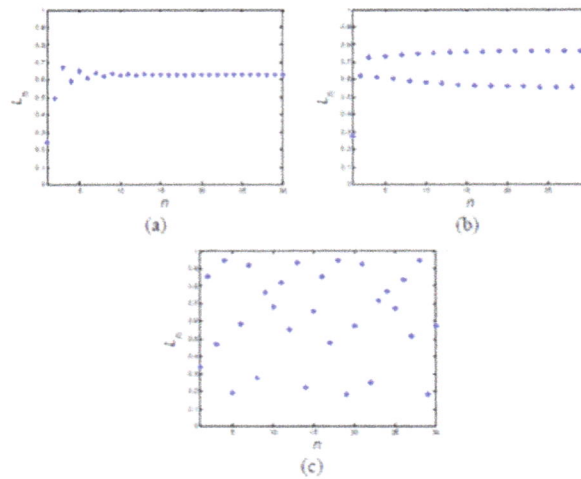

**Figure 3:** Property of logistic map with different bifurcation parameter with L0=0.1 (a) A=2.7 (b) A=3.1 (c) A=3.8.

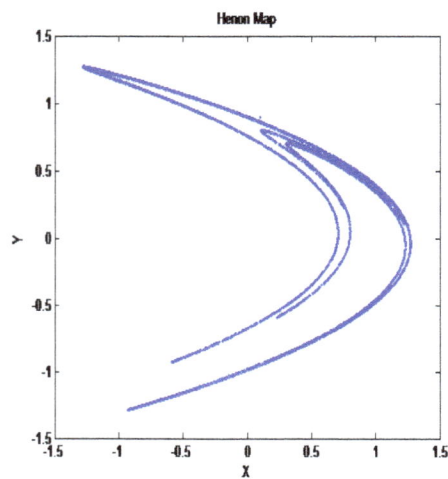

**Figure 4:** Attractors for the Henon map with a= 1.4 and b=0.3.

$$\overset{\bullet}{x} = s(y - x)$$
$$\overset{\bullet}{y} = rx - y - xz \tag{4}$$
$$\overset{\bullet}{z} = xy - pz$$

where $x$, $y$, and $z$ are dynamic variables and $s$, $r$, and $p$ are positive system parameters. The Lorenz system has a single positive Laypunov exponent, $\lambda_1 = 1.069$, while the other are $\lambda_2 = 0$ and $\lambda_3 = -12.73$ respectively. More detailed complex dynamics of the Lorenz system can be seen in [42]. For the typical values $s=10$, $r=28$, $p=8/3$ the time serials of the variables $X$, $Y$ and $Z$ are shown in Figure 5(a); the system has a 3-dimensional chaotic attractor as shown in Figure 5(b).

## Synchronization of two identical lorenz systems

Consider the following linear coupling of two identical Lorenz systems:

$$\overset{\bullet}{x_1} = s(y_1 - x_1) + d_1(x_2 - x_1)$$
$$\overset{\bullet}{y_1} = rx_1 - y_1 - x_1z_1 + d_2(y_2 - y_1) \tag{5}$$
$$\overset{\bullet}{z_1} = x_1y_1 - pz_1 + d_3(z_2 - z_1)$$

$$\overset{\bullet}{x_2} = s(y_2 - x_2) + d_1(x_1 - x_2)$$
$$\overset{\bullet}{y_2} = rx_2 - y_2 - x_2z_2 + d_2(y_1 - y_2) \tag{6}$$
$$\overset{\bullet}{z_2} = x_2y_2 - pz_2 + d_3(z_1 - z_2)$$

Where $x_i, y_i, z_i (i=1,2)$ are state variables, and $d_j(j=1,2,3)$ are coupling coefficients. The driver system consists of $x_1$, $y_1$ and $z_1$. The response system is described by $x_2$, $y_2$ and $z_2$. In particular, when $d_1 \neq 0$, $d_2=0$, $d_3=0$ the coupled systems and are x-coupled. Similarly, the systems with $d_2 \neq 0$, $d_1=d_3=0$ are y-coupled, and the systems are z-coupled when $d_3 \neq 0$, $d_1=d_2=0$. We can define the synchronization errors $e_x(t)$, $e_y(t)$, $e_z(t)$ as

$$e_x = x_1 - x_2,$$
$$e_y = y_1 - y_2, \tag{7}$$
$$e_z = z_1 - z_2,$$

then

$$-x_1z_1 + x_2z_2 = -z_1e_x - x_2e_z,$$
$$-x_1y_1 - x_2y_2 = y_1e_x + x_2e_y \tag{8}$$

From (5)-(8), the error dynamics is given by

$$\overset{\bullet}{e_x} = -(s+2d_1)e_x + se_y$$
$$\overset{\bullet}{e_y} = -(r-z_1)e_x - (1-2d_2)e_y - x_2e_z \tag{9}$$
$$\overset{\bullet}{e_z} = y_1e_x + x_2e_y - (p+2d_3)e_z$$

The coefficient matrix of this system is

$$A(t) = \begin{bmatrix} -(s+2d_1) & s & 0 \\ r-z_1 & -(1+2d_2) & -x_2 \\ y_1 & x_2 & -(p+2d_3) \end{bmatrix}, \tag{10}$$

Define

$$B(t) = \frac{A(t) + A^T(t)}{2}$$
$$= \begin{bmatrix} -(s+2d_1) & (r+s-z_1)/2 & y_1/2 \\ (r+s-z_1)/2 & -(1+2d_2) & 0 \\ y_1/2 & 0 & -(p+2d_3) \end{bmatrix} \tag{11}$$

Let $\alpha(t)$ and $\beta(t)$ be the minimum and maximum eigenvalues of matrix $B(t)$ respectively. According to the result in [43], we have the following lemma.

Lemma 1. The differential equation $\overset{\bullet}{X} = A(t)X$ has a solution X (t), then

$$\|X(0)\|\exp\left\{\int_0^t \alpha(t)dt\right\} \leq \|X(t)\| \leq \|X(0)\|\exp\left\{\int_0^t \beta(t)dt\right\} \tag{12}$$

It is easily proven by the following equation

$$\frac{d\|X(t)\|^2}{dt} = X^T(t)B(t)X(t)$$

Therefore, if $\exists\ \varepsilon > 0$ such that $\beta(t) < -\varepsilon$, then for any initial state $X(0)$, one has $X(t) \to 0$ exponentially. Note that $B(t)$ is a symmetric matrix, thus all eigenvalues of $B(t)$ are real for all t. Let the eigenvalues be $\lambda_i(i=1,2,3)$ with $\lambda_1 \leq \lambda_2 \leq \lambda_3$. For the two identical Lorenz systems, if $(x_1(0), y_1(0), z_1(0) \neq (x_2(0), y_2(0), z_2(0))$, then the state trajectories of the two identical Lorenz systems will separate as time goes by and become unrelated. When $d_j(j=1,2,3)$ satisfy $F(d_1,d_2,d_3) > 0$, the two identical chaotic systems will travel at the same orbit simultaneously.

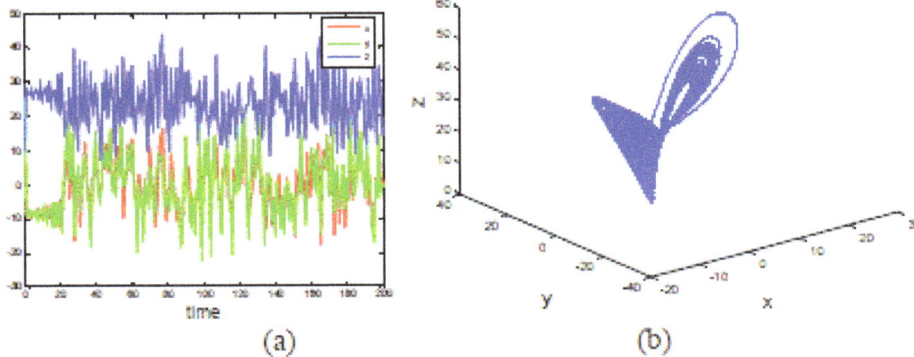

**Figure 5:** The Lorenz system (a) time profiles of the variables x, y and z (b) x-y-z phase trajectories.

That is, the two identical Lorenz systems with linear coupling will be synchronized. On the contrary, if the coupling coefficient $F(d_1,d_2,d_3) < 0$, the two identical chaotic systems will operate independently at their own orbits, i.e., they are not synchronized.

Theorem 1: Given the coupling coefficients $d_i>0$, i = 1, 2, 3 if $d_1$, $d_2$, $d_3$ satisfy the following condition

$$\gamma_0 = (s + 2d_1)(1 + 2d_2)(p + 2d_3) - \frac{p^2(s+r)}{16(p-1)}M > 0,$$

$$\sigma_0 = (s + 2d_1)(1 + 2d_2)(1 + s + 2d_1 + 2d_2)$$
$$+ (p + 2d_3)(1 + s + 2d_1 + 2d_2)(1 + s + p + 2(d_1 + d_2 + d_3))$$
$$- \frac{p^2(s+r)}{16(p-1)}(s + 2d_1 + M) > 0,$$

Where $M$ = max $\{1+2d_2, \quad p+2d_3\}$ then for any $(x_1(0),y_1(0),z_1(0),x_2(0),y_2(0),z_2(0))$, the two coupled Lorenz systems will be synchronized as $t \rightarrow +\infty$, provided that the orbit is close enough to the basin of attraction.

Pf: See Appendix for the details.

For $s=10$, $r=28$, $p=8/3$, the initial states $x_{10}=10$, $y_{10}=25$, $z_{10}=10$, $x_{20}=20$, $y_{20}=11$, $z_{20}=5$, and the coupling coefficients $d_1=1.2$, $d_2=0.8$, $d_3=2.1$, the numerical simulation of the corresponding chaotic phase trajectories and state errors versus time are illustrated in Figures 6 and 7.

## System Design and Secure Data Transmission

### ECG acquisition

Traditionally, ECG signals are recorded through more than three electrodes attached to the human body and manipulated in a complex data management system. This is not suitable for the current purpose. Instead of the way, this research proposes to use a convenient handheld device, developed by our research team, to collect physiological signals from only two leads [44], as shown in Figure 8(a). Each lead is attached to an electrode. The required signals are acquired when two electrodes are simultaneously touched. Figure 8(b) shows the device's structure, which comprises two sensing electrodes. The two active sensor electrodes are connected to the pulse measurement device and the pulse measurement device comprises a negative feedback difference common mode signal and a buffer/balanced circuit for providing a circuit with a self-common point electrode potential. The first bio-potential signal is detected by the first active sensor electrode and the common

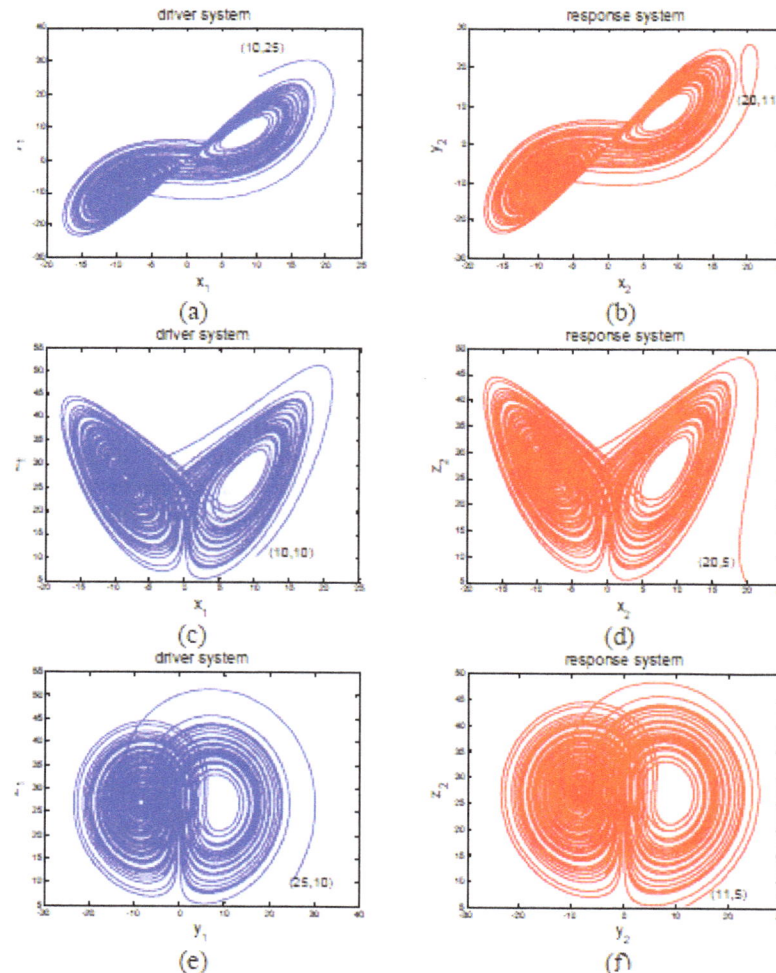

**Figure 6:** Chaotic phase trajectories for the two Lorenz systems (a) $x_1 - y_1$ plane (b) $x_2 - y_2$ plane (c) $x_1 - z_1$ plane (d) $x_2 - y_2$ plane (e) $y_1 - z_1$ plane (f) $y_2 - z_2$ plane.

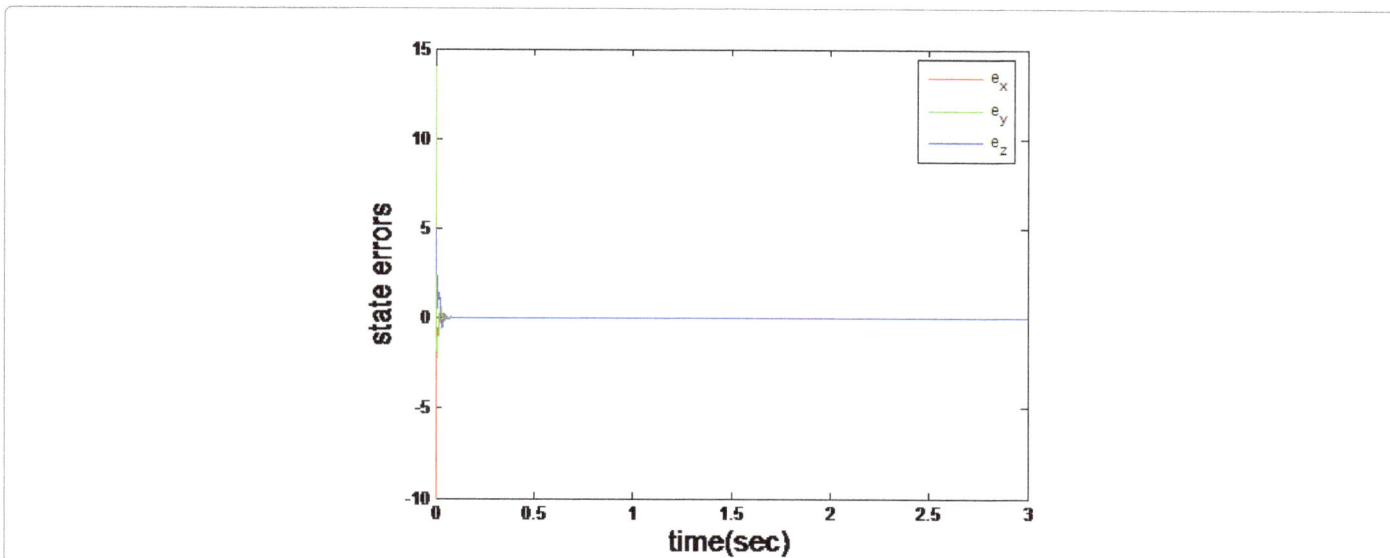

**Figure 7:** Synchronization errors $e_x(t)$, $e_y(t)$, $e_z(t)$ for the two Lorenz systems.

(a)

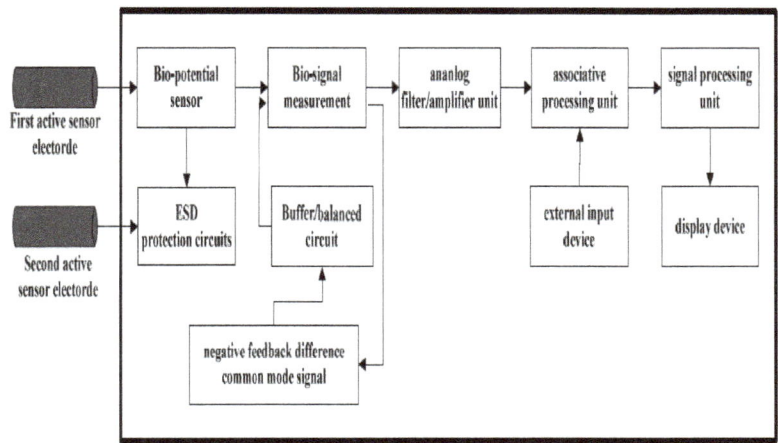

(b)

**Figure 8:** Self-developed handheld ECG acquisition device (a) measurement device (b) structure of the patented portable instrument ET-600.

point electrode. The second bio-potential signal, possesses the same magnitude, but with a different phase as the first bio-potential signal detected by the second active sensor electrode and the common point electrode. The associative processing unit receives the signal which is

processed by an analog filter/amplifier unit with the operational frequency from 0.5 to 40Hz.

The self-developed ECG management device accompanied with a

digital signal processing unit (NI USB6211) and the ECG data acquisition in the LabVIEW environment. The signals measured are then used to reconstruct ECG signals and extract the features by our feature extraction program.

## Secure data transmission

The structure of the proposed secure information transmission system based on the two Lorenz circuits is proposed in Figure 9. The encryption person's ECG data are collected and saved as a private key. The processed secret information is transmitted via the proposed chaotic encryption system, which is activated by the private key. To decrypt the secret information, the recipient should possess both of the

ECG plot and the chaotic decryption algorithm. In addition, the ECG extraction program must be used to extract the features as the initial key for the proposed chaotic decryption algorithms.

## Implementation of synchronization circuit for secure communication

An electric circuit is designed to realize the Lorenz-based synchronized circuit for secure data communication, as illustrated in Figure 10. The voltages at the nodes labeled $x_1$, $y_1$ and $z_1$ correspond to the states of  and $x_2$, $y_2$ and $z_2$ to the states of, respectively. The operational amplifier LF412 and associated circuitry perform the basic operations of addition, subtraction, and integration. The nonlinear

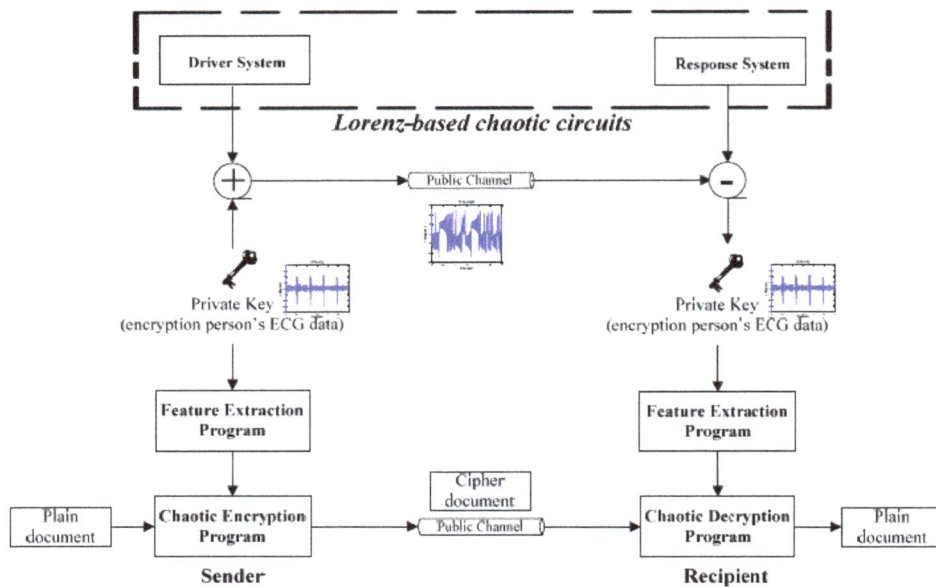

**Figure 9:** Structure of the secure information transmission based on the chaotic masking Lorenz circuits.

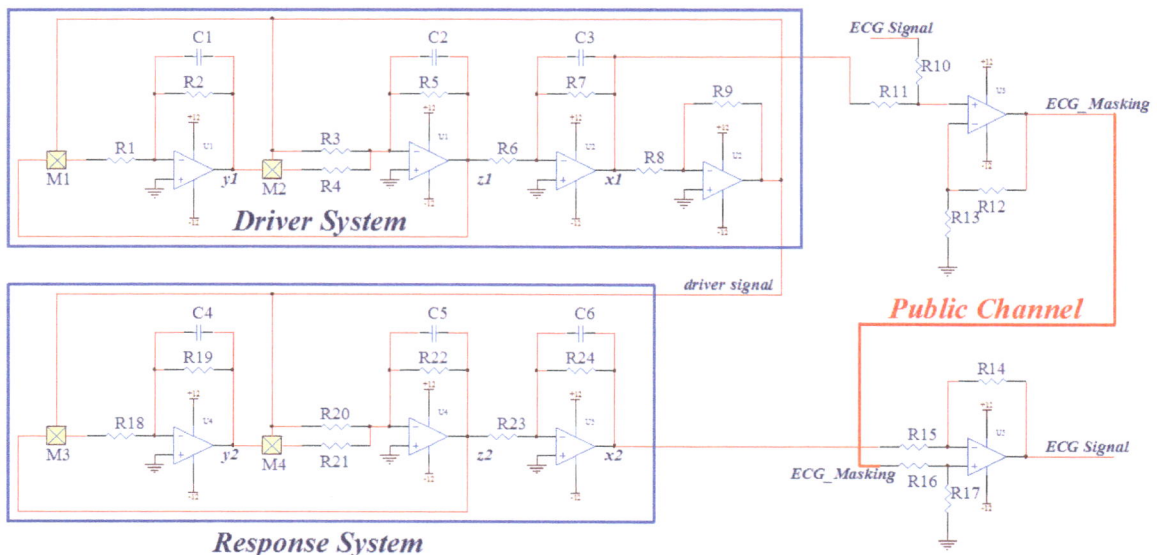

**Figure 10:** Lorenz-based chaotic masking communication circuit.

| Device | Description | Value | Tolerance |
|---|---|---|---|
| U1~U5 | Op Amp (LF412) | | |
| $R_1,R_4,R_8{\sim}R_{18},R_{21},R_{23}$ | 1/4W Resistor | 10 KΩ | ±0.05% |
| $R_2,R_{19}$ | 1/4W Resistor | 374 KΩ | ±0.05% |
| $R_3,R_{20}$ | 1/4W Resistor | 35.7 KΩ | ±0.05% |
| $R_5,R_{22}$ | 1/4W Resistor | 1 MΩ | ±0.05% |
| $R_6,R_7,R_{23},R_{24}$ | 1/4W Resistor | 100 KΩ | ±0.05% |
| C1~C6 | Capacitor | 0.1μF | ±0.1% |
| M1~M4 | Analog multiplier | | |

**Table 1:** Components of the chaotic masking communication circuits.

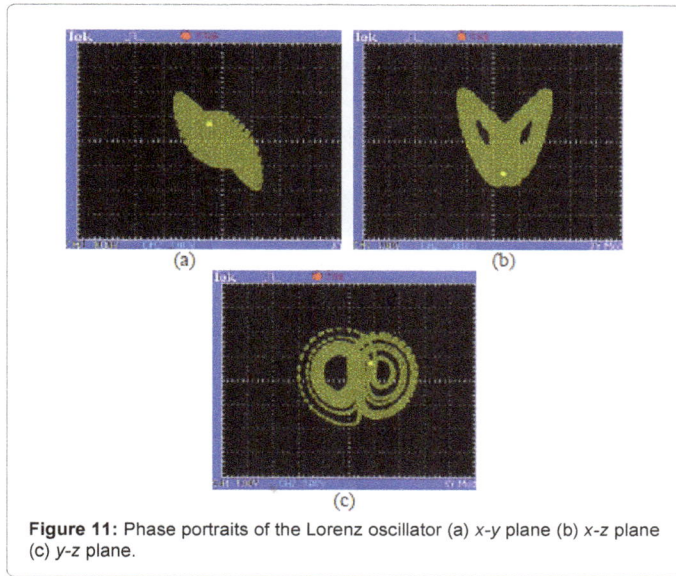

**Figure 11:** Phase portraits of the Lorenz oscillator (a) x-y plane (b) x-z plane (c) y-z plane.

terms in the system and are implemented with the analog multiplier AD633. The component list of the Lorenz-based chaotic masking communication circuit is given in Table 1. The system parameters *s, r,* and *p* can be implemented by resistors $R_2,R_3,R_5$ and $R_7$ as follows

$$s = \frac{R_5}{R_7}, \; r \approx \frac{R_5}{R_3}, \; p \approx \frac{R_5}{R_2} \tag{13}$$

The private key ECG signal was masked by chaotic signal of the driver system and is presented as ECG_masking, which is sent out through a public channel. On the other side, the ECG_masking signal is received and the private key is recovered by synchronized chaotic signal of the response system.

Experimental results for synchronization and secure communication are given to demonstrate the performance of the proposed scheme. Figure 11 shows the Lorenz-based circuit's attractor projected onto the x-y plane, x-z plane, and y-z plane, respectively. Figure 12 shows the phase portrait in $x_1$-$x_2$ plane illustrating synchronization of the Lorenz-based circuits. Figure 13(a) shows practical implementation of the proposed secure data communication system. Figure 13(b) depicts the scrambled private key ECG signal, the transmitted chaotic signal ECG_masking, and the recovered private key ECG_signal in the response system.

### Encryption/decryption algorithms

We now explain the procedure of the proposed information encryption/decryption system using ECG signals with a chaotic logistic map for text encryption and chaotic Henon map for image encryption.

Figure 14 presents the block diagram of the information encryption/decryption scheme. The chaotic functions depicted in Section 2 are employed in the information encryption/decryption algorithm using the logistic map, Henon map, ECG extraction program, and Wolf algorithm. The ECG extraction program extracts the individual features of the users as the initial key ($\lambda_i$) for the logistic map and Henon map, and subsequently uses these chaotic functions to generate an unpredictable random orbit. The unpredictable random orbit is used as a private encryption key serial to replace pixel values, images coordinates, and ASCII codes. Conversely, the chaotic decryption algorithm fulfills the inverse operation.

Figure 15 shows the flow chart of the chaotic encryption algorithm for the document with blended Figure and text. First, text and images of the encrypted document are separated. Set the encrypted grayscale image to be S, whose size is M×N and the pixel related to the coordinates (i, j) is denoted I(i,j), $1{\leq}i{\leq}M$ and $1{\leq}j{\leq}N$. The new coordinates of the pixel I(i,j), after replacement, denoted (i', j') with I'(i,j) representing the replaced I(i,j). To enhance undetectability, the new coordinates and the pixel I'(i', j') are produced using the chaotic Henon map . The format of encrypted text is transformed into Text file (T.txt). We obtain the strings with the line terminators and convert characters into ASCII codes until the end of the file (T.text), and then the ciphertext (T_mask. text) is converted into ASCII codes by using the chaotic logistic map and the converted ASCII codes into characters accordingly.

### Experimental Results

Table 2 lists key parameters of logistic map and Henon map for testing the encryption and decryption algorithms.

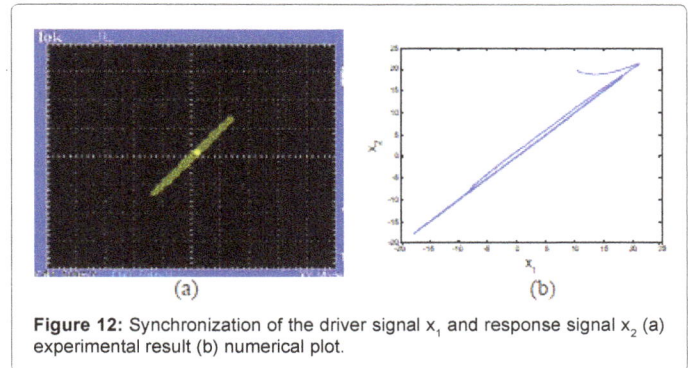

**Figure 12:** Synchronization of the driver signal $x_1$ and response signal $x_2$ (a) experimental result (b) numerical plot.

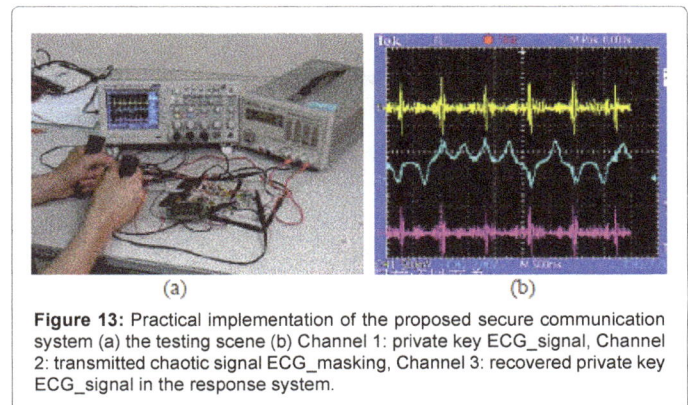

**Figure 13:** Practical implementation of the proposed secure communication system (a) the testing scene (b) Channel 1: private key ECG_signal, Channel 2: transmitted chaotic signal ECG_masking, Channel 3: recovered private key ECG_signal in the response system.

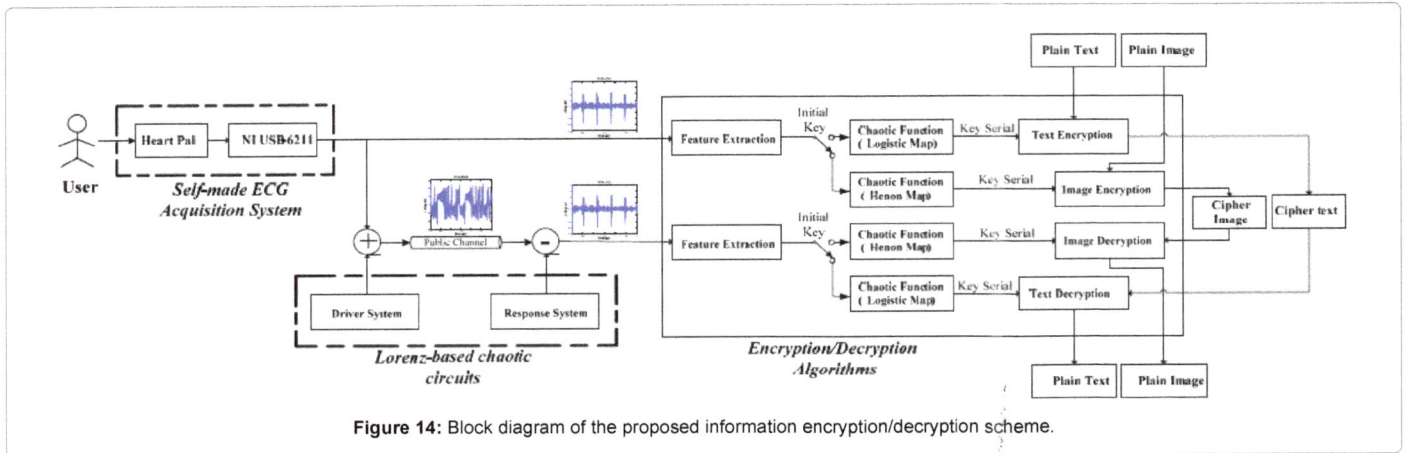

**Figure 14:** Block diagram of the proposed information encryption/decryption scheme.

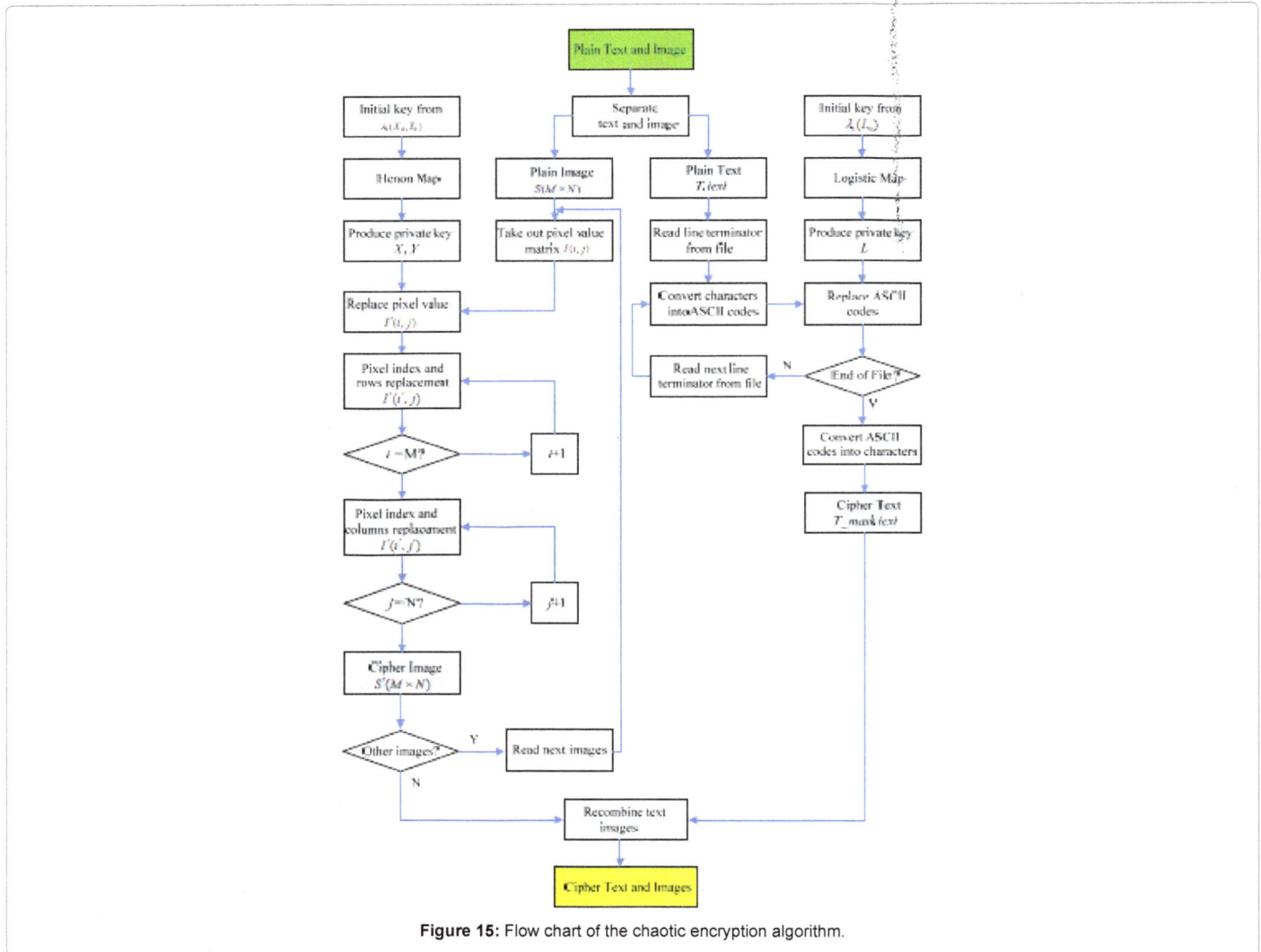

**Figure 15:** Flow chart of the chaotic encryption algorithm.

## Case 1: Figureure encryption and decryption

The physiological signals of the users were collected from the self-made portable instrument accompanied with a digital signal processing unit and analyzed in the LabView environment. For a qualified encryption system, the key serial should be able to Figureht against the brute-force attack. It should also be sensitive to the private key. A variety of simulation studies were conducted to test robustness of the proposed encryption system. Table 3 lists three representative images supported in the MATLAB image processing toolbox. Table 4 reveals

| Items | Value | Description |
|---|---|---|
| N | 1500 | number of iterations |
| $\lambda_1(X_0, Y_0, L_0)$ | 0.01573 | initial value formed by $\lambda_1$ of the encryption person |
| a | 1.4 | system parameter of Henon map |
| b | 0.3 | system parameter of Henon map |
| A | 4 | system parameter of logistic map |

**Table 2:** Parameters of the chaotic functions for encryption and decryption.

| Filename | Size | Color type |
|---|---|---|
| Liftingbody.png | 512 × 512 | 8 bits grayscale |
| Canoe.tif | 346 × 207 | 8 bits indexed |
| pears.png | 732 × 486 | 24 bits RGB |

**Table 3:** Different kinds of images.

| Items | Value | Description |
|---|---|---|
| n | 1500 | number of iterations |
| $\lambda_1(X_0, Y_0, L_0)$ | 0.01487 | initial value formed by $\lambda_1$ of the non-encryption person |
| a | 1.4 | system parameter Henon map |
| b | 0.3 | system parameter Henon map |
| A | 4 | system parameter of Logistic map |

**Table 4:** Parameters of chaotic functions for decryption.

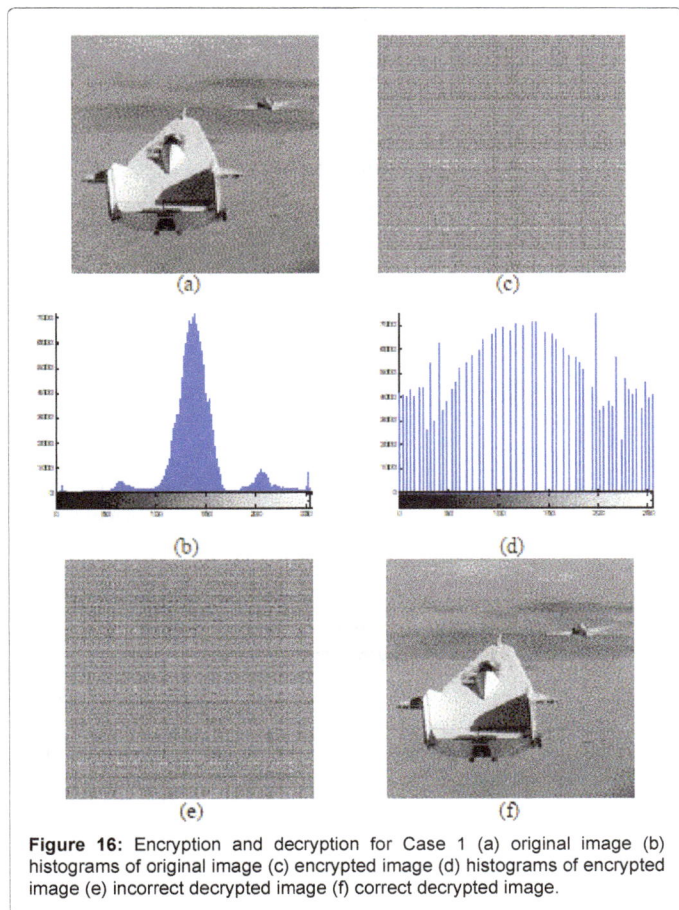

**Figure 16:** Encryption and decryption for Case 1 (a) original image (b) histograms of original image (c) encrypted image (d) histograms of encrypted image (e) incorrect decrypted image (f) correct decrypted image.

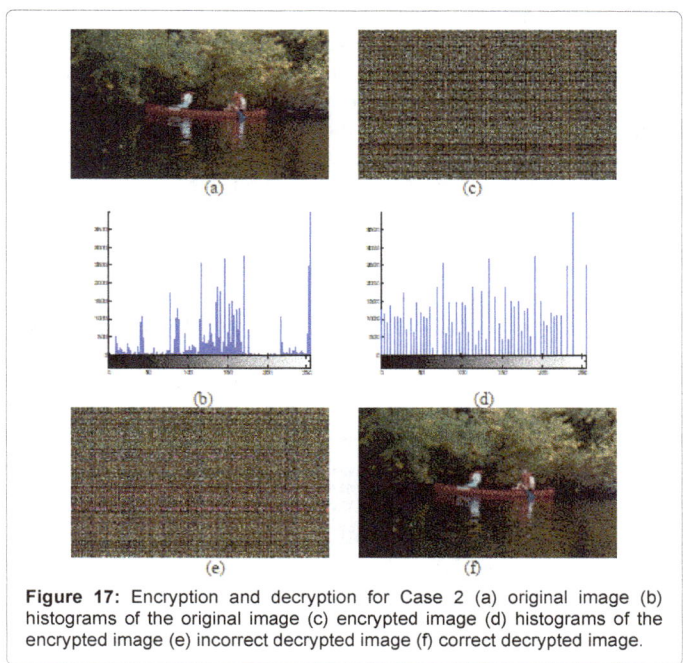

**Figure 17:** Encryption and decryption for Case 2 (a) original image (b) histograms of the original image (c) encrypted image (d) histograms of the encrypted image (e) incorrect decrypted image (f) correct decrypted image.

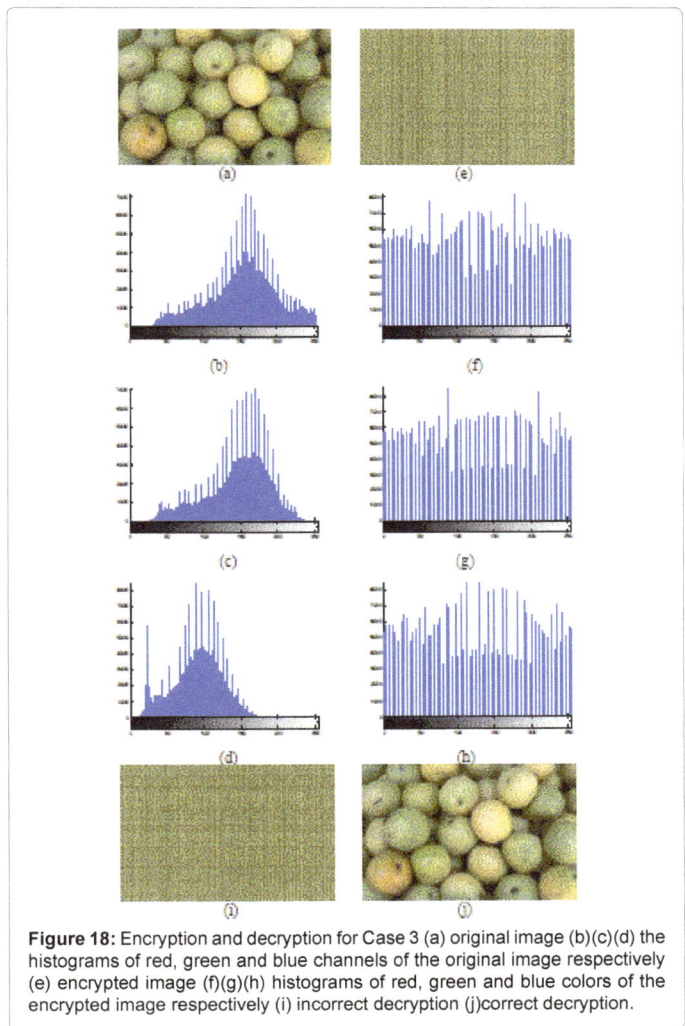

**Figure 18:** Encryption and decryption for Case 3 (a) original image (b)(c)(d) the histograms of red, green and blue channels of the original image respectively (e) encrypted image (f)(g)(h) histograms of red, green and blue colors of the encrypted image respectively (i) incorrect decryption (j) correct decryption.

that when the initial values changed to $\lambda_i$ for the non-encryption person, the decryption scheme generated a completely different decrypted result. Figures 16-18 display simulation results and histograms for three kinds of images. The image histogram illustrates how pixels in an image are distributed by graphing the number of pixels at the intensity level of color. The results of histogram analysis show an extremely different content in the original and encrypted images.

### Case 2: Document blended with figures and text

We take Page 2 of this paper as the object of experiment, which contains text and Figures to be encrypted. We transform the formats of text and Figures into Text (.txt) file and Image (.png) file simultaneously. Figures. 19-21 show the demonstration that it incorporates the text encryption algorithm with Logistic map and the image encryption

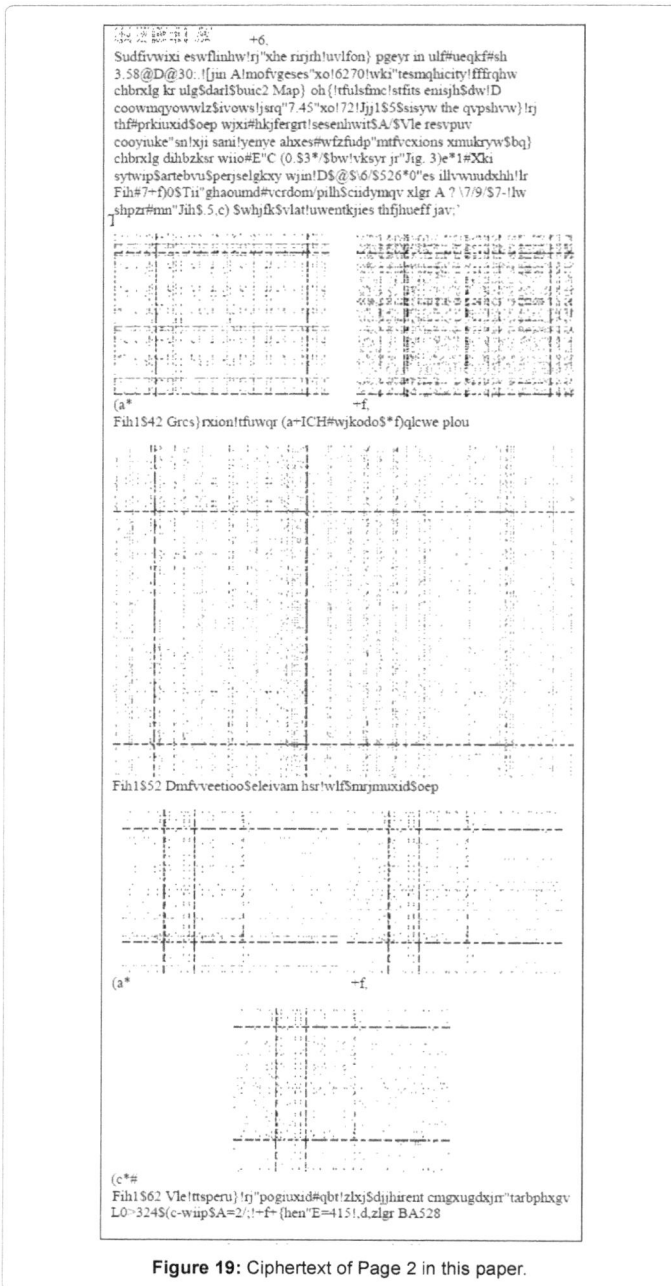

**Figure 19:** Ciphertext of Page 2 in this paper.

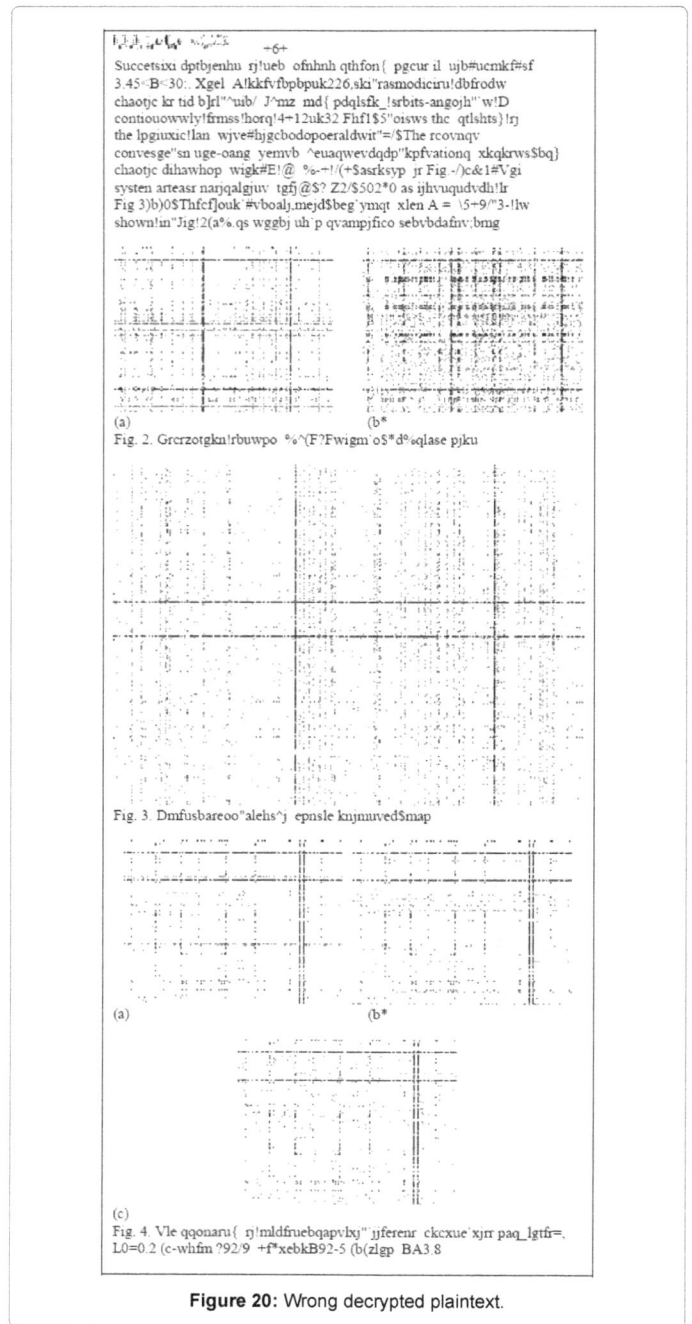

**Figure 20:** Wrong decrypted plaintext.

algorithm with Henon map. The encrypted plaintext is obviously non-readable, as shown in Figure 19. To compare the decrypted result of the chaotic encryption system, we chose an incorrect key and a correct one to activate the decryption algorithm. Figures. 20 and 21 show the results of decryption indicating that the proposed encryption system is quite sensitive to the key chosen and thus is appropriate for secure communication.

## Conclusions

This paper has presented theoretical and experimental studies on chaos synchronization and masking of data communication using electronic devices that are described by the Lorenz equations, and showed

$$L_{n+1} = AL_n(1-L_n) \qquad (2)$$

where $n=0,1,2,...$, $0 \leq L \leq 1$, $0 \leq A \leq 4$, $A$ is a (positive) bifurcation parameter. Fig. 2 shows the bifurcation diagram of the logistic map in the range $1 \leq A \leq 4$. When the vertical slice $A=3.4$, the iteration sequence splits into two periodic oscillations, which continues until $A$ is slightly larger than 3.45. This is called periodic-doubling bifurcation in chaos theory. Successive doublings of the period quickly occur in the range of $3.45 < A < 3.6$. When $A$ increases to 3.6, the periodicity becomes chaotic in the dark area. Many new periodic orbits emerge as $A$ continuously grows from 3.45 to 4. Fig. 3 shows the property of

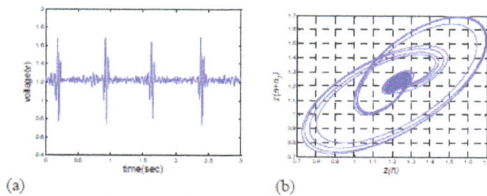

(a)                              (b)
Fig. 1. Encryption person (a)ECG signal (b)phase plot

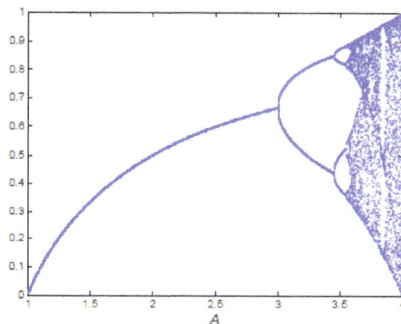

Fig. 2. Bifurcation diagram for the logistic map

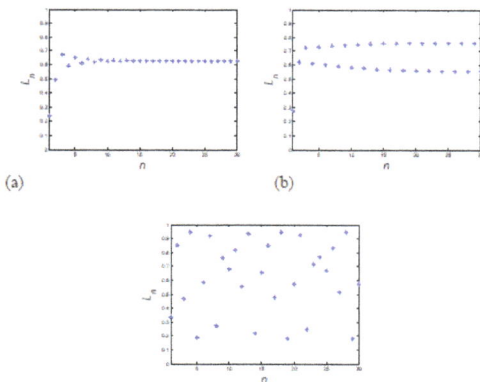

(a)                              (b)

(c)
Fig. 3. The property of logistic map with different bifurcation parameter A. L0=0.1 (a)when A=2.7 (b)when A=3.1 (c)when A=3.8

**Figure 21:** Resulting (correct) decrypted plain text.

that the private key created by ECG signals can be recovered from a chaotic carrier using a response system whose chaotic dynamics is synchronized with a driver system. The use of ECG signal's features from nonlinear dynamic modeling for information encryption is investigated. A personalized encryption scheme based on the individual-specific features of ECG as a personal key is proposed. To decrypt the encrypted message that one needs a specific ECG message accompanied with our proposed encryption algorithm. The blended functionality yields a doubly encrypted scheme, which is extremely hard to be decrypted.

Unlike traditional cryptographic algorithms, the presented approach features an infinite key space. This makes it an ideal key generator for encryption algorithms. Experimental results have proved feasibility and effectiveness of the proposed design. Moreover, the encryption time shows its potential applicability in real-time applications.

### Acknowledgment

This research was sponsored by Ministry of Science and Technology, Taiwan, ROC under the grant 104-2622-E-005-011.

### References

1. Chen TH, Wu CS (2010) Compression-unimpaired batch-image encryption combining vector quantization and index compression. Information Science 180: 1690-1701.

2. Kao YW, Huang KY, Gu HZ, Yuan SM (2013) uCloud: a user-centric key management scheme for cloud data protection. IET Information Security 7: 144-154.

3. Lu J, Wei Y, Fouque PA, Kim J (2012) Cryptanalysis of reduced versions of the Camellia block cipher, IET Information Security 6: 228-238.

4. Salleh M, Ibrahim S, Isnin IF (2003) Image encryption algorithm based on chaotic mapping, Journal Teknologi 39: 1-12.

5. Usama M, Khan MK (2008) Classical and chaotic encryption techniques for the security of satellite image. Proceedings of IEEE International Conference Biometrics and Security Technologies.

6. Alexopulos C, Bourbakis NG, Ioannou N (1995) Image encryption method using a class of fractals. Journal Electronic Imaging 4: 251-259.

7. Chen CK, Lin CL, Chiang CT, Lin SL (2012) Personalized information encryption using ECG signals with chaotic functions. Information Science 193: 125-140.

8. Chen CK, Lin CL (2010) Text encryption using ECG signals with chaotic logistic map. Proceedings of IEEE Conference Industrial Electronics and Applications.

9. Chen CK, Lin CL, Chiu YM (2010) Data encryption using ECG signals with chaotic Henon map. Proceedings of IEEE Conference Industrial Electronics and Applications.

10. Behnia S, Akhshani A, Mahmodi H (2008) A novel algorithm for image encryption based on mixture of chaotic maps. Chaos, Solitons and Fractals 35: 446-471.

11. Oliveira LPL, Sobottka M (2008) Cryptography with chaotic mixing. Chaos, Solitons and Fractals 35: 408-419.

12. Solak E, Cokal C (2011) Algebraic break of image ciphers based on discretized chaotic map lattices. Information Science 181: 227-233.

13. Zhu ZL, Zhang W, Wong KW, Yu H (2011) A chaos-based symmetric image encryption scheme using a bit-level permutation. Information Science 181: 1171-1186.

14. Jakimoski G, Kocarev L (2001) Chaos and cryptography: block encryption ciphers based on chaotic maps. IEEE Trans. Circuit and Systems-I: Fundamental Theory and Applications 48: 163-169.

15. Chen CK, Lin CL, Lin SL, Chiu YM, Chiang CT, et al. (2014) A chaotic theoretical approach to ECG-Based identity recognition. IEEE Computational Intelligence Magazine? 9: 53-63.

16. Lin SL, Chen CK, Lin CL, Yang WC, ChiangC T, et al. (2014) Individual identification based on chaotic electrocardiogram signals during muscular exercise. IET Biometrics 3: 257-266.

17. Biel L, Pattrsson O, Philipson L, Wide P (2001) ECG analysis: a new approach in human identification. IEEE Trans. Instrumentation and Measurement 50: 808-81.

18. Chiu CC, Chuang CM, Hsu CY (2009) A novel personal identity verification approach using a discrete wavelet transform of the ECG signal. International Journal of Wavelets, Multi-resolution and Information Process. 7: 341-355.

19. Israel SA, Irvine JM, Cheng A, Wiederhold MD, Wiederhold B K, et al. (2005) ECG to identify individuals, Pattern Recognition. 38: 133-142.

20. Singla SK, Sharma A (2010) ECG as biometric in the automated world. International Journal of Computer Science & Communication. 1: 281-283.

21. Loong J L C, Subari K S, Besar R, Abdullah M K (2010) A new approach to ECG biometric systems: a comparative study between LPC and WPD systems. Word Academy of Science, Engineering and Technology 68: 759-764.

22. Chen CK, Lin CL, Chiu YM (2011) Individual identification based on chaotic electrocardiogram signals. Proceedings of IEEE International Conference on Industrial Electronics and Applications.

23. Casalegio A, Braiotta S (1997) Estimation of Lyapunov exponents of ECG time series-the influence of parameters. Chaos, Solitons, and Fractals 8: 1591-1599.

24. Jovic A, Bogunovic N (2007) Feature extraction for ECG time-series mining based on chaos theory. Proceedings of International Conference Information Technology Interfaces.

25. Owis MI1, Abou-Zied AH, Youssef AB, Kadah YM (2002) Study of features based on nonlinear dynamical modeling in ECG arrhythmia detection and classification. See comment in PubMed Commons below IEEE Trans Biomed Eng 49: 733-736.

26. Pecora LM, Carroll TL (1990) Synchronization in chaotic systems. See comment in PubMed Commons below Phys Rev Lett 64: 821-824.

27. Pecora LM, Carroll TL (1991) Driving systems with chaotic signals. See comment in PubMed Commons below Phys Rev A 44: 2374-2383.

28. Lian KY1, Chiang TS, Chiu CS, Liu P (2001) Synthesis of fuzzy model-based designs to synchronization and secure communications for chaotic systems. See comment in PubMed Commons below IEEE Trans Syst Man Cybern B Cybern 31: 66-83.

29. Shih-Yu Li, Zheng-Ming Ge (2011) Fuzzy Modeling and Synchronization of Two Totally Different Chaotic Systems via Novel Fuzzy Model. See comment in PubMed Commons below IEEE Trans Syst Man Cybern B Cybern 41: 1015-1026.

30. Cheng CJ, Liao TL, Yan JJ, Hwang CC (2006) Exponential synchronization of a class of neural networks with time-varying delays. See comment in PubMed Commons below IEEE Trans Syst Man Cybern B Cybern 36: 209-215.

31. Cao J1, Chen G, Li P (2008) Global synchronization in an array of delayed neural networks with hybrid coupling. See comment in PubMed Commons below IEEE Trans Syst Man Cybern B Cybern 38: 488-498.

32. Xu WG, Shen HZ, Hu DP, Lei AZ (2005) Impulse tuning of Chua chaos. International Journal of Engineering Science 43: 831-844.

33. Kunin, I., Chemykh, G., Kunin, B: Optimal chaos control and discretization algorithms, Int. J. Eng. Sci. 44, 59-66 (2006).

34. Tsay SC, Huang C K, Qiu D L, Chen W T (2004) Implementation of bidirectional chaotic communication systems based on Lorenz circuits. Chaos, Solitons, and Fractals 20: 567-579.

35. Nana B, Woafo P, Domngang S (2009) Chaotic synchronization with experimental application to secure communications. Communications Nonlinear Science Numerical Simulation 14: 2266-2276.

36. Pehlivan I, Uyaroglu Y, Yogun M (2010) Chaotic oscillator design and realizations of the Rucklidge attractor and its synchronization and masking simulation. Scientific Research and Essays 5: 2210-2219.

37. Nana B, Woafo P (2011) Synchronized states in a ring of four mutually coupled oscillators and experimental application to secure communications. Communications Nonlinear Science Numerical Simulation 16: 1725-1733.

38. Wolf A, Swift J B, Swinney H L, Vastano J A (1985) Determining Lyapunov exponents from a time series. Physics Letter 16: 285-317.

39. May RM (1976) Simple mathematical models with very complicated dynamics. See comment in PubMed Commons below Nature 261: 459-467.

40. Henon M (1976) A two-dimensional mapping with a strange attractor. Communication in Mathematical Physics 50: 69-77.

41. Lorenz EN (1963) Deterministic nonperiodic flow. Journal of Atmospheric Sciences 20: 130-141.

42. Ge ZM, Tsen PC (2008) Chaos synchronization by variable strength linear coupling and Lyapunov function derivative in series form. Nonlinear Analysis 69: 4604-4613.

43. Dialecii JL, Krein MG (1974) Stability of differential equations in Banach space, AMS, New York.

44. Chiang CT (2005) Contact Type Pulse Measurement Device. USA Patent NO. US6945940B1.

45. Leonov G, Bunin A, Koksch N (1987) Attractor localization of the Lorenz system. ZAMM 67: 649-56.

# 11

# Improving the Network Life Time of Wireless Sensor Network using EEEMR Protocol with Clustering Algorithm

**Nayak JA\*, Rambabu CH and Prasad VVKDV**

*ECE Department, Gudlavalleru Engineering College, Gudlavalleru, AP, India*

## Abstract

Energy efficient routing is a one of the major trusted area in Wireless Sensor Networks (WSNs). The wireless sensor network composed of a large number of sensor nodes which has limited energy resource. The sensor nodes are working through the battery, energy saving becomes more vital issue in WSNs. The routing algorithms assure the concept of energy saving without affecting the Quality of Service (QoS) Parameters like Throughput, End to End Delay, Overhead and Packet Delivery Ratio. In the existing system the Enhanced Energy Efficient Multipath Routing (EEEMR) Protocol is implemented. The EEEMR Protocol is modification of AOMDV Protocol. In this paper, we are implementing Clustering algorithm in EEEMR Protocol. The development of cluster based sensor networks have recently shown to decrease the system delay, overhead and increase the system throughput and packet delivery ratio. Simulation is performed using NS2 and results shows that the proposed system is better than the existing system. The proposed system energy consumption is decreased by 13% compared to the existing system.

**Keywords:** Wireless Sensor Networks (WSNs); Quality of Service (QoS); Energy Efficient Multipath Routing (EEEMR); Clustering algorithm

## Introduction

The progression of wireless sensor networks is initially motivated by military applications. Wireless sensor networks are used in numerous civilian application areas like detecting, monitoring the movement of enemies, chemical, and biological, radiological, tracking, and automation, nuclear and health care applications. The Wireless sensor network consists of hundreds or thousands of low powered sensor nodes that have ability to communicate either directly to the base station or among each other. These nodes are integrated with micro sensing, computing wireless communication capabilities. Which are capable of detecting various events related to its surrounding environment such as speed, temperature, pressure, light etc., the WSN nodes are operate in ad-hoc manner, limited hardware and limited energy resource because it's small size. The energy source of sensor nodes in wireless sensor networks is usually powered by battery. This is insufferable, even impossible to be recharged or replaced. The energy efficiency and maximizing the life time of the network are major challenges in wireless sensor network. In wireless sensor networks the sensor nodes are grouped into individual disjoint sets called a cluster. Clustering is considered as one of the method to reduce energy consumption. The clustering is used in WSNs; it provides network scalability and energy saving attributes. Clustering schemes offer reduced communication overheads, decreases the overall energy consumption and reducing the interferences among sensor nodes.

## Related Work

In Raj and Sumathi; Singh and Sharma; Dave and Dala; Aliouat and Harous; and Tian et al. [1-5], the authors proposed a new Enhanced Energy Efficient Multipath Routing (EEEMR) protocol and compares the EEEMR Protocol with Flat routing protocols such as DSDV, DSR and AODV. The EEEMR Protocol is a modification of existing AOMDV protocol with the Bio inspired Cuckoo Search Algorithm (CSA). The AOMDV routing protocol is an extension of Flat routing protocol AODV. The EEEMR Protocol is slightly improves the QoS Parameters like throughput, delay, overhead and packet delivery ratio when compared to DSDV, DSR and AODV.

In Younis and Fahmy; Mallapur and Terdal [6,7], the authors presented an energy-efficient distributed clustering approach for ad-hoc sensor networks. The approach is hybrid: Cluster heads are randomly selected based on their residual energy and nodes join clusters such that communication cost is minimized.

In Ahmadi et al.; Maisra [8,9], the authors presented a new method which using the parameters of distance and remaining energy of each node in the process of cluster head selection, using the algorithm to find the shortest path between cluster head and base station.

In Al-Karaki and Kamal [10], the authors presented a new method is proposed, which is using K-means algorithm to forming the clusters and genetic algorithm to select the cluster head in each cluster.

## Proposed Work

In this paper, we are implementing Clustering algorithm in Enhanced Energy Efficient Multipath Routing (EEEMR) Protocol. In our existing work we have implemented the EEEMR Protocol. EEEMR Protocol is extension of AOMDV routing protocol with the Bio inspired Cuckoo Search Algorithm. The EEEMR Protocol uses the distance vector concept and hop-by-hop routing approach. The EEEMR Protocol also uses a route request broadcasted between source to destination and route discovery process to find the on demand routes. It also offers intermediate nodes with alternate paths, which are reducing the route discovery rate. Clustering is a good method in wireless sensor networks for effective data communication and towards energy efficiency. Cluster based operations consists of rounds. These involve cluster heads selection, cluster formation and transmission

**\*Corresponding author:** Nayak JA, PG Student, ECE Department, Gudlavalleru Engineering College, Gudlavalleru, AP, India
E-mail: jarapala.nayak123@gmail.com

of data to the base station. The Figure 1 shows that the cluster based wireless sensor network.

## Proposed algorithm

The clustering algorithm proposed for energy efficient technique for WSNs consists of fixed number of sensor nodes that improve the Cluster Head selection approach to prolong the lifetime of networks. The Cluster Head selection in WSNs is based on the decision taken from the residual energy and certain threshold value of the respective nodes. The threshold value is:

$$T[n] = \begin{cases} \left(\left(\dfrac{P}{1-P*\left(r \mod\left(\frac{1}{P}\right)\right)}\right)\dfrac{E_{residual}}{E_{initial}} * K_{optimal}, & n \in G \\ 0 & otherwise \end{cases}$$

Where P is the desired percentage of cluster head, r is the current round number and G is the set of nodes that have not been selected as cluster heads in last 1/P rounds. Using this threshold, each node will be moderately selected as cluster head at some point within 1/P rounds of the cluster head selection process.

Where $K_{optimal}$ is the optimal number of cluster head during the state of cluster formation. It is defined as follows:

$$K_{optimal} = \sqrt{\dfrac{N}{2\pi}} * \sqrt{\dfrac{E_{fs}}{E_{amp}}} * \sqrt{\dfrac{M}{d^2_{to\ BS}}}$$

Where N is the number of nodes and M is the network area and Efs and Eamp are the amplification power losses and d is the distance between the selected cluster head to the base station. The desired percentage of cluster heads depends upon different networks parameters like average distance between the sensor nodes to the base station, number of the sensor nodes deployed by the field and area of the field. The desired percentage varies at each round of cluster head selection [11].

After this each node that is selected as a cluster head will send a broadcast advertisement message to the all the nodes in the wireless sensor network. The each non-cluster head node decides the cluster to which it will belong for its round depending on the signal strength or distance. The node will send a message to the cluster head informing that it will be a member of that cluster. We will choose the nearest cluster head. The cluster head receives all the messages from nodes that would like to be in its cluster. Once the cluster head know the

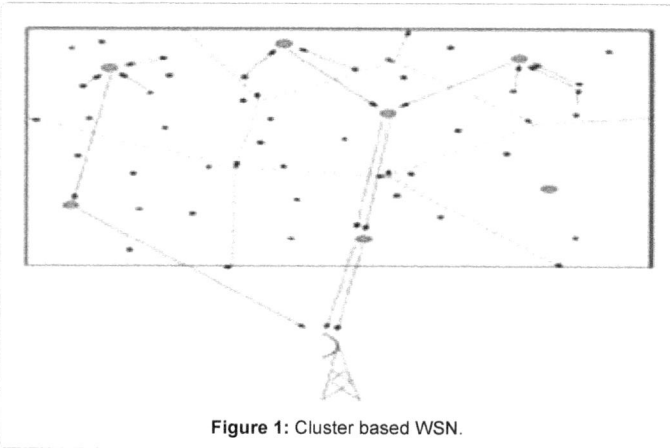

**Figure 1:** Cluster based WSN.

number of members in cluster it can create a TDMA schedule for data transmission purpose. Here each node in the cluster send their sensed data to the cluster head in one hop transmission and the cluster head send data to the base station by multi-hop transmission.

1. The algorithm takes into following assumptions:

2. The base station is far away from the sensor nodes.

3. The cluster head selection, cluster formation and transmission of data to the base station via cluster heads.

4. The selection of cluster head depends on the residual energy and certain threshold value, calculated by cluster head instead of calculating it by base station to reduce overhead and energy consumption at base station.

5. The cluster member nodes transmit their sensed data to their cluster head in one-hop transmission and cluster head to base station in multi-hop transmission.

6. The sensor nodes in the network infrastructure are forbid from being involved in the cluster head selection process to increase the stability in the network.

The major steps of the protocol are follows:

1. The algorithm is basically divided into the number of rounds.

2. For the first round the nodes with the highest energy node are selected as cluster head randomly for that particular cluster and data transmission is performed.

3. At the start of the second round the cluster head aggregates the residual energy of the particular members and calculates the threshold at that cluster head.

4. All the cluster heads do the same with their cluster members and effective clustering is performed to reach the base station by selecting optimal cluster head.

5. Every node has calculated the threshold value. If the threshold value of a node is greater than threshold value, the node will be candidate for the cluster head of that cluster for the next round.

6. If the cluster head threshold value is below the threshold value of network the cluster head is removed and again the cluster head selection process is performed in that cluster.

7. If the cluster head is below the threshold value in that time the cluster members are send their sensed data to the nearest cluster head. This process is continuous until the new cluster head is selected in that cluster.

8. The optimal cluster head at each round will transmit the information to the base station and do not involve base station to select cluster head at each round and to reduce energy consumption at each round.

## Simulation Setup

In this paper, we proposed and implemented EEEMR Protocol with adding the clustering algorithm, by the altering AOMDV in NS-2.34 simulator. The implemented EEEMR Protocol can be evaluated by the number of qualitative metrics such as Packet Delivery Ratio, Overhead, Delay, Throughput and Energy. Finally the simulated results are compared. Table 1 shows the simulation parameters.

Performance metrics: The EEEMR Protocol by using clustering

algorithm should address the following performance metrics such as increase the Packet Delivery Ratio and also Throughput, Minimization of Delay and also overhead, decrease the energy consumption of the wireless sensor network.

1. Throughput: It is the rate of successfully delivered data packets per second in the network between sources to destination.

2. End to end delay: It is the time taken by the data packets for the transmission between sources to destination across a wireless sensor network. This duration is caused by buffering, queuing and also the transmission delay at MAC.

3. Packet delivery ratio: It is the ratio between the received packets by the destination to the generated packets by the sources.

4. Overhead: It is calculated by the ratio of the total number of control packets sent by the sources to the number of data packets delivered to destination successfully.

5. Energy: It is calculated by the [Final Energy=Initial energy–Consumed energy].

## Simulation Results

We have done our research analysis in wireless sensor networks by using NS2. Comparative analysis done between proposed system and existing system.

The Overhead comparison is shown in Figure 2. Overhead is decreases when compared to the existing system. In first 20 s the existing system is better than the proposed system. If the simulation time is increase the proposed system is better than the existing system.

The Delay comparison is shown in Figure 3. Delay is decreases when compared to the existing system. In first 20 s the existing system is better than the proposed system. If the simulation time is increase the proposed system is better than the existing system.

The Throughput comparison is shown in Figure 4. Throughput is increase when compared to the existing system. In first 20 s the existing system is better than the proposed system. If the simulation time is increase the proposed system is better than the existing system.

The PDR comparison is shown in Figure 5. PDR is increase when compared to the existing system. In first 20 s the existing system is better than the proposed system. If the simulation time is increase the proposed system is better than the existing system.

The Energy usage comparison is shown in Figure 6. Energy usage is decrease when compared to the existing system.

| Parameter | Value |
|---|---|
| Routing Protocols | EEEMR Protocol |
| Algorithm | Clustering Algorithm |
| MAC Layer | 802.11 |
| Terrain Size | 840 × 840 |
| Number of nodes | 100 |
| Channel Type | Wireless Channel |
| Antenna Model | Omni Antenna |
| Radio Propagation Model | Two Ray Ground |
| Interface Queue Length | 50 |
| Interface Queue Type | Drop Tail/Pri Queue |
| Simulation Time | 100 s |
| Network Simulation | NS-2.34 |

Table 1: Simulation parameters.

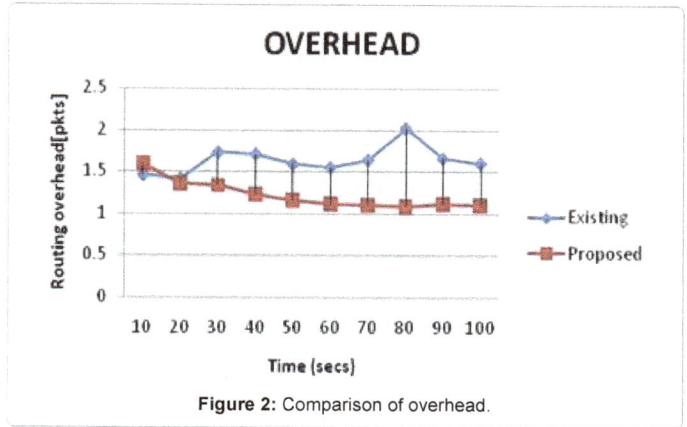

Figure 2: Comparison of overhead.

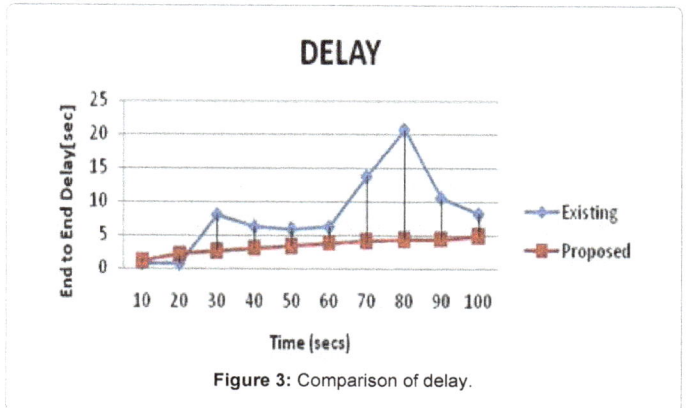

Figure 3: Comparison of delay.

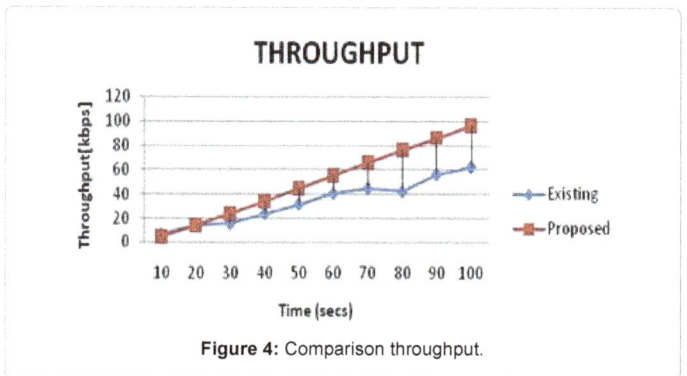

Figure 4: Comparison throughput.

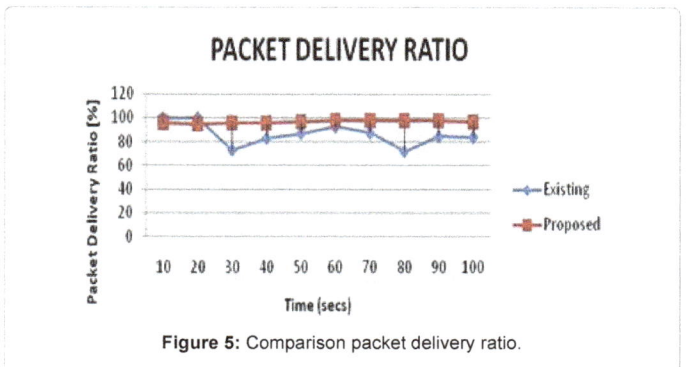

Figure 5: Comparison packet delivery ratio.

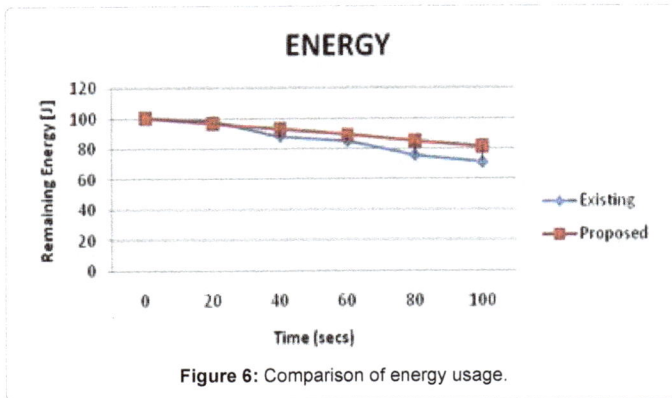

**Figure 6:** Comparison of energy usage.

## Conclusion

In this paper, the EEEMR Protocol is implemented by using Clustering algorithm. By using this method we improve the quality of service parameters like Throughput, Packet Delivery Ratio, Delay, Overhead and Energy of wireless sensor networks. When compared to the existing system the Throughput is around 35% increase, Packet Delivery Ratio is around 13% increase, Delay is around 40% decrease, Overhead is around 40% decrease and Energy consumption is around 13% decreases. The network life time of wireless sensor network is increases based up on Quality of Service parameters.

### References

1. Raj DAA, Sumathi P (2016) Analysis and comparison of EEEMR protocol with the flat routing protocols of wireless sensor networks. 2016 International Conference on Computer Communication and Informatics (ICCCI -2016), Coimbatore, India.

2. Singh SP, Sharma SC (2015) A survey on cluster based routing protocols in wireless sensor network. International Conference on Advanced Computing Technologies and Applications (ICACTA-2015) 45: 687-695.

3. Dave PM, Dala PD (2013) Simulation and performance evaluation of routing protocols in wireless sensor network. Int J Adv Res Comp Commun Eng 2: 1405-1412.

4. Aliouat Z, Harous S (2012) An efficient clustering protocol increasing wireless sensor networks lifetime. Innovations in Information Technology (IIT), IEEE International Conference on Communication, Networking & Broadcasting pp: 194-199.

5. Tian H, Shen H, Roughan M (2008) Maximizing networking lifetime in wireless sensor networks with regular topologies. Ninth International Conference on Parallel and Distributed Computing, Applications and Technologies PDCAT.

6. Younis O, Fahmy S (2004) Distributed clustering in ad-hoc sensor networks: a hybrid, energy-efficient approach. Twenty-third Annual Joint Conference of the IEEE Computer and Communications Societies, INFOCOM.

7. Mallapur SV, Terdal S (2010) Enhanced Ad-Hoc on Demand Multipath Distance Vector Routing Protocol (EAOMDV). Int J Comput Sci Inf Secur 7: 166-170.

8. Ahmadi M, Faraji H, Zonrevand H (2012) Clustering algorithm to reduce power consumptions for wireless sensor networks. Adv Mat Res 433-440: 5228-5232.

9. Maisra P (2010) Routing protocols for ad hoc mobile wireless networks.

10. Kazemi A, Akhtarkavan E (2014) Clustering algorithm to reduce power consumption in wireless sensor network. ACSIJ 3.

11. Al-Karaki JN, Kamal AE (2004) Routing techniques in WSN: a survey. IEEE Wireless Commun 11: 6-28.

# 12

# Light-Weight Energy Consumption Model and Evaluation for Wireless Sensor Networks

**Andrew Richardson, Jordan Rendall and Yongjun Lai\***

*Mechanical and Materials Engineering, Queens University, Kingston, Ontario, K7L 3N6, Canada*

**Abstract**

Wireless sensor networks are comprised of low power devices with fixed energy stores. They often require long term operation for successful deployment so it is important to efficiently manage and track their energy usage. To effectively accomplish this across distributed networks requires methods which have low energy cost with minimal error. In this paper we present a straightforward model for energy consumption in wireless sensor networks which is light-weight and accurate. The model has been applied to a wireless sensor network developed by the Queen's University MEMs lab and is evaluated with a custom testbed. Through testing, the model is exposed to realistic disturbances of communication loss, battery effects and variable voltage supplies. It was shown that with 99% packet reception rates in the network, the model accurately estimates end node energy consumption with less than 5% error. These results were demonstrated across varying data rates, battery supply capacities, and runtimes up to full network lifetime.

**Keywords:** Wireless sensor networks; Energy model; Energy consumption

## Introduction

Wireless sensor networks (WSNs) have shown promise in support of a wide range of applications from wildlife telemetry tags [1] to structural health monitoring of civil structures [2]. Additionally, WSNs have shown prospective use in industrial applications such as industrial machine monitoring where they can potentially reduce system cost and provide flexible testing platforms [3]. WSN are commonly comprised of battery powered nodes which require extended deployment durations to be successful. Because of this, lowering power consumption of WSNs is of major importance.

One method to lower WSN power consumption this is through the monitoring of node energy. Through this monitoring, researchers have shown possible reductions in WSN energy usage with techniques such as component-aware dynamic voltage scaling [4] and duty-cycle reconfigurable sensor electronics [5]. Monitoring node energy consumption across a distrusted WSN adds additional challenges. It requires a method which is able to wirelessly track energy usage with high accuracy while imposing minimal additional energy consumption on the system.

There has been some energy aware WSNs methods proposed in research that have shown promise. These methods include energy aware frame work with a focus on fault tolerance [6], a protocol for energy-aware LED lighting system control [7] and a stochastic model for a gradient based routing protocol [8]. The testing of these methods has been primarily through simulation, presenting the need for experimental validation of an energy consumption model to be performed.

Here we present a light-weight, energy consumption model and test its accuracy with experimental measurements. To remain light-weight, the model uses end node source voltage measurements, which are commonly taken and transmitted in many WSNs, and timing information taken by mains powered gateways. Through testing with a WSN developed by the Queen's University MEMs Lab, (referred to as QML-WSN), the model is shown to accurately represent the end node energy consumption while being exposed to communication issues, battery effects such as rate capacity and recovery, and variable supply voltages over extended test durations.

The rest of this paper is presented in the subsequent format. Section of testbed and experimental methods details the experimental setup used and further describes QML-WSN. WSN Energy Characterization section explains the energy characterization steps required for implementing the model for a WSN, while Energy Model section describes how the model functions and the initial tuning required. Model testing details the testing performed to validate the model and later important results are discussed. Last section summarizes the main conclusions of the work.

## Testbed and Experimental Methods

To evaluate the energy consumption model, a testbed for controlled WSN operation with simultaneous measurement of end node energy consumption was created. The testbed consists of two groups of components: the WSN under test and the measurement equipment (Figure 1). Time stamped current measurements and collected input data for the model allows the characterization of a WSN's energy usage, tuning of the model, and evaluation of the model's performance.

The QML-WSN is a star network designed for industrial monitoring which consists of one gateway and end nodes that communicate wirelessly over 2.4 GHz band. The gateway schedules communication timeslots and data requests for the end nodes, while the end nodes periodically collect and transmit data to the gateway. A controlled DC power supply or batteries can be used as the end node power supply allowing flexibility in testing. The gateway is connected to the computer over an Ethernet switch for user control of the WSN and the storage of WSN collected data into a MySQL database, which is used as input to the model. Only one end node is used during the testing demonstrated

**\*Corresponding author:** Yongjun Lai, Mechanical and Materials Engineering, Queens University, Kingston, Ontario, K7L 3N6, Canada
E-mail: lai@queensu.ca

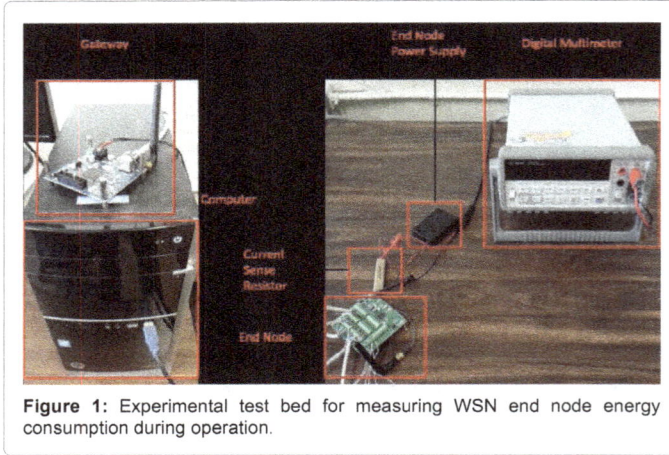

**Figure 1:** Experimental test bed for measuring WSN end node energy consumption during operation.

in the paper. To emulate the application requirements of QML-WSN, 5 STMicroelectronics LIS3DHTR accelerometers and 3 US Sensor USP11491 thermistors were connected to the end node sensor ports for the collection of acceleration and temperature data.

The measurement equipment is comprised of: one Key sight 34401A digital multi meter, a 10 Ω 25 W current sense resistor, and a computer. To measure end node current with minimal added loading, the digital multi meter measures the voltage drop across the current sense resistor which is in series between the end node power supply and the end node. The digital multi meter is controlled by the computer over GPIB through Lab VIEW script to store and capture voltage measurements at 226 Hz with +/- 0.1 μV resolution, or 1 μA of end node current.

For each test the desired data rate was set on the gateway, the end node was connected to the power supply and current measurements from the digital multi meter were started as the WSN operation began. A test was ended after either a set period of time elapsed or the end node supply voltage dropped below the functional range. During all testing, QML-WSN was operated with normal application behavior while current was measured from the digital multi meter.

QML-WSN's end node function is to remain primarily in a low energy usage sleep mode and periodically transition into a high energy usage active mode at the allocated timeslot. When in active mode, the end node first transmits a 'wake up' message to the gateway to ensure proper timing and then receives a data request message from the gateway. The end node samples its supply voltage, collects acceleration and temperature data from its sensors and transmits this data back to gateway in a 'data' message. A typical 'data' message is 1424 bits, including headers. The end node then transitions back to sleep mode. Through control of the gateway time slot scheduling the data rate can be set for the end node.

The time stamped current data measured during end node operation was processed in MATLAB to provide behavior specific end node energy consumption over a test's duration. The behaviors were isolated by identifying the transitions between sleep and active modes through derivative peak detection of the current measurements. The current measurements for each active and sleep period were integrated, resulting in end node energy consumption for each active and sleep period. The resulting measured end node energy information is used to characterization the WSN energy consumption. Additionally, since the measured energy information is time synchronized with the

MySQL database, which is used as an input to the model, it can provide experimental measurements to directly compare with the model.

## WSN Energy Characterization

The initial step to applying an energy consumption model is to characterize the WSN nodes energy usage [9]. This involves measuring the node energy usages for behaviors of interest to system operation across the working supply voltage range. Researchers have demonstrated highly detailed WSN profiling [9,10], but for QML-WSN, the behaviors of interest can be more plainly modeled as fixed sleep periods and fixed active period.

Using the testbed and techniques described in the above section, QML-WSN's current draw was measured for supply voltages of 5 V - 2.5 V from a controlled DC power supply. Measurements for each supply voltage were taken over 18 h - 24 h tests and the currents observed for sleep and active periods over the testing duration were each averaged. The resulting measured average currents, per supply voltage, for active and sleep behaviors can be seen in Figures 2a and 2b respectively. It can be seen that the active current is approximately three orders of magnitude larger than the sleep current, as the active periods are where we expect the majority of end node energy consumption. Linear fits have been applied to the measured data to form piecewise functions for active and sleep current dependent on supply voltage. Looking more closely at Figure 2, sections 1 of both the active and sleep current are the result of voltage regulation across the external TPS62740 converter [11]. At the supply range of 3.5 V - 2.5 V, the system is regulated by internal voltage converters of the MCU [12]. This results in linear relationships governed by the internal converters seen in active current section 2 and sleep current section 3. There is also a transitional period between converters for sleep current shown in section 2. Proper system operation does not occur below 2.5 V supply voltage resulting in sleep current spiking (sleep current section 4 in Figure 2b), but this is not included for the summarizing equations or for the model. The piecewise equations 1 and 2 are the result of profiling QML-WSN, providing linear voltage-current relationships for the end node behaviors which are necessary to the proposed energy consumption model.

$$I_{Active}(V_S) = \begin{cases} -2.6912V + 27.508, & V \geq 3.5 \\ 1.2597V + 14.157, & V < 3.5 \end{cases} \tag{1}$$

$$I_{Sleep}(V_S) = \begin{cases} -0.0011V + 0.0103, & V \geq 3.567 \\ -0.0536V + 0.1985, & 3.567 < V \geq 3.25 \\ -0.0029V + 0.0324, & 3.25 < V \geq 2.5 \end{cases} \tag{2}$$

## Energy Model

The energy model leverages the known voltage-current relationships gained from WSN profiling along with measured node supply voltages and timing information, collected by the gateway, to estimate energy usage with a minimal computational load required from end nodes. These terms are used to provide energy consumed for each period of node behavior, which are summed. The general model equation is given in (3) where the measured supply voltage $V_i$ is used with the voltage-current relationships I ($V_i$) to approximate the average current of node periods, while the timing information $\Delta t_i$ estimates the period durations and $E_{sum}$ is the summation of consumed energy.

$$E_{sum} = \sum_{i=0}^{i=number\ of\ periods} I(V_i)\Delta t_i \tag{3}$$

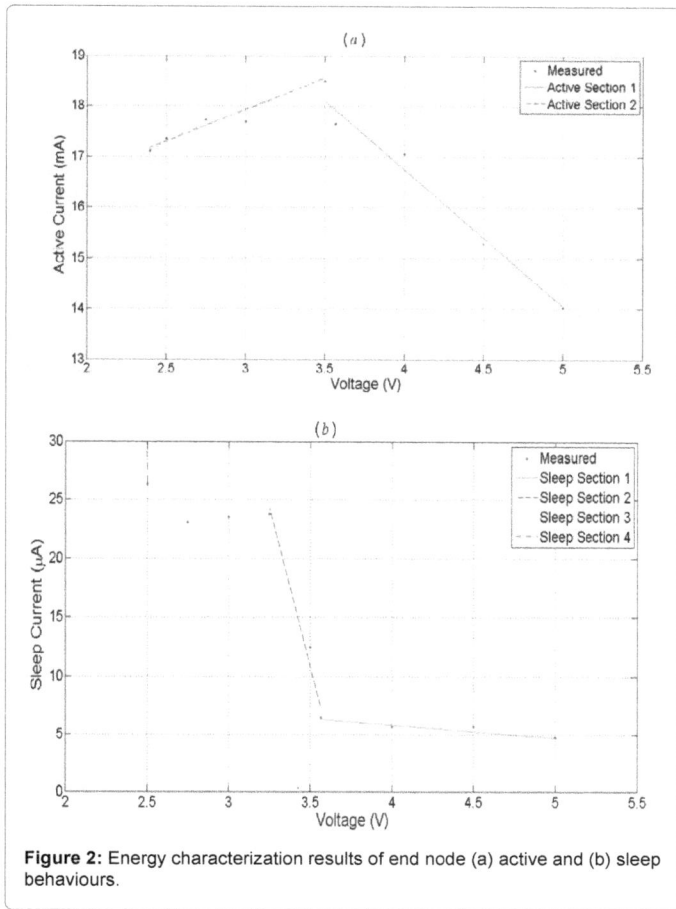

**Figure 2:** Energy characterization results of end node (a) active and (b) sleep behaviours.

The generic model is modified to allow a minimal computational load required rom nodes. Node information is only sent during node active periods while timing is tracked by the gateway around the active communication intervals, resulting in equation (4).

$$E_{sum} = \sum_{i=0}^{i=number\ of\ periods} I_{Active}(V_i)(t_{data_i} - t_{wake_i}) + I_{Sleep}(V_i)(t_{wake_{i+1}} - t_{data_i}) \quad (4)$$

Where $I_{Active}(V_i)$ and $I_{Sleep}(V_i)$ are the voltage-current relationships for active and sleep node behavior respectively, $t_{wake_i}$ is the time the node 'wakeup' message is receive by the gateway while $t_{data_i}$ is the time the node 'data' message is receive by the gateway, and Vi is the supply voltage measured by the node during the active period. The number of active periods is used for indexing because information collected by the node is only sent during active periods. Using this method, the active periods are directly tracked while the sleep periods occur in the duration between active periods. For calculation of the sleep period current, the supply voltage sampled during the previous active period is used which should show minimal change.

## Model development

The model was then applied to a controlled experimental scenario of QML-WSN while true node energy usage was measured from the digital millimeter. Using the test bed with node supply voltage fixed at 3.567 V from a DC power supply, a single node with 11.8 bps data rate was run for 24 hours with no packet losses. The resulting running energy summations plotted against testing time can be seen in Figure 3a, where the model and measured energy summations appear as constant linear

energy consumption rates. The model's slope appears lower than that of the measured slope by a constant value, causing an underestimate of node energy consumption. This suggests the underestimate is attributable to a small consistent error in energy calculations of each end node period. It was found that by calculating the weighted mean of the distribution for active period durations over the test, the measured active period duration was 609.8 ms while the modeled was 518.3 ms. This would lead to a 15% underestimate in the models end node active energy from the measured. The timing difference in the model can be attributed to a short portion of the end node active period occurring before the wakeup message is transmitted and after the data message is received by the gateway. This underestimate of active duration can be corrected by computing the difference between the weighted means and adding this calculated correction factor constant to each active period duration in the model, while also shifting the sleep durations. The resulting model is plotted as 'model correction factor' in Figure 3a, where the model and measured energy summations align much more closely.

The model's performance with and without the correction factor can be better seen in Figure 3b, which displays the error in modeled energy summation from the measured results. The error without the correction follows a linear trend with a slope of approximately 1.40 × $10^{-2}$ mA, which is expected from the slope difference of energy summation previously observed. It results in a final error of $3.291 × 10^{-1}$ mAh or a 14.48% error over the 24 hr test duration. For the model with the correction factor, the error stays very close to zero and is reflective of the close alignment to the measured energy summation. It results in a final error of $1.2 × 10^{-2}$ mAh or a 0.528% error over the 24 hr test duration.

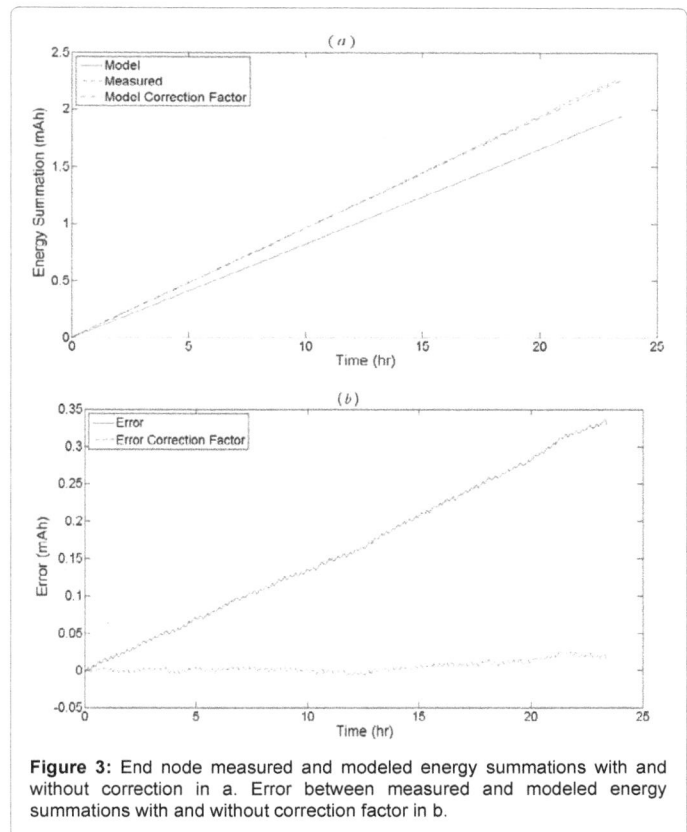

**Figure 3:** End node measured and modeled energy summations with and without correction in a. Error between measured and modeled energy summations with and without correction factor in b.

The model thus demonstrates it is able to effectively predict the energy consumption of the node within 15% error without the correction factor and this is improved to less than 1% error when including the correction factor. The correction factor will remain constant for QML-WSN since we expect the active period timing to be unchanged. During further use of the model with QML-WSN, the correction factor calculated from this test will be applied.

## Model Testing

The proposed energy consumption model has demonstrated the ability to accurately predict node energy under controlled experimental conditions, but this is not always the case with real world WSNs. In WSN deployments nodes can experience communication issues, battery effects such as rate capacity and recovery, and variable supply voltages throughout their lifetime. Two trials were undertaken using the testbed to better examine the impact of these factors on the model and its accuracy. The parameters of each trial are summarized in Table 1 and described in detail below.

In the first trial, the node's data rate was increased to 282.8 bps, the power source was changed to low capacity alkaline batteries, and the node was run until the batteries were fully depleted achieving network lifetime. Network lifetime was deemed to be reached when sustained proper end node function no longer occurred. The increased data rate and low capacity batteries were chosen to accelerate the battery depletion and reduce the testing time required while still achieving full network lifetime. This allowed three tests under the same conditions to be performed. Three Energizer A76 alkaline button batteries in series were used as the power source which has a rated capacity of 153 mAh at a discharge rate of 191 µA [13].

In the second trial, the power source was changed to standard capacity alkaline batteries; the batteries were only partially depleted by fixing the testing length to one week due to time constraints. Two tests were performed with differing data rates of 282.8 bps and 11.8 bps. This trial better reflects realistic application power sources and the associated battery effects, submits the energy consumption model to longer duration testing, and explores the impact of data rates on the model. Three Duracell Procell PC1500 AA alkaline batteries in series were used as the end node power source, which have a rated capacity of 3280 mAh at a discharge rate of 5 mA [14].

### Low capacity battery full depletion

For trial 1, three separate tests were conducted under the described experimental conditions. Figure 4a displays the modeled energy summation for each test, and Figure 4b displays the measured energy summation. From Figure 4a, it can be observed that the modeled energy summations of the three tests consist of closely matching linear trends that deviate in slope throughout the tests runtime. The gradual deviations in slope are likely due to end node current draw changing as supply voltage drops from 4.44 V - 2.5 V with battery depletion over the test runtime, as dictated by the profiled voltage-current relationship. Additionally, there is variation between the tests' network lifetime durations and their final modeled energy which ranged from 64.5 h - 75.2 h and 114 mAh - 135.8 mAh respectively. These differences are clarified when looking at the tests' measured energy summations in Figure 4b. It can be observed that the final measured energy for all tests is notably closer, ranging from 129.5 mAh - 136.4 mAh with differing energy consumption rates leading to the variations in test network lifetimes. The final measured energy is still lower than the rated 153 mAh specified by the battery manufacture; this could be attributed to

the rate capacity effect from discharging well above the listed rate of 191 µA.

From the measured energy summations in Figure 4b, linear trends similar to the modeled energy are evident but with significantly more variation throughout each of the tests. The variations appear to be due to regions of high energy consumption, between 15 h - 25 h for all three tests and between 0 h - 2.5 h for test 1. It should be noted that communication issues between the end node and gateway were observed during testing around these regions. For each of the three tests the measured energy is higher than the modeled energy, indicating that the model is underestimating the end node energy consumption. The high energy consumption regions are not apparent in the modeled energy summation, suggesting they are introducing error into the model and may be causing an underestimate of end node energy consumption.

### Standard capacity battery partial depletion

In trial 2, two separate tests were performed under the experimental conditions previously described. Test 1 used a data rate of 282.8 bps and tests 2 used 11.8 bps. The results of test 1 and 2 are presented in Figures 5a and 5b respectively, displaying the measured and modeled energy summations for each of the tests. Over the 167 h test duration,

| Trial number-test number | Data rate | End node power supply | Test duration |
|---|---|---|---|
| 1-1 | 282.8 bps | 3x Energizer A76 | Full battery depletion |
| 1-2 | 282.8 bps | 3x Energizer A77 | Full battery depletion |
| 1-3 | 282.8 bps | 3x Energizer A78 | Full battery depletion |
| 2-1 | 282.8 bps | 3x Duracell PC1500 | 1 week |
| 2-2 | 11.8 bps | | |

**Table 1:** Summary of parameters for each trail and test.

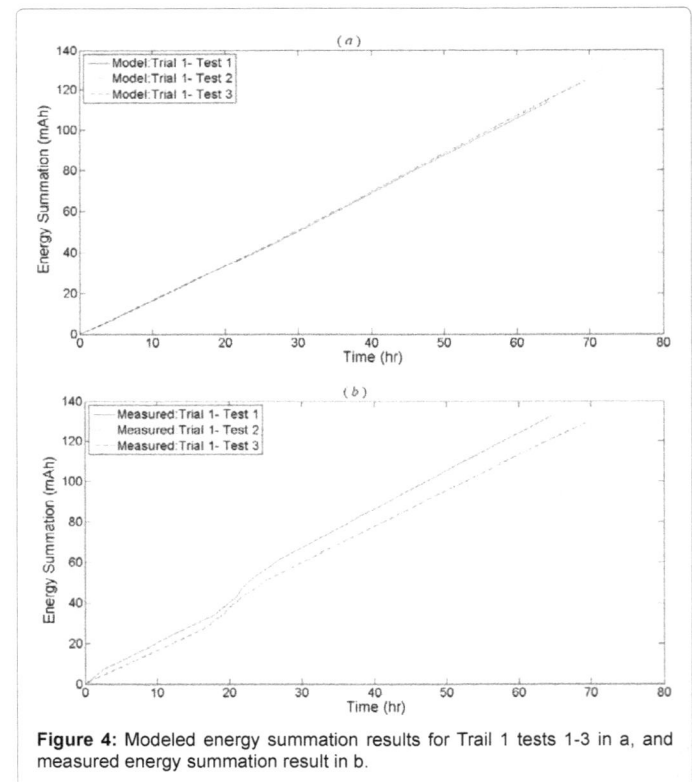

**Figure 4:** Modeled energy summation results for Trail 1 tests 1-3 in a, and measured energy summation result in b.

the final measured and modeled energy summations for test 1 were 284.2 mAh and 260.4 mAh respectively, while for test 2 they were 11.89 mAh and 11.4 mAh. The difference in final energy between tests 1 and 2 is due to end node data rate. In test 2 the data rate is 24 times lower than the data rate used in test 1.

The modeled energy summations appear to follow similar linear trends to those seen in trial 1with less slope deviations. The steadier slopes may be attributed to more stable supply voltages over the testing duration caused by a lower depletion of the overall battery capacity. The test 1 supply voltage changed from 4.71 V - 4.34 V while test 2 remained even more stable, only changing from 4.68 V - 4.66 V. In the measured energy summations of both tests, multiple regions of high energy usage were again observed. These same trends are observed in both tests of differing data rates, suggesting the model error apply evenly across data rates. The longer test duration with higher data rate of test 1 allowed the observation of a periodic like nature of the high energy usage regions at approximately: 70 h, 90 h, 115 h, 135 h, 160 h (Figure 5a). Also in Figure 5a, the model and measured energy summations appear to intersect at 70 h suggesting there may be competing factors in the energy model for over and undercompensation.

## Discussion

To better discern the effects observed during testing and the overall accuracy of the proposed model, the error between the measured and modeled energy summations has been calculated for both trials. The errors for trial 1 test 1-3 and trial 2 tests 1 are displayed in Figure 6a. The error for trial 2 test 2 is plotted in Figure 7a, due to the lower magnitude of energy usage compared to the previous tests. Increases in error can be seen for all tests which coincide closely in time with the high energy usage regions previously observed in the measured

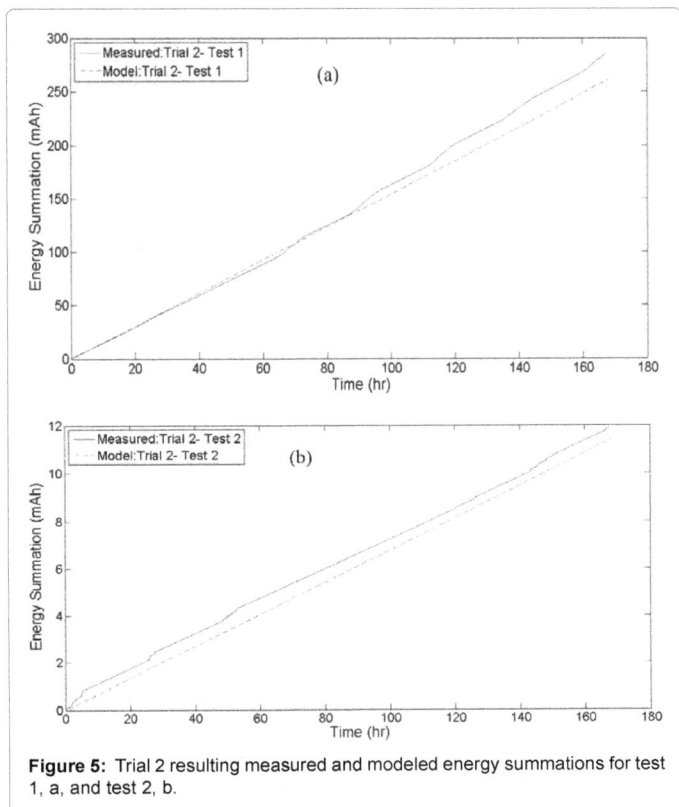

**Figure 6:** Error between modeled and measured energy summations of trial 1 tests 1-3 and trial 2 test 1 in a and each tests' packet loss in b.

energy summations. The increases appear to be the main source of error introduced into the model and thus require further investigation into their cause.

### Packet loss

End node communication interruptions were observed during testing near the high energy usage regions. To clarify the communication interruptions, the packet loss for each test is examined. The cumulative packet loss for trial 1 tests 1-3 and trial 2 tests 1 are shown in Figure 6b, while trial 2 test 2 packet losses is shown in Figure 7b. It can be observed for all tests that regions of high packet loss occur in the same time intervals as every large increase in error. Additionally, little to no packet loss can be observed in the duration between the jumps in error, with the exception of trail 1 test 1 which shows consistent packet losses and corresponding building error. Error introduced by packet loss can be explained by the presented model's reliance on proper communication to account for end node active period energy usage. When proper communication is not received by the gateway, the model estimates the end node to be in the low energy usage sleep mode. This causes the potential for the model to miss high energy usage periods where the node does not successfully communicate with the gateway. Few instances of packet loss without missed active periods can be seen, such as trial 2 test 2 at 80 h in Figure 7, but the majority appear to coincide with missed active periods. The periodic nature of the packet loss observed in Figure 6b for trial 2 test 1 and the high density of the packet loss regions for all tests suggests that packet loss is the main error introduced into the model.

### Model overcompensation

A smaller source of error in the model can be observed in Figures 6a and 7a as regions with negative slope in the error plots, which represent overcompensation of energy consumption by the model.

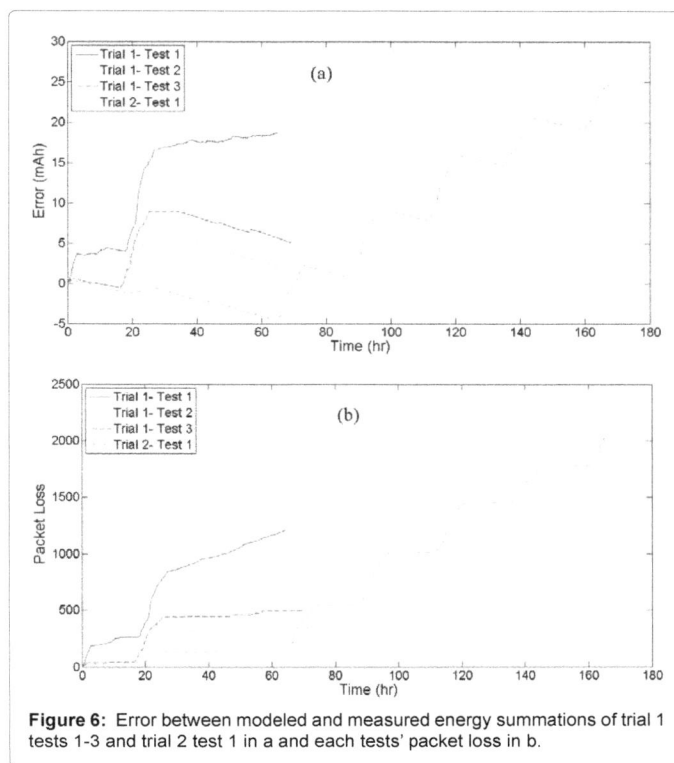

**Figure 5:** Trial 2 resulting measured and modeled energy summations for test 1, a, and test 2, b.

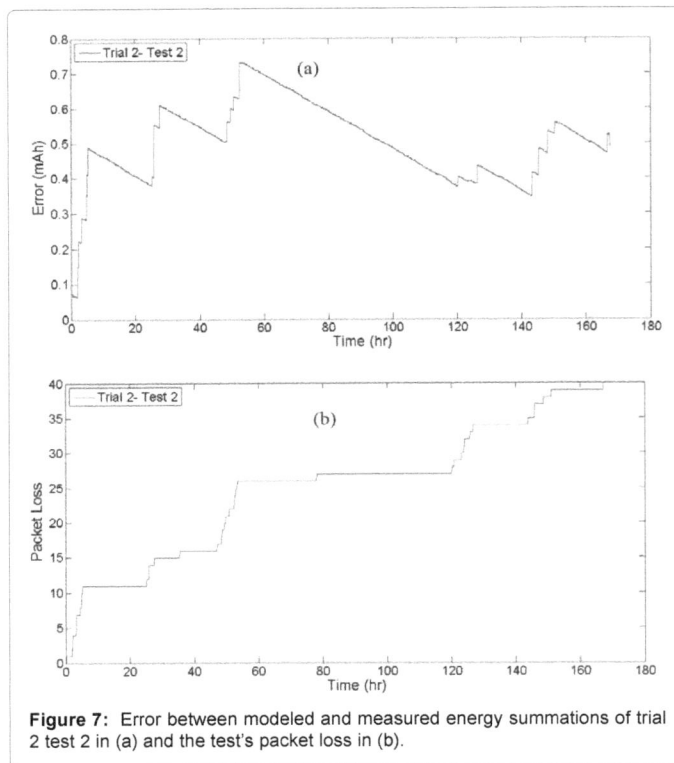

**Figure 7:** Error between modeled and measured energy summations of trial 2 test 2 in (a) and the test's packet loss in (b).

| Trial number - test number | Data rate | Final error (%) | Packet reception rate (%) |
|---|---|---|---|
| 1-1 | | 14.12 | 97.4 |
| 1-2 | 282.8 bps | 0.469 | 99.36 |
| 1-3 | | 4.107 | 99.00 |
| 2-1 | | 8.386 | 98.25 |
| 2-2 | 11.8 bps | 4.089 | 99.21 |

**Table 2:** Summary of test accuracy.

The overcompensation appears to occur during regions where packet loss remains stable, as seen in Figure 6b and Figure 7b. As the change in error is gradual, it suggests that it is the result of small miscalculations of energy in each end node period. It can be seen that the overcompensation appears constant for trial 2 test 1, 2 in Figures 6a and 7a, while it appears to change over the duration of trial 1 tests 2, 3. The change in slope can be noted in Figure 6a trial 1 tests 2, 3 as the slope of the error becomes near zero from 25 h - 35 h, which corresponds to source voltages between 3.75 V - 3.55 V for both tests. This suggests that source voltage influences this overcompensation, as it varies significantly more over the duration of the trail 1 tests than the trial 2 tests and it is used in the calculation of each end node period's energy consumption. The cause of the overcompensation could be attributed to an inaccuracy sampling the end node source voltage or misalignment in the current-voltage relationship creating during the energy characterizing. Both of these issues could be improved with calibration of the end node voltage sampling and more detailed energy characterizing. Ultimately, the overcompensation of node energy consumption shown is a minor source of error in the model compared to that caused by the communication issues.

### Model accuracy

The overall accuracy of the model can be clarified by examining the final error in each test as a percentage of final measured energy.

Table 2 summarizes the final errors and includes each test's packet reception rate. Examining Table 2, the model is seen across all tests to accurately predict energy consumption within 15% error for packet reception rates of 97.4% or greater. The model's accuracy was shown to increase for packet reception rates of 99% or greater, predicting energy consumption with 5% or less error. As packet reception rates of 99% have been feasibly demonstrated in WSN applications [15], it suggests that the model presented could predict energy consumption with 5% or less error in WSN applications. The overall accuracy of the model has been demonstrated to be high with consistency across: testing duration, network lifetime, data rates, battery depletion, and voltage supply capacity.

### Conclusion

The focus of this paper is to demonstrate an energy consumption model for WSN nodes that is accurate and light-weight. The model was implemented using a WSN developed by the Queen's University MEMs Lab, QML-WSN. This model is useful for tracking the energy usage throughout a distributed WSN, estimating the realistic network lifetime of a WSN deployment, and enabling energy aware networking protocols. The model was demonstrated and validated through controlled testing while exposing QML-WSN to communication issues, battery effects such as rate capacity and recovery, and variable supply voltages throughout the network lifetime.

To accurately evaluate the model a testbed has been created and detailed. QML-WSN's energy consumption was first characterized and the model was implemented on the platform. QML- WSN was tested under a range of conditions to ensure the model's performance. Two trails were undertaken with differing data rates, battery supply capacities, runtimes up to full battery depletion and network lifetime.

The model's performance was shown to be heavily influenced by packet loss. Despite this, the model still achieved high accuracy with errors lower than 5% for packet reception over 99%. These results were repeated for a range of battery capacities, data rates, network lifetime, and battery depletion. Additionally, a minor source of error of end node energy overcompensation was shown and could be reduced with a more detailed WSN energy characterizing. This model shows promise for a wide range of WSNs that have high packet reception rates. Further work of interest to this model is additional testing with an installed WSN in application environment and added compensation into the model for the error introduced by packet loss.

### References

1. Toledo S (2015) Evaluating batteries for advanced wildlife telemetry tags. IET Wirel Sens Syst 5: 235-242.

2. Chintalapudi K, Fu T, Paek J, Kothari N, Rangwala S, et al. (2006) Monitoring civil structures with a wireless sensor network. IEEE Internet Comput 10: 26-34.

3. Hou L, Bergmann NW (2012) Novel industrial wireless sensor networks for machine condition monitoring and fault diagnosis. IEEE Trans Instrum Meas 61: 2787-2798.

4. Hoermann LB, Glatz PM, Steger C, Weiss R (2011) Energy efficient supply of wsn nodes using component-aware dynamic voltage scaling. Wireless conference sustainable wireless technologies (European wireless), Austria.

5. Lutz K, Konig A (2010) Minimizing power consumption in wireless sensor networks by duty cycled reconfigurable sensor electronics. 8th Workshop on Intelligent Solutions in Embedded Systems, Crete.

6. Mitra S, De Sarkar A (2014) Energy aware fault tolerant framework in wireless sensor network. Applications and innovations in mobile computing, Kolkata.

7. Huynh TP, Tan YK, Tseng KJ (2011) Energy-aware wireless sensor network with ambient intelligence for smart LED lighting system control. 37th Annual Conference of the IEEE Industrial Electronics Society, Melbourne.

8. Migabo ME, Djouani K, Kurien AM, Olwal TO (2015) A stochastic energy consumption model for wireless sensor networks using GBR techniques. AFRICON, Addis Ababa.

9. Raghunathan V, Schurgers C, Park SPS, Srivastava MB (2002) Energy aware wireless micro sensor networks. IEEE Signal Process Mag 19: 40-50.

10. Antonopoulos C, Prayati A, Stoyanova T, Koulamas C, Papadopoulos G (2009) Experimental evaluation of WSN platform power consumption. IEEE International symposium on parallel and distributed processing, Rome.

11. Texas instruments (2014) TPS6274x 360nA I Q step down converter for low power applications.

12. Atmel Corporation (2014) ATmega256/128/64RFRS microcontroller with low power 2.4GHz transceiver.

13. Energizer (2015) Energizer A76 Alkaline battery.

14. Duracell (2015) Dyracell ID1500 alkaline manganese dioxide battery.

15. Geng D, Zhao Z, Fang Z, Xuan Y, Zhao J (2010) GRDT: group based reliable data transport in wireless body area sensor networks. IET International conference on wireless sensor network, Beijing.

# Extending the Functionality of Pymote: Low Level Protocols and Simulation Result Analysis

**Farrukh Shahzad***

*Information and Computer Science, King Fahd University of Petroleum and Minerals, Dhahran, Saudi Arabia*

**Abstract**

Wireless sensor networks (WSNs) are utilized in various applications and are providing the backbone for the new pervasive Internet, or Internet of Things. The development of a reliable and robust large-scale WSN system requires that the design concepts are checked and optimized before they are implemented and tested for a specific hardware platform. Simulation provides a cost effective and feasible method of examining the correctness and scalability of the system before deployment. In this work, we study the performance of Pymote, a high level Python library for event based simulation of distributed algorithms in wireless ad-hoc networks. We extended the Pymote framework allowing it to simulate packet level performance. The extension includes radio propagation, energy consumption, mobility and other models. The extended framework also provides interactive plotting, data collection and logging facilities for improved analysis and evaluation of the simulated system.

**Keywords:** Wireless sensor network; Simulation; Python; Distributed event modeling

## Introduction

Wireless Sensor Networks (WSNs) have been employed in many important applications such as intrusion detection, object tracking, industrial/home automation, smart structure and several others. The development of WSN system requires that the design concepts are checked and optimized using simulation [1].

The simulation environment for WSN can either be an adaptive development or a new development. The adaptive development includes simulation environments that already existed before the idea of WSNs emerged. These simulation environments were then extended to support wireless functionality and adapted for the use with WSNs. In contrast, new developments cover new simulators, which were created solely for simulating WSNs, considering sensor specific characteristics from the beginning [2].

Recently, several simulation tools have appeared to specifically address WSNs, varying from extensions of existing tools to application specific simulators. Although these tools have some collective objectives, they obviously differ in design goals, architecture, and applications abstraction level [3,4].

Simulators can be divided into three major categories based on the level of complexity:

- algorithm level,

- packet level, and

- instruction level.

Our work is based on Pymote [5] which an algorithm level simulator. Some other algorithm level simulators are NetTopo [6], Shawn [7], AlgoSensim [8] and Sinalgo [9].

In this work, our main contribution is the design and implementation of packet level modules for propagation, energy consumption and mobility models to extend the Pymote framework. Secondly, we also added graphing and data collection modules to enhance the Pymote base functionality and modified existing modules for node, network, algorithm and logging.

The rest of the paper is organized as follows: In section 4, we provide some background related to existing simulation tools and we discuss the python based simulation tool Pymote. We present our extension in section 5 and provide simulation examples and analysis in section 6. We conclude in section 7.

## Background and Related Work

Simulation has always been very popular among network-related research. A large number of simulators have been proposed in literature in which algorithms for wireless ad hoc networks can be implemented and studied. These simulators have different design goals and largely vary in the level of complexity and included features. They support different hardware and communication layers assumptions, focus on different distributed networks implementations and environments, and come with a different set of tools for modeling, analysis, and visualization. Classical simulation tools include NS-2, OMNeT++, J-Sim, OPNET, TOSSIM, and others [2-4].

Pymote is a high level Python library for event based simulation of distributed algorithms in wireless networks [5]. The library allows the user to make implementation of their ideas using Python-a popular, easy to learn, full featured, and objects oriented programming language. Functionalities provided by the library are implemented without additional layer of abstraction, thus harnessing full power of Python's native highly expressive syntax. Using the library, users can quickly and accurately define and simulate their algorithms. The library particularly focuses on fast and easy implementation of ideas and approaches at algorithm level without any specification overhead using formally defined distributed computing environment.

***Corresponding author:** Farrukh Shahzad, Information and Computer Science, King Fahd University of Petroleum and Minerals, Dhahran, Saudi Arabia
E-mail: farrukhshahzad@kfupm.edu.sa

## Extending the Functionality

### Propagation model

We implemented two basic radio propagation models and the commonly used shadowing model for WSN in the Pymote framework. These models are used to predict the received signal power of each packet. At the physical layer of each wireless node, there is a receiving threshold (P_RX_THRESHOLD). When a packet is received, if its signal power is below the receiving threshold, it is marked as error and dropped by the MAC [10-12] layer. The free space propagation model assumes the ideal propagation condition that there is only one clear line-of-sight path between the transmitter and receiver, while the two-ray ground reflection model considers both the direct path and a ground reflection path which gives more accurate prediction at a long distance than the free space model. However, in reality, the received power is a random variable due to multipath propagation or fading (shadowing) effects. The shadowing model consists of two parts: path loss component and a Gaussian random variable with zero mean and standard deviation σDB, which represent the variation of the received power at certain distance. Table 1 lists parameters available for propagation module. The propagation model type (free space, two-ray ground or shadowing) is a network level attribute, which should be selected before starting the simulation.

### Energy consumption model

In our extended framework, the energy model object is implemented as a node attribute, which represents the level of energy in a node. Each node can be conFigured to be powered by external source (unlimited power), Battery (default) or energy harvesting (EH) sources [13]. The energy in a node has an initial value which is the level of energy the node has at the beginning of the simulation. It also has a given energy consumption for every packet it transmits and receives which is a function of packet size, transmission rate and transmit (receive) power. The model also supports idle or constant energy discharge due to hardware/microcontroller consumption and energy charge for energy harvesting based WSN. During simulation, each node's available energy is recomputed every second based on the charging and/or discharging rate. If it drops below minimum energy required to operate ($E_{min}$) then that node assumed to be dead (not available for communication) until energy reaches above $E_{min}$ again later in simulation (for EH nodes). Table 2 lists parameters available for energy module which can be set differently for each node. The energy object keeps track of the energy available (for battery-operated or energy harvested nodes) and total energy consumption.

### Mobility model

Our extended framework allows nodes to be mobile during simulation. Each node can be conFigured as fixed or mobile. The mobility module support three types of motion as summarized in Table 3. During simulation, each mobile node location is recomputed every second.

### Plotting and data collection

These modules allow real-time plotting and data collection during and after simulation for interactive analysis and comparisons of useful information. The modules implements generic helper methods. The simulation script is responsible for utilizing these methods to plot/chart and collect/log appropriate information as required by the simulated algorithm and application scenario. The output files are managed by utilizing separate folder for each type of files within the current working path (Table 4). Also for each simulation run, a separate folder, prefixed with the current date time is used for all files created during that simulation run.

### Modified node module

Enhanced framework requires significant modification in the Node module. The Node object now contains node type, energy model object and mobility object. The modified send and receive methods check before transmission or reception whether node has enough energy to perform the operation. Also the propagation model dictates whether a packet is received without errors (i.e. when received signal power is greater than the threshold based on the distance between the sender and receiver nodes). The object also keeps track of number of messages transmitted, received, or lost.

### Simulation Example

We consider Internet of Things (IoT) application scenario where an energy [14] harvesting WSN (EHWSN) node is installed/embedded within the 'Thing' (object that need to be monitored). Several of such objects with EHWSN nodes form a cluster (in virtual star topology)

| Description | Parameter | Default |
|---|---|---|
| Transmit Antenna gain | G_TX | 1 |
| Receive Antenna gain | G_RX | 1 |
| System Loss (>=1.0) | L | 1 |
| Min. Received signal power threshold | P_RX_THRESHOLD | -70 dbm |
| Frequency | FREQ | 2.4 Ghz |
| Path loss exponent | BETA | 2.5 |
| Gaussian noise standard deviation | SIGMA_DB | 4 dbm |

**Table 1:** Propagation Model Parameters.

| Description | Parameter | Default |
|---|---|---|
| Transmit power | P_TX | 84 mW |
| Receive power | P_RX | 73 mW |
| Initial Energy | E_INIT | 2.0 Joules |
| Min. Energy required for operation | E_MIN | 0.5 Joules |
| Charging rate (EH nodes) | P_CHARGING | 2 mW/sec |
| Discharging rate | P_IDLE | 0.1 mW/sec |
| Transmission rate | TR_RATE | 250 kbps |

**Table 2:** Energy Model Parameters.

| Type | Parameters | Default |
|---|---|---|
| 0: Fixed | | None |
| 1: Mobile (uniform velocity) | VELOCITY HEADING | 20 m/s 45 deg |
| 2: Mobile (uniform velocity, random heading) | VELOCITY | 20 m/s |
| 3: Mobile (random motion) | MAX_RANDOM_ MOVEMENT | 30 m |

**Table 3:** Mobility Parameters.

| Type | Folder Name | Examples |
|---|---|---|
| Data files | /data | CSV files containing energy consumption for each node, Message received/lost counts, etc. |
| Charts/plots | /charts | Line plots and/or bar charts of energy levels. |
| Topology | /topology | Topology map of all nodes used for simulation (before/after simulation) |
| Logging | /logs | Simulation run and module level logging |
| Combined | /yyyy-mm-ddThh-mm-ss | All files generated during a specific run |

**Table 4:** File Management.

around a high power coordinator node (or cluster head). EHWSN nodes can only communicate to its own coordinator (when they have enough energy). Coordinators are special wireless nodes which have sufficient power available and can send data to base station directly or via other coordinators (multi-hop) in a typical converge-cast application as illustrated in Figure 1. These objects are mobile and can move around its neighborhood or move to another neighborhood (within the range of a different coordinator). The coordinators are installed at strategic fixed locations throughout the facility.

Figure 2 illustrates the scheme on time scale and can be summarized as follows:

- The coordinator periodically (period=Tc=t4-t0) broadcasts a beacon pulse with 10% duty cycle. The pulse contains the MAC address (48 or 64 bytes) of the radio, which is universally unique, and the number of registered nodes. After transmitting the beacon, it goes in the listening mode to receive messages from any EHWSN mode which have any packets to send.

- The neighboring EHWSN nodes (which have enough power to operate) periodically wake up (period=Tn>Tb) with a certain duty cycle to receive the beacon pulse from the coordinator (t1 in Figure 2). If the received coordinator's MAC address is different than the last communicated coordinator or the node has not communicated recently, then the node will send a registration message containing its MAC [15] address and the power status (t2 in Figure 2); otherwise node will go back to sleep or send the data packet if any. The destination address will be set to the coordinator address so that any other neighboring nodes which are listening will ignore it.

- On receiving the node' registration message, the coordinator record the registration information and increment the number of registered nodes (t3 in Figure 2). If coordinator receives a data message then it will buffer it for future transmission to base station or for aggregation. The coordinator will acknowledge the received packet in both cases.

- The case of multiple child nodes transmitting in response to the same beacon pulse is also shown in Figure 2 during the second beacon period. The contention is avoided by using a random back off time before transmission. The wining node will transmit first (node A transmits at t7 in Figure 2) and other

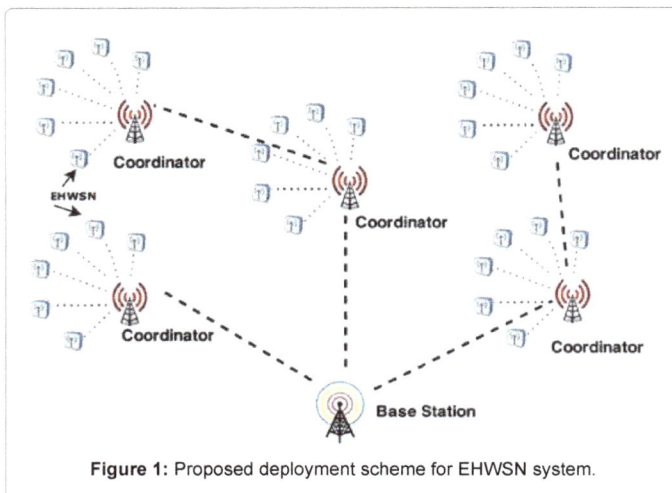

**Figure 1:** Proposed deployment scheme for EHWSN system.

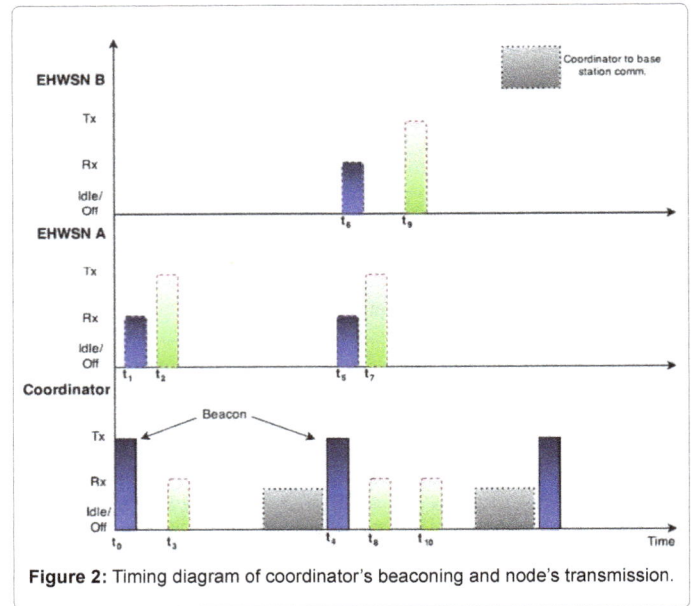

**Figure 2:** Timing diagram of coordinator's beaconing and node's transmission.

| Parameter | Name | Value |
|-----------|------|-------|
| Tb | Beacon period | 1 sec |
| Δb | Beacon duty-cycle | 10% |
| n | No. of nodes | 5 - 100 |
| Sd | Data packet size | 100 bytes |

**Table 5:** Simulation Parameters.

nodes will wait for the channel (e.g., node B transmits at t9 in Figure 2).

### Simulation setup

We only need to simulate the communication performance of one cluster formed by a coordinator and its children EHWSN nodes. The coordinator is placed in middle of *n* randomly deployed EHWSN nodes over a 600 m by 600 m area. We assumed that these nodes are constantly being charged during simulation and nodes are mobile (type=2, see Table 3). We consider beacon, registration and data packet sizes of 100 bytes while the acknowledgment packet size of 15 bytes. We used default parameters for different modules as listed in Tables 1-3. Some other parameters are shown in Table 5. The simulation script utilizes the plotting and data collection modules to generate image and data files for easy visualization and analysis of simulation results (Figure 3).

### Simulation results

Figure 4 shows a simple topology generated for simulation using the Pymote. The center node (#1) acts as the coordinator for the EHWSN nodes (numbered 2 to 26). The node in lighter color means that its available energy [16,17] is below E_MIN (=0.5 J).

We arbitrarily selected node 5 and node 10 as borderline in terms of energy available (i.e., the initial energy at start of simulation). Node 5 doesn't have enough energy to transmit in the beginning but charged up above E_MIN (Table 2) during the simulation and start communicating. On the other hand, Node 10 just has enough energy [18] to send few messages before its energy level dropped below E_MIN. We set the charging rate to 0 for Node 10. The energy level change during the simulation run is shown in Figure 5.

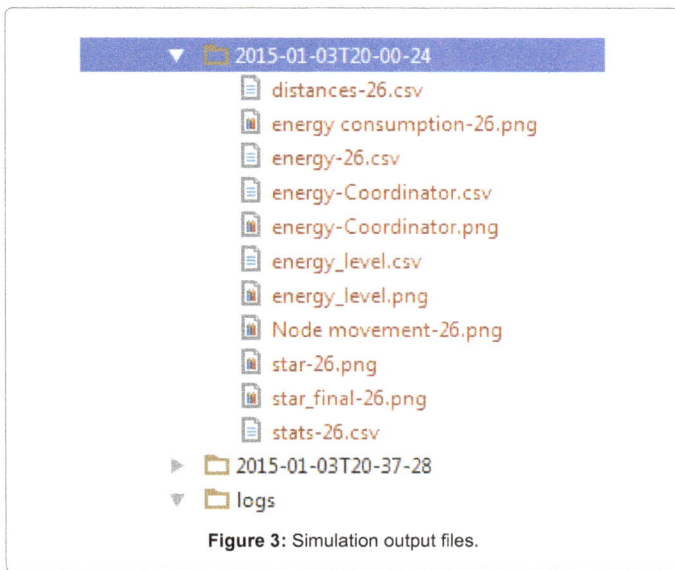

**Figure 3:** Simulation output files.

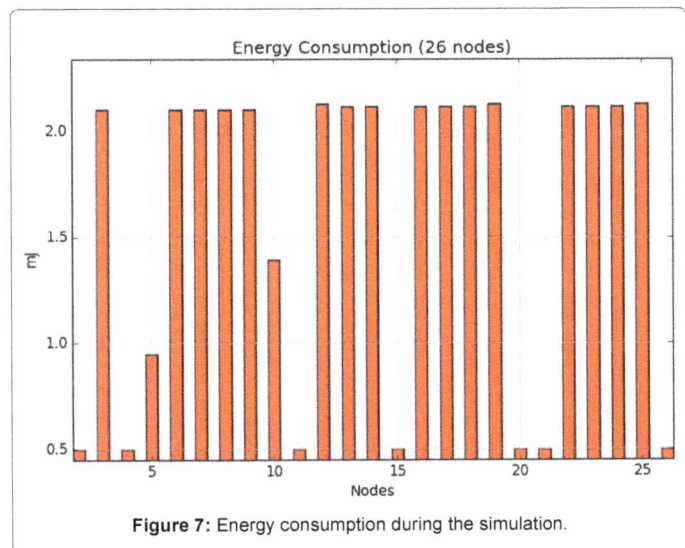

**Figure 4:** 25 EHWSN nodes around a coordinator.

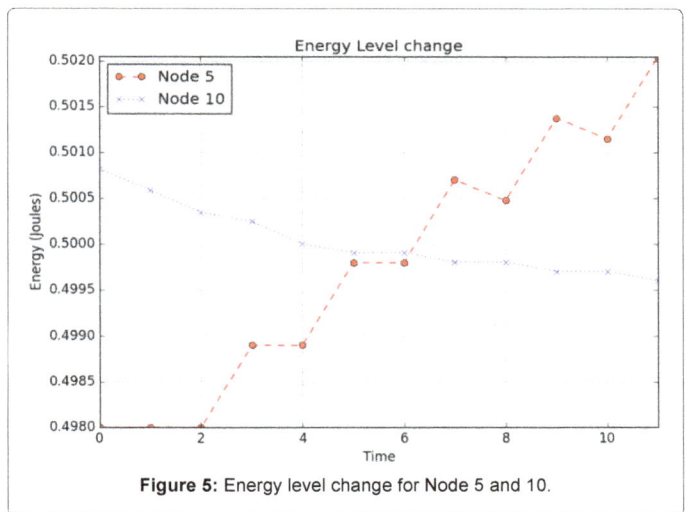

**Figure 5:** Energy level change for Node 5 and 10.

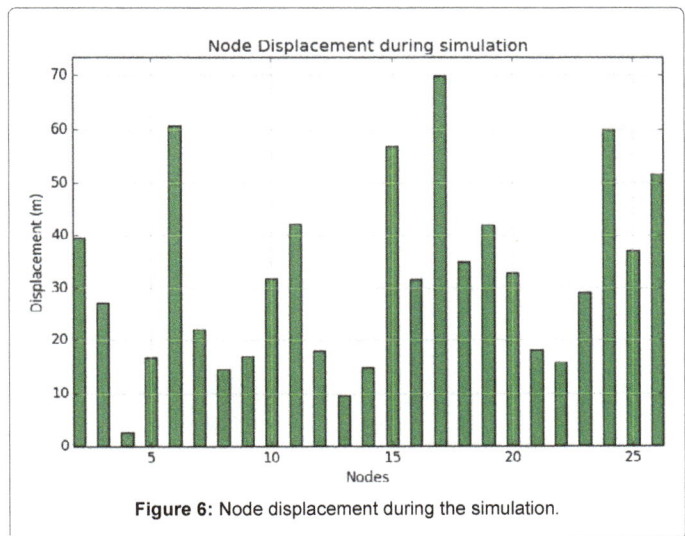

**Figure 6:** Node displacement during the simulation.

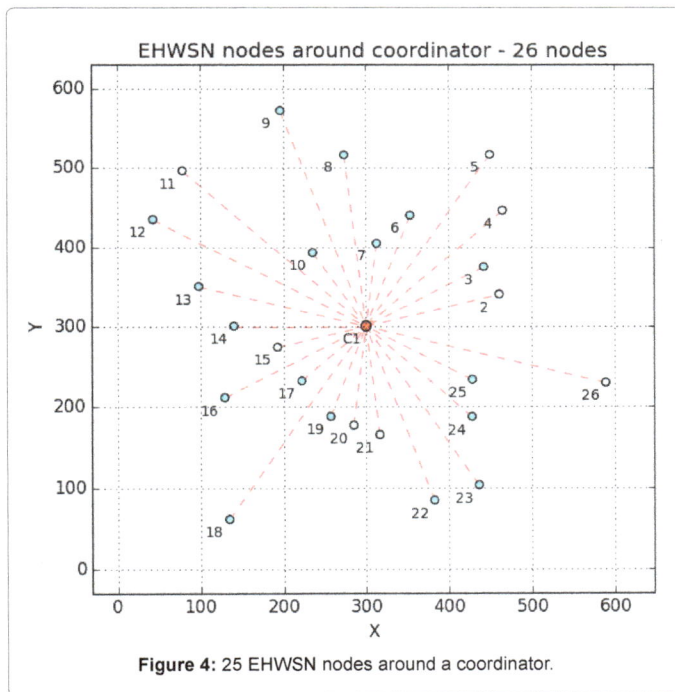

**Figure 7:** Energy consumption during the simulation.

Figure 6 shows the net node displacement during the simulation as they move around with constant speed but in random direction. Figure 7 illustrates the energy consumption of all EHWSN nodes. We can notice that some nodes never communicated due to low energy (like node 2, 4, 11, 15, 20, 21 and 26) whereas node 5 and 10 were only active during some part of the simulation as we discussed earlier. Finally Figure 8 shows the location of nodes at the end of simulation.

Secondly, we vary the number of EHWSN nodes in the network from 10 to 100 in the increment of 5. The extended framework generates simulation output files for each iteration. The output files also include the overall summary. Figure 9 shows the generated topology for 100 EHWSN nodes (2 to 101). Figure 10 shows energy consumption plots for coordinator and other nodes combined (sum of energy consumption for all EHWSN nodes). The chart also shows

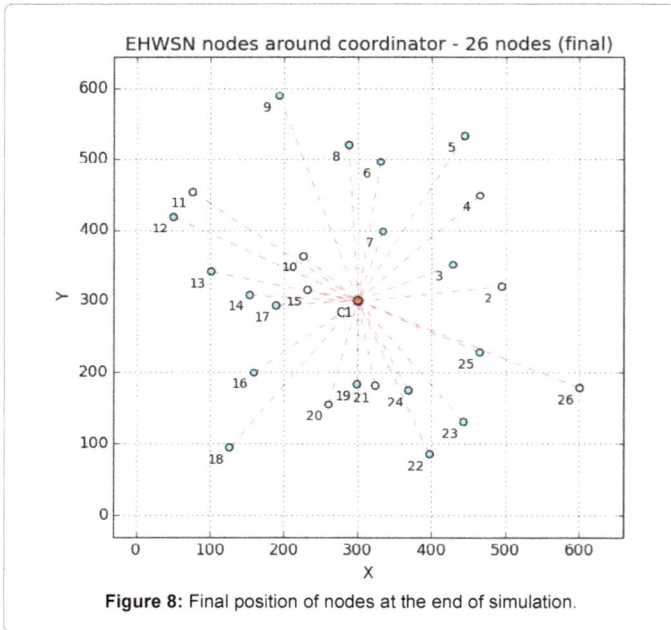

Figure 8: Final position of nodes at the end of simulation.

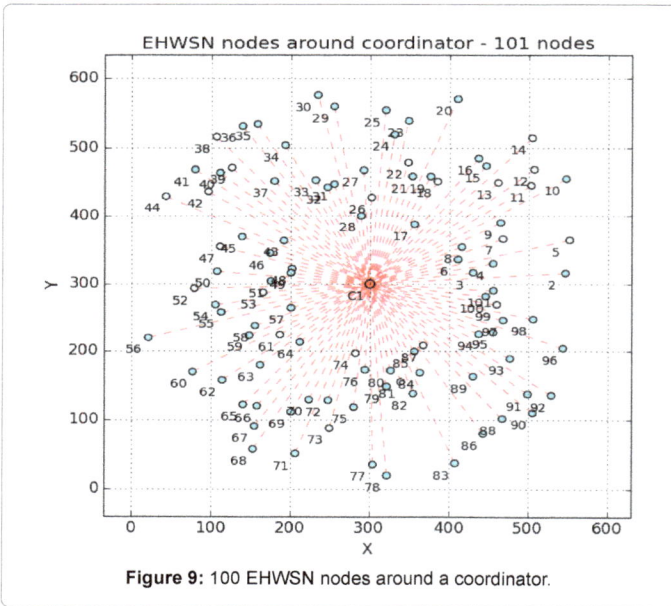

Figure 9: 100 EHWSN nodes around a coordinator.

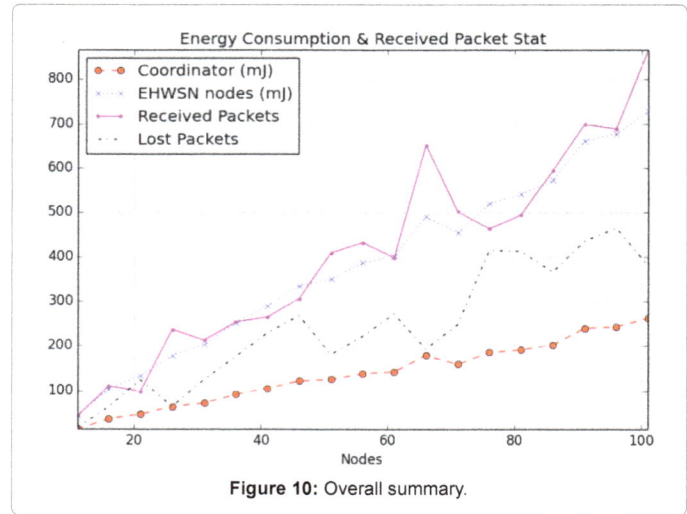

Figure 10: Overall summary.

| Nodes | Coordinator Energy consumption (mJ) | Total EHWSN consumption (mJ) | Received Packets at Coordinator | Lost Packets at Coordinator |
|---|---|---|---|---|
| 11 | 15 | 45 | 44 | 18 |
| 16 | 38 | 104 | 110 | 64 |
| 21 | 48 | 134 | 98 | 124 |
| 26 | 65 | 178 | 237 | 65 |
| 31 | 73 | 203 | 213 | 125 |
| 36 | 92 | 253 | 254 | 176 |
| 41 | 105 | 290 | 265 | 229 |
| 46 | 122 | 335 | 306 | 268 |
| 51 | 126 | 349 | 409 | 181 |
| 56 | 139 | 386 | 432 | 222 |
| 61 | 142 | 401 | 398 | 272 |
| 66 | 180 | 491 | 652 | 194 |
| 71 | 159 | 454 | 502 | 248 |
| 76 | 186 | 521 | 464 | 414 |
| 81 | 192 | 541 | 495 | 412 |
| 86 | 203 | 573 | 594 | 365 |
| 91 | 240 | 662 | 699 | 435 |
| 96 | 244 | 678 | 689 | 464 |
| 101 | 264 | 729 | 866 | 380 |

Table 6: Simulation Statistics.

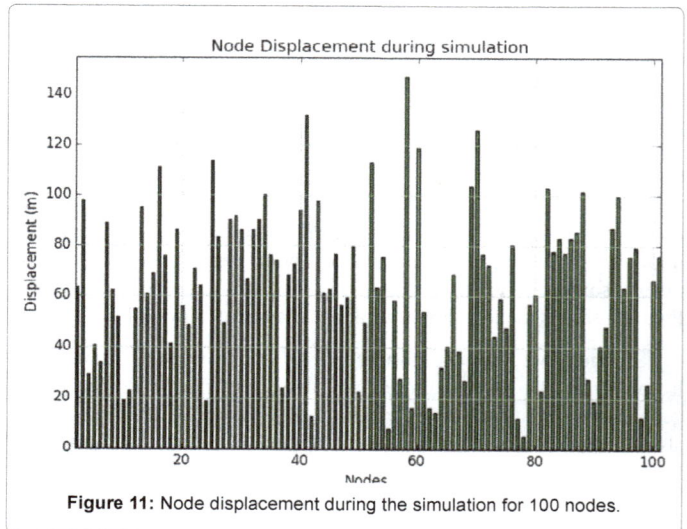

Figure 11: Node displacement during the simulation for 100 nodes.

number of messages (packets) received and lost at coordinator for each iteration.

Table 6 presented the overall simulation summary for all iterations. Node displacement and energy level change during the simulation for 100 nodes are shown in Figure 11 and 12 respectively.

## Conclusion

The development of a reliable and robust large-scale WSN system requires that the design concepts are checked and optimized before they are implemented and tested for a specific hardware platform. Simulation provides a cost effective and feasible method of examining the correctness and scalability of the system before deployment.

In this work, we utilized and extended the Python based Pymote

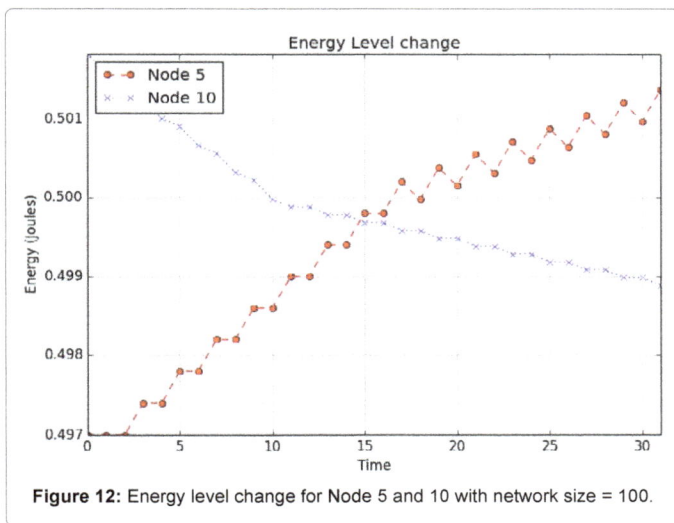

**Figure 12:** Energy level change for Node 5 and 10 with network size = 100.

framework to allow packet level simulation. We implemented modules for propagation, energy consumption and mobility models. We also added graphing and data collection modules to enhance the Pymote base functionality and modified existing modules for node, network, algorithm and logging to support the extended framework. Finally, we performed an example simulation for a scheme to efficiently utilize EHWSN in an IoT application. The simulation results presented include topology maps, plots for available energy, bar charts for node displacement and energy consumption and comparison of received and lost packets at the coordinator node.

### Acknowledgment

I would like to acknowledge the support provided by College of Computer Science and Engineering (CCSE) and the Deanship of Scientific Research at King Fahd University of Petroleum and Minerals (KFUPM).

### References

1. Abuarqoub, Abdelrahman (2012) Simulation issues in wireless sensor networks: A survey. SENSORCOMM 2012, The Sixth International Conference on Sensor Technologies and Applications.

2. Ali Q, Abdulmaojod A, Ahmed, H (2010) Simulation & Performance Study of Wireless Sensor Network (WSN) Using MATLAB. IJEEE Journal.

3. Sobeih A, Hou JC, Lu-Chuan Kung, Li N, Honghai Zhang, et al. (2006) J-Sim: a simulation and emulation environment for wireless sensor networks. Wireless Communications IEEE 4: 104&119.

4. Tselishchev Y, Boulis A, Libman L (2010) Experiences and Lessons from Implementing a Wireless Sensor Network MAC Protocol in the Castalia Simulator. Wireless Communications and Networking Conference (WCNC) IEEE pp: 1, 6, 18-21.

5. Damir Arbula, Kristijan Lenac (2013) Pymote: High Level Python Library for Event-Based Simulation and Evaluation of Distributed Algorithms. International Journal of Distributed Sensor Networks pp: 1-12.

6. Shu L, Hauswirth M, Chao HC, Chen M, Zhang Y (2011) Nettopo: a framework of simulation and visualization for wireless sensor networks. Ad Hoc Networks 9: 799-820.

7. Kroeller A, Pfisterer D, Buschmann C, Fekete S, Fischer S (2005) Shawn: a new approach to simulating wireless sensor network in Proceedings of theDesign, Analysis, and Simulation of Distributed Systems (DASD '05), San Diego, Calif, USA.

8. Algosensim. http://tcs.unige.ch/doku.php/code/algosensim/overview.

9. Sinalgo, http://dcg.ethz.ch/projects/sinalgo/.

10. Suriyachai P, Roedig U, Scott A (2012) A survey of MAC protocols for mission-critical applications in wireless sensor networks. IEEE Communications Surveys & Tutorials 14: 240-264.

11. Huang P, Xiao L, Soltani S, Mutka M, Xi N (2013) The evolution of MAC protocols in wireless sensor networks: A survey. IEEE Communications Surveys & Tutorials 15: 101-120.

12. Bachir AM, Dohler T, Watteyne KK, Leung (2010) MAC essentials for wireless sensor networks. IEEE Communications Surveys & Tutorials 12: 222-248.

13. Grady S, Cymbet JM (2014) The Design Secrets for Commercially Successful EH-Powered Wireless Sensors. Energy Harvesting for Powering WSN Symposia.

14. Eu ZA, Tan HP, Winston KG, Seah (2011) Design and performance analysis of MAC schemes for Wireless Sensor Networks Powered by Ambient Energy Harvesting, Ad Hoc Networks 9: 300-323.

15. Nintanavongsa P, Naderi MY, Chowdhury KR (2013) Medium access control protocol design for sensors powered by wireless energy transfer. INFOCOM, 2013 Proceedings IEEE pp: 150,154, 14-19.

16. Basagni, Stefano (2013) Wireless sensor networks with energy harvesting." Mobile Ad Hoc Networking: The Cutting Edge Directions: 701-736.

17. Seah, Winston KG, Tan YK, Alvin TSC (2013) Research in energy harvesting wireless sensor networks and the challenges ahead. Springer Berlin Heidelberg 13:73-93

18. Shahzad F (2013) Satellite monitoring of wireless sensor networks (WSNs). Procedia Computer Science 21: 479 – 484.

# 14

# A Novel Application of Sensor Networks in Biomedical Engineering

**Sabzpoushan SH[1]\*, Maleki A[2] and Miri F[2]**

[1]Biomedical Engineering Department, School of Electrical Engineering, Iran University of Science and Technology (IUST), Iran
[2]School of Computer Engineering, Iran University of Science and Technology (IUST), Iran

### Abstract

Born on other applications, wireless sensor networks (WSNs) grew on the promise of biomedical engineering applications. In this research we suggest system architecture for smart healthcare, based on a novel WSN. Our system particularly targets assisted-living residents and others who may help from continuous remote health monitoring. We present the objectives, advantages, and status of the design. An experimental livelihood space has been constructed at the Department of Biomedical Engineering (DBE) at Iran University of Science and Technology (IUST) for assessment of our system. A ten days monitoring and experimental results suggest a physically powerful potential for WSNs to open new research area in biomedical engineering, i.e. for low-cost, ad hoc use of multimodal sensors for a better quality of medical care.

**Keywords:** Healthcare; Wireless sensor network

## Introduction

Sensor networks permit data gathering and computation to be deeply embedded in the physical environment.

WSN is built by connecting a group of nodes together to perform the required tasks. These nodes cooperate and automatically create a network among themselves. Wireless sensor node, also commonly known as mote.

Today, WSN technologies have the potential to change the way of living with many applications in industry, travel, trade, environment monitoring, medicine, care of the dependent people, and emergency management and many other areas.

Wireless sensors, sensor networks, information technology (IT) and artificial intelligence, together have built a new interdisciplinary branch of biomedical engineering in order to overcome the challenges we face in everyday life. One of the major challenges of the world is the continuous elderly population increase in the developed countries. Population reference bureau [1] forecasts that in the next 20 years, the 65-and-over population in the developed countries will be nearly 20% of the overall population. It means the need of delivering quality care to a rapidly growing population of elderly while reducing the healthcare costs is an important issue as well. One hopeful application in this area is the integration of sensing and consumer electronics technologies which would allow people to be constantly monitored. In-home pervasive networks may assist residents and their caregivers by providing continuous medical monitoring, memory enhancement, control of home appliances, medical data access, and emergency communication [2]. Continuous monitoring will increase early detection of emergency conditions and diseases for at risk patients and also provide wide range of healthcare services for people with various degrees of cognitive and physical disabilities [3]. Not only the elderly and persistently ill but also the families in which both parents have to work will derive advantage from these systems for delivering high-quality care services for their babies and little children. Researchers in computer, networking, biomedical engineering and medical fields are working together in order to make the broad vision of smart healthcare possible. The importance of integrating large-scale wireless telecommunication technologies such as 3G, Wi-Fi Mesh, and WiMAX, with telemedicine has already been addressed by some researchers. Further improvements will be achieved by the coexistence of small-scale personal area technologies like radio frequency identification (RFID), Bluetooth, ZigBee, and wireless sensor networks, together with large scale wireless networks to provide context-aware applications [4].

In addition providing pervasiveness with existing and relatively more mature wireless network technologies, the development of small or wearable sensor devices is a mater of today researches. These researches may lead to enabling not only accurate information but also reliable data delivery.

Furthermore, the integration of all these technologies is the application, which is the coordinator between the caregivers and the caretakers and between the sensor devices and all of the actors in the overall system cycle. Since the application is the core of the high-quality healthcare service concept, the need for intelligent, context-aware healthcare applications will be increased.

Given the importance of the subject, there are already several applications and prototypes on the issue. For example, some of them are devoted to continuous monitoring for cognitive disorders like Alzheimer's, Parkinson's or similar cognitive diseases. Some focus on fall detection, posture detection and location tracking and others make use of biological and environmental sensors to identify patients' health status. There is also significant research effort in developing tiny wireless sensor devices, preferably integrated into fabric or other substances and be implanted in human body.

Some other studies [5-10] have either only smart home perspective or limited information about the design issues and challenges. The interested reader may find a wonderful survey on wireless sensor networks for health care [11].

In our paper we provide discussions not only from a smart home perspective but rather from a more healthcare related perspective. We

---

**\*Corresponding author:** Sabzpoushan SH, Biomedical Engineering Department, School of Electrical Engineering, Iran University of Science and Technology (IUST), Iran, E-mail: sabzposh@iust.ac.ir

also discuss benefits that will be achieved and challenges that will be faced while designing the future healthcare applications.

## Methods

In this research we developed network architecture for smart healthcare that will open up new prospects for continuous monitoring of assisted and independent-living people. While preserving resident comfort and privacy, the network manages a continuous medical history. No interfering area and environmental sensors combine with wearable interactive devices to evaluate the health of spaces and the people who inhabit them. Authorized care providers may monitor residents' health and life habits and watch for chronic pathologies. Multiple patients and their resident family members as well as visitors are differentiated for sensing tasks and access rights.

It is noteworthy that wearable sensors and systems have evolved to the point that they can be considered ready for clinical application. This is due not only to the tremendous increase in research efforts devoted to this area in the past few years but also to the large number of companies that have recently started investing forcefully in the development of wearable products for clinical applications. Stable trends showing a growth in the use of this technology suggest that soon wearable systems will be part of routine clinical evaluations. The interest for wearable systems originates from the need for monitoring patients over extensive periods of time. This case arises when physicians want to monitor individuals whose continual condition includes risk of sudden acute events or individuals for whom interventions need to be assessed in the home and outdoor environment. If observations over one or two days are satisfactory, ambulatory systems can be utilized to gather physiological data [12].

Another important point is that the patients under monitoring does not lose their mobility by the wire connections, it is why we considered wireless systems in our research as a practical solution.

We avoided high costs of installation and retrofit avoided by using ad hoc, self-managing networks. Based on the fundamental elements of future medical applications (integration with existing medical practice and technology, real-time and long term monitoring, wearable sensors and assistance to chronic patients, elders or handicapped people), our wireless system will extend healthcare from the traditional clinical hospital setting to nursing and retirement homes, enabling telecare without the prohibitive costs of retrofitting existing structures. The advantages of a WSN are numerous for smart healthcare with the main following important properties:

### a) Portability and desirability

Small devices collect data and communicate wirelessly, operating with minimal patient input. They may be carried on the body or deeply embedded in the environment. Unobtrusiveness helps with patient acceptance and minimizes confounding measurement effects. Since monitoring is done in the living space, the patient travels less often; this is safer and more convenient.

### b) Ease of deployment and measurability

Devices can be deployed in potentially large quantities with considerably less complexity and cost compared to wired networks. Existing structures, particularly decrepit ones, can be easily augmented with a WSN network whereas wired installations would be expensive and impractical. Devices are placed in the living space and turned on, self-organizing and calibrating automatically.

### c) Real-time and always-on

Physiological and environmental data can be monitored continuously, allowing real-time response by emergency or healthcare workers. The data collected form a health journal, and are valuable for filling in gaps in the traditional patient history. Even though the network as a whole is always-on, individual sensors still must conserve energy through smart power management and on-demand activation.

### d) Reconfiguration and self-organization

Since there is no fixed installation, adding and removing sensors instantly reconfigures the network. Doctors may re-target the mission of the network as medical needs change. Sensors self-organize to form routing paths, collaborate on data processing, and establish hierarchies.

## System architecture

Figure 1 shows the layout of simulated experimental area in our research.

The medical sensor network system integrates heterogeneous devices, some wearable on the patient and some placed inside the living space.

The architecture is multi-layer oriented (in this discipline is called multi tiered) with heterogeneous devices ranging from lightweight sensors, to mobile components, and more powerful stationary devices.

Data is collected, aggregated, pre-processed, stored, and acted upon using a variety of sensors and devices in the architecture (RFID tags, pressure sensor, floor sensor, dust sensor, environmental sensor, etc.). Multiple body networks may be present in a single system. Traditional healthcare provider networks may connect to the system by a gateway, or directly to its database. Some elements of the network are mobile, while others are stationary. Some can use line power, but others depend on batteries. If any fixed computing or communications infrastructure is present it can be used, but the system can be deployed into existing structures without retrofitting.

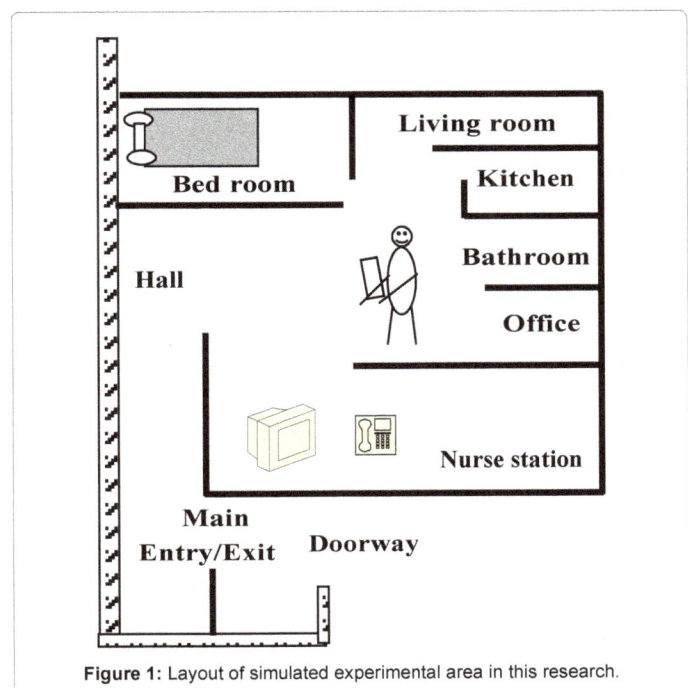

**Figure 1:** Layout of simulated experimental area in this research.

The components of the proposed architecture are shown in Figure 2, dividing devices into layers based on their roles and physical interconnection. Each layer of the architecture is described below:

### a) Network of body and its Subsystems

This network consists of small portable devices equipped with a variety of sensors such as heart-rate, heart-rhythm, temperature, accelerometer, oximeter. This layer performs biophysical monitoring, patient identification, location detection, and other desired tasks. These devices are small enough to be worn comfortably for a long time. Their energy consumption should also be optimized so that the battery is not required to be changed regularly. They may use "kinetic" recharging. Actuators notify the wearer of important messages from an external unit. For example, an actuator can remind an early Alzheimer patient to check the oven because sensors detect an abnormally high temperature. Or, a tone may indicate that it is time to take medication. The sensors and actuators in the body network are able to communicate among themselves. A node in the body network is designated as the gateway to the emplaced sensor network. Due to size and energy constraints, nodes in this network have little processing and storage capabilities.

### b) Emplaced Sensor Network

This network includes sensor devices installed in the environment like rooms, hallways, furniture and so on to support sensing and monitoring, including: temperature, humidity, motion, acoustic, camera, etc. It also provides a spatial context for data association and analysis. All devices are connected to a more resourceful backbone. Sensors communicate wirelessly using multi-hop routing and may use either wired or battery power. Nodes in this network may vary in their capabilities, but generally do not perform extensive calculation or store much data. The sensor network interfaces to multiple body networks, faultlessly managing handoff of reported data and maintaining patient presence information.

### c) Backbone

A backbone network connects traditional systems, such as PCs, and databases, to the emplaced sensor network. It also connects no

neighboring sensor nodes by a high-speed relay for efficient routing. The backbone may communicate wirelessly or may overlay onto an existing wired infrastructure. Nodes possess significant storage and computation capability, for query processing and location services. Yet, their number is minimized to reduce cost.

### d) Back-end Databases

One or more nodes connected to the backbone are dedicated databases for long-term archiving and data mining. If unavailable, nodes on the backbone may serve as in-network databases.

### e) Human Interfaces

Patients and caregivers interface with the network using PDAs, PCs, or wearable devices. These are used for data management, querying, object location, memory aids, and configuration, depending on who is accessing the system and for what purpose. Limited interactions are supported with the on-body sensors and control aids. These may provide memory aids, alerts, and an emergency communication channel. PDAs and PCs provide richer interfaces to real-time and historical data. Caregivers use these to specify medical sensing tasks and to view important data.

## Hardware and devices

We implemented our proposed system by using MicaZ motes and some other following hardware and devices. MicaZ is easy to implement and can be used to build WSN.

Existing motes typically use 8- or 16-b microcontrollers with tens of kilobytes of RAM, hundreds of kilobytes of ROM for program storage, and external storage in the form of Flash memory. These devices operate at a few mill watts while running at about 10 MHz [13]. Most of the circuits can be powered off, so the standby power can be about 1W. If such a device is active for 1% of the time, its average power consumption is just a few microwatts enabling long-term operation with two AA batteries.

Motes are usually equipped with low-power radios such as those compliant with the IEEE 802.15.4 standard for wireless sensor networks [14]. Such radios usually transmit at rates between 10 and 250 Kb/s, consume about 20-60 mW, and their communication range is typically measured in tens of meters [15,16]. Finally, motes include multiple analog and digital interfaces that enable them to connect to a wide variety of commodity sensors.

## Data acquisition

### a. Motion sensor

We have adapted a sensor module that is capable of detecting motion and ambient light levels. The module also has a simple one-button and LED user interface for testing and diagnostics. It is interfaced to a MicaZ wireless sensor node that processes the sensor data and forwards the information through the wireless network. A set of such modules is used to track human presence in every room of the simulated smart health home.

### b. Body network

We have implemented a wearable WSN service with MicaZ motes embedded in a jacket, which can record human activities and location using a 3-axis accelerometer and GPS. The recorded activity data is subsequently uploaded through an access point for archiving, from which past human activities and locations can be reconstructed.

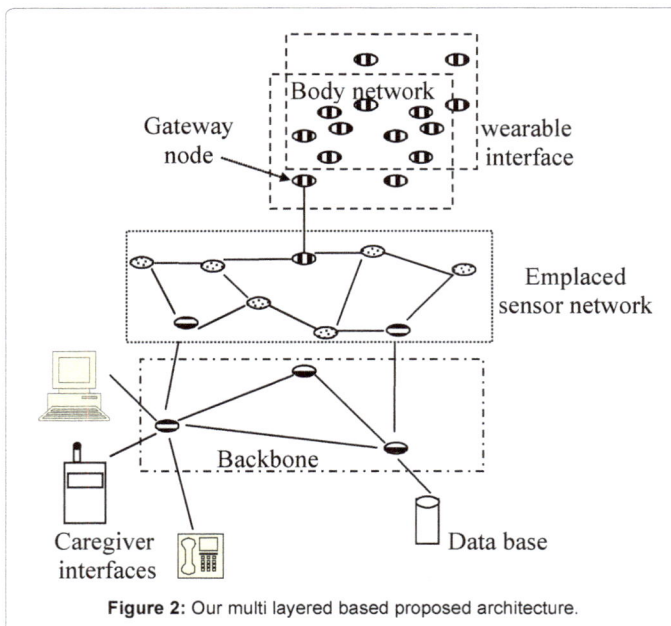

**Figure 2:** Our multi layered based proposed architecture.

### c. **Indoor temperature and light sensor**

These sensors give the environmental conditions of the habitat and are also connected to the backbone via MicaZ. Bed sensor. The bed sensor is based on an air bladder strip located on the bed, which measures the breathing rate, heart rate and agitation of a patient.

### d. **Pulse-oximeter and ECG**

These sensors are wearable, connecting to MicaZ and Telos devices, and collect patient vital signs. Heart rate (HR), heartbeat events, oxygen saturation (SpO2), and electrocardiogram (ECG) are available.

**Backbone infrastructure:** The backbone of our system is a single Stargate serving as a gateway between the motes deployed in the home environment and the nurse station. Motes use a Zigbee compliant (802.15.4) wireless protocol for communication. The Stargate runs Embedded Linux and possesses more power and capabilities than the motes.

**Database management and data mining:** A MySQL database serves as a backend data store for the entire system. It is located in a PC connected to the backbone, and stores all the information coming from the infrastructure for longitudinal studies and offline analysis.

**The user graphical interfaces (UGA):** Interfaces with residents, healthcare providers, and technicians have different requirements. Each must present an appropriate interface for performing the intended tasks, while conforming to the constraints imposed by form factor and usability. Currently, the system offers four different GUIs. The first is located on the local nurse station, and it tracks the motion of the resident using motion activations. A second GUI (Figure 3), which can run on a PDA, permits a caregiver to request real time environmental conditions of the living space and the vital signs of the resident.

It uses a query management system distributed among the PDA, Stargate and the sensor devices. The interface graphically presents requested data for clear consumption by the user. An LCD interface board was also designed for the MicaZ for wearable applications. It presents sensor readings, reminders and queries, and can accept rudimentary input from the wearer. A final GUI, from a direct medical application based on motion sensors, exists to study the behavioral profile of the user's sleep/wake patterns and life habits, and to detect some pathologies in the early stages. Figure 4 depicts some devices used in our system.

## Results

Validation of the system as a whole requires several complementary efforts. These include calculation of correct sampling of the signal, and assessment of system performance in real-world settings. The system is single hop, as the radio range covers all of the facility. A multi-hop protocol will be necessary for access of multiple floors, or if transmission power is reduced (we consider multi-hop routing algorithm in our next research). Data communication is bi-directional between the motes and the Stargate. Time-stamping is done by the PC when motion events are received.

A first experiment based on eight MicaZ motes, programmed to send motion events over the network containing the location of the user, was performed with no activity in the lab for five days. We observed no false detections in the system under these conditions. However, this experiment showed the necessity of enhancing the power management scheme to prolong the lifetime of the sensors. In another experiment, the supervision program located at the control

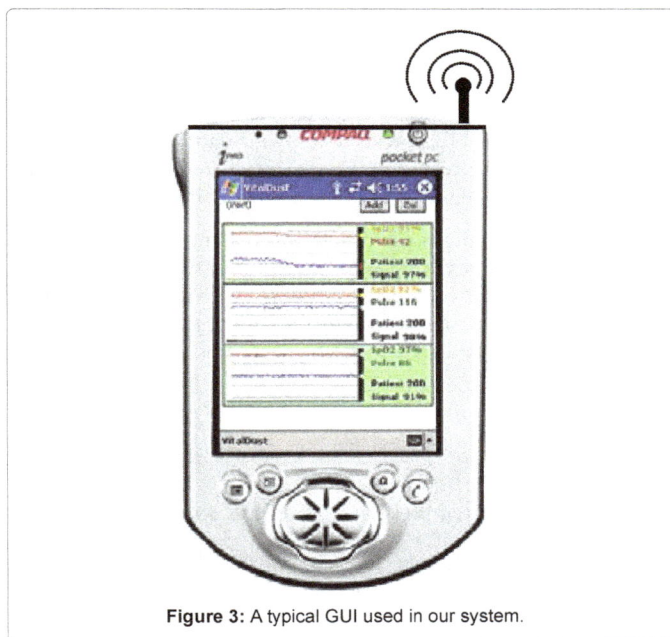

**Figure 3:** A typical GUI used in our system.

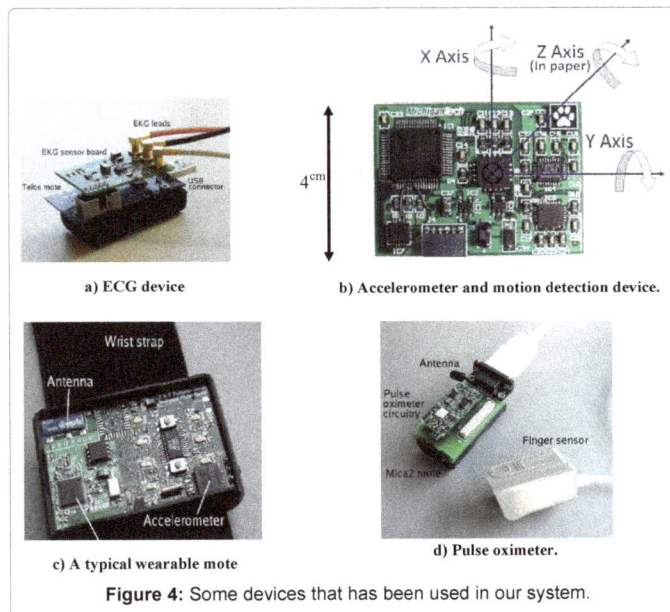

a) ECG device

b) Accelerometer and motion detection device.

c) A typical wearable mote

d) Pulse oximeter.

**Figure 4:** Some devices that has been used in our system.

station correctly displays the location of a mobile resident by polling the MySQL database for motion events.

A ten days experiment showed a robust system with some straight forward communications from front to backend of the system. We hope our system will greatly enhance quality of life, health, and security for those in assisted-living communities.

## Discussion

Data association is a way to know "who is doing what?" in a system without biometric identification and with multiple actors present, such as an assisted-living community. It permits us to recognize the right person among others when he is responsible for a triggered event. This is indispensable for avoiding medical errors in the future and properly attributing diagnostics. Consequently, dedicated sensors and data association algorithms must be developed to increase quality of data.

When the data association mechanisms are not sufficient, or integrity is considered critically important, some functionalities of the system can be disabled this preserves only the data which can claim a high degree of confidence. In an environment where false alarms cannot be tolerated, there is a tradeoff between accuracy and availability.

The system is monitoring and collecting patient data that is subject to privacy policies. For example, the patient may decide not to reveal the monitored data of certain sensors until it is vital to determine a diagnosis and therefore authorized by the patient at the time of a visit to a doctor. Security and privacy mechanisms must be throughout the system.

## References

1. Kinsella K, Phillips DR (2005) Global aging: the challenge of success. Population Bulletin 60: 1-44.

2. Stanford V (2002) Using pervasive computing to deliver elder care. IEEE Pervasive Computing 1: 10-13.

3. Stankovic JA, Cao Q, Doan T, Fang L, He Z et al. (2005) Wireless sensor networks for in-home healthcare: potential and challenges. In: High Confidence Medical Device Software and Systems (HCMDSS) Workshop.

4. Ng HS, Sim ML, Tan CM, Wong CC (2006) Wireless technologies for telemedicine. BT Technology Journal 24: 130-137.

5. Meyer S, Rakotonirainy A (2003) A survey of research on context-aware homes. In: Australasian Information Security Workshop Conference on ACSW Frontiers, Darlinghurst, Australia, pp: 159-168.

6. Orwat C, Graefe A, Faulwasser T (2008) Towards pervasive computing in health care – a literature review. BMC Medical Informatics and Decision Making 8: 26.

7. 7. Chan M, Esteve D, Escriba C, Campo E (2008) A review of smart homes-present state and future challenges. Computer Methods and Programs in Biomedicine 91: 55-81.

8. Sneha S, Varshney U (2009) Enabling ubiquitous patient monitoring: model, decision protocols, opportunities and challenges. Decision Support Systems 46: 606-619.

9. Koch S, Hagglund M (2009) Health informatics and the delivery of care to older people. Maturitas 63: 195-199.

10. Shin KG, Ramanathan P (1994) Real-time computing: a new discipline of computer science and engineering. Proceedings of the IEEE 82: 6-24.

11. Hande Alemdar, Cem Ersoy (2010) Wireless sensor networks for healthcare: A survey. Computer Networks 54: 2688-2710.

12. Bonito P (2003) Wearable sensors/systems and their impact on biomedical engineering. IEEE engineering in medicine and biology magazine 18-20.

13. Polastre J, Szewczyk R, Culler D (2005) BTelos: Enabling ultra-low power wireless research. In Proc. 4th Int. Conf. Inf. Process.Sensor Netw, pp: 364-369.

14. IEEE Standard for information technology (2003) telecommunications and information exchange between systems, local and metropolitan area networks, specific requirements-Part 15.4: Wireless medium access control (MAC) and physical layer (PHY) specifications for low-rate wireless personal area networks (LR-WPANs).

15. Atmel Corporation AT86RF230: Low power 2.4 GHz transceiver for ZigBee, IEEE 802.15.4, 6LoWPAN, RF4CE and ISM applications.

16. Texas Instruments (2006) 2.4 GHz IEEE 802.15.4/ZigBee-ready RF transceiver.

# 15

# Probabilistic Approach to Scheduling Divisible Load on Network of Processors

**Manar Arafat[1], Sameer Bataineh[2]\* and Issa Khalil[3]**

[1]*Department of Computer Science, An-Najah National University, Nablus, Palestine*
[2]*Faculty of Computer and Information Technology, Jordan University of Science and Technology, Jordan*
[3]*Qatar Foundation, Qatar*

## Abstract

Divisible Load Theory (DLT) is a very efficient tool to schedule arbitrarily divisible load on a set of network processors. Most of previous work using DLT assumes that the processors' speeds and links' speeds are time- invariant. Closed form solution was derived for the system under the assumption that the processors' speed s and the links' speeds stay the same during the task execution time. This assumption is not practical as most of distributed systems used today have an autonomous control. In this paper we consider a distributed system (Grid) where the availability of the processors varies and follows a certain distribution function. A closed form solution for the finish time is derived. The solution considers all system parameters such as links' speed, number of processors, number of resources (sites), and availability of the processors and how much of power they can contribute. The result is shown and it measures the variation of execution time against the availability of processors.

**Keywords:** Divisible load theory; Scheduling; Availability; Distributed systems

## Definitions and Notations

$N$: The number of available data sources (sites)

$M$: The number of available nodes to process the grid tasks

$S_i$: Data source (site). $i=1...N$

$P_i$: Available nodes to process the grid tasks. $i=1...M$

$L$: Total load at the originating load node

$cs_i$: Computational speed of node $i$ in the network

$w_i$: A constant that is inversely proportional to the computational speed of node $_i$. $w_{i=}1/cs_i$

$z_{ij}$: A constant that is inversely proportional to the link speed between site $i$ and node $j$.

$f_i$: The fraction of the load $L$ assigned to data source $i=1 ... N$

$L_i$: The load assigned to data source $i=1, ... N$

$\alpha_{ij}$: The fraction of load assigned from site $i$ to node $j$.

$T_{cp}$: The time it takes the $i^{th}$ node to process the entire load when $w_i=1$. The entire load can be processed on the $i^{th}$ processor in time $w_i T_{cp}$.

$T_{cm}$: The time it takes to transmit the entire load over a link when $z_i=1$. The entire load can be transmitted over the $i^{th}$ link in time $z_i T_{cm}$.

$T_{ij}$: Time to finish processing $\alpha_{ij}$.

$Pj(t)$: The probability of finding processor j is willing to contribute its whole computing power at time $t$ during $\alpha_{ij} w_j T_{cp}$, where $(i=1...N)$ and $(j=1...M)$.

$T_{ij}^c$: The time it takes to send $\alpha_{ij}$.

$T_{ij}^p$: The time it takes to process $\alpha_{ij}$.

$T_{finish}$: The minimum expected finish time for the total load.

$C_i$: The equivalent communication power for site $i$, $(i=1,2, ... N)$.

$C_T$: The total communication power in the system.

## Introduction

Geographically distributed heterogeneous systems become very powerful and popular in the past decade. Good examples of such systems are Clusters, Grids and Clouds. Grid can be viewed as a distributed large-scale cluster computing. From another perspective, it constitutes the major part of Cloud Computing Systems in addition to thin clients and utility computing [1-4]. Hence, Grid computing has attracted many researchers [5]. The interest in Grid computing has gone beyond the paradigm of traditional Grid computing to a Wireless Grid computing [6].

There are many attempts to find an analytical solution for scheduling load on the nodes (processors) on those systems. Queuing theory is a very famous tool participated to find analytical solution on such system [7,8]. Divisible Load Theory (DLT), deterministic in nature, was also used and proved that it is very much the same as Markov Chain Modeling [9]. However, DLT has shown that it is an excellent tool to schedule independent jobs on a Grid originated from multiple resources [7,10-14].

As Scheduling plays an important role in determining the performance of Grids, there are many algorithms in literature that discuss the scheduling on Grids [15-23]. It was shown that finding an

**\*Corresponding author:** Sameer Bataineh, Faculty of Computer and Information Technology, Jordan University of Science and Technology, Jordan
E-mail: samir@just.edu.jo

analytical solution for general scheduling problem in a Grid is a very difficult task [24].

There are several attempts to use the DLT to model scheduling arbitrarily divisible load on the Grid [25-27]. However, communication time is rarely considered [11]. In [7], communication time is studied but not in dividing the load, so the transfer input time of the load was not part of the model. Though all parameters of the system were considered in [10], the paper did not provide a closed form analytical solution for the finish time.

In our previous work in [28], we managed to come up with closed form solution for the minimum finish time of executing an arbitrarily divisible application on the Grid taking into consideration the communication time and the computation time simultaneously. In [28], we assumed that the nodes are always available to execute the grid task. In distributed environments, this assumption is unrealistic. A node may not be willing to contribute its whole computing power during the time span of a grid task execution.

The objective of this paper is to develop an analytical model to distribute the grid load from multiple sites to all nodes in the Grid such that the load is executed in a minimum time. The model deals with the dynamic availability of each node to serve the Grid tasks. We will derive closed form solutions for the load fraction of each node, and the minimum expected finish time of the total load. The solution considers all system parameters such as the links' speed, number of processors, number of resources (sites), and availability of the processors and how much of their power they can contribute.

The rest of the paper is organized as follows: In section 5 we present the model of the system, in section 3 we present the notations used throughout the paper. The system equations are in section 6. The results and discussion are in section 7. Finally the concluding remarks are in section 8.

## System Model

The system has three types of nodes: originating load node, N data sources (or sites) $S_i$, $i=1,... N$, and $M$ available nodes $P_i$, $i=1...M$ to process the grid tasks. Nodes have different speeds $w_i$, $i=1,2,...M$. The links that connect the data sources with the nodes have also different speeds $z_{ij}$, $i=1,2,...N$, $j=1,2,...M$.

The system works as shown in Figure 1. The originating node receives the total load L and distributes it to N available sites. The fraction of the load to be assigned to each site is $f_i$, $i=1 ... N$. It follows that the share of the load L to be assigned to data source $i$ is given by $L_i=f_iL$, $i=1...N$. Then, each data source will distribute its load $L_i$ to the M available nodes for processing. The fraction of the load to be assigned to each node $j$ from site $i$ is $\alpha_{ij}$ ($i=1, 2,.... N$) & ($j = 1, 2,...M$).

In section 6, first we obtain the fraction of load that has to be assigned to each site from the load originating node. Each site will be assigned a fraction that depends on the speed of the links in the network. Second, we obtain the fraction of load that has to be assigned from each site to each of the available nodes in the network. The fractions depend on the speed of the links, the speed of the nodes and the dynamic availability of each node in the grid.

Our analytical model guarantees that the total load is executed in a minimum time. We derive closed form solution for the minimum expected finish time of the total load.

## System Equations

Using the Equivalent Processor and Communication Link Concept, which was first introduced in [29], we can replace the set of $M$ nodes in Figure 1 with a single node, which has an equivalent computing power. In other words, if the $M$ nodes are all removed and replaced by the equivalent node, the performance of the system will be exactly the same. We can also replace the M links from each data source by one link which has an equivalent communication power $C_i$, $(i=1,2, ... N)$ as depicted in Figure 2. The value of $C_i$ is given by:

$$c_i = \sum_{j=1}^{M} \frac{1}{Z_{ij}} \qquad i=1 ... ... . N \tag{1}$$

The total communication power in the system is

$$c_T = \sum_{j=1}^{M} \frac{1}{Z_{1j}} + \sum_{j=1}^{M} \frac{1}{Z_{2j}} + ...... \sum_{j=1}^{M} \frac{1}{Z_{Nj}} = \sum_{i=1}^{N} C_i \tag{2}$$

The distribution of the total load L from the originating node to the N sites depends on the links speed to the equivalent node. Each site $S_i$ will be assigned a fraction that depends on the equivalent communication power $C_i$, $(i=1,2, ... N)$ and the total communication power in the system $C_T$. Consequently, the fraction of the load to be assigned to site $i$ is $f_i = \dfrac{c_i}{c_T}$ for $i=1 ... N$, such that $\sum_{i=1}^{N} f_i = 1$. It follows that the share of the load $L_i$ to be assigned to each site $i =1...N$ is given by:

$$L_i=f_iL \tag{3}$$

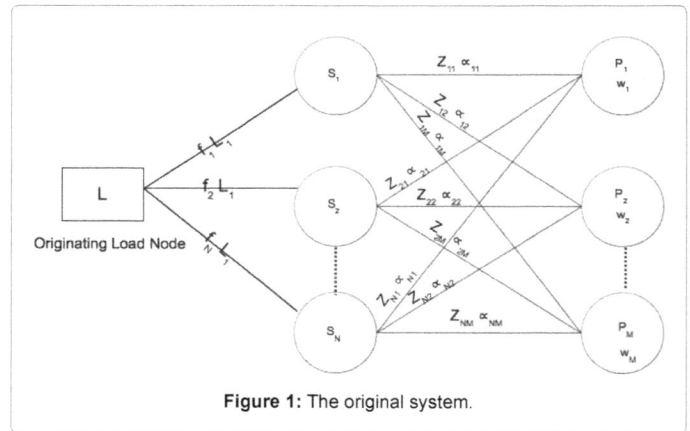

**Figure 1:** The original system.

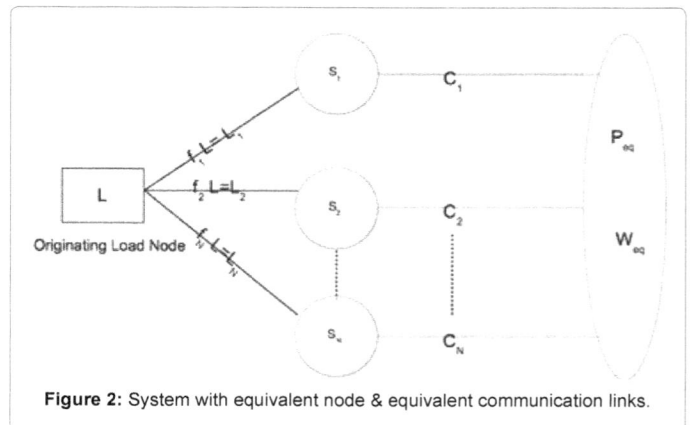

**Figure 2:** System with equivalent node & equivalent communication links.

Next, we obtain the fraction of load that has to be assigned from each site to each of the available nodes in the network. The assumption that computational resources at nodes are dedicated to Grid tasks is impractical. Each of the M nodes may not be willing to contribute its whole computing power during the time span of a grid task execution. Each node $P_j$ $(j=1...M)$ will be assigned a load fraction that depends on the speed of the links, the speed of the nodes and the dynamic availability of each node in the grid. So it is not possible to take advantage of the equivalent power concept at this stage.

In Figure 3, each data source has the illusion as if it is the only data source that is distributing its load to the $M$ available nodes. In other words, we can view the system as $N$ multiples of a single data source as shown in Figure 4.

In Figure 4, each site will distribute its load to M available nodes such that the total load is executed in a minimum time. The solution is based on the optimality principle [30]. Optimality solution assumes that all processors finish at the same time. It is analytically proved that a minimal solution time is achieved when the computation by each node finishes at the same time [31]. Intuitively, this is because otherwise, the processing time could be reduced by transferring some fractions of load from busy nodes to idle nodes [32-36].

We now derive closed form solution for the load fraction of each node, and the minimum expected finish time of the total load. The solution is based on the optimality principle [37]. Optimality solution assumes that all processors finish at the same time.

In general, the time to complete execution of $\alpha_{ij}$ on node $j$ consists of the communication time $T_{ij}^c$ and processing time $T_{ij}^p$ divided by the probability of finding node j available during $T_{ij}^p$

$$T_{ij} = T_{ij}^c + \frac{T_{ij}^p}{p_j(t)} \qquad \text{Where } i=1,2,.... \quad N; j=1,2,...M \qquad (4)$$

$$T_{ij} = \alpha_{ij} Z_{ij} T_{cm} + \frac{\alpha_{ij} w_j T_{cp}}{P_j(t)} \qquad (5)$$

Let $i=1$, Then the system equations are:

$$T_{ij} = \alpha_{ij}(Z_{1j} T_{cm} + \frac{w_j T_{cp}}{P_j(t)}) \qquad j=1,2,... M \qquad (6)$$

Applying the optimality criterion that all processors should stop

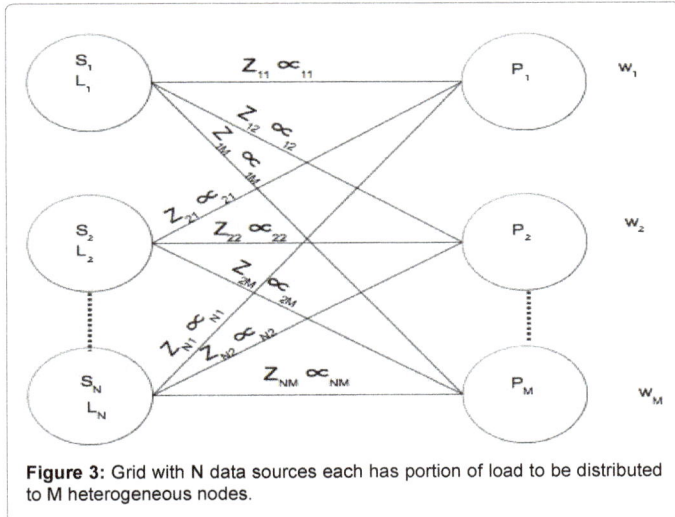

Figure 4: Grid system as viewed by a single data source.

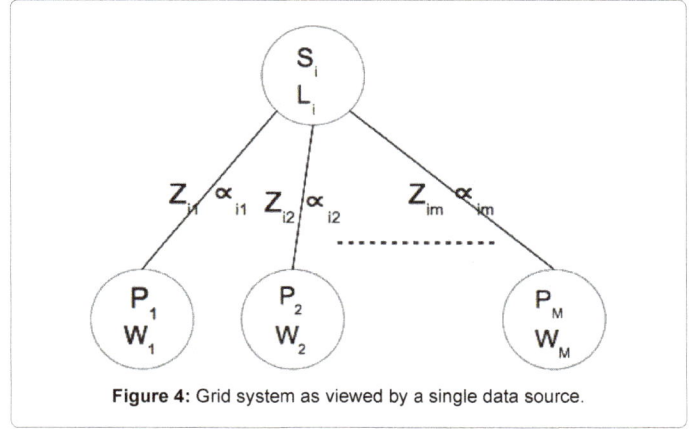

**Figure 5:** The expected finish time against the estimate of probability for different M.

computing at the same time [31,32]:

$$\text{Where } T_{1j} = T_{1(j+1)} \text{ where } j=1,2,....M\text{-}1 \qquad (7)$$

From equation (7), we obtain equations (8), (9), and (10) respectively.

$$\alpha_{12} = \alpha_{11} \frac{P_2(t)(Z_{11} T_{cm} P_1(t) + w_1 T_{cp})}{P_1(t)(Z_{12} T_{cm} P_2(t) + w_2 T_{cp})} \qquad (8)$$

$$\alpha_{1j} = \alpha_{11} \frac{P_j(t)(Z_{11} T_{cm} P_1(t) + w_1 T_{cp})}{P_1(t)(Z_{1j} T_{cm} P_j(t) + w_j T_{cp})} \qquad (9)$$

......

$$\alpha_{1(M-1)} = \alpha_{1M} \frac{P_{M-1}(t)(Z_{1M} T_{cm} P_M(t) + w_M T_{cp})}{P_M(T)(Z_{1(M-1)} T_{cm} P_{(M-1)}(t) + w_{(M-1)} T_{cp})} \qquad (10)$$

Now we can write all the above equations as a function of $\alpha_{11}$ and the parameters of the Grid as follows:

$$\alpha_{12} = \alpha_{11} \frac{P_2(t)(Z_{11} T_{cm} P_1(t) + w_1 T_{cp})}{P_1(t)(Z_{12} T_{cm} P_2(t) + w_2 T_{cp})} \qquad (11)$$

$$\alpha_{13} = \alpha_{11} \frac{P_3(t)(Z_{11} T_{cm} P_1(t) + w_1 T_{cp})}{P_1(t)(Z_{13} T_{cm} P_3(t) + w_3 T_{cp})} \qquad (12)$$

In general,

$$\alpha_{1j} = \alpha_{11} \frac{P_j(t)(Z_{11} T_{cm} P_1(t) + w_1 T_{cp})}{P_1(t)(Z_{1j} T_{cm} P_j(t) + w_j T_{cp})} \qquad \text{Where } j=1,2,....,M \qquad (13)$$

Obviously, the summation of all fractions of the load must equal 1.

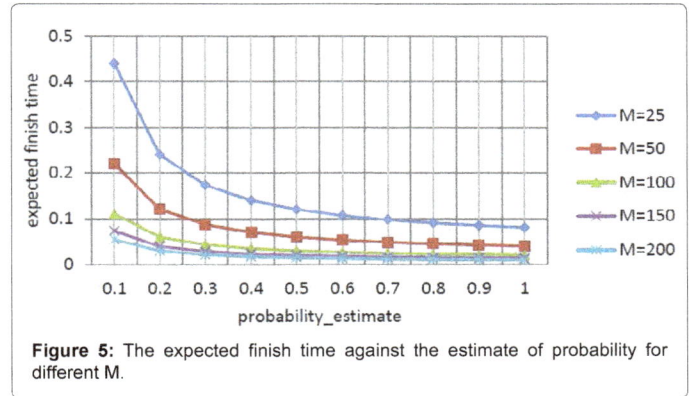

**Figure 3:** Grid with N data sources each has portion of load to be distributed to M heterogeneous nodes.

$$\alpha_{11} + \alpha_{12} + \alpha_{13} + \dots + \alpha_{1M} = 1 \qquad (14)$$

Using equations (13) and (14), we can find the exact value of $\alpha_{11}$ as a function of grid parameter

$$\alpha_{11} = \cfrac{1}{1 + \cfrac{Z_{11}T_{cm}P_1(t) + w_1 T_{cp}}{P_1(t)} \left[ \sum_{j=2}^{M} \cfrac{p_j(T)}{(Z_{1j}T_{cm}P_j(t) + w_j T_{cp})} \right]} \qquad (15)$$

Since $\alpha_{11}$ is found, $\alpha_{12}, \alpha_{13} \dots \alpha_{1M}$ can be calculated using equation (13) and the minimum expected finish time can also be determined using equations (6) and (15):

$$T_{finish} = T_{11} \cfrac{1}{\sum_{j=1}^{M} \cfrac{p_j(T)}{(Z_{1j}T_{cm}P_j(t) + w_j T_{cp})}} \qquad (16)$$

## Results and Discussion

In this section we demonstrate that the analytical model obtained is correct in that it produces results that are in agreement with intuitively expected results. The results are compared with the deterministic systems discussed in all previous work [7,10-14]. The total load in the system is found in $N$ load sources. Each load source will distribute its load fraction $f_k$ $(k=1\dots N)$ to the $M$ available nodes in the network. The availability of each node $i$ $(i=1\dots M)$ varies with time and is based on certain probability $P_j(t)$. If $P_j(t)=1$, the total computing power of node $j$ is available and if $P_j(t)= 0.5$, half of the total computing power of node j can be lend to the newly arriving job. In general $P_j(t)$ of the computing power of node j can be devoted to a newly arriving jobs. The probability function can be derived from a certain realistic distribution. An estimate of $P_j(t)$ can be measured over a reasonable period of time. As an example let us think of each processor as an M/M/1 queue where the customers arrive according to independent Poisson processes with rate $\lambda$ and the service times of all customers are exponentially distributed with mean $1/\mu$. We require that $\rho = \lambda/\mu < 1$, since, otherwise, the queue length will explode. The quantity $\rho$ is the fraction of time the server is working. From the equilibrium probabilities we can derive expressions for the mean number of customers in the system $E(l)$, which is:

$$E(l) = \frac{\rho}{1-\rho}$$

If the power of a processor $j$ is uniformly distributed among all jobs in the queue of processor $j$, then the speed of the processor $j$ at an instant of time devoted to each job served by processors $j$ is given by the following equation

$$p_j(t) = \frac{1}{E_j(l)} = \frac{1-\rho}{\rho}$$

One can further consider an M/M/1 system serving different types of customers. As a simple example assume that there are two types of jobs only, type 1 and 2, but the analysis can easily be extended the situation with more types of customers. Type 1 and type 2 customers arrive according to independent Poisson processes with rate $\lambda_1$, and $\lambda_2$ respectively. The service times of all customers are exponentially distributed with the same mean $1/\mu$. One can easily find a $p_j(t)$ even if a preemptive resume priority rule is applied.

In general an explicit solution for the probabilities $p_j(t)$ is

$$p_j(t) = \sum_{k=0}^{k=n} k \times p_{kj}(t)$$

Where $p_{kj}(t)$ is the probability that at time $t$ there are $k$ jobs in the system which can handle at most n jobs $p_{kj}(t)$ equation can be found in [38].

In the following we use the equations derived in the paper to study the effect of different parameters on the system performance. The results are compared with the deterministic case. Deterministic results are those obtained with $(t)=1$.

In Figures 5 and 6, we assume that the M available nodes in the network will have the same probability of availability to serve the grid loads ($0 < probability\_estimate \leq 1$). Figure 5 relates the expected finish time to the estimate of probability, for M=25, 100, 150 and 200, assuming that $z_{ij} =1$ and $wj=1$ (for $i=1\dots N$ and $j=1\dots M$), and $T_{cp}=T_{cm}=1$.

Figure 5 shows that as the number of available nodes in the grid increases, the expected finish time will decrease. For our parameters, there will be no significant improvement after M=150. The minimum expected finish time occurs when all the nodes are available to serve the grid loads. Comparing the results with the deterministic case, when the nodes devoted all power to one job, in other words when the probability estimate =1, we notice that the expected finish time could be 5 times slower when the system has a large number of processors.

Figure 6 relates the expected finish time to the estimate of probability for M=100 and $z_{ij}=0.1, 0.5, 1, 1.5$ and $2$ (for $i=1\dots N$ and $j=1\dots M$), assuming that $w_j=1$ (for $j=1\dots M$), and $T_{cp}=T_{cm}=1$.

Figure 6 shows that as each link speed increases, the expected finish will decrease.

Figure 7 relates the estimate of probability for node $k$, ($k$: may be any node $1\dots M$), to its load fraction $\alpha_{ik}$ assigned from load source $i$, for M=25, 50, 100, and 150. Assuming that the estimate of probability of each other available node is equal to one and that $z_{ij}=1$ and $w_j=1$ (for $i=1\dots N$ and $j=1\dots M$), and $T_{cp}=T_{cm}=1$.

From Figure 7, nodes that are available for serving the grid loads

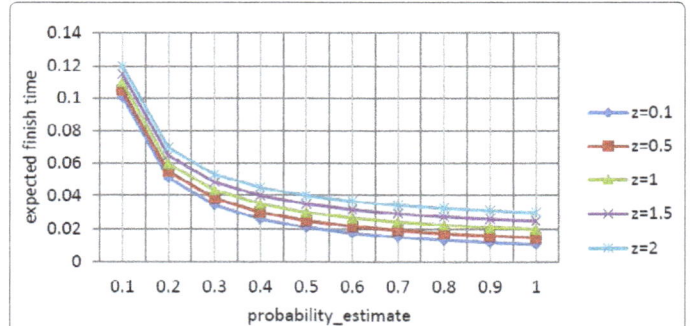

**Figure 6:** The expected finish time against the estimate of probability for different z.

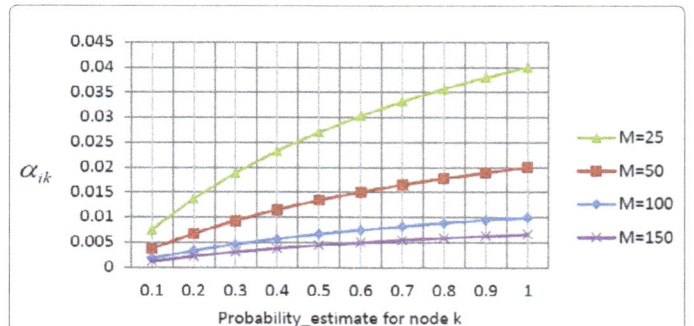

**Figure 7:** The load fraction against the estimate of probability for a node.

most of the time will be assigned larger load fractions. This is exactly what is expected because we assume that all processors must stop at the same time to abide with the optimality principle.

## Conclusion

Unlike all previous work, which using DLT, in this paper we propose a model where the processors speeds is a function of number of jobs that a node in the distributed system is in charge at time 't'. We assumed that the availability of a node is based on certain probability and it varies with time. Of course, the probabilistic function should be derived from a certain realistic distribution.

The load is found in N sites and has to be distributed and assigned to M nodes such that total load is executed in a minimum time. We derived closed form solution for the load fraction of each node, and the minimum expected finish time of the total load. The result is shown and it measures the variation of execution time against the availability of processors for different system parameters. Our next step is to find a distribution function that can apt the system behavior. This will generate an accurate estimate for the probabilities to precisely reflect the availability of the processors in the system.

## References

1. Armbrust M, Fox A, Griffith R, Joseph AD, Katz RH, et al. (2009) Above the Clouds: A Berkeley View of Cloud Computing. Technical Report No. U CB/EECS-2009-28.

2. Nandgaonkar SV, Raut AB (2014) A Comprehensive Study on Cloud Computing. International Journal of Computer Science and Mobile Computing 3: 733-738.

3. Kathrine GJW, Ilaghi MU (2012) Job Scheduling Algorithms in Grid Computing - Survey. International Journal of Engineering Research & Technology 1: 7.

4. Qureshi MB, Dehnavi MM, Allah NM, Qureshi MS, Hussain H, et al. (2014) Survey on Grid Resource Allocation Mechanisms. Journal of Grid Computing 12: 399-441.

5. Toma I, Iqbal K, Roman D, Strang T, Fensel D, et al. (2007) Discovery in grid and web services environments: A survey and evaluation. Multiagent and Grid Systems 3: 341-352.

6. Manvi SS, Birje MN (2009) Wireless Grid Computing: A Survey. IETE Journal of Education 50: 119-131.

7. Byuna EJ, Choia SJ, Baikb MS (2007) MJSA Markov job scheduler based on availability in desktop grid. Future Generation Computer Systems 23: 616-622.

8. Tian G, Xiao C, Xu X, Gao1 CQ, Nuslati, et al. (2010) Grid Workflow Scheduling Based on Time Prediction of Queuing Theory. Proceedings of the 2010 IEEE, International Conference on Information and Automation pp: 36-39.

9. Moges MA, Robertazzi TG (2009) Grid Scheduling Divisible Loads from Two Sources. Computers and Mathematics with Applications 58: 1081-1092.

10. Abdullah M, Othman M, Ibrahim H, Subramaniam S (2009) Closed form Solution for Scheduling Arbitrarily Divisible Load Model in Data Grid Applications: Multiple Sources. American Journal of Applied Sciences 6: 626-630.

11. Moges M, Robertazzi TG (2003) Optimal Divisible Load Scheduling and Markov Chain Models. Proceedings of the 2003 Conference on Information Sciences and Systems, The Johns Hopkins University, Baltimore, MD, USA.

12. Jia J, Veeravalli B (2010) Scheduling Multisource Divisible Loads on Arbitrary Networks. IEEE Transactions On Parallel And Distributed Systems 21: 520-531.

13. Li X, Veeravalli B (2010) PPDD: scheduling multi-site divisible loads in single-level tree networks. Cluster Computing 13: 31-46

14. Robertazzi TG, Yu D (2006) Multi-source grid scheduling for divisible loads. In: Proceedings of the 40th Annual Conference on Information Sciences and Systems (CISS'06) pp: 188-191.

15. Moise D, Moise I, Pop F, Cristea V (2008) Resource Co Allocation for Scheduling Tasks with Dependencies, in Grid. The Second International Workshop on High Performance in Grid Middleware HiPerGRID.

16. Fujimoto N, Hagihara K (2003) Near-optimal dynamic task scheduling of precedence constrained coarse- grained tasks onto a computational grid. Second International Symposium on Parallel and Distributed Computing 2003 Proceedings pp: 80-87.

17. Abdullah M, Othman M, Ibrahim H, Subramaniam S (2007) An integrated approach for scheduling divisible load on large scale data grids. In: Computational Science and Its Applications—ICCSA, 4705: 748-757.

18. Mamat A, Lu Y, Deogun J, Goddard S (2012) Scheduling real-time divisible loads with advance reservations. Real-Time Systems 48: 264-293.

19. Lin X, Mamat A, Lu Y, Deogun J, Goddard S (2010) Real-time scheduling of divisible loads in cluster computing environments. Journal of Parallel and Distributed Computing 70: 296-308.

20. Chuprat S (2010) Divisible Load scheduling of real-time task on heterogeneous clusters. In Proceedings of the IEEE International Symposium on Information Technology (ITSim '10) 2: 721-726.

21. Ghanbari S, Othman M, Leong WJ, Abu Bakar MR (2014) Multi-criteria based algorithm for scheduling divisible load. In: Proceedings of the 1st International Conference on Advanced Data and Information Engineering (DaEng '13), Lecture Notes in Electrical Engineering, pp: 547-554.

22. Othman SM, Ibrahim H, Subramaniam S (2012) New method for scheduling heterogeneous multi-installment systems. Future Generation Computer Systems 28: 1205-1216.

23. Yu C, Marinescu DC (2010) Algorithms for divisible load scheduling of data-intensive applications. Journal of Grid Computing 8: 133-155.

24. Bataineh S (2008) Divisible Load Distribution in a Network of Processors. Journal of Interconnection Networks 9: 31-51.

25. Shokripour A, Othman M (2009) Survey on Divisible Load Theory and its Applications. International Conference on Information Management and Engineering pp: 300-304.

26. Shokripour A, Othman M (2009) Survey on Divisible Load Theory and Its Applications. International Conference on Information Management and Engineering.

27. Ghanbari S, Othman M (2014) Comprehensive Review on Divisible Load Theory: Concepts, Strategies, and Approach. Mathematical Problems in Engineering Volume 2014: 460354

28. Bataineh S, Khalil I (2013) Scheduling divisible load on Wireless Grid with communication Delay. 3rd International Conference on Wireless Communications and Mobile Computing (MIC- WCMC 2013), Valencia, Spain

29. RichardsonP, Sieh L, Elkateeb AM (2001) Fault-tolerant adaptive scheduling for embedded real-time systems. IEEE Micro 21: 41-51.

30. Kleinrock L (1975) Queueing Systems: Theory. Wiley, New York.

31. Sohn J, Robertazzi TG (1996) Optimal Load Sharing for a Divisible Job on a Bus Network. IEEE Transactions on Aerospace and Electronic Systems 32: 34-40.

32. Mingsheng S (2008) Optimal algorithm for scheduling large divisible workload on heterogeneous system. Applied Mathematical Modelling 32: 1682-1695.

33. Yu D, Robertazzi TG (2003) Divisible Load Scheduling for Grid Computing. In: PDCS'2003, 15th Int'l Conf. Parallel and Distributed Computing and Systems.

34. Robertazzi TG (2003) Ten reasons to use divisible load theory. Computer 36: 63-68.

35. Bharadwaj V, Ghose D, Robertazzi TG (2003) Divisible load theory: a new paradigm for load scheduling in distributed systems. Cluster Comput. 6: 7-17.

36. Wang X, Wang Y, Meng K (2014) Optimization Algorithm for Divisible Load Scheduling on Heterogeneous Star Networks. JOURNAL OF SOFTWARE 9: 1757:1766

37. Bharadwaj V, Ghose D, Mani V (1992) A study of optimality conditions for load distribution in tree networks with communication delays. Dept. of Aerospace Engineering, Indian Institute of Science, Bangalore, India, Technical Report 423: 02-92.

38. Bataineh S, Hsiung T, Robertazzi TG (1994) Closed Form Solutions for Bus and Tree Networks of Processors Load Sharing A Divisible Job. IEEE Tran. on Computers 43: 1184-1197.

# 16

# Monitoring Internet Access along with Usage of Bandwidth Using Intrusion Detection System

**Rajagopal D\* and Thilakavalli K**

*Department of Computer Applications, K.S. Rangasamy College of Arts and Science, Tiruchengode, Namakkal, Dt-637 215, India*

## Abstract

New Approach to observe web Access beside Usage of information measure victimization Intrusion Detection System could be a comprehensive web use observation and news utility for company networks. It takes advantage of the very fact that the majority companies give web access through proxy servers, like MS ISA Server, MS Forefront TMG, WinGate, WinRoute, MS Proxy, WinProxy, EServ, Squid, Proxy Plus, and others. Whenever the user accesses several websites, transfer files or pictures, these actions were logged. The system processes these log files to supply system directors a good vary of report-building choices. It might build reports for individual users, showing the list of internet sites visited, beside elaborate classification of web activity (downloading, reading text, viewing footage, observation movies, paying attention to music, and working). This technique might produce comprehensive reports with analysis of overall information measure consumption, building easy-to-comprehend visual charts that show the areas wherever wasteful information measure consumption has eliminated. This new approach is employed to observation the web information measure employed by the user. victimization this technique will simply decide that user fill the information measure most heavily, when, and what specifically they transfer, what proportion time they pay on-line, and what knowledge transfer traffic they produce.

**Keywords:** Bandwidth; Intrusion detection system; Proxy server; Network traffic; Network security

## Introduction

Network managers and directors should get on guard against all types of unauthorized network use [1]. Intrusion Detection System observation network traffic for activity that falls inside the definition of prohibited activity for the network [2]. When found, the Intrusion can alert directors and permit them to require corrective action, interference access to vulnerable ports, denying access to specific science addresses, or move down services wont to enable attacks, this fast-alert capability makes an Intrusion Detection System the front-line weapon within the network directors war against hackers. The planned Intrusion Detection System put in on the server that serves native hosts and users over web. There are four actors within the system monitor, user, network and computer user. User sends request to the server over the web or native space Network and Intrusion Detection System can analyze the packets received by the server. This Intrusion Detection System detects each internal and external intrusion. If it detects any intrusion then it alerts computer user (Figure 1).

The planned approach permits centralized monitor of Users web access prevents personal usage of company information measure, reduces the web expenses, very easy-to-use. It will begin observation user's couple of minutes once, once the installation complete, works with all trendy proxy servers, permits the generation of a good range of reports and diagrams, that show the potency of proxy server usage, and it's a task computer hardware to automates the creation and delivery of reports to authorize personnel.

### Advantages of the approach

- Allows centralized observation of Users web access

- Prevents personal usage of company information measure and reduces the web Expenses

- Extremely easy-to-use; will begin observation users couple of minutes once, once the installation is complete

- Works with all trendy proxy servers and permits the generation

of a nice range of reports and diagrams, that show the potency of Proxy Server usage

- Task computer hardware to automates the creation and delivery of reports to authorize personnel.

## Intrusion Detection System

Intrusion Detection refers to the method of observation the system for unauthorized access incidents, which might be the violation of the protection policy, system use policy, or the other security standards [2-5]. On the opposite hand, An Intrusion Prevention System (IPS) prevents unauthorized access incidents from being prosperous. To safeguard the system from any attacks, Intrusion Detection and Prevention System (IDPS), that give an utterly machine-controlled observation service, deployed on the systems [3]. Most of the IDPS systems log the incident on every occasion an attack on the system is that observe and notifies the administration of the system so all necessary actions will taken to avoid such incidents once more within the future. The directors of the system also can put together the IDPS to observe the violations of the tip user policies and alternative unauthorized activities [3].

## Intrusion Detection System Types

### Network intrusion detection system

It is a freelance platform, which identifies intrusions by examining network traffic and monitors multiple hosts [2]. Network Intrusion

**\*Corresponding author:** Rajagopal D, Department of Computer Applications, K.S. Rangasamy College of Arts and Science, Tiruchengode, Namakkal Dt-637 215, India, E-mail: sakthiraj2782007@gmail.com

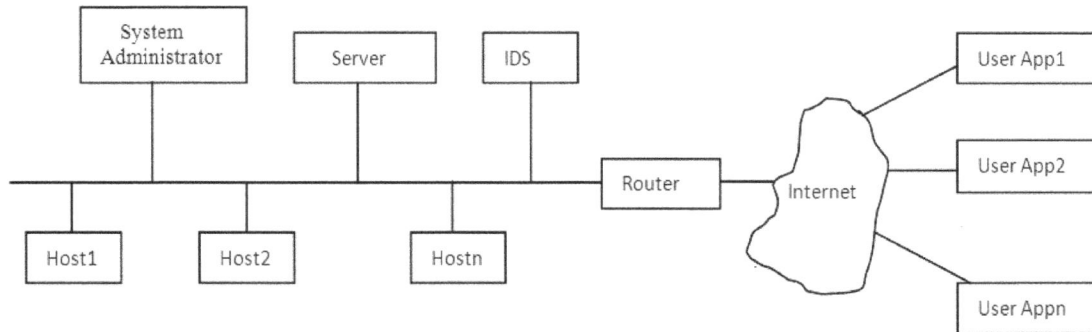

**Figure 1:** IDS working place in the network.

Detection System gain access to network traffic by connecting to a network hub, network switch designed for port mirroring, or network faucet [6,7]. A NIDS is place on a network to investigate traffic in search of unwanted or malicious events. Network traffic designed on varied layers; every layer delivers knowledge from one purpose to a different. The OSI model and transmission management protocol (TCP)/IP model show however, every layer stacks up. Inside the TCP/IP model, rock bottom link layer controls however, knowledge flows on the wire, like dominant voltages and the physical addresses of hardware, like Mandatory access Control (MAC) addresses (Figure 2).

The web layer controls address routing and contain the science stack. The transport layer controls knowledge flow and checks knowledge integrity. It includes the communications protocol and user datagram protocol (UDP). Lastly, the foremost sophisticated however most acquainted level is that the application layer, that contains the traffic employed by programs. Application layer traffic includes the online (hypertext transfer protocol [HTTP]), file transfer protocol (FTP), email, etc. Most NIDSs observe unwanted traffic at every layer; however concentrate totally on the applying layer.

Two main element sorts comprise a NIDS: appliance and software package solely. A NIDS appliance could be a piece of dedicated hardware: it is solely operated to be IDS. The Operating System (OS), software, and the network interface cards (NIC) are enclose within the appliance. The second element kind, software package solely, contains the entire IDS software package and generally the OS; but the user provides the hardware. Software-only NIDSs are usually more cost-effective than appliance-based NIDS because of they are doing not give the hardware; but, a lot of configuration is need, and hardware compatibility problems could arise.

**Component types:** Two main component types comprise a NIDS: appliance and software only. A NIDS appliance is a piece of dedicated hardware: its only function is to be an IDS. The operating system (OS), software, and the network interface cards (NIC) are included in the appliance. The second component type, software only, contains all the IDS software and sometimes the OS; however, the user provides the hardware. Software-only NIDSs are often less expensive than appliance-based NIDS because they do not provide the hardware; however, more configuration is required, and hardware compatibility issues may arise.

With an IDS, the "system" component is vital to efficiency. Often a NIDS is not comprised of one device but of several physically separated components. Even in a less complicated NIDS, all components may be present but may be contained in one device. The NIDS is usually made of components identified, but more specifically, the physical

components usually include the sensor, management sever, database server, and console.

**Sensor:** The sensor or agent is the NIDS component that sees network traffic and can make decisions regarding whether the traffic is malicious. Multiple sensors are usually placed at specific points around a network, and the location of the sensors is important. Connections to the network could be at firewalls, switches, routers, or other places at which the network divides.

**Management server:** The management server will make decisions based on what the sensor reports. It can also correlate information from several sensors and make decisions based on specific traffic in different locations on the network.

**Database server:** Database servers are the storage components of the NIDS. From these servers, events from sensors and correlated data from management servers can be logged. Databases are used because of their large storage space and performance qualities.

**Console:** As the user interface of the NIDS, the console is the portion of the NIDS at which the administrator can log into and configure the NIDS or to monitor its status. The console can be installed as either a local program on the administrator's computer or a secure Web application portal. Traffic between the components must be secure and should travel between each component unchanged and unviewed. Intercepted traffic could allow a hacker to change the way in which a network views an intrusion.

**Inline:** An Inline NIDS sensor is placed between two network devices, such as a router and a firewall [8]. This means that all traffic between the two devices must travel through the sensor, guaranteeing that the sensor can analyze the traffic. An inline sensor of an IDS can be used to disallow traffic through the sensor that has been deemed malicious. Inline sensors are often placed between the secure side of the firewall and the remainder of the internal network so that it has less traffic to analyze (Figure 3a and 3b).

As the above diagram on the left, the computer running snort is connected to the firewall, the firewall would be configured with a "mirror" or "spanning" port that would essentially copy all of the incoming and outgoing traffic to a particular interface for the snort software to monitor. This way, any suspicious traffic passing the border of the network would be subject to examination.

As the above diagram on the right, the traffic is passing directly through the snort machine, using two Ethernet interfaces. This is an excellent solution for environments where a mirror port is unavailable,

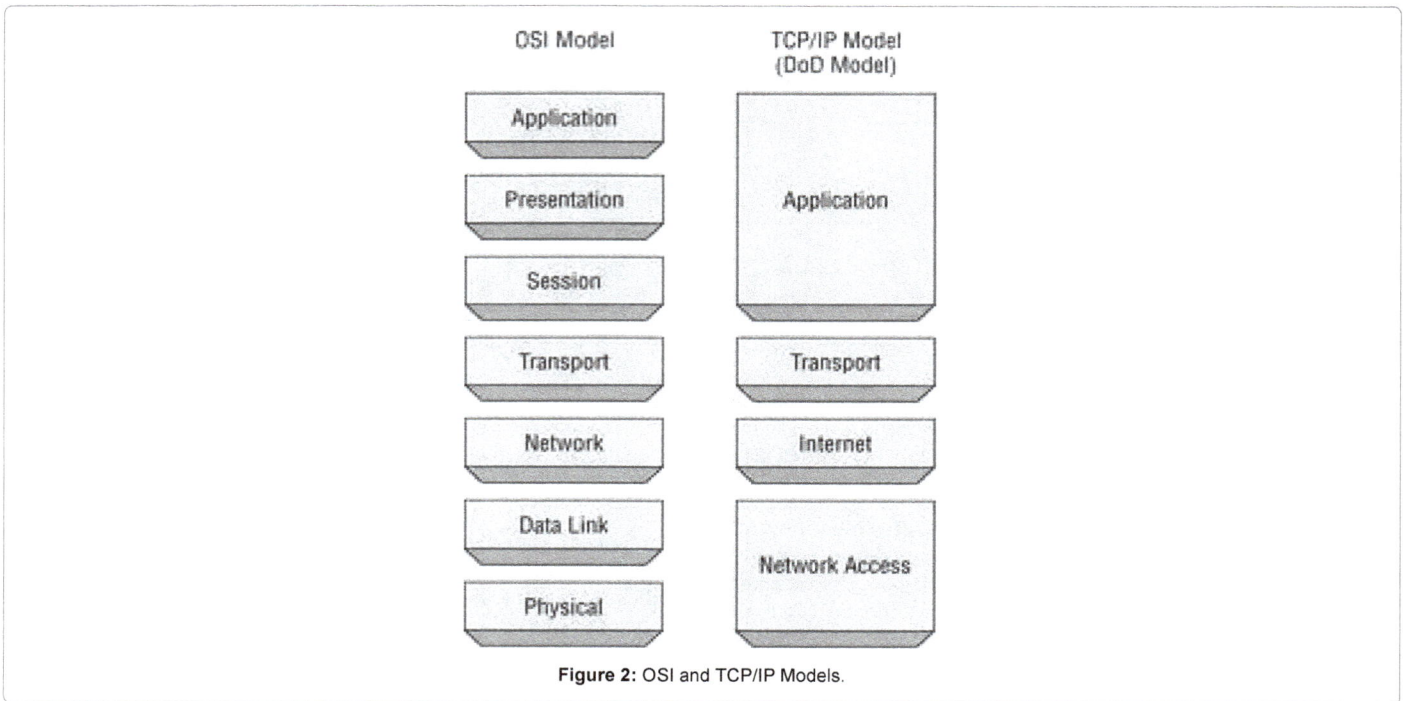

**Figure 2:** OSI and TCP/IP Models.

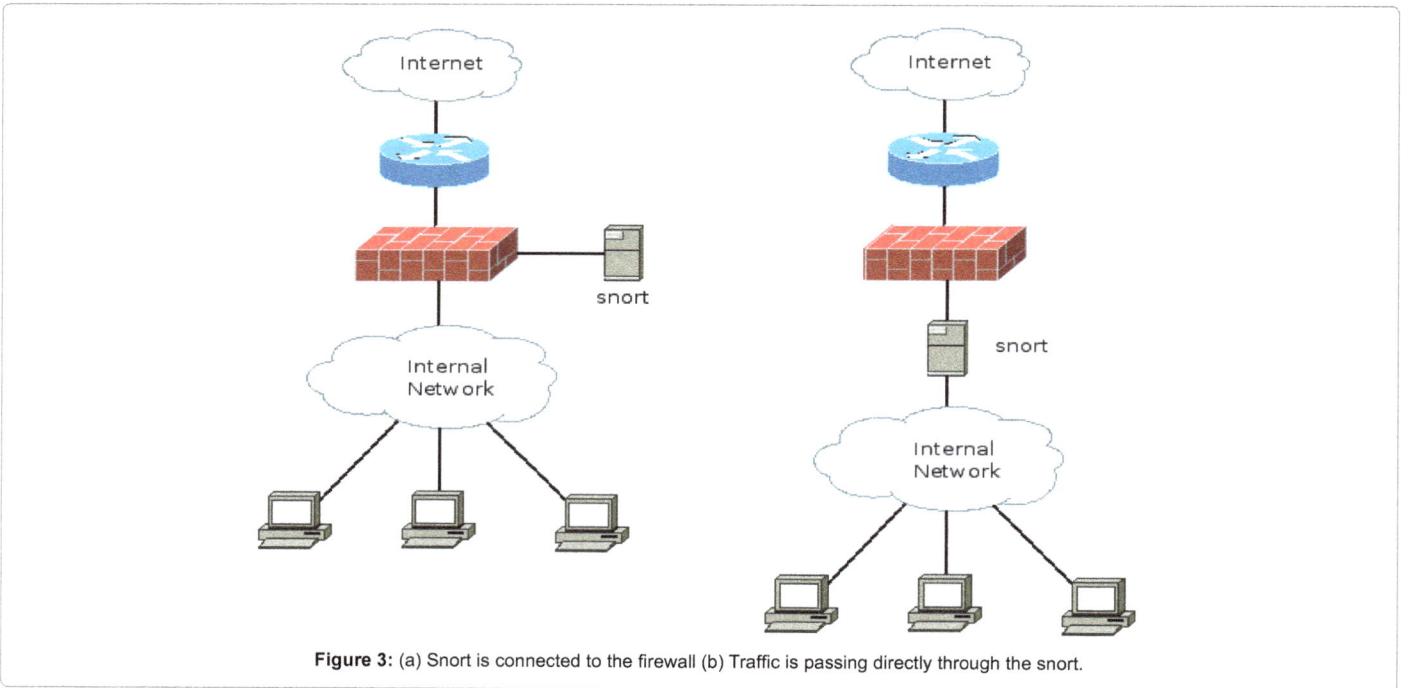

**Figure 3:** (a) Snort is connected to the firewall (b) Traffic is passing directly through the snort.

such as a branch office using low-end networking equipment that can't provide the additional interface. (It is important to note that a NIDS should be carefully placed within the network topology for maximum effectiveness. If two of the client machines in these diagrams are passing suspicious traffic between them, the snort machine will not notice; it only sees traffic destined for the Internet. It is always possible, of course, to run multiple NIDS systems and tie all of the alerts into one console for processing so as to eliminate these blind spots.)

Because of its large install base, rules for detecting new threats

are constantly being produced and published for free usage on sites like Emerging Threats. If the administrator want to be alerted when a host on the network is connecting a known botnet controller [9], for example, the up-to-the-minute rules for this can be downloaded from ET. The same goes for signatures of new worms and viruses, command-and-control traffic, and more. So a NIDS is an excellent tool for detecting when a host on the network has been compromised or is otherwise producing suspicious traffic.

**Passive:** A passive sensor analyzes traffic that has been copied from the network versus traffic that passes through it. The copied traffic can come from numerous places.

**Spanning port:** Switches often allow all traffic on the switch to be copied to one port, called a spanning port. During times of low network load, this is an easy way to view all traffic on a switch; however, as the load increases, the switch may not be able to copy all traffic. Also, if the switch deems the traffic malformed, it may not copy the traffic at all the malformed traffic that may be the type the NIDS sensor must analyze.

**Network tap:** A network tap copies traffic at the physical layer. Network taps are commonly used in fiber-optic cables in which the network tap is inline and copies the signal without lowering the amount of light to an unusable level. Because network taps connect directly to the media, problems with a network tap can disable an entire connection.

### Advantages of network based ids

- Monitor network for port scans.

- Monitor network for malicious activity on known ports such as http port 80.

- Identify varied varieties of spoofing attacks.

- Does not impact network performance.

- Increased tamper resistant.

- Operating systems independent.

### Drawbacks of network based ids

- Packets lost on flooded networks.

- Reassemble packets incorrectly

- No understanding of O/S specific application protocols like SMB.

- No understanding of obsolete network protocols.

- Does not handle encrypted data.

## Host-based intrusion detection system

It consists of an agent on a host that identifies intrusions by analyzing system calls, application logs, file-system modifications (binaries, secret files, capability databases, Access management lists etc.) and alternative host activities and state [7]. Whereas a NIDS watches the traffic on a network phase, HIDS watches the activities of a selected host. A common open-source HIDS system is OSSEC, named as a contraction of Open Source Security [10].

Much like a NIDS, the position of HIDS software package must arrange carefully. The user doesn't need to receive an alert on every occasion a file is reaching on a digital computer. The system has fastidiously to put together and the monitored behaviors cropped to on eliminate false alarms and make sure the true security problems are noticed and alerted properly [11].

### Advantages of host based ids

- Monitor events native to a host, and might observe prosperous or failure of attacks that cannot be seen by a network-based IDS.

- Operate in the setting during which network traffic encrypted.

- Unaffected by switched networks and is independent of topology.

- Monitor system specific activities like file access, user access, etc.

- Provide thorough data gathered via logs and audit; for instance, Kernel logs know who the user is.

- No extra hardware is required to implement Host based IDS solution.

- When Host-based IDSs operate on OS audit trails, they will facilitate observe attacks that involve software package integrity breaches.

### Drawbacks of host based ids

- Host based IDS are tougher to manage, as data should be designed and managed for each host singly.

- Host based IDS's network blind and cannot detect a network scans or other such surveillance that targets entire network.

- If the host is compromised, collected log by Host based IDS can be subverted.

- Disabled by bound denial-of-service attacks.

- Uses operating system audit trails as an information source. The number of information is large and might need extra native storage on the system.

- Inflict performance deficiency on monitored host.

## Stack-based intrusion detection system

This type of system consists of an evolution to the HIDS systems. The packets examined as they are going through the TCP/IP stack and, therefore, it is not necessary for them to work with the network interface in promiscuous mode. This reality makes its implementation to be dependent on the Operating System [12].

This can be latest IDS technology and varies dramatically from vendor to vendor. Stack Based IDS works by integration closely with the TCP/IP stack, allowing packets to be watch as they traverse their way up the OSI Layers. Observation the packets during this means permits the IDS to drag the packets from the stack. To be complete Stack- Based ID ought to watch each incoming and outgoing network traffic on a system. By monitoring network packets destined just for a simple host, the principle is to create the IDS have sufficiently low overhead so each system on the network will run Stack-Based IDS.

## Intrusion Detection Techniques

### Statistical anomaly-based ids

A statistical anomaly-based IDS determines traditional network activity like what type of bandwidth is mostly used, what protocols are used, what ports and devices usually connect to each other- and alert the administrator or user once traffic is detected that is abnormal(not normal) [13]. The anomaly-based detection model detects the attacks based on the profiles. The profiles contain the patters or the traditional behavior during which the system is used. The profiles supported specific users, networks, or the applications. They are making by monitoring the system use over a period, known as the evaluation period. This model compares the present activities with the profiles to get the abnormal activity in progress, which regularly is an attack. Since the system use and also the network use do not seem to be not static and always contain some variation over time, the profile should additionally modify consequently. Therefore, once the creation of the profiles in the evaluation period, an IDPS changes the profiles over time. The samples of the profiles mentioned below. A user profile contains an email activity of 5%. Once the IDPS victimization anomaly-based detection model senses that the e-mail activity on the system is over 5%, it will consider about it an attack. Over the past few weeks, on average a user performs that open, read, and write operations on the file system for 2% of the time. Once the IDPS detects a growth within the file access operations, it reports an attack incident. The advantage of the anomaly-based detection model is that it is able to detect even unknown attacks by comparison the present abnormal events with something that is considering traditional. Further, this model also can be more efficient than the signature-based model given that there are a large range of signatures to compare inside the signature-based detection model. On the other hand, the attack incidents that anomaly-based model produces don't seem to be terribly specific and it takes some efforts by the administrator to pin - purpose the basis of the attack. Additionally this model subject to a "slow attack". In this type of attack, the attacker first finds out the threshold between the normal and abnormal activities. The attacker then would slowly attack the system making sure that the activities during the attack do not reach the threshold which results into the anomaly-based detection model not detect the attack.

### Advantages of anomaly based ids

- Identify any potential attack.

- Identify attacks that have not seen before, or close variants to antecedently well-known attacks.

### Drawbacks of anomaly based ids

- Normal will amendment over time, introducing the requirement for periodic on-line preparation of the behavior profile, result either in inaccessibility of the intrusion detection system or in extra false alarms.

- Current implementations give high false alarms.

- Requires experience to work out what triggered an alarm.

## Signature-based ids

Signature based IDS monitor's packets within the Network and compares with pre-configured and pre-determined attack patterns called signatures [14]. The difficulty is that there will be lag between the new threat discovered and Signature being applied in IDS for detecting the threat. Throughout this lag time, IDS are going to be unable to spot the threat.

The signature refers to the pattern during which a antecedently well-known attack was performed. The signature-based detection methodology is that the method of comparison the present events with the signatures. The signature-based detection model produces terribly specific attack event reports as oppose to the anomaly-based detection model [5]. The disadvantage of a signature-based detection model is its inability to detect new unknown attacks since the system does not have any signature entry within the system for the new attacks.

### Benefits of signature based ids

- Provides terribly low false alarms as compare to Heuristic based IDS.

- Provides detail contextual analysis providing steps for preventive or corrective actions.

### Drawbacks of signature based ids

- It is tough to assemble data concerning well-known attacks and keeping up-to-date with new vulnerabilities.

- Signatures and corrective recommendations are generalized; so it makes it tougher to grasp them.

- Knowledge concerning attacks is extremely centered, keen about the operating system, version, platform, and application.

- Signature/Pattern based IDS are more popular and commercially used than Heuristic/Anomaly detection based IDS. Major vendors such as ISS offer network based and host based signature detection.

## Stateful protocol inspection

Stateful protocol inspection is similar to anomaly based detection, but it can also analyze traffic at the network and transport layer and vender-specific traffic at the application layer, which anomaly-based detection cannot do.

## Differences between Detection Techniques

### Misuse detection vs. Anomaly detection

A misuse detection system, also known as a Signature-Based Intrusion Detection System identifies intrusions by watching for patterns of traffic or application data presumed to be malicious. These types of systems are presumed to be able to detect only 'known' attacks. However, depending on their rule set, signature-based IDSs can sometimes detect new attacks which share characteristics with old attacks.

An Anomaly-Based Intrusion Detection System identifies intrusions by notifying operators of traffic or application content presumed to be different from 'normal' activity on the network or host [15,16]. Anomaly-based IDSs typically achieve this with self-learning. In anomaly detection, the system administrator defines the baseline, or normal, state of the network's traffic load, breakdown, protocol, and typical packet size. The anomaly detector monitors network segments to compare their state to the normal baseline and look for anomalies [17].

### Network-based vs. Host-based systems

In a network-based system, or NIDS, the sensors are located at choke points in the network to be monitored, often in the network borders. The sensor captures all network traffic flows and analyzes the content of individual packets for malicious traffic. In a host-based system, the sensor usually consists of a software agent which monitors all activity of the host on which it is installed. Hybrids of these two types of system also exist. A Network Intrusion Detection System is an independent platform which identifies intrusions by examining network traffic and monitors multiple hosts. Network Intrusion Detection Systems gain access to network traffic by connecting to a hub, network switch configured for port mirroring, or network tap. An example of a NIDS is Snort. A Host-based Intrusion Detection System consists of an agent on a host which identifies intrusions by analyzing system calls, application logs, file-system modifications (binaries, password files, capability/acl databases) and other host activities and state. A Hybrid Intrusion Detection System combines both approaches. Host agent data is combined with network information to form a comprehensive view of the network. An example of a Hybrid IDS is Prelude [18].

## Passive system vs. reactive system

In a passive system, the IDS sensor detects a potential security breach, logs the information and signals an alert on the console. In a reactive system, the IDS respond to the suspicious activity by logging off a user or by reprogramming the firewall to block network traffic from the suspected malicious source, either autonomously or at the command of an operator. Though they both relate to network security, IDS differs from a firewall in that a firewall looks out for intrusions in order to stop them from happening. The firewall limits the access between networks in order to prevent intrusion and does not signal an attack from inside the network. IDS evaluate a suspected intrusion once it has taken place and signal an alarm. IDS also watch for attacks that originate from within a system.

## Network Bandwidth

Network bandwidth refers to the amount of knowledge being transmitted across a network at any given purpose in time. Network bandwidth will decrease if devices that change networked communications fail. Network bandwidth might be forced by each hardware and software package limitations. Optimizing the out there Network information measure could be a primary responsibility of network administrators.

## Current Techniques in Network Security

A number of techniques have been invented in the past few years to help a system administrator in strengthening the security of a single host or the whole computer network.

## Audit trails

"A chronological record of system activities that is sufficient to enable the reconstruction, reviewing, and examination of the sequence of environments and activities surrounding or leading to an operation, a procedure, or an event in a transaction from its inception to final results." defined by the National Computer Security Center

Audit trail can be used in determining whether an unexpected or unauthorized behavior has occurred in a system.

## Fire wall

A recent trend in network security enhancement involves the use of firewall, which is a collection of filters and gateways that shield trusted networks within a locally managed security perimeter from the external untrusted networks [13].

## Screening router

Screening router is a router, which in addition to forwarding packets likes a normal router, also examines data in the packets, and applies some predefined access control policies on the packets to determine whether they can be forwarded to the next hop or should be discarded.

## Application gateway

Application gateways provide one or more of the following functionality: relay, proxy and server filter. Relay gateway, passes the data between the two sides of a firewall system. In some special environments, like a company using "local" IP addresses (i.e. visible only within the company) for internal network, a relay gateway should also provide the function for translating these addresses before they are sent out.

Proxy is of most importance to a firewall system, for most of access control policies are enforced through application proxies. Usually, a proxy gateway is application specific. When a client program inside the firewall requires a connection with an outside server, an application proxy on the firewall will handle the request first.

Server filter works in the opposite direction as an application proxy. It handles the incoming connection requests from external network to the internal servers. Similar to inetd under most UNIX systems, a server filter acts as a proxy for multiple internal application servers. When receiving a connection request, the server filter dispatches it to the corresponding application server.

As an application gateway examines more data in a network packet than a screening router does, it provides more power in network intrusion detection and prevention. On the downside, it requires more system resources and more processing time.

## Design and Methodology

There is a need for testing directly two different types of NIDS, in other words anomaly and signature detection. There is also a need for online evaluation using realistic simulated traffic generation, as opposed to offline evaluation using network traffic traces.

The initial objective of this experiment was to set up a testbed for two different types of NIDS and generate simulated background traffic as well as range of exploits. Such an experiment proved too generic since the choice of exploits ready to use was relatively small compared to the amount of existing exploits. Instead, the experiment was split in two: a first experiment on the learning window variation of an anomaly IDS, and a second experiment testing two different types of IDS in a specific, well-defined scenario. States that a valid computer security experiments should consist of only one varying component. The following sections define an overview of the testbeds used.

## Network architecture

The basic network architecture is composed of a router and switch. Since the background traffic is split into two IP address ranges according to whether it is client or server, two **V**irtual **L**ocal **A**rea **N**etwork ( VLAN ) are needed to mimic internal and external traffic. The router is used in this configuration in order to route traffic between both VLAN s. Another essential piece of configuration is setting up the **S**witched **P**ort **A**nalyser (SPAN) port on the switch in order to send all traffic crossing the switch to the IDS station for analysis.

## Training window experiment

This experiment aims at demonstrating any effects that a variation of training window length could have on an anomaly-based IDS .This station is linked to a switch and monitors all network traffic crossing this network device. The traffic generator is used to produce benign background traffic for anomaly system profile creation. Ideally, this station should produce this type of traffic with a traffic generation simulation tool. The exploit generator is used after the profile generation phase has been completed.

Finally, the router is used in this testbed in order to route or discard the generated traffic and make all the connections appear real to the IDS. To sum up, the anomaly-based IDS will be subjected to different learning periods. For each period, the profile created will be stored for the next experimental phase, being the attack detection. After this profile generation phase, the IDS will be subjected to a mix of benign background traffic and malicious traffic (Figure 4).

### Scenario

This experimental scenario is realized in order to focus this research on a specific type of threat rather than only available threats. For this scenario, the following background is to be considered. A renowned bank branch computer networks system includes an FTP server hosting highly sensitive data, such as bank account details for example.

Currently, the corporation uses the following tools as part of its security system: a firewall at the boundary with the untrusted network, antivirus on local machines and built-in IDS on the gateway router analyzing traffic going in and out of the trusted network, such as on a Cisco router. The security at the boundary of the corporate network is optimum, but the security staff is worried about threats present on the inside of their network. Insiders threats are multiple, although here

the main concern is data theft from the FTP server, only protected by a username and password combination.

In order to protect the branch from such a threat, the security staff would like to know which type of NIDS would be best suited in this case (Figure 5).

There are some considerations to take into account with regards to this scenario. Sensitive data would probably not be stored on a simple FTP server in a real case environment, and the access to such a server would probably be more securely controlled. The simplistic approach used in this scenario is chosen due to time considerations and testing focus: a FTP is faster to breach than a more secure server,

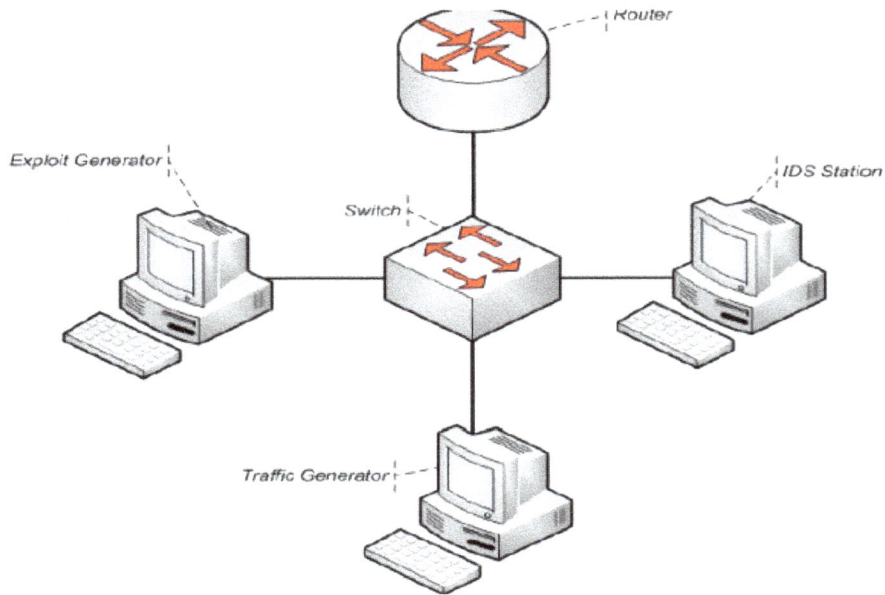

**Figure 4:** Training window experiment design.

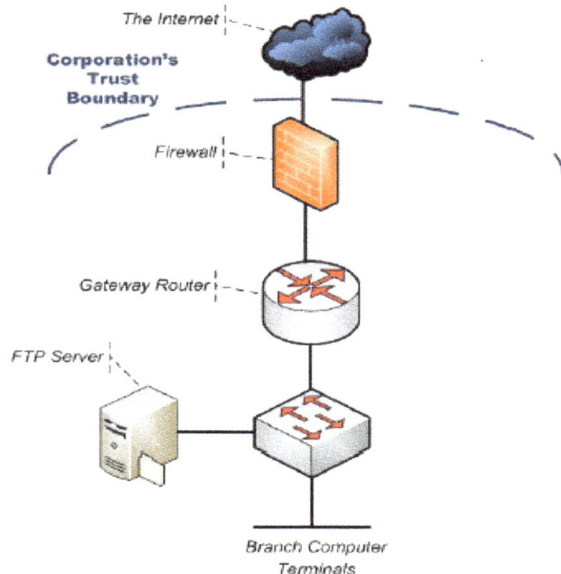

**Figure 5:** Scenario bank computer system.

and this experiment is focused on NIDS rather than server security. This scenario can show which IDS is best for such an environment. Adding additional security measures could not do any harm but make the whole system more secure (Figure 6).

The testbed is very similar to the one used in the learning window experiment, with the difference of an extra machine running an FTP server. The scenario threat is data theft. Data theft is usually composed of a collection of exploits following the steps highlighted. In this case, the data theft consists of:

- Live IP addresses scan
- Portscan on live addresses
- Brute force attack on FTP username/password
- Data theft

The exploit generation station will carry out every step of a data theft threat.

## Distributed intrusion detection system

The Distributed Intrusion Detection System (DIDS) is an intrusion detection mechanism which was developed jointly by the University of California at Davis, Lawrence Livermore Laboratory, Haystack Laboratory and the U.S. Air Force. DIDS combines attributes of a network monitoring system with the system-level capabilities of an audit record-based combined anomaly/misuse detector. DIDS incorporates a monitor on each host, a monitor on the local area network (LAN), and a DIDS director [19].

Each host monitor consists of a host event generator and a host agent. The host event generator reviews the audit data from the host for indications of events which may be part of an attack. The DIDS host event generators also utilize user and group profiles to identify anomalous behaviors in the audit record. The information identified by the host event generator is reported to the DIDS director by the host

agent. The LAN monitor is the network equivalent of the host monitor. It includes the LAN event generator and the LAN agent.

The DIDS director forms the heart of the intrusion detection mechanism. It is composed of three components, the communications manager, an expert system and a user interface. The communications manager is receives input from each of the host monitors and from the LAN monitor and forwards the information to the expert system for analysis. The communications manager is also capable of forwarding requests for additional information from the expert system to the host monitors and the LAN monitor.

The DIDS expert system is a rule-based system which is responsible to analyzing the information received from the monitors and reporting it to the security official. The final component of the DIDS system, the user interface, allows a security official to interactively review the status of the system, receive reports from the expert system, and request additional security-related information from the system. One of the essential elements of the DIDS system is the use of a Network-user Identification (NID). This is a process of establishing an identifier for each individual when they are initially logged into the network. This is especially important because many attackers use multiple accounts to attack a network or use the interconnectivity of computer networks to attempt to disguise their identity.

Once a user has logged into the network and been assigned a NID all subsequent activity conducted by that user is attributable through the NID. While the NID offers the potential to track an intruder through a variety of hosts and possible identities, there are ways to defeat the mechanism. By logging out of the monitored domain and then reentering under a different user id, an attacker can prevent DIDS from relating the two sessions. In addition, the DIDS system probably cannot attribute two related sessions to the same user if the user passes through an unmonitored domain. These difficulties aside, an initial DIDS prototype has successfully demonstrated the ability to track users through a monitored domain. Because of the complexity of the

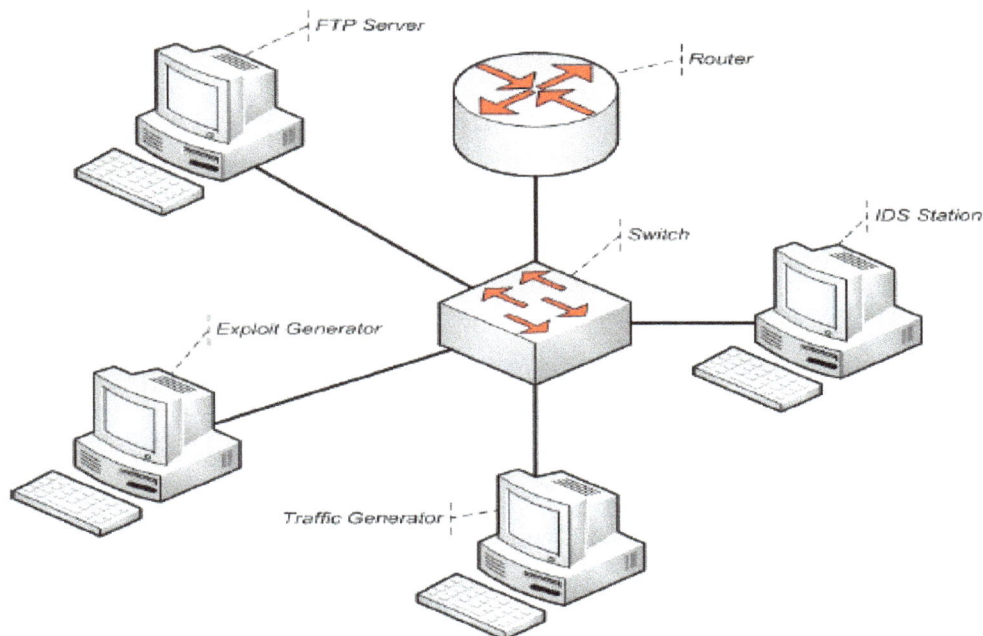

**Figure 6:** Scenario experiment design.

system and its use of audit data, DIDS retains the negative effect on the performance of the system which plagues most traditional intrusion detection systems. While this could be a significant disadvantage of the DIDS system, the innovative design of the system effectively addresses the difficulty in identifying intrusions in a networked environment.

### State transition analysis tool (stat/ustat)

The State Transition Analysis Tool (STAT) and USTAT, the variation of STAT which was designed specifically for the UNIX operating system environment, are rule-based penetration detection approaches which characterize the process of an attack on a computer system as a series of transitions from an initial state to a compromised state. The technique defines specific events, called signature actions, which occur between each of the intermediate transitions. The omission of any of the signature actions results in a failed attack on the system.

If the current pattern of activity matches an established intrusion scenario, STAT/USTAT has the ability to predict the future activities of an attacker. The ability to predict behavior offers the advantage of allowing the Security Administrator to be more confident that an actual attack is occurring prior to utilizing any countermeasures. As more of the established scenario's activities are matched, the confidence level that an attack is occurring increases. In addition, because this technique does not rely on possibly unrelated events to indicate a potential attack, the incidence of false alarms reported by the system should be significantly reduced.

Another advantage of this approach is that because STAT/USTAT selects specific audit data for confirmation of potential intrusion patterns, only a portion of the audit data is actually reviewed. This reduces the reliance of the system on the entire set of audit data, thereby reducing the required storage space and memory requirements necessary for processing an entire audit trail. While STAT/USTAT offers significant advantages in its approach to intrusion detection, the technique is unable to detect other attack-type behavior such as denial of service attacks and masquerading. Because these indications of an attack cannot be ignored by an effective intrusion detection system, a mechanism which employs STAT/USTAT would also require a complementary rule-based anomaly/misuse detection system.

### Tripwire

In November 1992, the COAST laboratory at Purdue University introduced Tripwire. Tripwire is an integrity checking program which permits a system administrator to monitor system files for addition, deletion, or modification.

Tripwire operates in one of four modes. In the database initialization mode, the program generate a database which contains all of the relevant information on the system files, including signatures. Because the baseline database is being generated based on the files which currently exist in the system, it is critical that the existing database is free of logic bombs, viruses, Trojan horses, or other attack programs. The integrity checking mode results in the creation of a new database from information contained in the configuration. The information in the new database is compared with the results contained in the original database. Any discrepancies are processed through a filter which determines which file attributes can be changed without adversely effecting the system. The remaining identified changes are then reported to the system administrator.

The final two operating modes are used to ensure that the information in the database is consistent. The database update mode calculates new signatures for those files which have been legitimately changed. In the interactive database update mode the program generates a list of those files which have been modified and updates those which are identified by the system administrator as legitimate.

Tripwire is a good tool for monitoring the status of system files. Tripwire makes no pretense of insuring the complete security of the computer system. It functions to notify system administrators of a very important indication of an intrusion. This information, combined with other security-related tools, should provide a more secure operating environment.

### Graph-based intrusion detection system (GrIDS)

Researchers in the COAST laboratory have recently proposed a novel approach to intrusion detection based on the analysis of activity graphs. The Graph-Based Intrusion Detection System (GrIDS) is designed to analyze network activity in large networks for the presence of attacks [19]. GrIDS aggregates the actions of a networks users into the activity graphs. Based on a review of the structure of these graphs the system can identify patterns which indicate intrusive behavior. Information received from other intrusion detection devices and network monitors can be included in the attributes of the activity graphs.

Individual types of graphs will be maintained in graph spaces with the GrIDS system. Because there are a number of possible attacks on the network, multiple graph spaces must be maintained. Each graph space is dependent on a specific rule set which modifies the graphs within it's graph space based on inputs to the system. GrIDS is able to analyze activity on large networks because of it's ability to model networks as a series of hierarchies. Each area within the hierarchy has a GrIDS module which is responsible for that area. Any activity which crosses area boundaries will be passed up to the GrIDS in the next higher level for resolution. The GrIDS in that level builds reduced graphs which model the underlying structure on a smaller scale. This ability to model subhierarchies allows GrIDS to monitor networks of increasing complexity. The true promise in the GrIDS system is in its ability to assist users in creating rule sets for the system. GrIDS includes a policy language which enables administrators to translate organizational policies and guidelines into rule sets which are used to analyze the network activity.

### Thumb printing

Thumb printing is a method of tracking intruders through a sequence of logins, referred to by the authors as a connection chain. While it is not intended to be an independent intrusion detection system, it could prove to be a valuable addition to other technologies. Thumb printing was developed by researchers at the University of California at Davis in response to a weakness in DIDS.

A current weakness in this approach is that it assumes that the content of the connections along the chain are the same. As a result, the use of different encryption techniques by two points would render the method useless.

### Cooperating security managers

While DIDS takes a centralized security approach to network intrusion detection, Cooperating Security Managers (CSM) decentralizes the process. A separate CSM is run on each computer which is connected to the network. Each CSM consists of six elements. The heart of the CSM is the Security Manager (SECMGR). The SECMGR receives input from the various CSM components and coordinates with CSM's on other hosts as users pass through the network. The command monitor (CMNDMON) intercepts the commands from the user and forwards them to the host intrusion detection system (IDS). While CSM requires the presence of an intrusion detection system on each host, the actual mechanism is separate from the CSM and can therefore be any intrusion detection tool. Any intrusions detected by the IDS are reported to the SECMGR.

CSM's ability to utilize a variety of intrusion detection systems also prevents the system from being limited by any of the specific approaches to intrusion detection. As new approaches are developed which more efficiently process user information, they can be incorporated into the CSM, effectively upgrading the CSM as a whole.

The network administrators, the senior management, and the users of their systems, on a scale of 1 (minimal) to 10 (very significant). The respondents considered security to be a major concern with an average score of 8.9. The reported levels of perceived significance of the other categories diminished to the typical users level of concern for security rating of 2.4.

The survey respondents to rate six types of threats in order of concern from 1 (lowest perceived threat) to 6 (most significant threat to their network). The survey results indicated that most of the respondents utilize a combination of security devices on their networks. Ninety percent of the respondents utilize the existing security features which are present in their host operating system. Some types of intrusion detection system and firewall mechanism are used in seventy-two percent of the networks. The respondents reported that they were utilizing their security mechanisms to defend their networks from external penetrations, masquerades, internal attacks, viruses and denial-of-service attacks. Seventy-seven percent of the respondents reported that their networks had been attacked in the past. A further breakdown of that group indicated that sixty-nine percent reported the attack to a superior or other authority. The same group responded that seventy-two percent were utilizing a firewall mechanism at the time that the attack occurred. Among those respondents who were using a security mechanism when attacked, sixty-three percent reported that the mechanism had reduced the severity of the attack.

This often consisted of a timely notification of the security administrator which allowed active defensive measures to be conducted before significant damage could occur to the system. Additional security measures were implemented after the attack by sixty-three percent of the group. Most of these measures consisted of improved security education and training of the users and the correction of well-known system flaws.

The results of that intrusion detection systems should be capable of identifying various types of threats, or be capable of being seamlessly incorporated with other security mechanisms which can defend against those threats not addressed by the intrusion detection system.

## Proposed Model

### Interacting with a pop3 server

Downloading an email from a POP3 server is rather straight forward. The communication with a POP3 server uses only few commands and is easily human readable. Once a connection, possibly with SSL, is established, the client needs to provide a user name and password to enter the POP3 state TRANSACTION, called 'connected' in Pop3MailClient (Figure 7).

### Error handling & tracing

To further help with the investigation of communication problems,

a Trace event is raised. It shows commands and responses exchanged between PopClient and PopServer, including warnings. It is strongly recommended to use this feature in the beginning of a project, because RFC1939 gives the server implementer great freedom. It often provides additional information which can be seen in the trace.

### Server settings

The Pop3MailClient requires server name, port, should SSL be used, username and password in the constructor and they cannot be changed. To get the demo code running, need to enter own credentials for username and password in the following line:

Pop3.Pop3MailClient DemoClient=new Pop3.Pop3MailClient("pop.gmail.com",995, true, Username@gmail.com, "password");

### Reading raw email

The method GetRawEmail returns the complete email content for one particular message number. RFC1939 specifies that only ANSI characters can be used and therefore the raw email can be easily displayed.

### Auto reconnect after server timeout

The isAutoReconnect property is set, the Pop3MailClient tries to reconnect exactly once after a timeout. That's all it usually takes, but notice that any emails marked for deletion are not deleted on the server.

### SMTP trace listener

The SMTP Trace Listener class is derived from the TraceListener class found in the System.Diagnostics library. The methods **MUST** override are:

- public override void Write(string Message)
- public override void WriteLine(string Message)
- public virtual void Fail(string);
- public virtual void Fail(string, string);
- public virtual void Close();
- public virtual void Flush();

The new flexible tracing architecture provides us the ability to

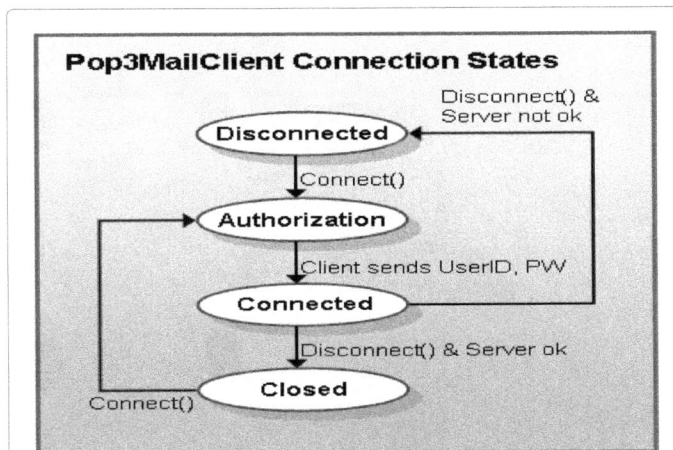

**Figure 7:** Interacting with POP3 Server.

Trace Listener. Other than TraceListener, new classes like TraceSource, TraceSwitch, and TraceFilter give us complete control over application tracing. The Trace Listener class, PortWriterTraceListener, that listens to Trace messages and sends the Trace messages to a UDP port.

**Trace listener class enumerators**

a. **Classes:**

TraceListener

TraceListenerCollection

TraceSource

TraceSwitch

TraceFilter

TraceEventCache

Trace

a. **Enumerators:**

TraceEventType

TraceLevel

TraceOptions

## Port writer trace listener

The UDP protocol is efficient and fast, but less reliable. Tracing is not a mission-critical task, and reliability is not the highest priority criteria. In fact, there are only two abstract functions in the TraceListener class, Write and Writeline. Following is a functional Trace Listener class that, if used, will log the Trace in the event log [20].

Now, PortWriterTraceListener has two custom attributes called 'destination' and 'port'. These two custom attributes will be used to make the UDP connection. Trace Listener sends Trace messages to a UDP port using the UdpClient class of the System.Net.Sockets namespace. Need to do two things, connect to the destination computer, and send data to the port.

## Traceview

The TraceView simply listens to the UDP port where PortWriterTraceListener sends the Trace messages and displays the messages in a ListView control. There are many useful features in TraceView, like logging Trace to a file, saving Trace to a file, search messages, filter messages, and copy messages to the Clipboard. SOAP Extensions to create reusable components to manage authorized users of Web Services and to track Web Services usage by those users (Figure 8). The figure below depicts the role of SOAP Extensions in the Web Services architecture:

The small solid circles in the above figure show the points in the serialization/de-serialization process where the incoming/outgoing messages can be tapped in SOAP Extensions (Figure 9).

**Architecture of SOAP extensions:** SOAP Extensions offer a rich and extensible mechanism for implementing reusable infrastructure components for Web Services. Their ease of implementation without losing flexibility has made them the "way to go" for the developer community. SOAP Extensions–based reusable components will be the backbone of Web Service offerings, and enforces business rules without distracting from the design goals defined at the start of this article. The following list provides a few broad categories where these types of components can be used:

- Data encryption
- Authentication and authorization
- Accounting/logging
- Monitoring system performance

The figure below depicts an example of this process flow (Figures 10-12):

## Bandwidth

Bandwidth performance is one of the critical requirements for every website. In today's time major cost of the website is not hard disk space but its bandwidth. So transferring maximum amount of data over the available bandwidth becomes very critical and how can use IIS compression to increase bandwidth performance. How does IIS compression work?

The user requests for a 'Home.html' page which is 100 KB size. IIS serves this request by passing the 100 KB HTML page over the wire to the end user browser (Figure 13).

When compression is enabled on IIS the sequence of events changes as follows:-

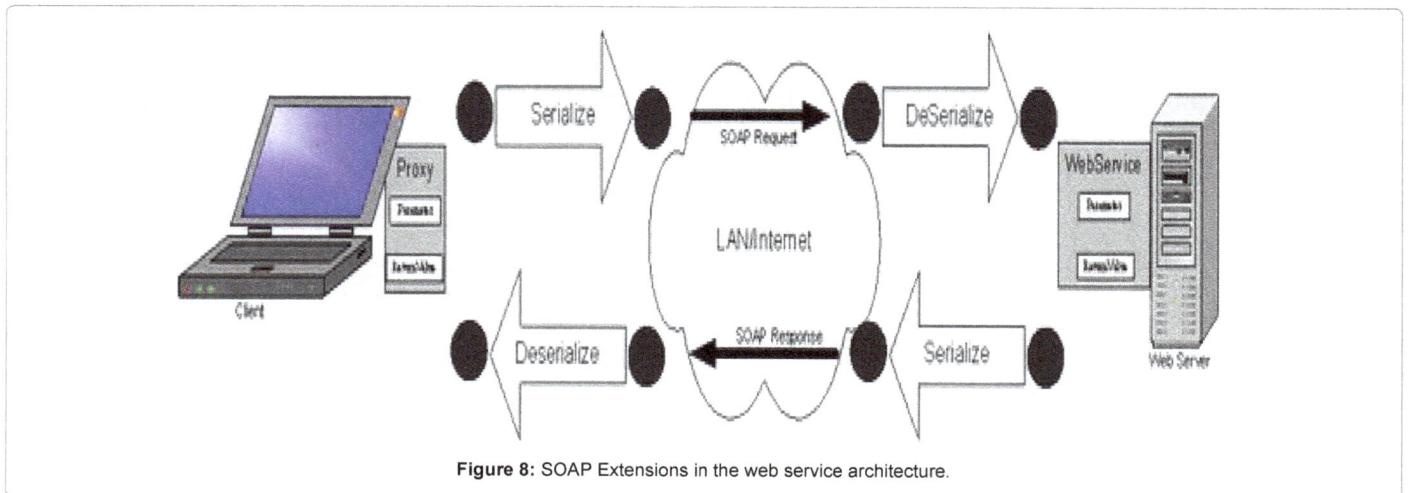

**Figure 8:** SOAP Extensions in the web service architecture.

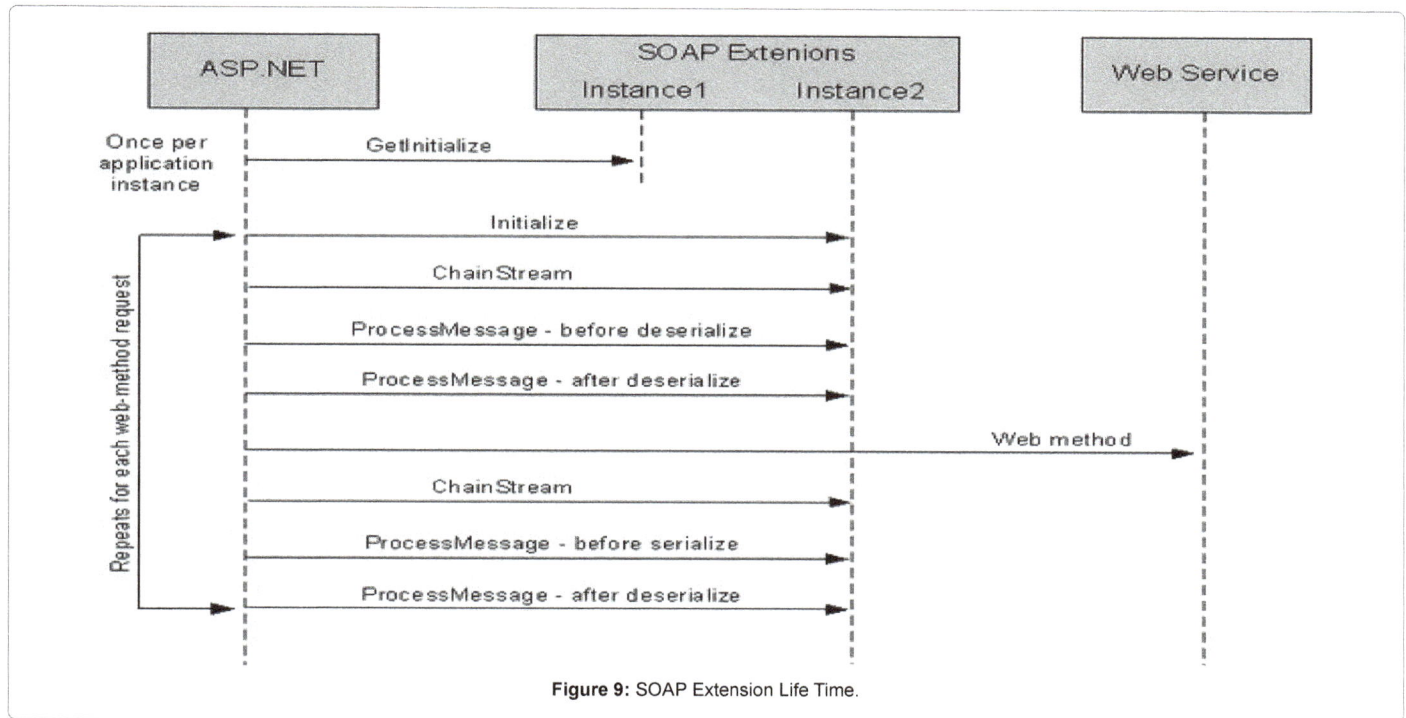

**Figure 9:** SOAP Extension Life Time.

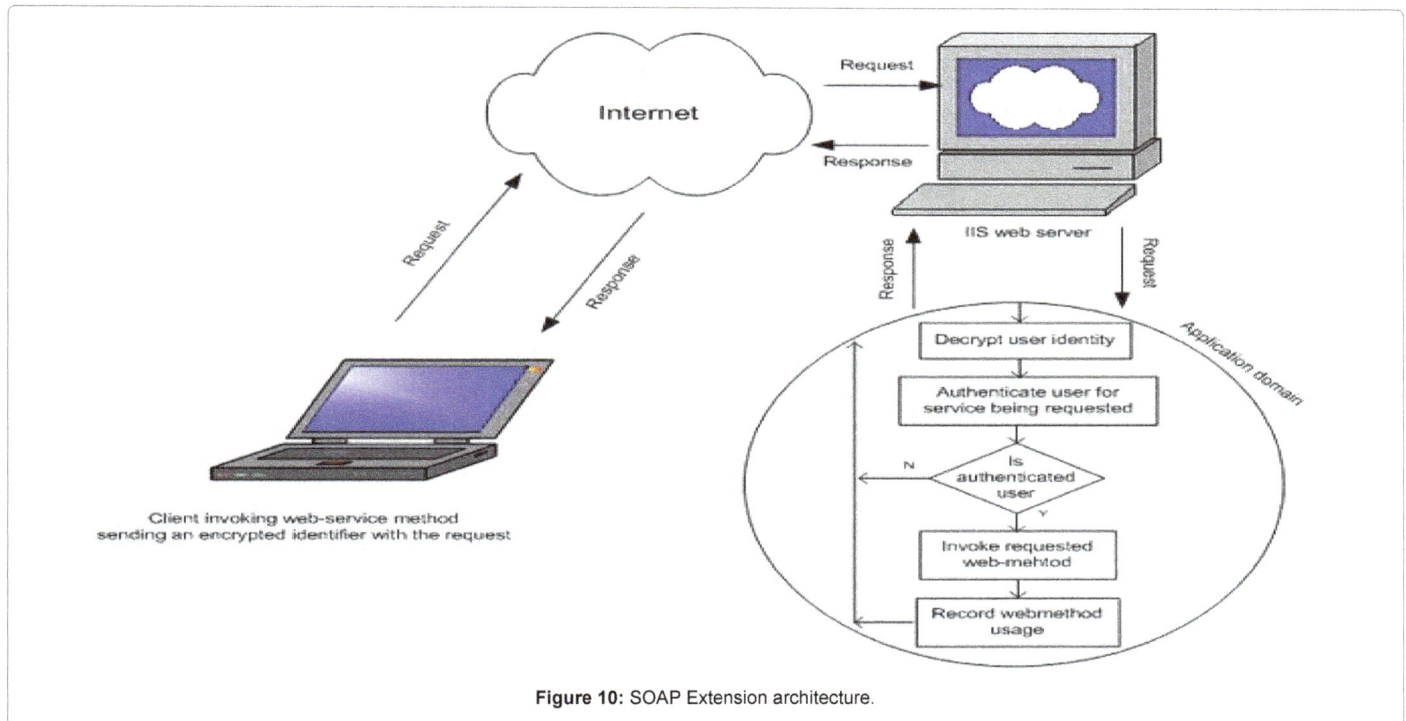

**Figure 10:** SOAP Extension architecture.

• User requests for a page from the IIS server. While requesting for page the browser also sends what kind of compression types it supports. Below is a simple request sent to the browser which says its supports gzip and deflate.

GET/questpond/index.htmlHTTP/1.1

Accept: image/gif,image/x-xbitmap, image/jpeg, image/pjpeg, application/x-shockwave-flash, application/vnd.ms-excel*/*

Accept-Language:en-us

Accept-Encoding:gzip,deflate

User-Agent:Mozilla/4.0

Host:www.questpond.com

Connection: Keep-Alive

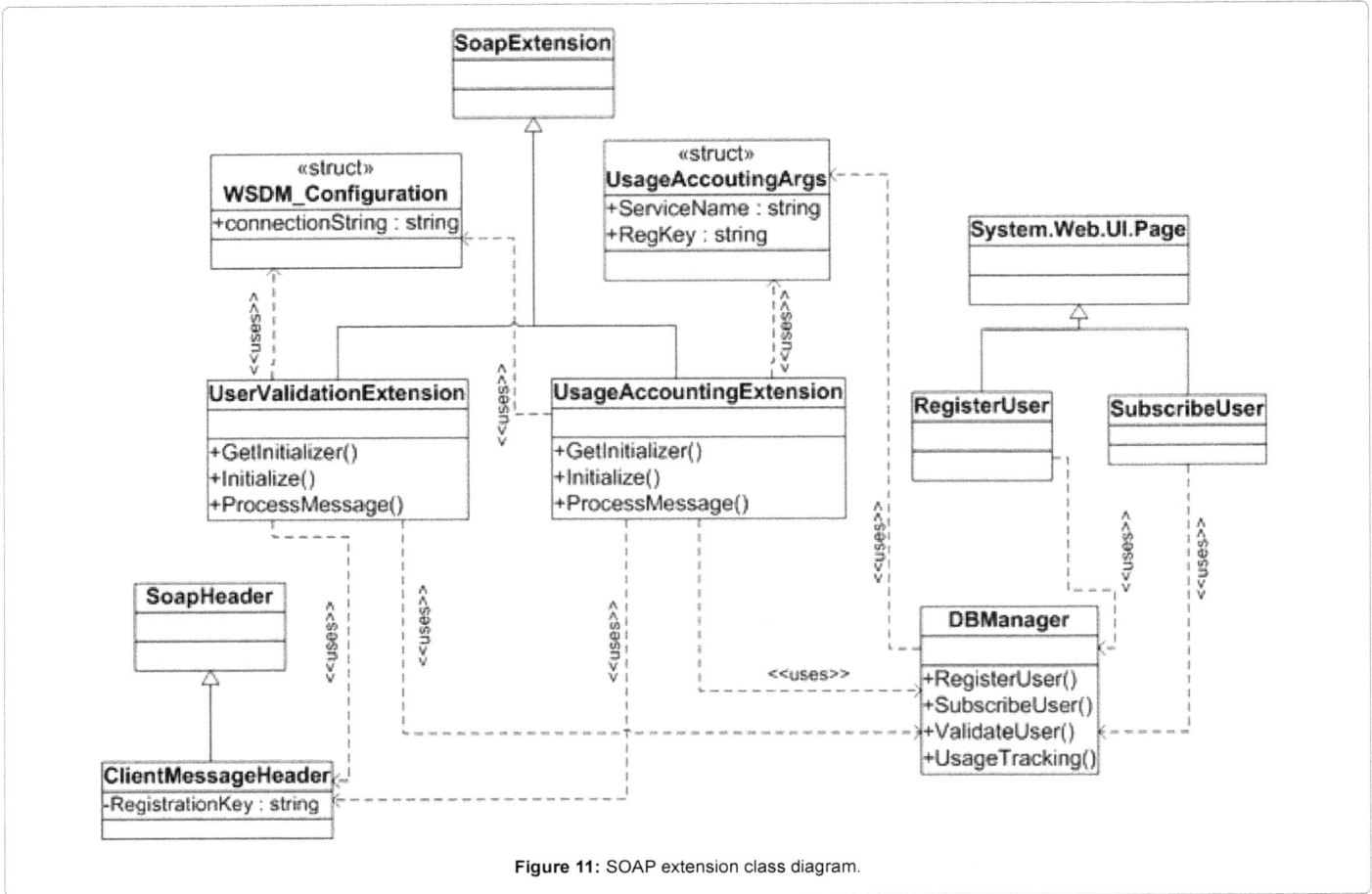

**Figure 11:** SOAP extension class diagram.

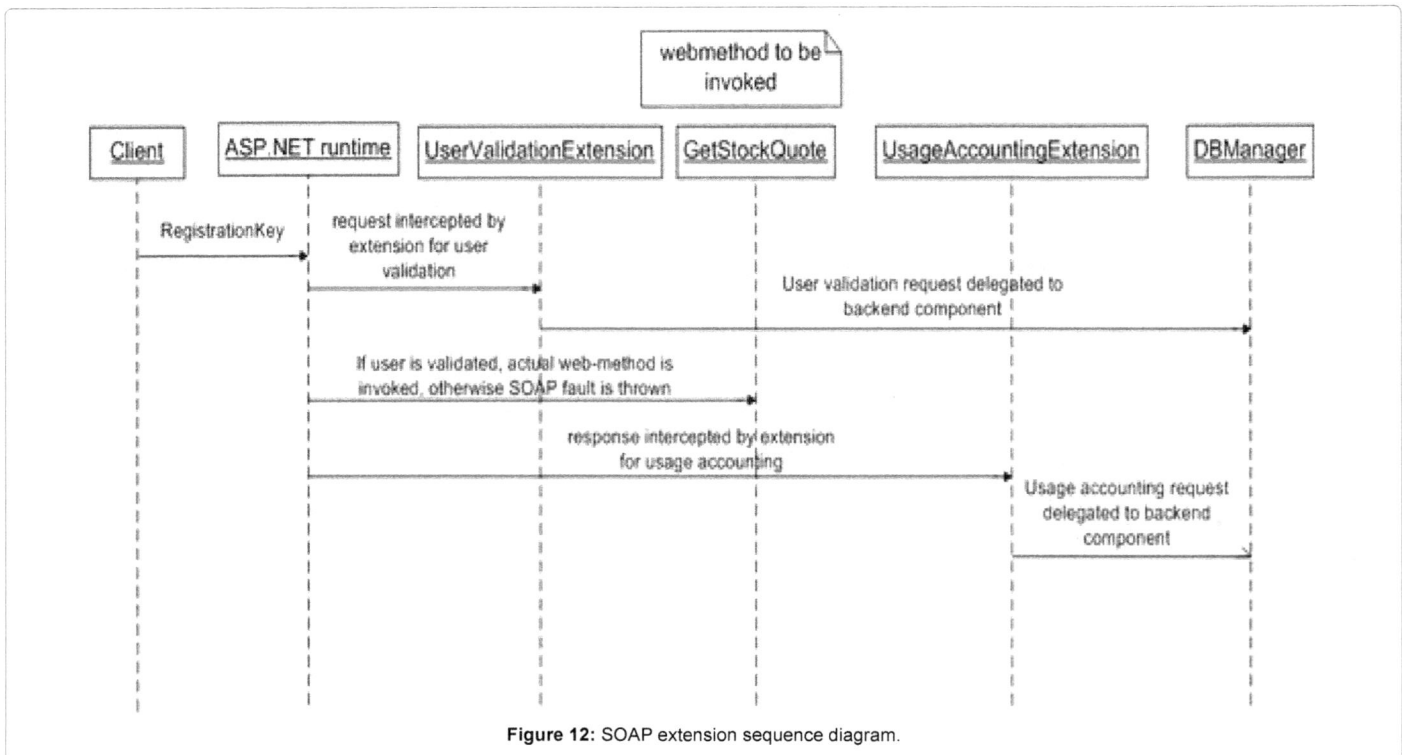

**Figure 12:** SOAP extension sequence diagram.

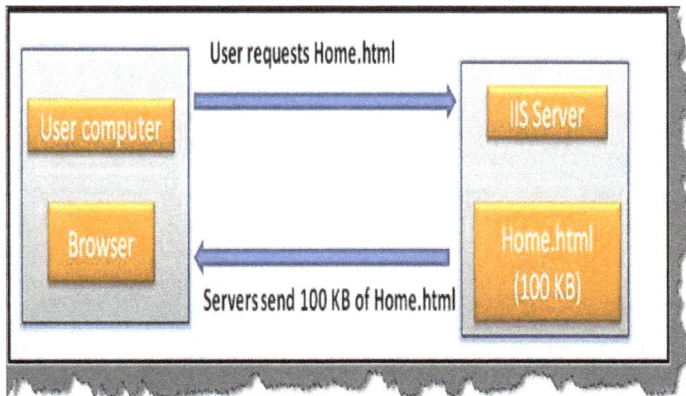

**Figure 13:** IIS compression.

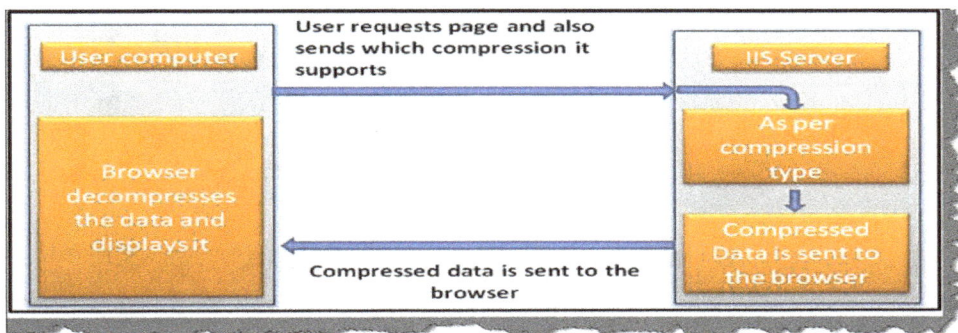

**Figure 14:** IIS compressed dataflow.

**Figure 15:** Extension over deflate.

- Depending on the compression type support sent by the browser IIS compresses data and sends the same over the wire to the end browser (Figure 14).

- Browser then decompresses the data and displays the same on the browser.

**Compression fundamentals: - Gzip and deflate:** IIS supports to kind of compressions Gzip and deflate. Both are more or less same where Gzip is an extension over deflate (Figure 15). Deflate is a compression algorithm which combines LZ77 and Huffman coding.

Gzip is based on deflate algorithm with extra headers added to the deflate payload (Figure 16).

Below are the headers details which is added to the deflate payload data. It starts with a 10 byte header which has version number and time stamp followed by optional headers for file name. At the end it has the actual deflate compressed payload and 8 byte check sum to ensure data is not lost in transmission (Figure 17).

**Enabling IIS compression**

a. **Step 1:- Enable compression**

The first step is to enable compression on IIS. So right click on websites à properties and click on the service tab. To enable compression need to check the below two text boxes from the service tab of IIS website properties. Below figure shows the location of both the checkboxes (Figure 18).

b. **Step 2:- Enable metabase.xml edit**

Metadata for IIS comes from 'Metabase.xml' which is located at "%windir%\system32\inetsrv\". In order to make changes to this XML file need to direct IIS to gives us edit rights. So right click on IIS server root à go to properties and check 'enable direct metabase edit' check box as shown in the below figure (Figure 19).

c. **Step 3:- Set the compression level and extension types**

**Figure 16:** Extension over Gzip.

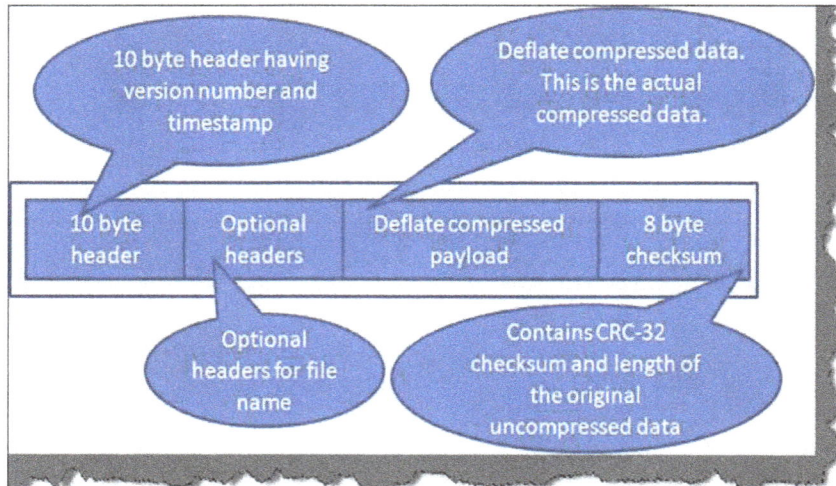

**Figure 17:** Deflate payload data.

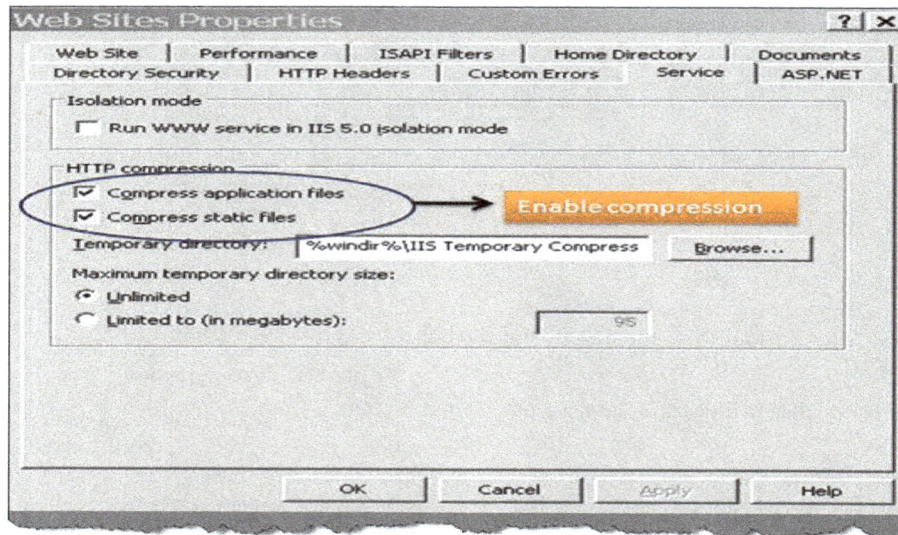

**Figure 18:** Enable compression.

Next step is to set the compression levels and extension types. Compression level can be defined between 0 to10, where 0 specifies a mild compression and 10 specifies the highest level of compression. This value is specified using 'HcDynamicCompressionLevel' property. There are two types of compression algorithms 'deflate' and 'gzip' (Figure 20).

Need to also specify which file types need to be compressed. 'HcScriptFileExtensions' help to specify the same. For the current scenario specified that need to compress ASPX outputs before sent to the end browser (Figure 21).

d. **Step 4:- Does it really work?**

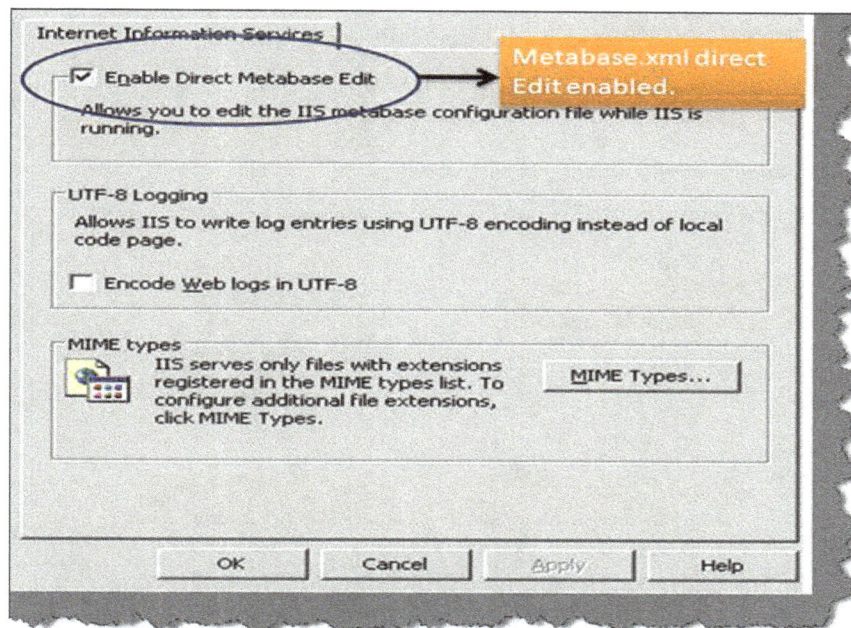

**Figure 19:** Enable matabase.xml edit.

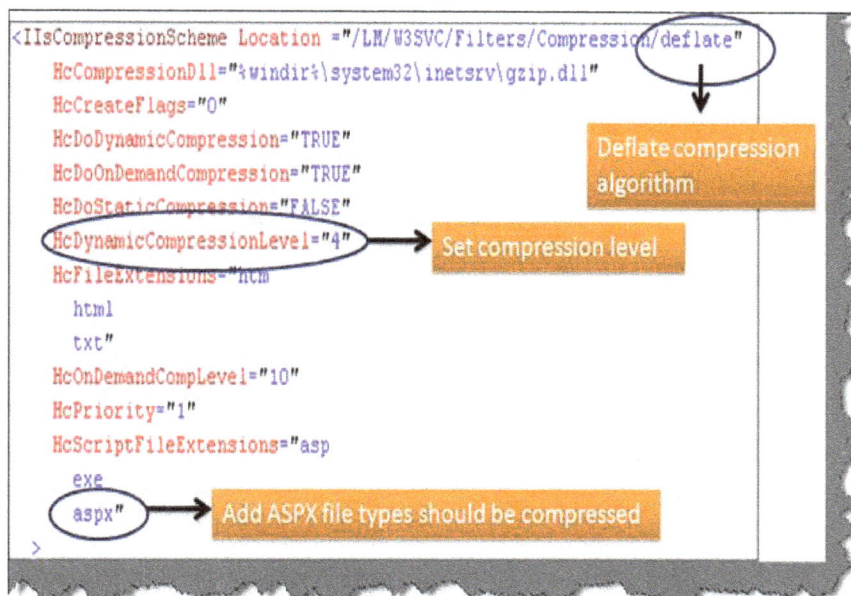

**Figure 20:** Set the compression level.

In order to see the difference before compression and after compression will run the fiddler tool as run ASP.NET loop page. Below screen shows data captured by fiddler without compression and with compression. Without compression data is "80501 bytes" and with compression it comes to "629 bytes". I am sure that's a great performance increase from bandwidth point of view (Figure 22).

If site is only serving compressed data like 'JPEG' and 'PDF', it's probably not advisable to enable compression at all as CPU utilization increases considerably for small compression gains. On the other side

need to balance compression with CPU utilization. The more increase the compression levels the more CPU resources will be utilized. Different data types needs to be set to different IIS compression levels for optimization. In the further coming section will take different data types, analyze the same with different compression levels and see how CPU utilization is affected. Below figure shows different data types with some examples of file types (Figure 23).

**Static data compression:** Let's start with the easiest one static content type like HTML and HTM. If a user requests for static page

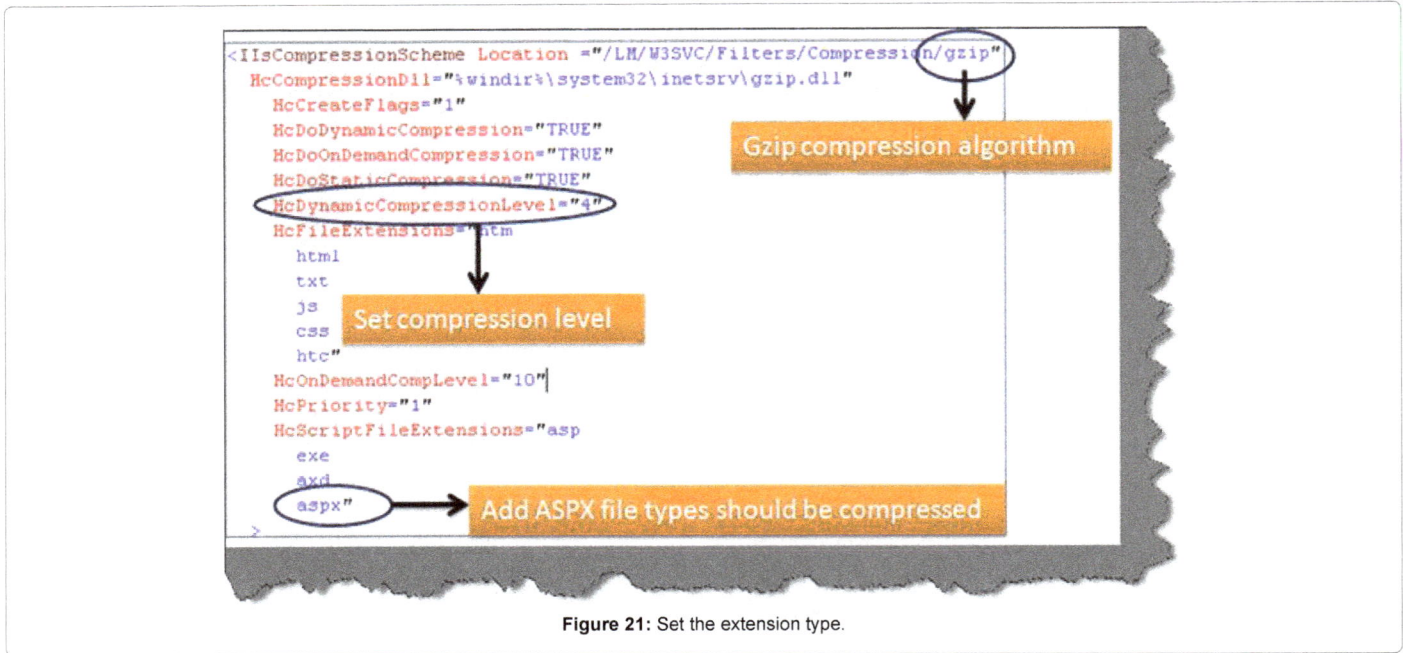

**Figure 21:** Set the extension type.

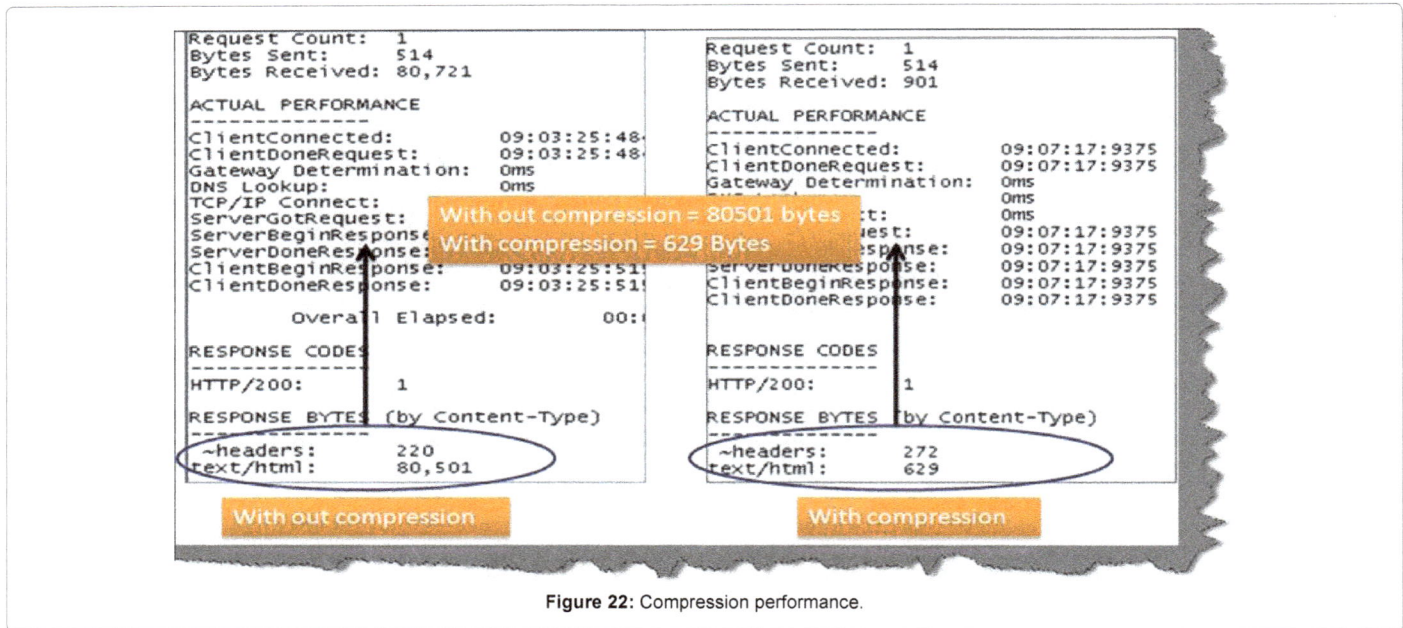

**Figure 22:** Compression performance.

from IIS who has compression enabled, IIS compresses the file and puts the same in '%windir%\IIS Temporary Compressed Files' directory. Below is a simple screen which shows the compressed folder snapshot. Compression happens only for the first time. On subsequent calls for the same compressed data is picked from the compressed files directory (Figure 24).

Below are some sample readings taken for HTML files (Table 1) of size range from 100 KB to 2048 KB and set the compression level to '0' (Figure 25).

**Dynamic data compression:** Dynamic data compression is bit different than static compression. Dynamic compression happens

every time a page is requested. So, the users need to balance between CPU utilization and compression levels (Table 2).

The above readings do not show anything specific, its bit messy. So plotted the graph using the above data and hit the sweet spot. Even after increasing the compression level from 4 to 10 the compressed size has no effect. So the conclusion from this is, setting value '4' compression level for dynamic data pages will be an optimized setting (Figure 26).

**Compressed file and compression:** Compressed files are file which are already compressed. For example files like JPEG and PDF are already compressed. The compressed files after applying IIS compression did not change much in size.

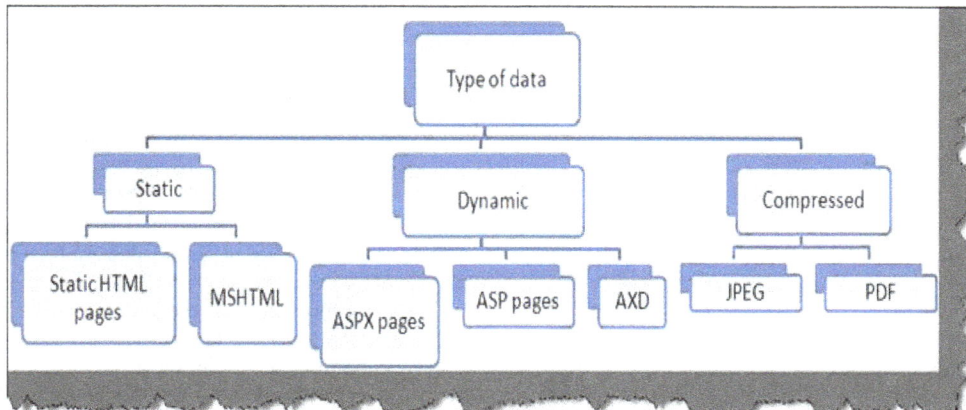

**Figure 23:** Different data types.

**Figure 24:** Static data compression.

The compression benefits are very small. End up utilizing more CPU processor resource and gain nothing in terms of compression (Table 3 and Figure 27).

So the conclusion can draw for compressed files is that can disable compression for already compressed file types like JPEG and PDF.

## Proposed System Result

The result of the Intrusion Detection Expert System (IDES) has become a regular in intrusion detection systems. Many current systems are based in partly on IDES prototype technology, The Next-Generation Intrusion Detection Expert System (NIDES) is the comprehensive enhancement to IDES. NIDES is a real-time intrusion detection application which integrates a statistical analysis -based anomaly detector and a rule-based misuse detection system [21]. The combination gives the flexibility to observe penetrations from internal and external attacks. NIDES additionally includes a comprehensive program that allows access to any or all the applications capabilities, similarly as a context -sensitive facilitate system.

While NIDES is considered the present progressive during a combined anomaly and misuse detection system, the applying retains the problem possessed by all similar models in detection cooperative attacks, long-term penetration situations and virus propagation. Another potential disadvantages is that NIDES retains the reliance on the system's audit record for input.

Future expansions of the rulebase and the development of profiles of entities aside from users ought to scale back the potential vulnerabilities that don't seem to be adequately addressed by the present system.

One of the main challenges to attempting implement and validate a new intrusion detection methodology, is to assess it and compare its performance therewith of alternative out there approaches. It is noticeable that this task is not restricted to A-NIDS, however is additionally applicable to NIDS (and even to IDS sometimes) generally. The requirement for test-beds that provide robust and reliable metrics to quantify NIDS has been prompt, for instance, by the National Institute for Standards and Technology (NIST). An advantage of assessment in real environments is that the traffic is sufficiently realistic; however, this approach subject to: (a) the risk of potential attacks, and (b) the possible interruption of the system operation due to simulated attacks. On the other hand, the evaluation of NIDS methodologies in experimental environments involves the generation of synthetic traffic as well as background traffic representing legal users, that is much from being a trivial endeavor.

| Actual KB | Compressed in KB |
|---|---|
| 100 | 24 |
| 200 | 25 |
| 300 | 27 |
| 1024 | 32 |
| 2048 | 41 |
| **Compression level set to '0'** | |

**Table 1:** File Compression range.

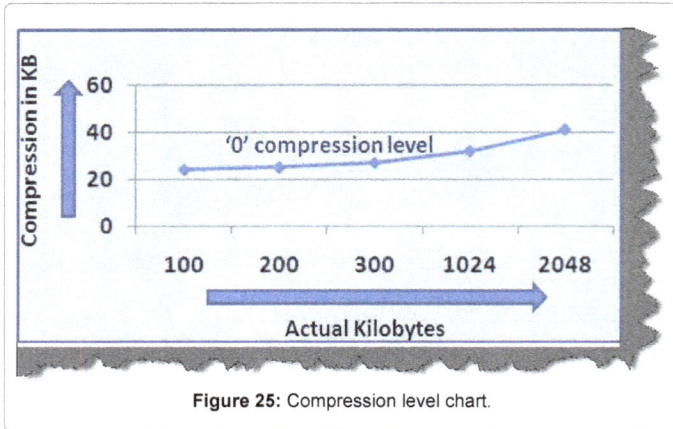

**Figure 25:** Compression level chart.

| Compression Levels | File size | | | | |
|---|---|---|---|---|---|
| | **100 KB** | **200 KB** | **300 KB** | **1 MB** | **2 MB** |
| 0 | 32,774 | 35,496 | 37,699 | 52,787 | 109,382 |
| 1 | 30,224 | 32,300 | 34,104 | 46,328 | 92,813 |
| 2 | 29,160 | 31,004 | 32,673 | 43,887 | 87,033 |
| 3 | 28,234 | 29,944 | 31,628 | 42,229 | 83,831 |
| 4 | 26,404 | 27,655 | 29,044 | 34,632 | 44,155 |
| 5 | 25,727 | 26,993 | 28,488 | 33,678 | 42,395 |
| 6 | 25,372 | 26,620 | 28,488 | 33,448 | 41,726 |
| 7 | 25,340 | 26,571 | 28,242 | 33,432 | 41,678 |
| 8 | 25,326 | 26,557 | 28,235 | 33,434 | 41,489 |
| 9 | 24,826 | 26,557 | 28,235 | 33,426 | 41,490 |
| 10 | 24,552 | 25,764 | 27,397 | 32,711 | 42,610 |

**Table 2:** CPU utilization and compression levels.

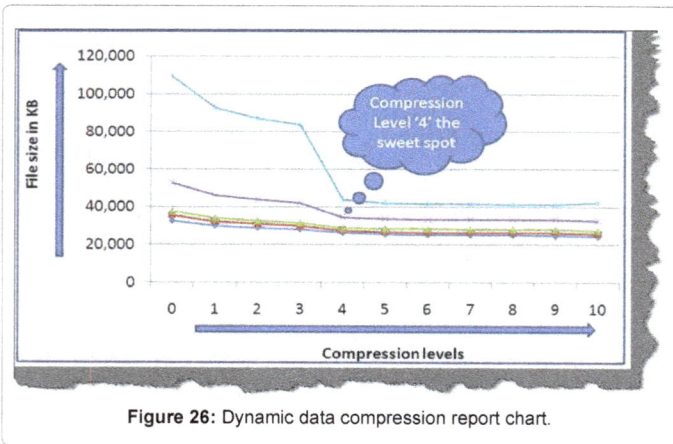

**Figure 26:** Dynamic data compression report chart.

Other network traffic related studies deal with the problem of standardizing the acquisition and use of real traffic for validating NIDS environments. During this respect, that contributes some proposals

on a general methodology to amass and organize traffic datasets, to outline AN analysis framework to check the performance of anomaly-based NIDS [19,22]. The significant try created up to now in NIDS assessment proof of its importance. However, it remains an open issue and a big challenge.

A new methodology that would achieve more accuracy than the existing six classification patterns (Gaussian Mixture, Radial Basis Function, Binary Tree Classifier, SOM, ART and LAMASTAR),called Hierarchical Gaussian Mixture Model[HMM] for IDM. Development of host-based anomaly intrusion detection, focusing on system call based HMM training. This was later enhanced with the inclusion of data pre-processing for recognizing and eliminating redundant sub sequences of system calls, resulting in less number of HMM sub models. Experimental results on three public databases showed that training cost can be reduced by 50% without affecting the intrusion detection performance. False alarm rate is higher yet reasonable compared to the batch training method with a 58% data reduction. An anomaly detection system comprising of detection modules for detecting anomalies in each layer. The anomaly detection results of the neighbor node(s) is taken by the current node and its result in turn is sent to the neighbor node(s).Experimental results revealed increased detection rate and reduced false alarm positives, compared to other methods [23].

The new framework builds the patterns of network services over datasets labeled by the services [22]. With the built patterns, the framework detects, attacks in the datasets. This approach is independent of attack-free training datasets, but assumes that each network service has its own pattern for normal activities.

The biometrics-based intrusion detector model to provide a light-

| Actual compressed file size | File size after IIS compression |
|---|---|
| 100 | 102 |
| 220 | 210 |
| 300 | 250 |
| 1024 | 980 |
| 2048 | 1987 |

**Table 3:** IIS compression.

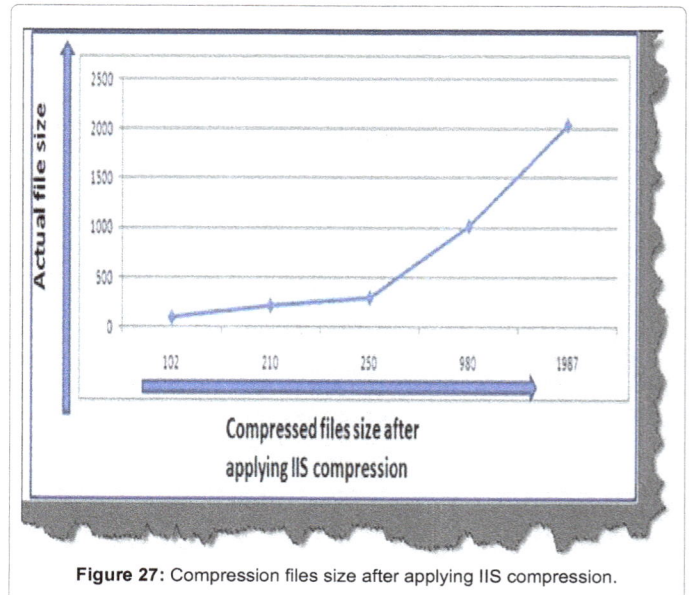

**Figure 27:** Compression files size after applying IIS compression.

weight and self-contained module for detection user identities misuse. System-calls and network traffic monitoring systems ought to be combined to the present detector to achieve the best solutions.

The proposed a technique to detect anomalies at all layers of a network stack in a sensor network, segregating the service at various levels. Physical layer intrusion is detected by using RSSI values of neighbors (dependant on background noise, weather conditions etc). Targeting MAC layer will work for schedule based and sleep/wake-up based MAC protocols whereas IASN protocol is geared toward the routing layer.

Experiments show that IASN is used for supply started routing protocols, table driven routing protocols and data dissemination mechanisms like directed diffusion. The probability of detection increases linearly with the amount of nodes running IASN. Nodes guard each other from masquerade at application layer. Depending on the resource availability, any combination of the above methods can be employed, as they are independent of one another. All technique are energy efficient as they have very low false positive rates (except RSSI and round trip time) and low overhead.

Combining multiple independent data sources and studying combined traditional intrusion attack and anomaly intrusion, the anomaly intrusion traffic detection provided the statistical wavelet based detection mechanism. The properties like attack length, packet count, packet rate, and dominant protocol kind match with the two data sets, as is showed by attack structure. At lean and significant traffic situations, the demand capability of the server was determined to administer higher clarity of anomaly intrusion detection though server period of time. Analysis of many traffic anomaly properties that is not possible victimization ancient intrusion measurements is performed by a brand new model that used anomaly intrusion attack measurements.

Windows Host Anomaly Detection System, which is used as a supplement for other security mechanisms under windows. It can only detect intrusions which invoke an anomaly sequence by programs. The Statistical anomaly detection technology called that HIDE with hierarchical multitier multi-observation window system to monitor network traffic parameters simultaneously, using a real-time probability distribution function (PDF) for each parameter, collected during the observation window. The similarity measurements of measured PDF and reference PDF are combined into an anomaly status vector classified by a neural network. This technique detects that attacks and soft faults with traffic anomaly intensity as low as 3 to 5 percent of typical background traffic intensity, thereby generating an early warning.

The anomaly based mostly intrusion detection system for mobile networks, supported simulation results of quality profiles for enhancing ABID in mobile wireless networks. If the quality behavior of users has not been accurately found, the choice of specific values for key parameters, like sequence length and cluster size is absurd. One potential strategy for enhancing the characterization of users and addressing construct drift (keeping official up-to-date), is to take care of a window of the fresh determined sequences (analogous to the exponential weighted moving average) which will then be wont to update the coaching patterns sporadically and, thus scale back the false positives.

An intrusion detection algorithm and its architecture (two layered, global central layer and a local layer, together performing data collection, analysis and response), based on data mining and useful in real time for network security, By filtering out the known traffic behavior

(intrusive and normal) the IDS focuses on analysis on unknown data thereby reducing false alarm rates. The model supported contiguous professional selection rule ways observe most anomalies, unsuccessful match does not imply AN abnormity, as traditional rules might not cowl all traditional data. Detection rates in this is not commendable but it has vast future scope for improvement.

In recent literature, anomaly detection through a Bayesian Support Vector Machine is found as interesting machine learning model for anomaly detection. Use of a SVM with one-class to detect the system anomalies at their early stage is studied along with drift output classification probabilities. Experimentally, absence of failure training data under one-class SVM leads to quick detection of unknown anomalies. Initially dividing the training data into multiple unrelated lower dimensional models, the test data will be evaluated on each model separately thereby revealing outliers in different capacities (as is used to evaluate the posterior class probabilities in Bayesian framework) [24-27].

## Results of the Various Algorithms

The experimental results that have obtained with the assorted algorithms. the assorted rules designed and whose results are bestowed the Brute-Force algorithm, the Karp-Rabin rule, the Boyer-Moore rule and also the Knuth-Morris-Pratt rule.

The results of the running time of these algorithms vary the input size, where the input is the words. The number of patterns to be matched remains the same. The running time (in milliseconds) for the various algorithms is recorded within the following Table 4.

From the Table 4 and the graph (Figure 28), there is no trend within the performance between the algorithms. Solely that KMP performs the worst and Boyer-Moore performs the simplest.

## Conclusion

The extension of the model with additional attributes can help to unearth further mistakes. The analysis of statistical properties of router configurations appears to be a promising approach to help operators in detecting mistakes. Unlike most of the current research, which use only one agent per engine for detection of various attacks, the proposed system is constructed by several agents in a single engine. The NIDS will broaden its read on completely different behaviors of the network traffic by every of the agents with its own strength on capturing a form of network behavior. Firewall policy rules are one in every of most significant part of network security system. It plays the very important role in management of any organization's network and its security infrastructure.

Thus the management of policy rule could be a vital task for the network security. There are many tools and techniques won't to perform anomaly detection and rule editing by using given set of existing policy rules. However, one in every of the idea and so its limitation is that

| Input Size | Running Time (in milliseconds) | | | |
|---|---|---|---|---|
| | Brute-Force | Karp-Rabin | Knuth-Morris-Pratt | Boyer-Moore |
| 20000 | 15 | 15 | 17 | 15 |
| 60000 | 46 | 45 | 46 | 40 |
| 100000 | 74 | 73 | 79 | 68 |
| 140000 | 102 | 102 | 108 | 102 |
| 180000 | 129 | 132 | 139 | 119 |
| 200000 | 144 | 144 | 159 | 133 |

**Table 4:** Running time for various algorithms.

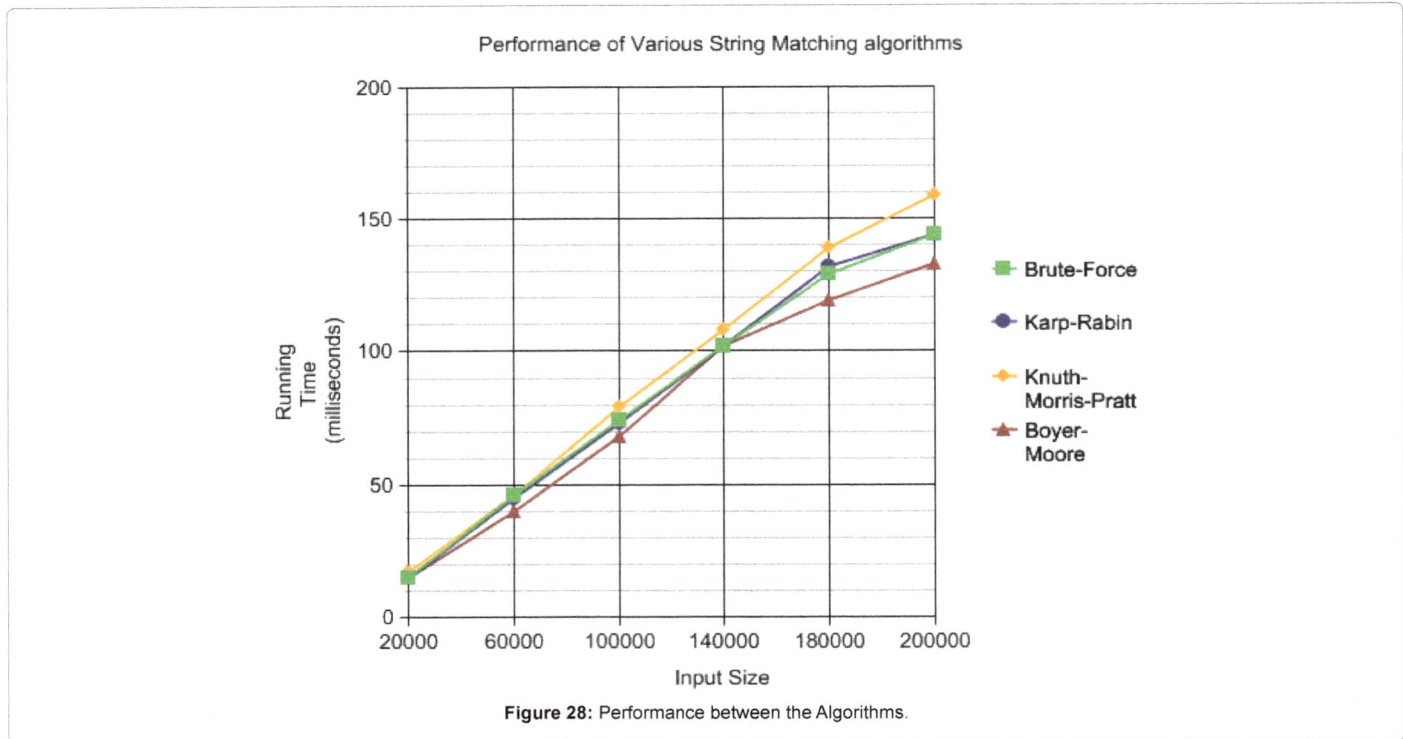

**Figure 28:** Performance between the Algorithms.

firewall and its rules are set to be static and so while not a capability to replicate the network behavior determined by firewall.

## References

1. Shirbhate RS, Patil PA (2012) Network Traffic Monitoring Using Intrusion Detection System. International Journal of Advanced Research in Computer Science and Software Engineering 2: 1-5.

2. Raju B, Srinivas B (2012) Network Intrusion Detection System Using KMP Pattern Matching Algorithm. International Journal of Computer Science and Telecommunications 3: 33-36.

3. Shrivastava N, Richariya V (2012) Ant Colony Optimization with Classification Algorithms used for Intrusion Detection. IJCEM International Journal of Computational Engineering & Management 15: 54-63.

4. Ranjan R, Sahoo G (2014) A New Clustering Approach for Anomaly Intrusion Detection. International Journal of Data Mining & Knowledge Management Process (IJDKP) 4: 29-39.

5. Kumar N, Angral S, Sharma R (2014) Integrating Intrusion Detection System with Network Monitoring. International Journal of Scientific and Research Publications 4: 1-4.

6. Shelokar ND, Ladhake SA (2010) Network Intrusion detection using correlation functional dependency. Oriental Journal of Computer Science & Technology 3: 185-188.

7. Sivakumar V, Yoganandh T, Mohan Das R (2012) Preventing Network From Intrusive Attack Using Artificial Neural Networks. International Journal of Engineering Research and Applications (IJERA) 2: 370-373.

8. Kabila R (2008) Network based Intrusion Detection and Prevention Systems in IP-Level Security Protocols. World Academy of Science, Engineering and Technology 2: 661-667.

9. Prasad KM, Mohan Reddy AR, Jyothsna V (2012) IP Traceback for Flooding attacks on Internet Threat Monitors (ITM) Using Honeypots. International Journal of Network Security & Its Applications (IJNSA) 4: 13-27.

10. Al-Dabagh NB, Fakhri MA (2014) Monitoring and Analyzing System Activities Using High Interaction Honeypot. International Journal of Computer Networks and Communications Security 2: 39-45.

11. Laing B (2000) Internet Security Systems, How to guide –Implementing a network based Intrusion Detection System.

12. Yadav MR, Kumbharkar PB (2014) Intrusion Detection System with Supervised Learning Algorithms. International Journal of Advanced Research in Computer Science and Software Engineering 4: 305-310.

13. Khari M, Gaur M, Tuteja Y (2013) Meticulous Study of Firewall Using Security Detection Tools. International Journal of Computer Applications & Information Technology 2: 1- 9.

14. Sharma S, Kumar S, Kaur M (2014) Recent trend in Intrusion detection using Fuzzy-Genetic algorithm. International Journal of Advanced Research in Computer and Communication Engineering 3: 6472-6475.

15. Gaidhane R, Vaidya C, Raghuwanshi M (2014) Survey: Learning Techniques for Intrusion Detection System [IDS]. International Journal of Advance Foundation and Research in Computer [IJAFRC] 1: 21-28.

16. Dixit U, Gupta S, Pal O (2012) Speedy Signature Based Intrusion Detection System Using Finite State Machine and Hashing Techniques. International Journal of Computer Science Issues 9: 387-391.

17. Goel R, Sardana A, Joshi RC (2012) Parallel Misuse and Anomaly Detection Model. International Journal of Network Security 14: 211-222.

18. Meenatchi I, Palanivel K (2014) Intrusion Detection System in MANETS: A Survey. International Journal of Recent Development in Engineering and Technology 3: 42-50.

19. Shuang-can Z, Chen-jun H, Wei-ming Z (2014) Multi Agent Distributed Intrusion Detection System Model Based on BP Neural Network. International Journal of Security and Its Applications 8: 183-192.

20. Jawale DR, Bhusari VK (2014) A Novel Approach for classification and Detection of attacks in Network Intrusion Detection System Using ANN. International Journal of Advanced Research in Computer Science and Software Engineering 4: 802-806.

21. Cannady J, Harrell J A Comparative Analysis of Current Intrusion Detection Technologies.

22. Jaisankar N, Saravanan R, Swamy KD (2009) Intelligent Intrusion Detection System Framework Using Mobile Agents. International Journal of Network Security & its Applications (IJNSA) 1: 72-88.

23. Saini P, Godara S (2014) Modelling Intrusion Detection System using Hidden Markov Model: A Review. International Journal of Advanced Research in Computer Science and Software Engineering 4: 542-548.

24. Das V, Pathak V, Sharma S, Sreevathsan, Srikanth MVVNS, et al. (2010) Network Intrusion Detection System based on Machine Learning Algorithms. International Journal of Computer Science & Information Technology (IJCSIT) 2: 138-151.

25. Mudholkar SS, Shende PM, Sarode MV (2012) Biometrics Authentication Technique for Intrusion Detection Systems Using Finger Print Recognition. International Journal of Computer Science, Engineering and Information Technology (IJCSEIT) 2: 57-65.

26. Golmah V (2014) An Efficient Hybrid Intrusion Detection System based on C5.0 and SVM. International Journal of Database Theory and Application 7: 59-70.

27. Khanbabapour H, Mirvaziri H (2014) An Intelligent Intrusion Detection System Based on Expectation Maximization Algorithm in Wireless Sensor Networks. International Journal of Information and Communication Technology Research 4: 1-10.

# 17

# A Rapid Prototyping Matlab Based Design Tool of Wireless Sensor Nodes for Healthcare Applications

**Rammouz R[1,2]\*, Labrak L[1], Abouchi N[1], Constantin J[2], Zaatar Y[2] and Zaouk D[2]**

[1]*Nanotechnology Institute of Lyon, Lyon, France*
[2]*Applied Physics Laboratory, Beirut, Lebanon*

**Abstract**

Remote patient monitoring enables medical facilities and personnel to monitor patients outside of conventional clinical settings. This technology is greatly affected by the available energy sources. The limited energy supply is the bottleneck for measurement, data transmission, network connection and lifetime. Thus, efficient power consumption estimation in the early stage of the design would give the user an insight on its feasibility within medical constraints. In this paper, a power-oriented Matlab based design tool is proposed. It will help the user to make an energy aware design. In fact, it will enable him to determine the power consumption of each node based on a set of selected components, the daily routine of the patient and many other factors. It will also allow him to test multiple configurations prior to the physical implementation. This will minimize the time and financial impact of any modification. Furthermore, it will provide guidance to properly adapt the design within the requirements of the application: Choosing components, energy sources, etc. A validation is proposed via a physical implementation in order to compare the results obtained through simulation with practical measurements. Future works aim to take into account memory access operations, include several sensors operating within a network, and adapt the tool to other applications (smart buildings, environmental monitoring, etc.).

**Keywords:** Remote patient monitoring; Power consumption; Matlab; Energy aware design; Validation

## Introduction

The ever increasing number of people who need continuous medical attention coupled with the rising costs of healthcare has triggered the concept of remote patient monitoring. This can be achieved using wireless sensor nodes attached to the human body. These nodes ensure periodic measurements of vital signs [1] such as heart rate, blood pressure, temperature, etc. Data acquired is first sent to a local gateway (cellphone, tablet, PDA …) through short distance wireless protocols. It is then uploaded to a medical database, analyzed and viewed by healthcare professionals (Figure 1) [2,3].

Remote patient monitoring is frequently used with the elderly and the chronically ill. It allows these patients and their physicians to closely monitor their medical condition and, if needed, intervene [4]. This technology is mainly limited by the amount of available energy. Thus estimating the power consumption of a node prior to its physical implementation will allow the designer to properly adapt its architecture (choosing energy sources, selecting components, etc.).

Some ready to use network simulators already have energy profiling functions. Nevertheless none of these can guide a designer in the decision process at electrical architectural level. The developed tool is power oriented and has two advantages. The main is an easy configurable component based environment. It allows fast creation of new component description based on the datasheets thus producing a flexible growing database. The second lies in many built-in solutions coupled with optimization processes that provide the user with guidance to properly adapt his design within the constraints of the application: choosing the more power efficient configuration, the required energy sources, etc.

The rest of the paper is organized as follows. Section 2 gives an overview on energy profiling and networked embedded systems simulators. In section 3, we will explain the methodology used to determine the power consumption of the node and implementation of the built-in functions. We will then present its application to a wireless temperature sensor node in section 4. Section 5 will be dedicated to the physical implementation of that node. The results are then presented in section 6 and the conclusion is in section 7.

## State of Art

Energy profiling has already been at the center of several studies.

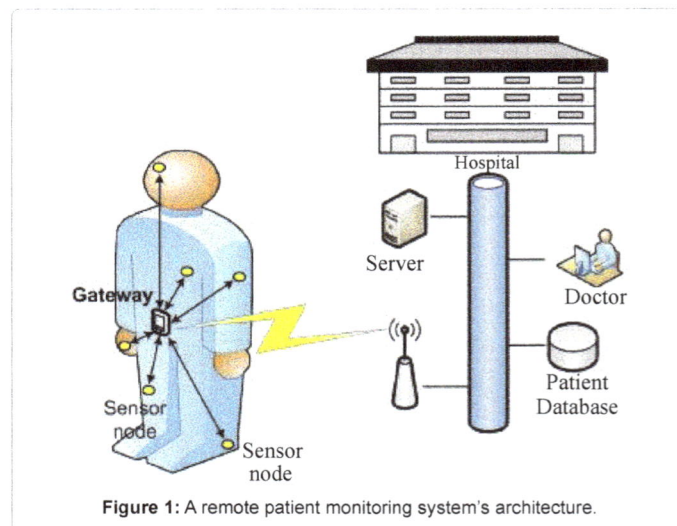

**Figure 1:** A remote patient monitoring system's architecture.

**\*Corresponding author:** Rammouz R, Nanotechnology Institute of Lyon, Lyon, France, E-mail: ramzy.rammouz@cpe.fr

Plenty of networked embedded systems simulators have been developed in the last decade. This is mostly due to the uncertainty faced by designers regarding the system's energy consumption.

A first example is the general network simulators such as NS-2, NS-3 [5] and OPNET [6]. These simulators lack in the infrastructure for easy and accurate hardware modelling despite having an important amount of protocol implementations.

A second example is specific ad hoc networks simulators. Those were created to overcome general network simulators deficiencies regarding system models for simulating sensor nodes. They included some specific models for radio propagation, sensors and hardware systems. SENSE (SEnsor Network Simulator and Emulator) is one of these simulators. It is built over a component based modelling approach. Yet it lacks in the implementation of commonly used wireless standards and does not include models for noise or interference [7].

A third example is that of virtual prototyping as in IDEA1. IDEA1 is based on System C and C++ language. It allows cycle-accurate energy consumption estimation. Moreover, it has an instruction set simulator that supports different microcontrollers [8]. However, accurate models require disclosure of details that are not always provided by microcontroller designers [9].

A last example is the multi-domain simulation based on simulators coupling. Interfacing simulators is a complex task taking into account that each is developed in a different programming language and has its own user interface.

These factors lead us to propose a new approach: creating a design tool that is generic enough to make it simple to implement new components without a loss of accuracy. This will significantly speed up the production cycle especially when designers have prior knowledge of each component's individual performances.

## Design Tool

In this section we aim to explain the concept we were based on to determine the power consumption of the node.

Only then, we will be able to elaborate on the built-in functions.

## Concept

Our goal is to simulate a wireless sensor node and estimate its power consumption taking into account a physical implementation. The first step was to select an efficient modeling approach. We used a Model Based Design (MBD) approach as it brings cost and time saving in the development. Moreover the function models can be easily reused for different setups. MATLAB/SIMULINK offers an adapted environment and many suitable features to implement our system as an executable specification.

Our purpose is to determine the power consumption of a wireless sensor node similar to the one shown in Figure 2. In order to do that, we need to identify the activity of each component. It is obtained from the time spent in each state (active, sleep, idle …) as well as state transitions [10]. This conducted us to establish the cycle of operation presented in Figure 3 [11] and explained here under:

- The processor wakes up periodically in accordance with a predefined measurements rate following an assessment of the patient's medical condition

- The processor activates the sensor to start converting the physiological signal into an electrical one

- Simultaneously, an analog to digital conversion is taking place in the ADC

- Once the measurement/conversion process is done, both the sensor and the ADC go to sleep mode

- The microprocessor processes the ADC output and data is sent to the RF transceiver in order to be transmitted to the gateway

- Meanwhile the processor becomes idle: the CPU is disabled during transmission, and the processor waits for an interruption from the wireless module

- The RF transceiver informs the processor of the outcome of the transmission and is put to sleep

- At the end, the whole node goes back to sleep mode

Based on this cycle, we have modeled the instantaneous current consumption of each component as well as the whole unit using Matlab. Figure 4 shows the blocks embedded in our model. Component and configuration related inputs are used to extract consumption waveforms as well as mean values [11].

## Built-in functions

In order to provide an efficient reconfigurable tool, we decided to adopt a model based approach. It consists in two simple built-in functions coded in Matlab. The first one computes current consumption of a given node, while the other provides the user with guidance during the design process.

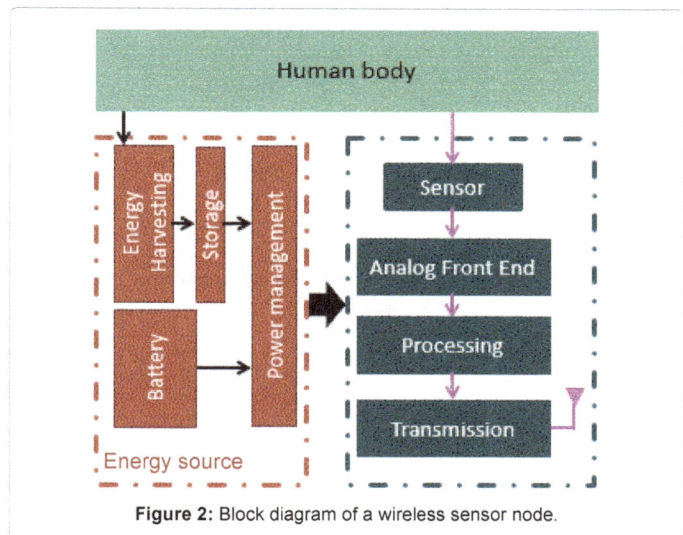

**Figure 2:** Block diagram of a wireless sensor node.

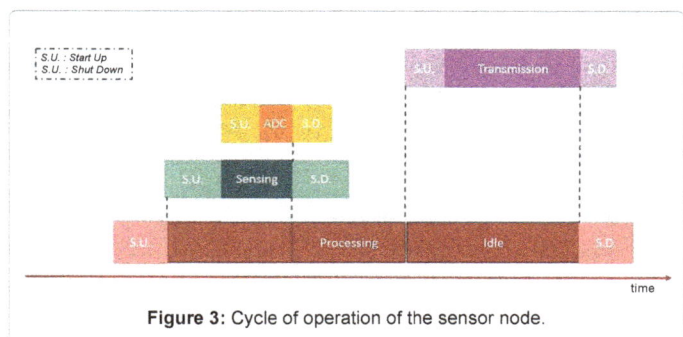

**Figure 3:** Cycle of operation of the sensor node.

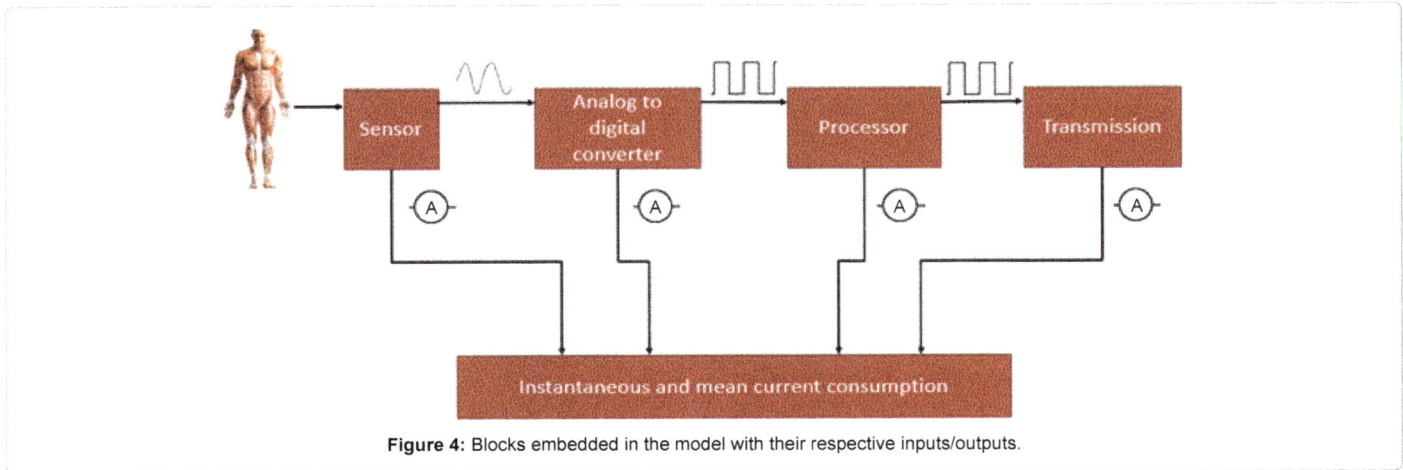

**Figure 4:** Blocks embedded in the model with their respective inputs/outputs.

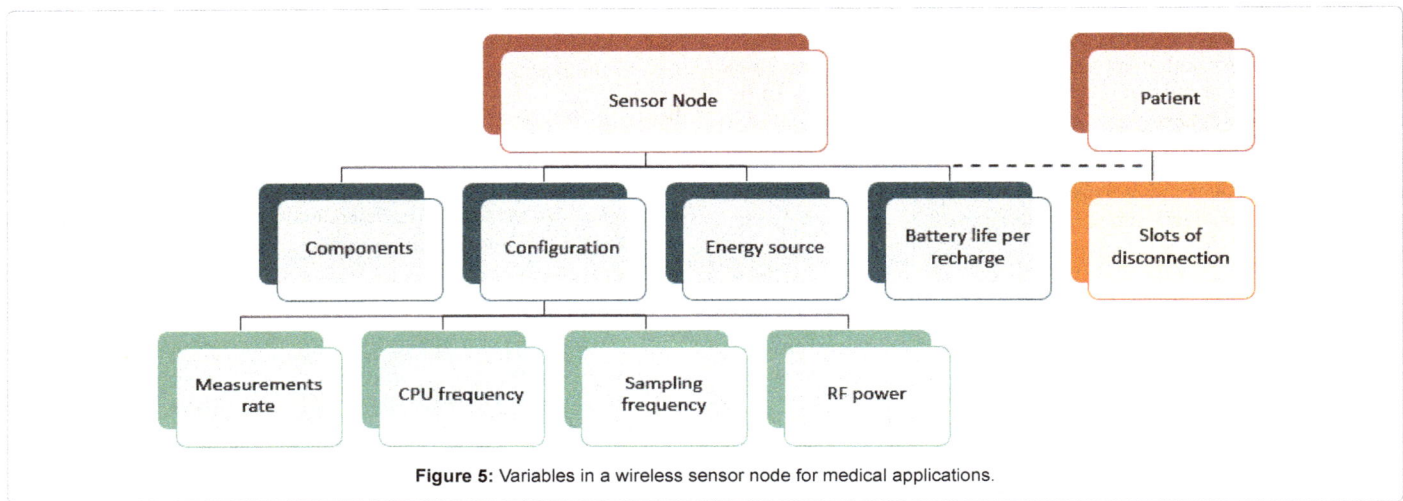

**Figure 5:** Variables in a wireless sensor node for medical applications.

| Variables \ Scenario | Battery life | Disconnection slots | Configuration of the node | | | | |
|---|---|---|---|---|---|---|---|
| | | | Energy source | Measurements rate | Sampling frequency | CPU frequency | RF power |
| Battery life per recharge | ■ | | | | | | |
| Picking energy sources | | | ■ | | | | |
| Disconnection time slots | | ■ | | | | | |
| Design | | | ■ | ■ | ■ | ■ | ■ |

Legend:　■ Variable　□ Parameter

**Table 1:** Variables/parameters combination for each scenario.

**Estimating the current consumption of the node:** Our goal is to determine for a given node, current consumption throughout the whole monitoring period which extends to weeks, maybe months. This consumption is sometimes affected by the outcome of the connection to the gateway. In fact, the RF module's operation may differ whenever we are connected or not. This is the case for some RF communication standards such as ZigBee [12], Bluetooth [13], and Bluetooth Low Energy [14]. Since these protocols are the most used in sensor network, calculating the current consumption of a node throughout the whole monitoring period has to take into account any disconnection imposed by the daily routine of the patient. And that's exactly what the first built-in function does.

**Guidance towards an energy aware design:** Another built-in function aims to guide the user to a power efficient design within the medical requirements of the application. It computes optimization routines to determine the best trade-off among several possible configurations. The diagram shown in Figure 5 displays the variables in a wireless sensor node. We can classify these variables in two main categories: patient related or node related. While designing a node, some of these variables can be set as parameters according to feasibility or medical constraints. This function includes solutions for four different scenarios. Each scenario represents a different configuration of these variables i.e., a different variables/parameters combination (Table 1).

**Battery life per recharge:** In this case, the architecture of the node is fixed, the energy source is chosen, and the patient has an established daily routine (disconnection slots). We aim to extract the mean current consumption of the node and determine, for a given battery, how long it will keep the sensor node operating without needing a recharge.

**Picking energy source:** This scenario assumes that a minimal battery life is required. Moreover, disconnection time slots are already established and the architecture of the node is set. However, we need to properly choose an energy source in order to comply with the above requirements. We will start by estimating the mean current consumption of the node, we will then determine the minimal battery size required. The solution given to the user is the one with the closest larger capacity.

**Disconnection time slots:** For this scenario, a minimal battery life per recharge is imposed. The node is already designed in terms of components and energy sources. Our goal is to determine the total amount of tolerated disconnection time. We will start by calculating the maximal mean current consumption allowed in order to respect the battery life constraint. We will then determine the disconnection period using a dichotomist search algorithm.

**Design:** In this case, a minimal battery life per recharge is to be respected. The daily routine of the patient is given. We have several commercial modules and energy sources to choose from. We would also need to properly configure these components. Our design tool will start by choosing, for each component, the most power efficient commercial module. It will then pick the smallest energy source that allows us to run the node with minimal performances. The last step, configuring these components, is done using an optimization algorithm.

Choosing an adequate optimization algorithm is crucial at this stage. Since the variables can be put in a discrete form, and the energy consumption increases with enhanced performances. We had to develop our own optimization algorithm. It is based on a brute-force search problem-solving technique. However, some features were added to make it more intelligent, thus reducing the amount of processing time. In fact, we defined an upper and lower limit of current consumption. The first is to respect the battery life imposed, while the second (-10%) aims to discard the solutions with the lowest performances. This algorithm will start by sorting the values taken by each variable from the one that produces the lowest consumption to the highest. It will then go through these values in order, keeping the ones that fit within the consumption limits. Whenever we reach the upper bound, the rest of the values are discarded without any calculation.

Another advantage is to take into account the user's preferences; for example, if a designer would rather work with a CPU frequency above 8 MHz, our algorithm will provide solutions that respect that constraint as well as the requirements of the application.

## Application

### Wireless body temperature sensor

In this section, we present an application. The sensor node we selected is a wireless body temperature sensor. Our choice was driven by the significance of body temperature in medical diagnosis. Table 2 matches each value with the corresponding sickness/symptoms.

The first step is to benchmark and look for components that provide low energy consumption coupled with good performances. We also took into account whether it is simple to assemble them in real life (supply voltage, size, etc.). We chose to work with Bluetooth Low Energy protocol because of its compatibility with laptops as well as smartphones. Table 3 matches each component with the commercial module we selected [8].

### Estimating the current consumption of the node

Figure 6 displays the configuration of the problem. We consider that we are taking a measurement every two minutes for a month (30 days). Each one is a double digit precision value coded in ASCII (five bytes).

Figure 7 presents the current consumption recorded for the sensor, the analog to digital converter as well as the microprocessor. The state of the connection to the gateway has a major influence on the current consumption of the RF module. To understand this effect it requires a brief explanation of the Bluetooth Low Energy communication protocol.

In this technology, the connection takes place between a master/client (the gateway) and a slave/server (the node). Whenever it is not

| Temperature | Symptoms |
|---|---|
| 44 | Death |
| 43 | Normally death, brain damage, cardio-respiratory collapse |
| 41-42 | Fainting, confusion, convulsion, Low/high blood pressure |
| 38-40 | Severe sweating, dehydration, weakness, vomiting, dizziness, fast heart rate |
| 37 | Normal temperature |
| 36 | Mild or moderate shivering |
| 34-35 | Intensive shivering, numbness, blue/gray skin, confusion, loss of movement in the finger |
| 29-33 | Confusion, sleepiness, slow heart rate, hallucinations |
| 24-28 | Breathing may stop, death |

**Table 2:** Symptoms according to body temperature.

| Component | Commercial module |
|---|---|
| Sensor | LM60 |
| Processor | MS430FR5969 |
| RF transceiver | Ble113 (Bluetooth Low) Energy) |

**Table 3:** Components alongside the commercial modules chosen.

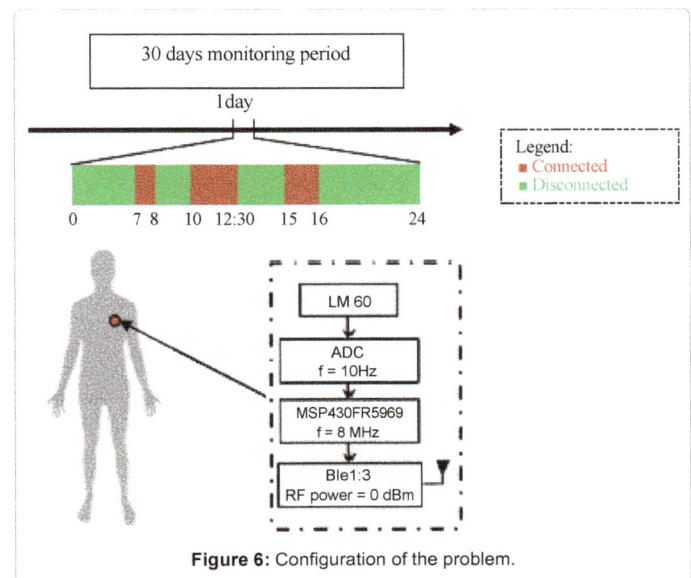

**Figure 6:** Configuration of the problem.

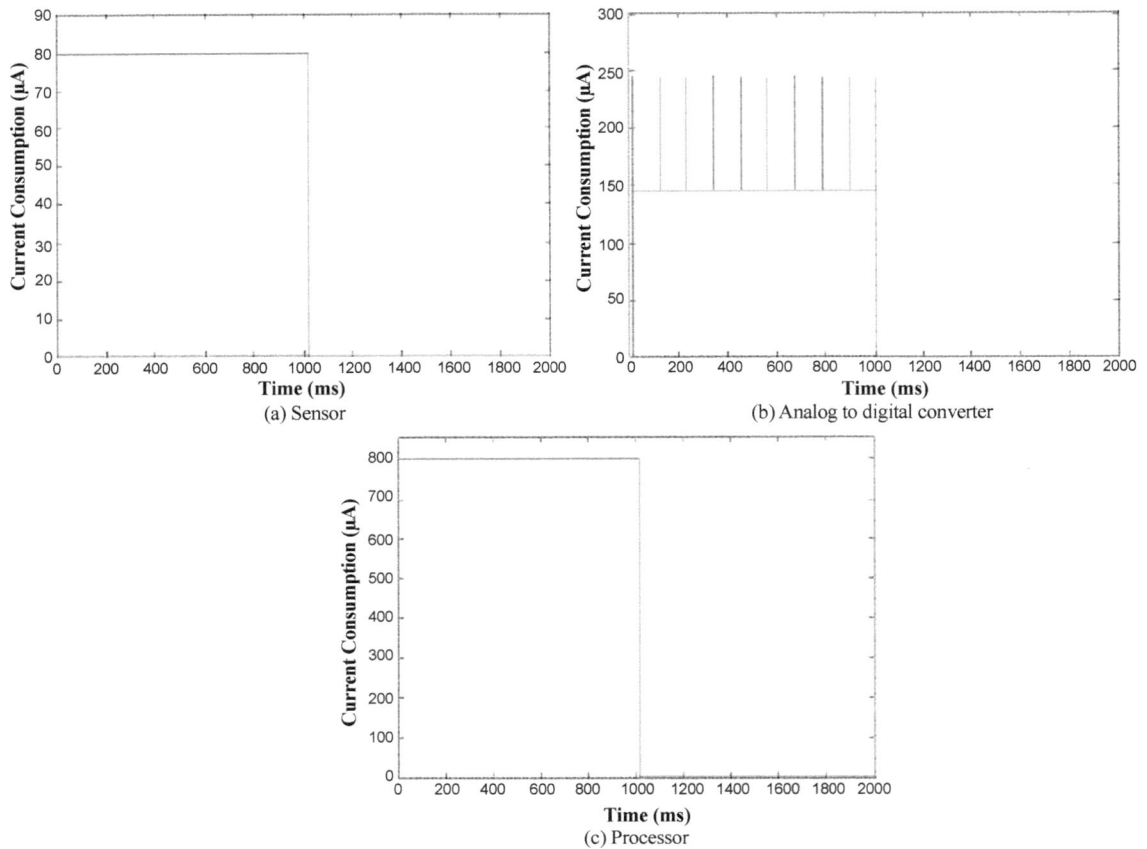

Figure 7: Current consumption registered in the sensor, the analog to digital converter and the processor.

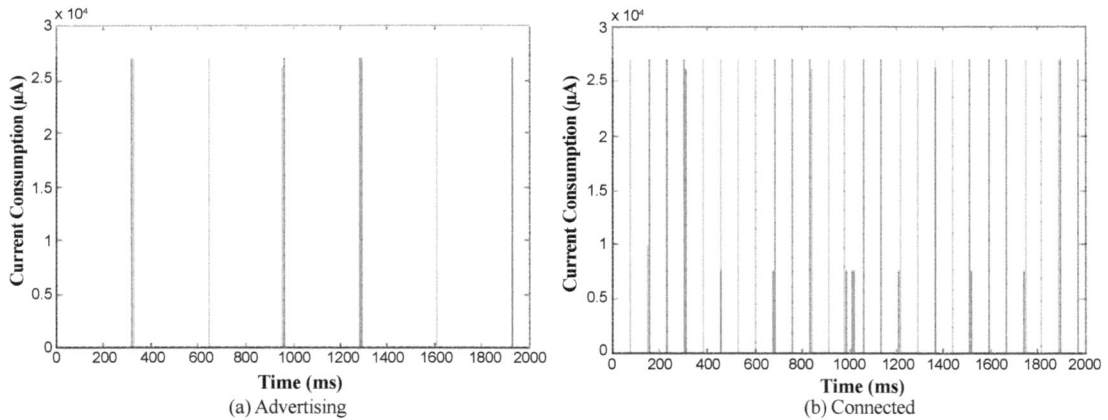

Figure 8: Current consumption registered in the RF module in advertising and connected modes.

connected, the server transmits data in advertising packets through three advertising channels. This transmission occurs in intervals of time called advertising events. After each sending of the advertising packets, the advertiser will be listening on the same channel for a while to check if there is a response coming from the gateway (i.e., the master). A connection starts whenever a master sends a connection request packet to the slave. The connection request packet can only be sent right after a successful reception of an advertising packet. It will contain connection

related information such as the connection interval. Data transfer occurs within connection events. Such an event starts when the master sends a packet to the slave at the defined connection interval. The slave can respond 150 µs after its reception. It will send a packet that may contain data or not (connection maintenance packet) [15,16].

Figure 8 shows the current consumption of the ble113 module in both advertising and connected modes. We assume an advertising interval of 320 ms (default value in the ble113 module) and a connection interval of 75 ms (default value whenever connected to a bled112 dongle). The mean current consumption registered throughout the whole monitoring period is equal to 320.12 μA.

## Guidance towards an energy aware design

Since the first three scenarios are quite straightforward, we will only be presenting an application for the Design solution.

**Design:** For this example, we assume we need to respect a 30 days minimal battery life as well as disconnection slots imposed by the daily routine of the patient. Moreover, we will be choosing between two microprocessors from TI's MSP430 series: the MSP430FR5969 and the MSP4305739. Table 4 summarizes their specifications as found in their respective datasheets. The MSP430FR5739 consumes less current in active mode than its counterpart but more in sleep mode. We will also be handpicking an energy source among the four batteries: CR2032 (230 mAh), CR2450 (620 mAh), 2xAAA (1150 mAh) and 2xAA (2100 mAh).

Our goal is to provide a user with the solution that includes the more power efficient components, the smallest possible battery and a configuration to efficiently use that energy source. This must be done according to the application requirements as well as the designer's preferences (Table 4). The proposed design included TI's MSP430FR5969 as the microprocessor and a CR2032 battery to power up the whole node. Several acceptable component's configurations are presented in Table 5. Figure 9 shows how our optimization algorithm brings us closer to the targeted consumption including each iteration.

## Physical Implementation

Having associated our design tool with an application to a wireless temperature sensor, the next step is to physically implement the design. This will allow us to compare the results obtained through simulation with practical measurements.

This process is divided into three major parts: Prototyping the node, creating an application to collect data, and measuring the current consumption.

### Prototyping a wireless sensor node

Adopting the Bluetooth Low Energy communication protocol required defining a Generic ATTribute profile (GATT). This profile mainly consisted of a "Body Temperature" service identified by a Universally Unique IDentifier (UUID). The service itself contains

**Figure 9:** Evolvement of the current consumption deviation with each iteration.

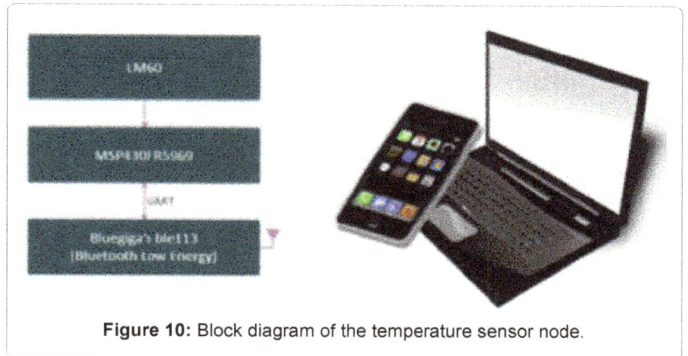

**Figure 10:** Block diagram of the temperature sensor node.

a "Body Temperature Measurement" characteristic. Whenever this characteristic's value is updated, the change will be indicated to the gateway. Indications are acknowledged.

Figure 10 shows the block diagram of the node we prototyped. Its operation is summarized by the following:

The measurement's rate is defined using the microprocessor's Real Time Clock;

The processor will wake up periodically according to this rate;

It will power up the sensor through one of its pins for 1 second;

During that time an analog to digital conversion takes place in the processor's analog to digital converter with a 10 Hz sampling frequency;

The temperature's initial value is retrieved in the processor;

It's then passed to the Bluetooth Low Energy module;

The Bluetooth Low Energy module updates the "Body Temperature Measurement" characteristic's value.

In order to achieve minimal current consumption, modules, timers and peripherals were put to sleep when not in use. Moreover, current leakage through pins had to be avoided.

### Collecting data

At this stage, we had to create a PC application in order to collect the incoming data. Such software must be able to scan for all the available connections. However it must only connect to the one advertising the desired service. The next step is to look for certain characteristics in

| Module | Current consumptions | |
|--------|-------------|-------------|
| | **Active mode** | **Sleep mode (LPM3)** |
| MSP430FR5739 | 81.4 μA/MHz | 6.3 μA |
| MSP430FR5969 | 100 μA/MHz | 0.4 μA |

**Table 4:** TI's MSP430 processors alongside their specifications.

| Configuration of the node | | | |
|---------------------------|--------------|--------------------|----------|
| Measurements' rate | CPU frequency | Sampling frequency | RF power |
| Users preferences | | | |
| ≥ One each 10 minutes | ≥ 8 MHz | ≥ 1 kHz | 0 dBm |
| Proposed configurations | | | |
| One each 5 minutes | 8 MHz | 10 kHz | 0 dBm |
| One each 5 minutes | 8 MHz | 1 kHz | 0 dBm |
| One each 10 minutes | 16 MHz | 10 kHz | 0 dBm |
| One each 10 minutes | 16 MHz | 1 kHz | 0 dBm |

**Table 5:** The proposed configurations with respect to the designer's preferences.

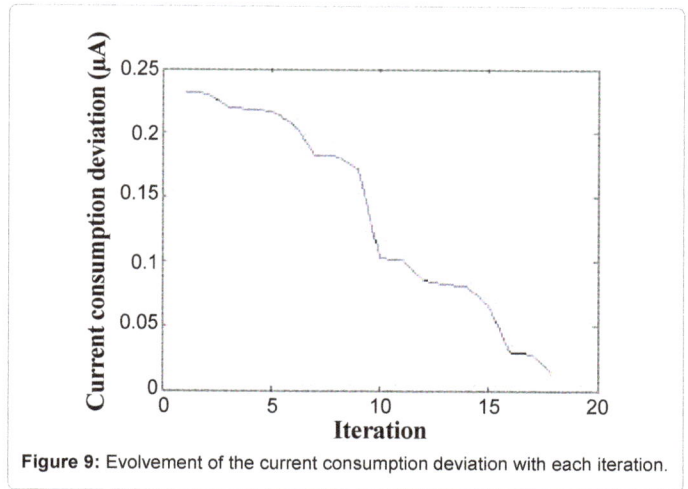

order to read, write, activate notifications/indications, etc. We chose Python as a programming language in order to benefit from multiple libraries allowing us to plot data, write it into files, access date and time, etc. The application we created will scan for Bluetooth Low Energy modules looking for the "Body Temperature" service. It will then connect to the corresponding ble113 module and activate indications for the "Body Temperature Measurement" Characteristic. Once this is done, it will start receiving temperature related information. Real time plots as well as a data log are updated with each value.

## Current measurements

Since both Texas Instruments and Silicon Labs prioritize power consumption, the developments kits were made in a way to make it easy for the user to determine the current consumption of the module. That's why we have decided to work with these kits for current measurements.

**Sensor (LM60), processor and analog to digital converter (MSP430FR5969):** These components are powered from the MSP430FR5969. Thus, measuring the latter's consumption will take into account theirs as well. Measurements were done using TI's Energy Trace [17] and an ampere meter [18]. The first outputs the power consumption's diagram as well as the mean value for current and voltage. The second will allow us to measure the current consumption in sleep mode without any influence from the debugger.

**Radiofrequency module (Ble113):** The first step would be to compare current consumption waveforms to the ones measured with a scope. This measurement, later referred to as $V_0$, is done over a 3 ohm resistor using an instrumental amplifier with a gain of 10. It will result in an inverted signal and the current consumption I can be calculated based on the basic following equation [19]:

$$I=(3.3-V_0)/30$$

This will allow us to measure the current registered during transmission and reception. It will also enable us to measure the connection and the advertising intervals. The next step is to accurately measure the module's current consumption in sleep mode using an ampere meter [19].

## Results

In this section, we compare results obtained through simulation with practical measurements in order to discuss relevance of our design tool.

## Sensor (LM60), processor and analog to digital converter (MSP430FR5969)

For experimental reasons, we are sending temperature data each 5 seconds. For this setup a mean current consumption of 153.7 µA was measured. Figure 11 matches the power measured in real life with the one simulated. Both have the same one second active period. However the consumption in active mode is much higher in simulation than in real life.

Since the results obtained do not match those of the simulation, additional measurements were made. The deviation recorded earlier was due to the current registered in active mode: this current depends on the cache hit ratio, i.e., the size and structure of the program (Table 6) [20].

We then took all these measurements and tuned our design tool. The mean current consumption obtained by simulation was 146.25 µA. Since Energy Trace technology requires a connected debugger, we had to take into account its consumption (4 µA) in idle mode and deduce it from the measured mean current consumption. The results are shown in Table 7.

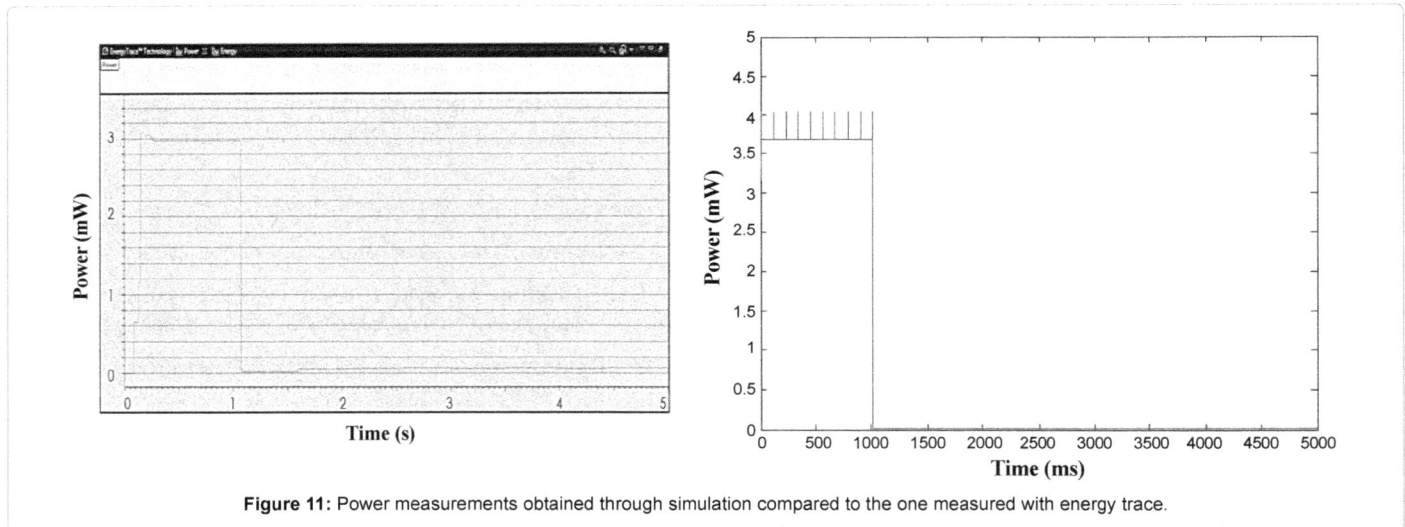

**Figure 11:** Power measurements obtained through simulation compared to the one measured with energy trace.

| Cache hit ratio (%) | Current consumption in active mode (µA) |
|---|---|
| 00 | 2510 |
| 50 | 1440 |
| 66 | 1070 |
| 75 | 890 |
| 100 | 420 |

**Table 6:** Current consumption registered in the MSP430FR5969 according to the cache hit ratio for an 8 MHz CPU frequency.

| Current consumption (µA) | Simulation | Real life measurements | Relative error |
|---|---|---|---|
| Mean value | 146.25 | 149.7 | -3.5 % |

**Table 7:** The mean current consumption measured compared to the one obtained through simulation for the sensor-ADC-processor sequence.

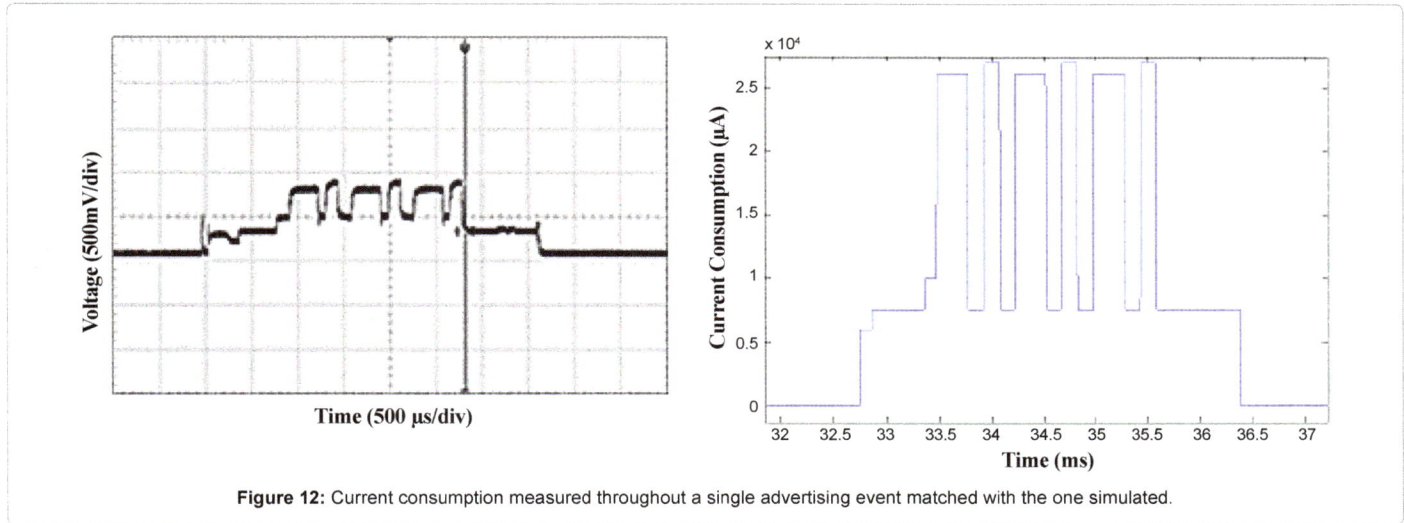

**Figure 12:** Current consumption measured throughout a single advertising event matched with the one simulated.

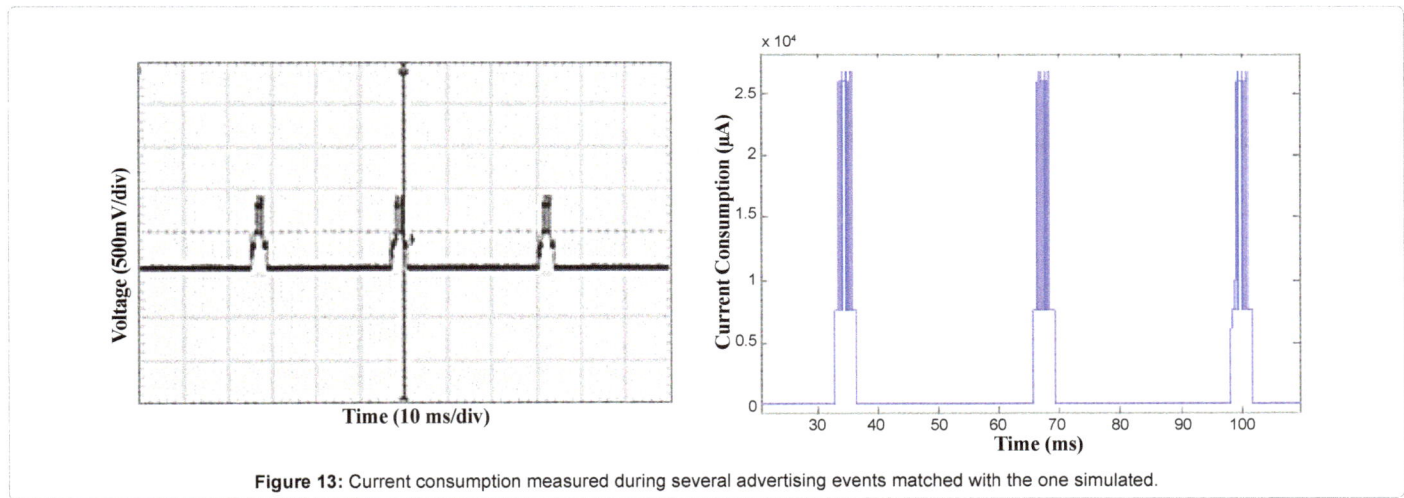

**Figure 13:** Current consumption measured during several advertising events matched with the one simulated.

## Radiofrequency module (Ble113)

We first matched the waveforms measured with the scope with the ones produced by simulation. Note that we reduced the advertising interval (32 ms). This will result in more accurate measurements although it will significantly increase the mean current consumption. Figure 12 is related to the advertising mode. It zooms in on a single advertising event allowing us to measure the currents in transmission as well as reception. It is clear that both waveforms are similar in terms of transitions and time spent in each state. Figure 13 also displays more than 1 advertising event allowing us to measure the advertising interval and compare it with the value we set. Figures 14 and Figure 15 provide the same information in connected state. Table 8 matches each measurement taken with its value as provided in the module's datasheet and used in our design tool [21].

Errors recorded over transmission, reception and active currents are slightly higher. This is due to the current consumption of the measurement setup (the 3 ohm resistor and the instrumental amplifier) [19]. This extra consumption is not a constant and depends on the current consumed by the ble113 module. Moreover, it cannot be accurately measured since the time spent in each state is relatively small. The measurement setup was disconnected when measuring the current in sleep mode (PM2).

Concerning the advertising interval, configuring it in the ble113 was done by defining a minimal and a maximal value which explains the minimal error observed. On the other hand, the connection interval was set by default as a single value in the Bluetooth Low Energy dongle. That is why the error was 0%.

Furthermore, the data obtained from simulation is similar to the one measured in the physical implementation of the temperature sensor node in terms of the consumption waveforms as well as the mean values. Thus, our design tool allows the user to predict the consumption of a node prior to its implementation with small error.

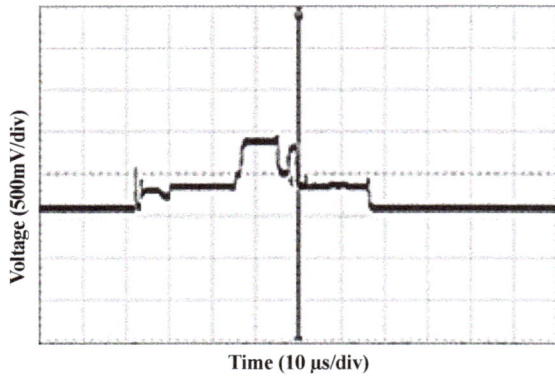

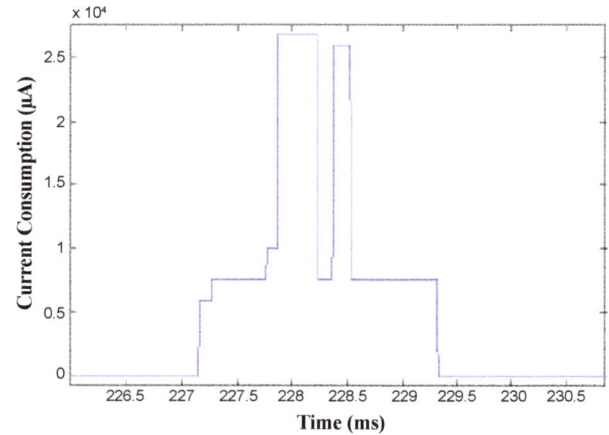

**Figure 14:** Current consumption measured during a single connection event matched with the one simulated.

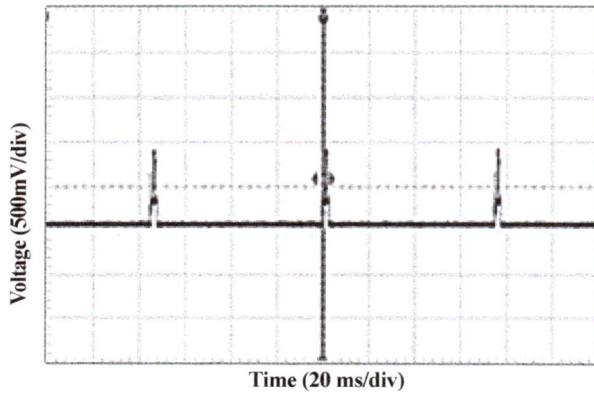

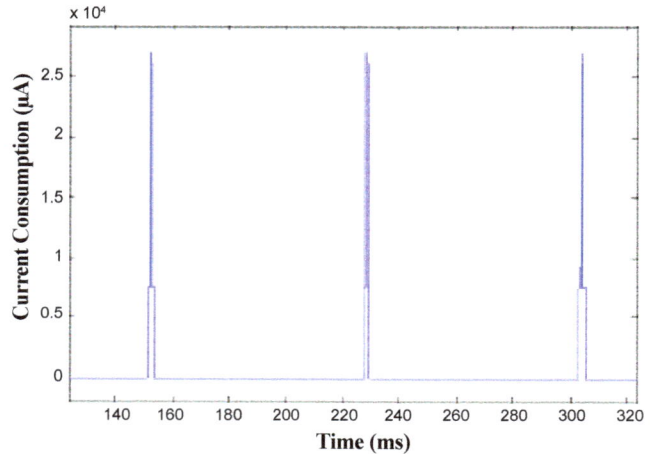

**Figure 15:** Current consumption measured during several connection events matched with the one simulated.

| Data measured | Real life measurements | Datasheet | Relative error |
|---|---|---|---|
| Transmission current (mA) | 28.3 | 26.1 | +7.77% |
| Reception current (mA) | 29.3 | 27 | +7.78% |
| Active current (mA) | 8 | 7.6 | +5.3% |
| Sleep current–PM2 (µA) | 0.93 | 0.9 | +3.22% |
| Advertising interval (ms) | 31.8 | 32 | -0.62% |
| Connection interval (ms) | 75 | 75 | 0% |

**Table 8:** Real life measurements compared to the values found in the datasheet of the ble113 module.

## Conclusion

This paper presented a Matlab based design tool for wireless sensor nodes used in remote patient monitoring. It is power oriented and enables early design explorations. It has many built-in solutions to guide the user during an energy aware design process. It can also be easily used for different setups. The deviation between simulation and experimental measurements recorded over the mean value is around 5%.

In order to guarantee a patient's wellbeing, it is vital not to lose any information. Storing data over a memory chip will ensure that and make the system more reliable. Thus it is mandatory to model memory access operations, guide the user to the data backup strategy that best suits his application and study its effect over the consumption of the node.

In addition, most medical conditions require monitoring simultaneously more than one vital sign. This requires several nodes operating within a network. However, the energy consumption as well as the quality of service depends on the number of sensors and the amount of data to transmit by each.

Furthermore, it would be interesting to include other applications such as industrial applications, civil structure monitoring, security and surveillance, smart buildings, etc.

Finally, creating a user interface would further reduce the time and effort required to design.

## Acknowledgement

Rammouz R, gratefully acknowledges the support of The Azm and Saade Association - Lebanon for funding his work. A part of this work is funded by the Lebanese University research program.

A part of this work is funded by the Lebanese University Research Program.

## References

1. Yuce MR, Khan J (2011) Wireless body area networks: technology, implementation, and applications. CRC Press.

2. Hanson MA, Powell Jr HC, Barth AT, Ringgenberg K, et al. (2009) Body area sensor networks: Challenges and opportunities in Computer, pp: 58-65.

3. Movassaghi S, Abolhasan M, Lipman J, Smith D, Jamalipour A (2014) Wireless body area networks: A survey in IEEE Communications Surveys & Tutorials. IEEE 16: 1658-1686.

4. Noury N, Fleury A, Nocua R, Poujaud J, Gehin C, Dittmar A (2009) Sensors for medical monitoring. Sensors, algorithms and networks in IRBM 30: 93-103.

5. Henderson TR, Roy S, Floyd S, Riley GF (2006) ns-3 project goals. Proceeding from the 2006 workshop on ns-2: the IP network simulator 13.

6. Tiwari V, Malik S, Wolfe A (1994) Power analysis of embedded software: a first step towards software power minimization, Very Large Scale Integration (VLSI) Systems, IEEE Transactions on 2: 437-445.

7. Chen G, Branch J, Pflug M, Zhu L, Szymanski B (2005) SENSE: a wireless sensor network simulator. In Advances in pervasive computing and networking Springer US, pp: 249-267.

8. Galos M, Navarro D, Mieyeville F, O'Connor I (2012) A cycle-accurate transaction-level modelled energy simulation approach for heterogeneous Wireless Sensor Networks. In New Circuits and Systems Conference (NEWCAS) 2012 IEEE 10th International, pp: 209-212.

9. Molina JM (2010)Energy profiling of networked embedded systems, Ph.D thesis, Vienna University of Technology, Vienna.

10. Terrasson G (2008) Contribution à la conception d'un émetteur-récepteur pour microcapteurs autonomes, Ph.D Dissertation, Université Sciences et Technologies-Bordeaux I, Bordeaux.

11. Rammouz R, Labrak L, Abouchi N, Constantin J, Zaatar Y, et al. (2015) A generic Simulink based model of a wireless sensor node: Application to a medical healthcare system. Proceedings International Conference on Advances in Biomedical Engineering (ICABME), pp: 154-157.

12. Casilari E, Cano-García JM, Campos-Garrido G (2010) Modeling of current consumption in 802.15. 4/ZigBee sensor motes. Sensors 10: 5443-5468.

13. Negri L, Beutel J, Dyer M (2006) The power consumption of Bluetooth scatternets. In Proc. IEEE Consumer Communications and Networking Conference (CCNC), pp: 519-523.

14. Gomez C, Oller J, Paradells J (2012) Overview and evaluation of bluetooth low energy: An emerging low-power wireless technology in Sensors 12: 11734-11753.

15. Dementyev A, Hodges S, Taylor S, Smith J (2013) Power consumption analysis of Bluetooth Low Energy, Zigbee and ANT sensor nodes in a cyclic sleep scenario. Proceedings IEEE International Wireless Symposium (IWS), pp: 1- 4.

16. Liu J, Chen C, Ma Y, Xu Y (2013) Energy analysis of device discovery for Bluetooth Low Energy networks. Proceedings IEEE 78th Vehicular Technology Conference (VTC Fall), pp: 1-5.

17. Texas Instruments (2014) MSP430TM Advanced Power Optimisations: ULP AdvisorTM and EnergyTraceTM Technology. Application Report.

18. Texas Instruments (2014) MSP430FR5969 LaunchPadTM Development Kit (MSP-EXP430FR5969) User's Guide.

19. Silicon Labs (2013) Ble113 Development Kit 2.0 datasheet.

20. Texas Instruments (2012) MSP430FR59xx Mixed Signa Microcontrollers datasheet.

21. Silicon Labs (2015) Bluegiga Bluetooth Smart Software v.1.4 API Documentation.

# Handcrafted Microwire Regenerative Peripheral Nerve Interfaces with Wireless Neural Recording and Stimulation Capabilities

**Ali Ajam[1], Ridwan Hossain[1], Nishat Tasnim[1], Luis Castanuela[1], Raul Ramos[1], Dongchul Kim[2] and Yoonsu Choi[1*]**

[1]*Department of Electrical Engineering, University of Texas Rio Grande Valley, Edinburg, Texas, 78539, USA*
[2]*Department of Computer Science, University of Texas Rio Grande Valley, Edinburg, Texas, 78539, USA*

## Abstract

A scalable microwire peripheral nerve interface was developed, which interacted with regenerated peripheral nerves in microchannel scaffolds. Neural interface technologies are envisioned to facilitate direct connections between the nervous system and external technologies such as limb prosthetics or data acquisition systems for further processing. Presented here is an animal study using a handcrafted microwire regenerative peripheral nerve interface, a novel neural interface device for communicating with peripheral nerves. The neural interface studies using animal models are crucial in the evaluation of efficacy and safety of implantable medical devices before their use in clinical studies. 16-electrode microwire microchannel scaffolds were developed for both peripheral nerve regeneration and peripheral nerve interfacing. The microchannels were used for nerve regeneration pathways as a scaffolding material and the embedded microwires were used as a recording electrode to capture neural signals from the regenerated peripheral nerves. Wireless stimulation and recording capabilities were also incorporated to the developed peripheral nerve interface which gave the freedom of the complex experimental setting of wired data acquisition systems and minimized the potential infection of the animals from the wire connections. A commercially available wireless recording system was efficiently adopted to the peripheral nerve interface. The 32-channel wireless recording system covered 16-electrode microwires in the peripheral nerve interface, two cuff electrodes, and two electromyography electrodes. The 2-channel wireless stimulation system was connected to a cuff electrode on the sciatic nerve branch and was used to make evoked signals which went through the regenerated peripheral nerves and were captured by the wireless recording system at a different location. The successful wireless communication was demonstrated in the result section and the future goals of a wireless neural interface for chronic implants and clinical trials were discussed together.

**Keywords:** Neural interface; Microchannel scaffold; Nerve regeneration; PDMS

## Introduction

Neural interface technologies are envisioned to facilitate direct connections between the nervous system and external technologies such as limb prosthetics or data acquisition systems for further processing. Although cultured *in vitro* neuronal networks have shown a variety of mechanisms of neuronal functionality, the major role of behavioral control by the nervous system cannot be incorporated with the *in vitro* neuronal networks system. Neuronal interface signals captured from awake, freely behaving animals are crucial for the next level of clinical applications. In amputees, such technologies would provide direct neural control of prosthetic movements and restore sensory feedback by functionally reconnecting damaged efferent motor and afferent sensory pathways. The peripheral nerve has been one target for bidirectional interfacing, with renewed interest generated by reports that peripheral nerve tissue is viable for interfacing even years after injury or amputation [1-4]. Several designs, such as cuff electrodes, flat interface nerve electrodes (FINE) [5-7], longitudinal intrafascicular electrodes (LIFE) [5,8-10], Utah Slanted Electrode Arrays (USEA) [11-13], and regenerative sieve and microchannel electrodes [14-20] demonstrated selective recording and stimulation. However, the devices have limited electrode sites and recordings can only be obtained from the limited number of nerve fascicles.

A regenerative peripheral nerve interface, developed here, can be utilized to address these goals and is designed to communicate with the brain through the peripheral nervous system. Previously, we developed 4-electrode and 8-electrode microwire regenerative peripheral nerve interfaces (µPNI) [21-26]. Here we report an advanced generation of the 16-electrode µPNI and even further advanced the µPNI with wireless communication capabilities. The whole implantable microdevice consists of a µPNI for recording placed on the transection site of the sciatic nerve, and three µCuff electrodes, one for stimulation placed on the proximal site of the transection site of the sciatic nerve and the other two for recording placed on the tibial nerve and the common peroneal nerve. Additionally, two electromyography electrode pairs were implanted on the tibialis anterior (TA) and soleus (SOL) muscles on the right hind leg to record the muscle signals during animal's locomotion tests. It gives us the capability of both electrophysiological recording and stimulation to develop a communication pathway from the brain to the endings of peripheral nerves. Peripheral nerve stimulation from one end of the µPNI initiates a neural signal pathway. Animal locomotion on a treadmill was tested in the animal facility at UTRGV and the µPNI has enabled us to analyze any sophisticated behavioral patterns.

There are three fundamental neuroscience backgrounds correlated with the µPNI (Table 1). Independent microchannel neural interfaces will be creatively achieved by microwires embedded inside the microchannel scaffolds which can be occupied by regenerated nerve

*****Corresponding author:** Yoonsu Choi, Department of Electrical Engineering, University of Texas Rio Grande Valley, Edinburg, Texas, 78539, USA
E-mail: yoonsu.choi@utrgv.edu

| Neuroscience Fundamentals | µPNI |
|---|---|
| Peripheral nerves regenerate (like hair and nails) | Custom designed Microchannel Scaffolds support nerve regeneration |
| Action potentials are recordable every 1mm from nodes of Ranvier | 1.0 mm length open microwire Electrodes in 3 mm length microchannel scaffolds |
| CNS-PNS neurons are connected from the brain to peripheral nerves | Peripheral Nerve Interface will collect details of the Brain making the process Noninvasive Brain-Machine Interfaces |

**Table 1:** Supports from fundamental neuroscience.

and develop an isolated neural signal communication. Along with peripheral nerve regeneration, the microwires on the nodes of Ranvier can cover and record neural signals selectively from an isolated neural signal source. The microchannel and microwire are long enough to cover and record neural signals from the isolated nerve branch by structural selectivity during nerve regeneration.

## Methods

### Fabrication of PDMS scaffolds

Microfluidic channel scaffolds were developed to direct peripheral nerve growth. 50 wires (160 µm in diameter) were tightly packed into Silastic® tubes (OD 1.96 mm, ID 1.47 mm; Cat. No. 508-006, Dow Corning, MI) and then were cast in liquid PDMS (Sylgard® 184, Dow Corning, MI) with a 10:1 base to curing agent ratio. They were placed in a vacuum chamber until all air dissipated and then were placed in an oven for 2 hours at 90°C to allow the liquid PDMS to solidify. The Silastic® tube and Sylgard 184® are composed of the same PDMS material and became a single structure as the liquid PDMS solidified. The solidified PDMS was soaked in chloroform, causing it to expand. The wires were henceforth, removed leaving behind a long, flexible scaffold with an array of microchannels within it. Chloroform is a highly volatile solvent, leaving no residue when it is evaporated. No special process is required for the fabrication process to clean chloroform. Chloroform-swollen PDMS was switched into 70% ethanol to clean the device while shrinking down PDMS and run sterilization process together. Then PDMS scaffolds were placed in an oven at 100°C for 20 minute to make any remaining Chloroform evaporate.

### Embedding microwires

The 75 µm diameter microwires (Stablohm 800A, California fine wire, Grover Beach, CA) were inserted in the 160 µm diameter microchannels to record the neural signals from the regenerated nerves inside microchannels. No special micromachining equipment was required and commercially available microwires were efficiently used to implement the µPNI structures. Once the PDMS scaffolds were fabricated, commercially available microwires (75 µm diameter) were embedded within their microchannels. The scaffold was cut 3 mm lengthwise and PDMS tubes, used as suture guides, were placed at both proximal and distal ends of the scaffold, hereafter referred to as proximal and distal tubes, respectively. The proximal tube was cut 3.5 mm long and placed on one end of the scaffold, thereby covering 1.4 mm of the scaffold. It was secured to the scaffold by placing a drop of liquid PDMS solution where they make contact (on the outer surface of the scaffold) and placing in an oven at 90°C for 10 minutes. The distal tube was cut 4 mm long and placed so that it covered 1.4 mm of the other end of the scaffold. To assist in the process of embedding microwires in the scaffold, a small circle with a slit leading to it was made in the distal tube. The gap between the tubes was filled with a dental cement and was subjected to ultraviolet (UV) light for 8 seconds

to cause it to harden. The distal tube was opened along the slit to facilitate embedding of microwires into the microchannel.

16 microwires were cut into 8 inch segments and 1 mm of insulation was trimmed off at the tips. They were then folded at 90° angles, 1 mm away from the uninsulated parts. The exposed wires were placed one-by-one into the microchannels through the circle made in the proximal tube and glued to the dental cement previously applied using the same technique. Dental cement was then used to seal the circle in the proximal tube, being careful not to allow any dental cement into the regenerating path of axons. All 16 wires were then braided together to make them as compact as possible and were connected subcutaneously to a head stage connector which was attached to the skull. Microwires are beneficial because they are easy to implant, permit smaller wounds, and create minimal obstruction to the regenerative path.

### Animal implantation

Surgical procedures were performed under aseptic conditions at the UTRGV Animal Facility (Figure 1). Prior to implantation, a Lewis rat was placed into an induction chamber and subjected to gas anesthesia (Isoflurane) until unconscious. The surgery locations (right thigh and top of head) were shaved and cleaned using a betadine scrub and isopropyl alcohol. Its maxillary central incisors were hooked into a gas mask through which it continued to receive small doses of anesthesia. It was secured to a surgery table and its body temperature was regulated with a hot pad. Incisions were made along the right thigh to expose the sciatic nerve, tibialis anterior (TA), and soleus (SOL) muscles. The nerve was severed, proximal to the tibial and fibular nerves, and the µPNI was implanted by suturing both the distal and proximal ends of the nerves to the guides of the device (Figure 2). EMG signals were obtained by implanting pairs of microwires (Stablohm 800A, California Fine Wires, CA) (75 µm diameter) into the TA and SOL. All electrodes were guided subcutaneously to an incision made at the top of the head and henceforth attached to a connector (Nano Strip Connector, A79022-001, Omnetics, MN), which was secured to the skull using dental cement and stainless steel screws. All procedures conformed to the Guide for the Care and Use of Laboratory Animals of the Institute of Laboratory Animal Resources, Commission on Life Sciences, National Research Council (National Academy Press, Washington, DC, 1996) and were reviewed and approved by the Institutional Animal Care and Use Committee UTRGV.

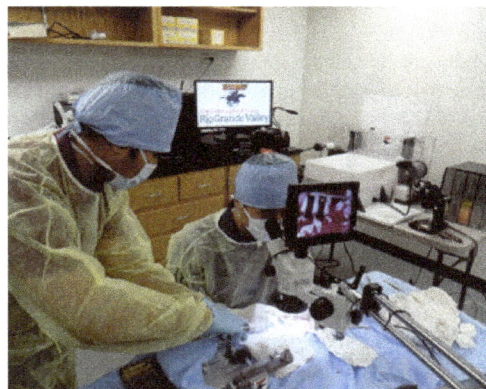

**Figure 1:** Surgery setup and implementation. All surgical procedures were done under stringent ethical standards at the UTRGV animal facility.

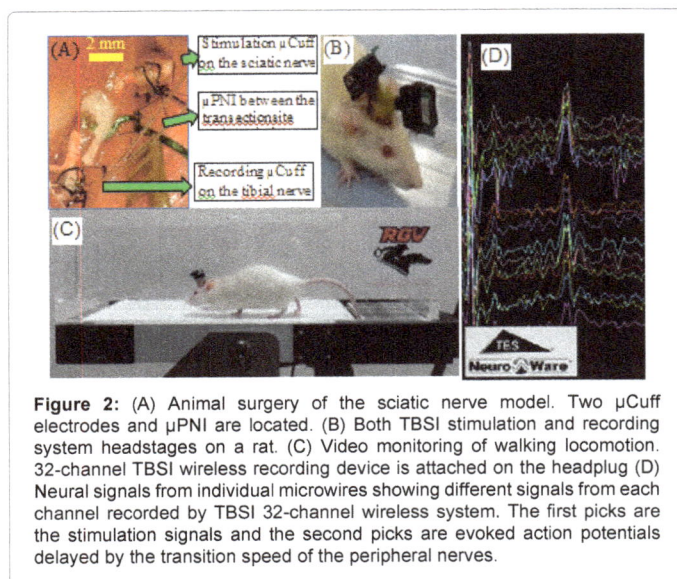

**Figure 2:** (A) Animal surgery of the sciatic nerve model. Two µCuff electrodes and µPNI are located. (B) Both TBSI stimulation and recording system headstages on a rat. (C) Video monitoring of walking locomotion. 32-channel TBSI wireless recording device is attached on the headplug (D) Neural signals from individual microwires showing different signals from each channel recorded by TBSI 32-channel wireless system. The first picks are the stimulation signals and the second picks are evoked action potentials delayed by the transition speed of the peripheral nerves.

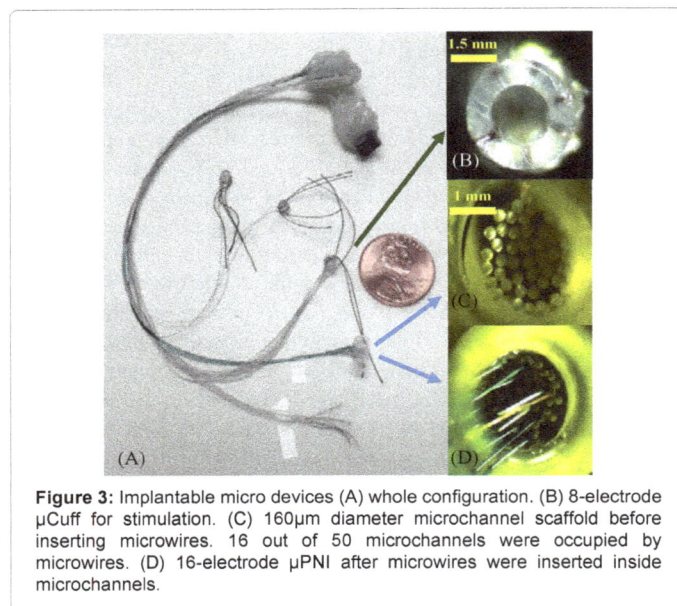

**Figure 3:** Implantable micro devices (A) whole configuration. (B) 8-electrode µCuff for stimulation. (C) 160µm diameter microchannel scaffold before inserting microwires. 16 out of 50 microchannels were occupied by microwires. (D) 16-electrode µPNI after microwires were inserted inside microchannels.

## Results

A manually fabricated implantable micro device ready for the surgery is shown in Figure 3. Both an omnetics connector (Nano Strip Connector, A79022-001, Omnetics, MN) for TBSI neuroware and a sullins connector (S9009E-04-ND, 8 Position 050, Dual Row, Digi-key) for TBSI stimware were placed at one end of the microwire bundle. The other end of the microwire bundle was connected with a µPNI, three µCuffs, and two electromyography electrodes. Peripheral nerve axons were targeted in the µPNI using microchannels that isolate different groups of axons. Since the developed fabrication technique is simple and adjustable, the scaffold parameters (length and microchannel diameter) can be modified to fit different applications. Figure 3C indicates that each microchannel (160 µm diameter) is individually separated and completely sealed. This feature aims to improve the design by reducing the crosstalk between adjacent microchannels and increasing the signal-to-noise ratio. This is a significant advantage of

the µPNI because other electrodes that are near each other can create crosstalk due to parasitic capacitances [27].

Figure 2A shows the successful surgery result. The µPNI was implanted between the transected sciatic nerve stumps at the location marked as 'transection' of the schematic design in Figure 4. All embedded electrodes were routed subcutaneously and connected to a head-mounted plug (Figure 2B). The nerve stumps were sutured on each side of the µPNI. The stimulation µCuff was placed on the proximal sciatic nerve from the µPNI. Once the microchannel scaffolds were occupied by the regenerated nerve, the stimulation signal from the µCuff was recorded from the recording µCuff on the tibial and the common peronial nerves. This confirmed the successful nerve regeneration through the µPNI. The µCuff electrodes were used as supplementary recording devices and contributed as a part of neural networks in the sciatic nerve branches. The electrophysiological locomotion data of the sNI was captured by Triangle BioSystems International (TBSI, Durham, NC) wireless system. Figure 2B shows both TBSI w-32 wireless recording system and TBSI S2W stimulator. While an animal was walking on a treadmill for behavioral pattern analysis, TBSI w-32 system recorded the electrophysiological signals from all implanted micro devices (Figure 2C). We used a rodent treadmill system that has the slop angle control capability (760306, Harvard Apparatus, South Natick, Massachusetts), which was installed in a procedure room at the animal facility at UT-RGV. TBSI S2W stimulator was used to generate the evoked signal to analyze the neural pathways and electrophysiological properties of the sciatic nerve branches, SOL, and TA muscles. Neural recording and stimulation signals were forced to flow longitudinally within the microchannel scaffolds which make each microchannel independent from all other microchannels, making it possible to retrieve specific signals. The implantable devices of the µPNI, the µCuff, and the EMG electrodes were implanted in the animal and neural signal recordings were obtained, while the animal was running on a treadmill. The TBSI wireless recording system gave the maximum flexibility for the locomotion studies. Due to the robust nerve regeneration of the sciatic nerve model, all channels were occupied with regenerated nerves. Figure 2D shows the electrophysiological signals captured by the 16-electrode microwire µPNI using TBSI wireless recording system three weeks after implantation. The neural signals through the regenerated nerves in the µPNI were recorded and analyzed to retrieve data corresponding to animal behavior patterns. Electrophysiological signals were recorded from all 16 electrodes. Although some signals showed were identical, these also suggested possible axonal branching from the regenerated nerves. With further analysis in the future work, we could determine if the multiple axons were originated from the same parent neuron to make a same neural signal pattern. It could be confirmed by histology analysis at the end of the procedure after harvesting regenerated nerve tissues. The unique neural signal patterns of the µPNI, depending on the animal behavior patterns, will not only confirm the brain-controlled neural singnals at the µPNI, but also pioneer the delicate neuronal networks in the brain linked to the sensory and motor feedback of peripheral nerves. Action potentials with similar waveforms were identified in the locomotion microelectrode recordings and extracted using a time-amplitude window discriminator routine. The average amplitude of the action potentials extracted from microchannels was about 100 µV with amplitudes ranging from 40~200 µV. Selected and repeated action potentials comparing the µPNI, TA muscle, and cuff data were clearly demonstrating the step cycles. A neural signal combination of all microwires of the µPNI, or part of them, will express a behavioral pattern at a specific temporal moment. A repeatable behavioral pattern

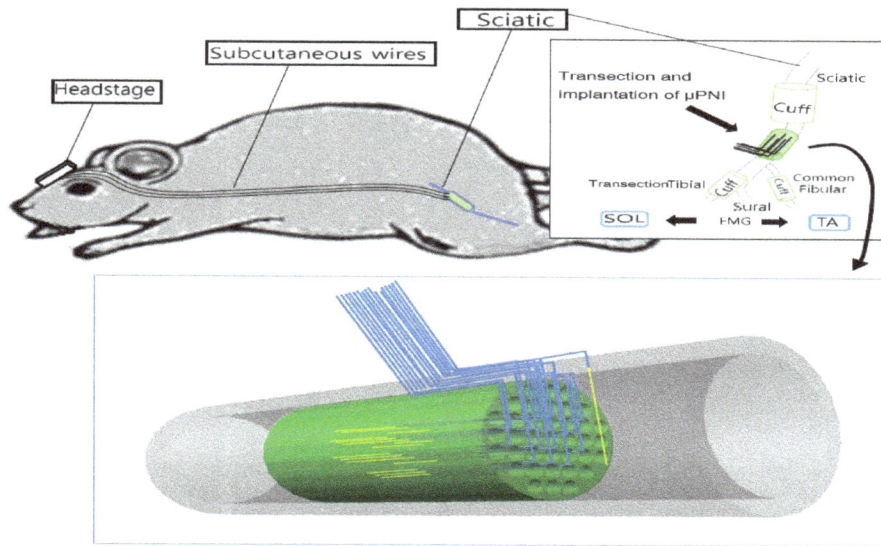

**Figure 4:** Schematic view of the animal model for μPNI, μCuff, and EMG electrodes on the sciatic nerve braches.

may express a series of temporal neural signal patterns. Thin scar tissue formation covering outside PDMS scaffolds was observed from the harvested μPNI. However, no obstructing inflammation responses was observed inside microchannels with a two-month regeneration period. PDMS is an FDA-approved biomaterial for several clinical applications. As a biocompatible material, PDMS has been used in a wide range of applications, such as a structure itself as part of the device and an insulator. PDMS cuff electrodes have been used on the extradural sacral root to sense the bladder response to stimulation in patients [7].

## Discussion

For the chronic animal studies, the microdevices need to be implanted securely inside the animal body without biological rejection and mechanical failure. Moreover, minimally invasive surgery is always required. A biological reaction to foreign materials could be significant in any chronic animal study requiring implantation, especially those that require the implanted device to be kept for more than three months. We recorded electrophysiological signals with the μPNI for two months period, since the axonal reinnervation was achieved after one month and stable muscle signals could be captured afterward. Many implantable devices have been used in the everyday clinical practice and decades of implantation is not a significant issue anymore, which makes us confident about the chronic studies. We used PDMS as a base material for all components of the μPNI. PDMS has been widely used as a major material of the implantable devices for both research and clinical purposes [28-33], due to its easy fabrication technique and biocompatibility. To achieve translational capabilities, PDMS could be replaced by biodegradable materials, such as PCL, PLGA, and PGA [34-39]. After the nerve regeneration, the biodegradable microchannel will be dissolved to give the structures as close to a natural nerve as possible. Each biodegradable material needs to be tested for its own biocompatibility and degradation rate in the peripheral nerve model.

Though we have developed the μPNI targeting the sciatic nerve model with gait analysis of the somatic nervous system, it can be easily adapted for other nerve models including the modulation of the autonomic nervous system. The developed fabrication technique of the μPNI is not dependent on the nerve size as any size of the μPNI components can be developed, ranging from few hundred micrometers to several millimeters in diameter which covers almost all major peripheral nerves and their branches. This allows for flexibility in choosing from a variety of design configurations that target specific nerve fibers.

The μPNI has a significant potential as neuroscience research test beds, if it is combined with biochemical neurotropic factors. The microchannels of the μPNI can be coated with different neurotropic factors to separate the growth of sensory or motor specific axons into the microchannels. They will encapsulate multiple neurotropic factors, such as nerve growth factor (NGF) [40], neurotrophin-3 (NT-3) [41-43], brain-derived neurotrophic factor (BDNF) [44,45], and neurotrophin 4/5 (NT-4/5) [46], ciliary neurotrophic factor (CNTF), and glial cell line-derived neurotrophic factor (GDNF). Inducing the specific axonal growth from a microchannel structure to biochemically infused microchannels could provide data for a more in-depth analysis of axon growth behavior.

## References

1. Dhillon GS, Lawrence SM, Hutchinson DT, Horch KW (2004) Residual function in peripheral nerve stumps of amputees: implications for neural control of artificial limbs. J Hand Surg Am 29: 605-615.

2. Kuiken TA, Li G, Lock BA, Lipschutz RD, Miller LA, Stubblefield KA, et al. (2009) Targeted Muscle Reinnervation for Real-Time Myoelectric Control of Multifunction Artificial Arms. JAMA 301: 619–628.

3. Rossini PM, Micera S, Benvenuto A, Carpaneto J, Cavallo G, et al. (2010) Double nerve intraneural interface implant on a human amputee for robotic hand control. Clin Neurophysiol 121: 777-783.

4. Agnew SP, Schultz AE, Dumanian GA, Kuiken TA (2012) Targeted reinnervation in the transfemoral amputee: a preliminary study of surgical technique. Plast Reconstr Surg 129: 187-194.

5. Badia J, Boretius T, Andreu D, Azevedo-Coste C, Stieglitz T, et al. (2011) Comparative analysis of transverse intrafascicular multichannel, longitudinal intrafascicular and multipolar cuff electrodes for the selective stimulation of nerve fascicles. J Neural Eng 8: 036023.

6. Tyler DJ, Durand DM (2002) Functionally selective peripheral nerve stimulation

with a flat interface nerve electrode. IEEE Trans Neural Syst Rehabil Eng 10: 294-303.

7. Grill WM, Mortimer JT (2000) Neural and connective tissue response to long-term implantation of multiple contact nerve cuff electrodes. J Biomed Mater Res 50: 215-226.

8. Serra J, Bostock H, Navarro X (2010) Microneurography in rats: a minimally invasive method to record single C-fiber action potentials from peripheral nerves in vivo. Neurosci Lett 470: 168-174.

9. Lawrence SM, Dhillon GS, Jensen W, Yoshida K, Horch KW (2004) Acute peripheral nerve recording characteristics of polymer-based longitudinal intrafascicular electrodes. IEEE Trans Neural Syst Rehabil Eng 12: 345-348.

10. Rossini PM, Rigosa J, Micera S, Assenza G, Rossini L, et al. (2011) Stump nerve signals during transcranial magnetic motor cortex stimulation recorded in an amputee via longitudinal intrafascicular electrodes. Experimental Brain Research 210: 1-11.

11. Branner A, Stein RB, Fernandez E, Aoyagi Y, Normann RA (2004) Long-term stimulation and recording with a penetrating microelectrode array in cat sciatic nerve. IEEE Trans Biomed Eng 51: 146-157.

12. Thurgood BK, Warren DJ, Ledbetter NM, Clark GA, Harrison RR (2009) A wireless integrated circuit for 100-channel charge-balanced neural stimulation. IEEE Trans Biomed Circuits Syst 3: 405-414.

13. Normann RA (2007) Technology insight: future neuroprosthetic therapies for disorders of the nervous system. Nat Clin Pract Neurol 3: 444-452.

14. Navarro X, Calvet S, Rodríguez FJ, Stieglitz T, Blau C, et al. (1998) Stimulation and recording from regenerated peripheral nerves through polyimide sieve electrodes. J Peripher Nerv Syst 3: 91-101.

15. Ceballos D, Valero-Cabré A, Valderrama E, Schüttler M, Stieglitz T, et al. (2002) Morphologic and functional evaluation of peripheral nerve fibers regenerated through polyimide sieve electrodes over long-term implantation. J Biomed Mater Res 60: 517-528.

16. Lago N, Ceballos D, Rodríguez FJ, Stieglitz T, Navarro X (2005) Long term assessment of axonal regeneration through polyimide regenerative electrodes to interface the peripheral nerve. Biomaterials 26: 2021-2031.

17. Castro J, Negredo P, Avendaño C (2008) Fiber composition of the rat sciatic nerve and its modification during regeneration through a sieve electrode. Brain Res 1190: 65-77.

18. Cho SH, Lu HM, Cauller L, Romero Ortega MI, Lee JB, et al. (2008) Biocompatible SU-8-Based Microprobes for Recording Neural Spike Signals From Regenerated Peripheral Nerve Fibers. IEEE Sensors Journal 8: 1830-1836.

19. Garde K, Keefer E, Botterman B, Galvan P, Romero MI (2009) Early interfaced neural activity from chronic amputated nerves. Front Neuroeng 2: 5.

20. FitzGerald JJ, Lago N, Benmerah S, Serra J, Watling CP, et al. (2012) A regenerative microchannel neural interface for recording from and stimulating peripheral axons in vivo. J Neural Eng 9: 016010.

21. Gore RK, Choi Y, Bellamkonda R, English A (2015) Functional recordings from awake, behaving rodents through a microchannel based regenerative neural interface. J Neural Eng 12: 016017.

22. Choi Y, Gore RK, English AW, Bellamkonda RV (2012) Multilumen PDMS scaffolds for peripheral nerve repair and interface. In: the 40th Neural Interfaces Conference, Salt Lake City, UT, 2012.

23. Choi Y, Park S, Chung Y, Gore RK, English AW, et al. (2014) PDMS microchannel scaffolds for neural interfaces with the peripheral nervous system. In: The 27th IEEE International Conference on Micro Electro Mechanical Systems (MEMS 2014), San Francisco, CA.

24. Choi Y, Shafqat F, Heo H, Bellamkonda RV (2012) Development of micro channel nerve grafts and their application in peripheral nerve regeneration. In: 2012 International Annual Symposium on Regenerative Rehabilitation, Pittsburgh, PA.

25. Gore RK, Choi Y, English AW, Bellamkonda RV (2012) Peripheral neural recordings in awake and behaving rats after sciatic regeneration through a microchannel based neural interface. In: Neuroscience 2012, Society for Neuroscience 42nd annual meeting, New Orleans, LA.

26. Hossain R, Kim B, Pankratz R, Ajam A, Park S, et al. (2015) Handcrafted multilayer PDMS microchannel scaffolds for peripheral nerve regeneration. Biomed Microdevices 17: 109.

27. Takei K, Kawano T, Kawashima T, Sawada K, Kaneko H, et al. (2010) Microtube-based electrode arrays for low invasive extracellular recording with a high signal-to-noise ratio. Biomed Microdevices 12: 41-48.

28. Kurstjens GaM, Borau A, Rodriguez A, Rijkhoff NJM, Sinkjaer T (2005) Intraoperative recording of electroneurographic signals from cuff electrodes on extradural sacral roots in spinal cord injured patients. The Journal of Urology 174: 1482–1487.

29. Stöver T, Lenarz T (2009) Biomaterials in cochlear implants. GMS Curr Top Otorhinolaryngol Head Neck Surg 8: Doc10.

30. Malcolm RK, Edwards KL, Kiser P, Romano J, Smith TJ (2010) Advances in microbicide vaginal rings. Antiviral Res 88 Suppl 1: S30-39.

31. Grill WM, Craggs MD, Foreman RD, Ludlow CL, Buller JL (2001) Emerging clinical applications of electrical stimulation: opportunities for restoration of function. J Rehabil Res Dev 38: 641-653.

32. Martens FM, Heesakkers JP (2011) Clinical results of a brindley procedure: sacral anterior root stimulation in combination with a rhizotomy of the dorsal roots. Adv Urol 2011: 709708.

33. Andriot M, Degroot JV, Meek J, Meeks R, Gerlach E, et al (2007) Silicones in Industrial Applications. Inorganic Polymers, RD Jaeger and M. Gleria, Eds ed: Nova Science Publishers pp. 61-161.

34. Sivak WN, Bliley JM, Marra KG (2014) Polymeric Biomaterials for Nerve Regeneration: Fabrication and Implantation of a Biodegradable Nerve Guide. In: Axon Growth and Regeneration A. J. Murray, Ed ed: Springer pp. 139-148.

35. Liu JJ, Wang CY, Wang JG, Ruan HJ, Fan CY (2011) Peripheral nerve regeneration using composite poly (lactic acid-caprolactone)/nerve growth factor conduits prepared by coaxial electrospinning. Journal of Biomedical Materials Research Part A 96A: 13-20.

36. Nectow AR, Marra KG, Kaplan DL (2012) Biomaterials for the development of peripheral nerve guidance conduits. Tissue Eng Part B Rev 18: 40-50.

37. Reid AJ1, de Luca AC, Faroni A, Downes S, Sun M, et al. (2013) Long term peripheral nerve regeneration using a novel PCL nerve conduit. Neurosci Lett 544: 125-130.

38. Kehoe S, Zhang XF, Boyd D (2012) FDA approved guidance conduits and wraps for peripheral nerve injury: a review of materials and efficacy. Injury 43: 553-572.

39. Mobasseri A, Faroni A, Minogue BM, Downes S, Terenghi G, et al. (2015) Polymer scaffolds with preferential parallel grooves enhance nerve regeneration. Tissue Eng Part A 21: 1152-1162.

40. Lotfi P, Garde K, Chouhan AK, Bengali E, Romero-Ortega MI (2011) Modality-specific axonal regeneration: toward selective regenerative neural interfaces. Frontiers In Neuroengineering 4: 11-11.

41. Sterne GD, Brown RA, Green CJ, Terenghi G (1997) Neurotrophin-3 delivered locally via fibronectin mats enhances peripheral nerve regeneration. Eur J Neurosci 9: 1388-1396.

42. Yamauchi J, Chan JR, Miyamoto Y, Tsujimoto G, Shooter EM (2005) The neurotrophin-3 receptor TrkC directly phosphorylates and activates the nucleotide exchange factor Dbs to enhance Schwann cell migration. Proc Natl Acad Sci U S A 102: 5198-5203.

43. Sahenk Z, Oblinger J, Edwards C (2008) Neurotrophin-3 deficient Schwann cells impair nerve regeneration. Exp Neurol 212: 552-556.

44. Utley DS, Lewin SL, Cheng ET, Verity AN, Sierra D, et al. (1996) Brain-derived neurotrophic factor and collagen tubulization enhance functional recovery after peripheral nerve transection and repair. Arch Otolaryngol Head Neck Surg 122: 407-413.

45. Sendtner M, Holtmann B, Kolbeck R, Thoenen H, Barde YA (1992) Brain-derived neurotrophic factor prevents the death of motoneurons in newborn rats after nerve section. Nature 360: 757-759.

46. Liu X, Jaenisch R (2000) Severe peripheral sensory neuron loss and modest motor neuron reduction in mice with combined deficiency of brain-derived neurotrophic factor, neurotrophin 3 and neurotrophin 4/5. Dev Dyn 218: 94-101.

# 19

# Enhanced Algorithms for Fault Nodes Recovery in Wireless Sensors Network

**Darwish IM\* and Elqafas SM**

[1]Institute of Postgraduate Studies and Research, Selangor, Malaysia
[2]Arab Academy for Science, Technology and Maritime Transport, Giza Governorate, Egypt

## Abstract

An integration of sensing environment with the numerous deployments of sensor nodes in Wireless Sensor Network (WSN) causes the severe security threats and hence the trust assurance mechanisms are required. For the large scale WSN, the existence of a number of intermediate nodes is responsible for the data forwarding to the sink node. Due to the battery operated sensors, the recharge and replace mechanisms suffer from the energy conservation and minimum network lifetime. The identification of fault nodes on the transmission path plays the major role in energy conservation. With the dense deployment of sensor nodes, the failures in node and link are high that disrupts the entire communication. This paper proposes the suitable alternative fault-free path prediction model to perform the communication among the nodes. Initially, the sensor nodes are deployed in the WSN environment. Once the initialization of source and destination nodes are over, the path between them is predicted through the Hamiltonian path prediction model. During the failure scenario, this paper estimates the node and link parameters such as Received Signal Strength Indicator (RSSI), queue size, response time, and bandwidth are individually estimated and group them into the Quality Factor (QF). Based on the QF, the proposed work predicts the fault-free link to alleviate the unnecessary transmissions to the fault node and reduces the energy consumption. The comparison between the proposed Hamiltonian Path-based Hyper Cube (HPHC) network with the existing fault detection mechanisms regarding the performance measures such as Packet Delivery Ratio (PDR), fault node detection rate, throughput and end-to-end delay assures the effectiveness of HPHC in WSN communication.

**Keywords:** Link failure handling; Link quality factor; Reliable data delivery; Routing protocol; Stable routing

## Introduction

Large scale sensing technologies are integrated with the several wireless communication links open up the various research issues regarding the minimum energy and network lifetime in Wireless Sensor Networks (WSN) [1]. The existence of shared unreliable transmission medium among the nodes induces the security threats during the communication. There are numerous schemes are developed to address the security issue in which they are limited by four factors as follows: architectural differences, limitations of sensor nodes, network density and size. Besides, the high computation, overhead, energy and memory requirement are high for the necessary solutions of WSN. For large scale WSN, the intermediate nodes perform the routing of data packets towards the sink with the following characteristics:

- Due to the limited resource availability and the large size nodes deployment induces the difficulties in the trust-based scheme.

- Traffic routed from base station to all nodes is high.

- Effective resource utilization against the number of constraints is essential.

The organization of routing is split-up into two stages such as single path routing and multi-path routing [2].

### Single path routing

The simple and scalable routing established between the source and destination nodes for the specific period. The selection of an intermediate node by the source node is the repetitive task which increases the power depletion and minimizes the network lifetime. Besides, the data manipulation is corrupted with the presence of malicious node on the routing path and the lack of fault tolerant mechanism.

### Multi-path routing

In this routing, a multiple number of paths are extracted to deliver the data from source to destination. Due to the intensive number of multiple paths, the assurance of reliability, integrity and load balancing is the major requirement alternative to the single path routing.

The maintenance of powerfully connected topology at all the times is the major requirement for an effective communication. The generation of faults induces the failures in WSN due to the following major issues [3]:

- Problems in fabrication process.

- Environmental factors.

- Battery power depletion

- Enemy attacks.

The failure in node causes the partition of the whole network into disjoint blocks and breaks the network connectivity adversely. The Quality of Services (QoS) also affected by such type of failure nodes. The deviations from the normal behavior of nodes due to the random faults disrupt the network functions. The unaware of structure and

---

**\*Corresponding author:** Darwish IM, Researcher, Institute of Postgraduate Studies and Research, Arab Academic of Science and Technology, Selangor, Malaysia, E-mail: imsaad73@gmail.com

state of WSN by the faulty nodes leads to the ignorance of other faults generated in a network. Besides, the misbehaving nodes have the capability to collide with another node with the required knowledge about structure. In both cases, the structural knowledge is the major requirement to identify the faulty nodes in WSN. Recently, the evolution of linear consensus algorithms considered the misbehaving or faulty nodes and sends the information to the neighbors. The characterization of resilience properties [4] of linear consensus strategies also required analyzing the connectivity among the nodes.

The route establishment comprises two steps transmission of Routing Request (RREQ) and replies (RREP) among the neighbors which lead to the power consumption. Due to the extensive power consumption, the battery life is degraded and the nodes in the network will become no longer available. Hence, the evolution of directed diffusion algorithms alleviates the power consumption problem by transmitting the neighbors to the first set. The nodes available in network elect the neighbor based on the hop count or rules with the exchange of request and replies [5]. The number of diffuser tower is increased drastically for large size WSN. The evolution of clustering approaches [6] addresses the issues in energy/power consumption minimization. The communication among the head branches reduce the diffusion tower that leads to reduction of power consumption effectively.

The fault detection approaches [7,8] are categorized into two as follows: architectural and methodological dimensions. On the basis of the architectural view, the detection approaches are classified into centralized and distributed. In centralized approaches, the fusion center is responsible for the collection of all the sensor measurements and selects the reliable subset which eventually sent to the base station. In distributed approaches, the exchange of sensor measurements with nearby nodes without any fusion center. The majority voting scheme, threshold scheme and Bayesian methods are available for methodological point of view. Among them, the Bayesian formulation considers the background information like sensor fault to classify the measurement as corrupted or not. The major deficiency of Bayesian formulation is the exponential growth of computational complexities and energy consumption. Hence, the suitable framework is needed to provide the trade-off between the effective communication and minimum energy consumption in WSN. This paper proposes the suitable fault-free path selection to achieve the effective communication with minimum computational resources. The technical contributions of proposed work are listed as follows:

- Hamiltonian-hyper cube construction predicts the necessary fault-free path on the basis of the link quality measures. The quality factor estimation depends on the three major parameters such as Received Signal Strength Indicator (RSSI), bandwidth and queue size.

- The simultaneous inclusion of node and link failures in link quality factor estimation validates the link and switch over to the adjacent link if it is unstable.

- The integration of Hamiltonian Path with the Hyper Cube (HPHC) enables the self-configuration of WSN environment with the best link information and the immediate update of the link failure to the nodes.

- The table maintenance is associated with all the nodes is responsible for the faulty node detection.

The paper organized as follows: The detailed description about the related works on fault-free path selection mechanism is discussed in section 4. The implementation process of Hamiltonian Path-based Hyper Cube (HPHC) is described in detail in section 5. The comparative analysis of proposed approach with existing fault detection framework is provided in section 6. Finally, the conclusions about the application of cube-extension of path prediction presented in section 7.

## Related Work

This section discusses the traditional fault-discovery schemes to achieve among the WSN sensor nodes under failure conditions. Due to the powering of sensor nodes by the battery, the replacement and recharge of them are difficult to issue and hence the energy-efficient routing is the major concern in WSN applications. Jain [9] presented the brief review of the flat and data centric routing techniques with the necessary comparison. The brief survey concluded that more efficient, scalable and robust routing schemes are to be required to reduce the energy consumption and improves the battery lifetime effectively. The proper functioning of network is the necessary task under failure conditions observed in some components. Chouikhi et al. [10] presented the overview of mechanisms that improved the fault tolerance property. The solutions highlighted in this survey focused on detection and prevention of fault occurrence in energy aware routing and data aggregation mechanisms. The occurrence of selfish or malicious nodes disrupts the communication among the nodes by considering the multi-dimensional trust attributes. Bao et al. [11] proposed the highly scalable cluster-based hierarchical trust management protocol for WSN. The utility of hierarchical trust management protocol was demonstrated by applying the trust-based geographic routing and intrusion detection approaches. The existence of optimal trust threshold value effectively minimized the false positives and false negatives. The maintenance of high-quality WSN dependent on the delay requirement. Duche and Sarwade presented the new method to detect the sensor node failure or malfunctioning in the WSN environment. The utilization of Round Trip Delay (RTD) provided the accurate measure of confidence factor of RTD path. On the basis of confidence factor, the faulty nodes were detected [12,13].

Lee and Choi [14] presented the distributed fault detection algorithm that identified the faulty sensor nodes on the basis of the comparisons between the neighboring nodes and decision. During the sensing of the communication process, the time redundancy was used to tolerate the transient faults in the system. The employment of sliding window eliminated the delay in redundancy measurement. The extension of sensor nodes and their operation caused the inactive nodes were unaware of communication strategy. Hence, the split-up of the network (normal and inactive) was the difficult stage during the communication. To avoid this issue, Akbari et al. [15] designed the suitable techniques to maintain the cluster structure during the two scenarios such as faulty condition and energy drained cases. On the basis of the residual energy, the Cluster Head (CH) and secondary CH were selected. The energy consumption and the remaining energy available during the cluster formation were measured effectively. The lifetime improvement, fault discovery and recovery were the major constraints in effective communication. Paradkar et al. [16] applied the grad diffusion and genetic algorithms to find the lost node information and the recover the routing path effectively. The time-out mechanism was used to detect the hard faults effectively. The dissemination of all diagnostic information was the major requirement to assure the global view of fault status of WSN. Mahapatro and Khilar [17] assumed the cluster-based routing mechanism where the nodes were organized into one-hop clusters. The spanning tree construction spanned all the CHs disseminated the local diagnostics effectively. The binary variable was employed for distributed fault detection. When the event was detected, the binary variable assigned to be 1 and it was 0 for the undetected events. Ould-Ahmed-Vall et al. [18] proposed the new approach which considered the different failure probabilities, drift overtimes, and calibration related failures in different accuracy levels. High spatial correlation was the major requirement to analyze the failure conditions. With the increase in number of sensors, the failure in sensor nodes also increased drastically. Lo et al. [19] presented the distributed, reference free fault detection mechanism on the basis of the local pair-wise verification between the sensors. Based on the relationship, the faulty nodes were detected. The decentralized fashion based fault detection mechanism ensured the energy saving capability effectively. Research studies turned their works into energy-aware distributed fault tolerant topology control algorithm namely Adaptive Disjoint Path Vector (ADPV). Deniz et al. [20] ensured the secure super connectivity among the nodes through the ADPV algorithm. The ADPV comprised the two phases such as single initialization and restoration. They obtained the two-fold increase of super connectivity among the nodes compared to the conventional DPV algorithm. Ding et al. [21] proposed the Multi-Particle Swarm Immune Cooperative Algorithm (MPSICA) to provide the intelligent route discovery and improve the fault tolerant capability. Lau et al. [22] proposed the Centralized Naïve Bayes Detector (CNBD) that analyzed the end-to-end transmission time and collected the necessary information. The minimum power burden to the battery of each sensor nodes was achieved. Chanak and Banerjee [23] performed Fuzzy-rule formation for the fault detection and uncertainties prediction that provided high effectiveness in fault recovery.

## Hamiltonian Path Based Hyper Cube for Route Discovery

This section discusses the implementation of proposed Hamiltonian Path based Hyper Cube (HPHC) for the routing path estimation to deliver the data to the sink node by avoiding the node and link failures. The routing path comprises the set of tiny immediate sensor nodes to accept the data from the source node and forward them to sink node through the number of neighbor nodes. The failures in node and link level cause the disruption in communication. The failure report to the base station creates as many as routing requests and retransmissions that cause the overhead and time complexity. Besides, the energy consumed by the sensor nodes is increased due to the following factors:

- Communication in failure path.

- No. of retransmissions due to the node/link failures.

To alleviate these issues, this paper focuses on the fault-free communication among the nodes under node and link failures. The proposed work namely HPHC constructs the hyper cube structure on the basis of the Hamiltonian path with the Harmony search algorithm. Figure 1 shows the workflow of proposed HPHC in WSN applications. Initially, the WSN environment is constructed with the number of sensor nodes and base station on the suitable locations. The presence of faulty nodes violates the link and affects the communication adversely. The failures observed during the data transmission are observed in node and link levels. In node level, the sequential measurements such as Received Signal Strength Indicator (RSSI), velocity and relative velocity are performed.

### Background: Ad-hoc on demand vector

With the infrastructure less and the existence of unreliable links, the routing of data to the sink node induces the energy saving requirements. The scarcity in energy consumption and the kept of

sensors in operational stages throughout the simulation period are the major issues to be considered during the development of routing protocols. The manual assignment of unique identifiers unique identifiers is infeasible. The utilization of potentially unique identifier through the MAC address and GPS coordinates increase the payload in the messages. To alleviate these issues, the Ad-hoc On Demand Vector (AODV) [16] is used as the background framework for the proposed cube structure.

The set of tables is maintained in the AODV is used to find the information regarding the paths available, position, energy and speed. When a node wishes to transmit the message (source node) to the other node, it initiates the route discovery process to identify the location of destination node. The source node floods the query packet to all the nodes in the network. The acknowledgement reply from the node to the source node is responsible for the creation of route between them. Once the route request (RREQ) is initiated, the intermediate nodes update their routing table during forward and reverse route. Figure 2 shows the AODV workflow among the source (S) and destination (D) nodes.

Initially, the AODV utilizes the sequence numbers to determine

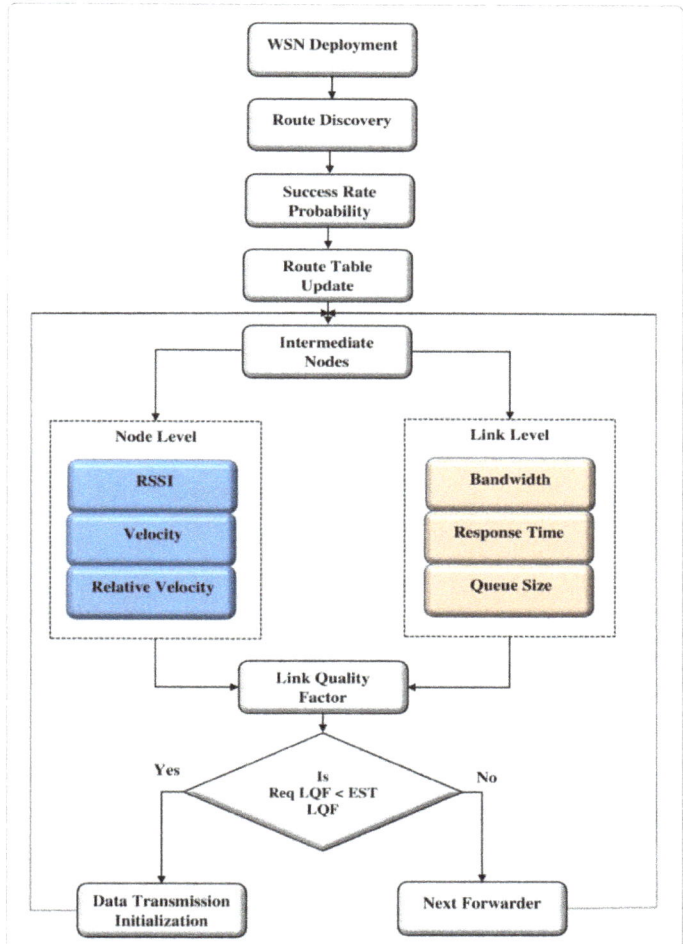

**Figure 1:** Workflow of HPHC scheme for faults tolerant WSN. In the link level, the bandwidth, response time and the queue size are computed. Based on these measurements, the Link Quality Factor (LQF) is computed. If the estimated LQF is less than the actual LQF, then the data transmission is initiated. Otherwise, the next forwarder is selected and the controller is transferred to the intermediate nodes.

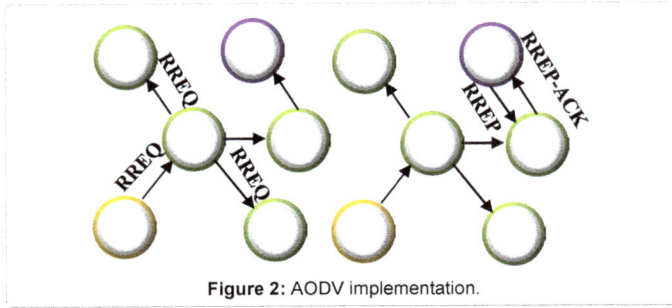

**Figure 2:** AODV implementation.

the timeliness of each packet with the avoidance of loops. The kept of all the routes is fresh with the help of expiry timers. The maintenance of advantages of fundamental distance vector routing components with the evaluation philosophy to check the operation. The addition of malicious node to the AODV will induce the fault model (failure scenario) in WSN formation. The node which is declared as malicious node drop out the packets during the communication. The steps modified in the AODV protocol in this paper are listed as follows:

1. Initialize the malicious variable with the value 'false' and this declaration inside the constructor is listed as follows:

*AODV::AODV(nsadrr_t id):Agent(PT_AODV)...*

{

.....

Malicious=false;

}

2. The TCL script is updated as follows

If (strcmp(argv (1) "malicious")==0)

{

Malicious-true;

Return TCL-OK

}

3. Then, implement the behavior of node on the basis of the rt_ resolve function used in AODv protocol.

If (malicious==true)

{

drop (p_DROP_RTR_ROUTE Loop)

}

With these modifications, the AODV is selected in this paper and integrate with the Hamiltonian path to estimate the relevant path for data transmission and minimum energy communications.

## Hamiltonian path estimation

The number of tiny sensor nodes deployed in WSN environment has the capability to collect the information about the zone of interest and track the target over the specific region to deliver the services. The major tasks performed in WSN are sensing and transmission. The fault node present in the network affects the sensing and transmission tasks. The failures observed in WSN model are categorized into two levels as follows:

**Node Level:** The failures occurred in node level are caused either by hardware (sensing unit, CPU, memory, network interface and battery etc.) or software (routing, MAC and application) malfunctioning. If the battery energy falls below the level, the sensing unit provides the incorrect reading that leads to improper data acquisition. The major fault tolerance solution is to implement the any mechanism that minimizes the energy consumption and improves the network lifetime. During the node failure conditions, the RSSI, velocity and relative velocity are measured respectively.

**Link level:** The interferences between the WSN nodes and the packet collisions lead to loss of transmitted data. The paths built by the routing protocol lead to the dropping and loss in transmission data. During the selection of routes, the routing protocol considers the requirements of applications. During the link failure conditions, the bandwidth, queue size and response time are estimated accordingly.

In both cases, the occupation of data packets in the queue within the frame length limits the link capacity. Then, the time required for the data forwarding and the acknowledgement is more. The existence of obstacles among the sender and receiver nodes makes the RSSI as unsuitable for link formation. The ratio of difference between the total queue size and occupied queue size to the frame length is referred as queue size (Qs) and it is formulated as follows:

$$Qs = \frac{Total \text{ Queue size - Occupied Queue size}}{Frame \text{ length}} \quad (1)$$

Then, the mathematical formulation of response time is expressed as follows:

$$Rt_i = \left(T_{DR} - T_{DDT}\right) + \left(T_{ACKT} - T_{ACKDR}\right) \quad (2)$$

Where, $T_{DR}$–Time of data packet accumulated in queue of received port.

$T_{DDT}$–Time of data removed from the queue of transmission port.

$T_{ACKT}$–Time of acknowledgement packet accumulated in transmission port.

$T_{ACKDR}$–Time of acknowledgement packet removed from receiver port.

In the minimal energy consumption mode, the RSSI value is estimated in this paper. The distance between the nodes is calculated using a constant value 'k' and power required for transmission/ reception (Pt, Pr) as follows:

$$d = \sqrt[4]{k.Pt/Pr} \quad (3)$$

Relative velocity is calculated by:

$$\overline{v} = \Delta d / \Delta t \quad (4)$$

Finally, the LQF is estimated with the above formulations as follows:

$$LQF\left(S, D\right) = \sum_{i=0}^{n} \left(QS_i + Rt_i + BW_i\right) \quad (5)$$

With the above formulation, the quality factor of each link is estimated. If the estimated value is less than the required value, then the transmission is initiated. Otherwise the next forwarder is selected on the basis of Hamiltonian fault-free path estimation [24]. The algorithm used to estimate the path between the nodes is described as follows:

| Hamiltonian path |
|---|
| **Input**: Number of nodes (n), Location n(x, y), Node Failure Rate (NFR) (Z)<br>**Output:** Fault free path |
| Step 1: Deployment of sensor nodes in field of in range 1000 X 1000<br>Step 2: Placing the Base Station in Center of Region (Mid (Rx, Ry))<br>Step 3: Calculate distance between the neighbor nodes to reach the base station<br>$D = \sqrt{\left(X_i - X_j\right)^2 + \left(Y_i - Y_j\right)^2}$<br>Step 4: if (D < TxRange200) Then ALM (i, j) =1 //Adjacency Link Matrix (ALM)<br>Step 5: Calculate the Node Failure Rate $NFR(i, j)=((Z_i)+(Z_j))/2$<br>Step 6: Find the intermediate hop by hop node estimation (Xi, NFR(i, j))<br>// To Choose hop by hop high reliable node in path<br>Step 7: Create Graph using the Node as edges and calculated NFR value as vertices<br>Step 8: Use Hamiltonian Path A utility function to check if the vertex v can be added at index 'pos' in the Hamiltonian Cycle constructed so far (stored in 'path[]') |

The interconnection among the nodes in WSN is modelled as the simple graph [25] called G=(V, E) which contains set of edges (E) and nodes (V). The travelling cycle in which each node in the graph G traverses exactly once refers the Hamiltonian cycle and such type of graph refers Hamiltonian. The faulty nodes ($f_v$) and edges ($f_e$) are referred as the faulty set (f). Then, the graph is said to be fault-free only if the set $f$ is not equal to 0. The Hyper cube (HQ) construction through the Hamiltonian formulation isolates the faulty set from the normal set effectively. The node is labeled with the unique n-bit string as its address and is adjacent to the n distinct nodes. The adjacency or neighbor nodes modeling are performed by n-dimensional HQ with the complementary in bits. Let us consider the two four dimensional HQs namely $0\text{-}HQ_n$ and $1\text{-}HQ_n$ as shown in Figure 3.

The set of crossing edges in the hyper cube is described as:

$$E^c = \left\{ (x,y) \middle| (x,y) \in E\left(HQ_{n+1}\right), x \in HQ_n^0 \ and \ y \in HQ_n^1 \right\} \quad (6)$$

Where, x, y–Arbitrary nodes in hyper cube.

If the arbitrary nodes are located in the same hyper cube, then the first Hamiltonian path (p) and cycle (c) are constructed. By merging of path and cycle, the fault-free Hamiltonian path is constructed. The major characteristics of the fault free Hamiltonian path are described as follows:

- Hamiltonian path is present in HQ1.

- Fault free Hamiltonian path is described as $HQ_2\text{-}f$, where, $F$ defines only one faulty node and edge.

- Fault-free Hamiltonian path in $HQ_3\text{-}f$, Where $f \leq 2$.

- Fault-free Hamiltonian path $HQ_{n+1}$ with $f_v + f_c \leq n$.

With the above characteristics, the fault-free path is selected to provide the effective communication among the nodes in WSN.

## Recursive hyper cube network

The popular technology in nowadays is the interconnection of the nodes in communication is hypercube network. The one-dimensional hypercube network is modeled as the graph with two vertices ($v_0$, $v_1$) with the following conditions [26]:

- Vertex set $V=v_0 U v_1$.

- The matched edge set $M \sqsubset E(G)$.

The hypercube network is initialized with the vertices and matched edges as follows:

$$HQ_n = (v_0 U v_1, E_0 \cup E_1 \cup M) \quad (7)$$

The position of the network is extracted with the place including high LQF. The additional vertex added to the Hamiltonian cycle by using the following algorithm.

| Addition of vertex to Hamiltonian cycle |
|---|
| For each vertex<br>If (safe(v, G, P, pos)<br>Path[pos]=v; // Recursive process to construct the fault-free path<br>If(HamcycleUtil((G, P, pos+1)==true)<br>Return true;<br>If(added v does not lead to the solution)<br>P(pos)=-1;<br>Endif<br>End For |

During the demanding conditions, the Hamiltonian path is formed on the basis of hypercube network construction in the previous section. During the communication, some of the nodes or links underwent the failure that leads to the disruption in communication. The node

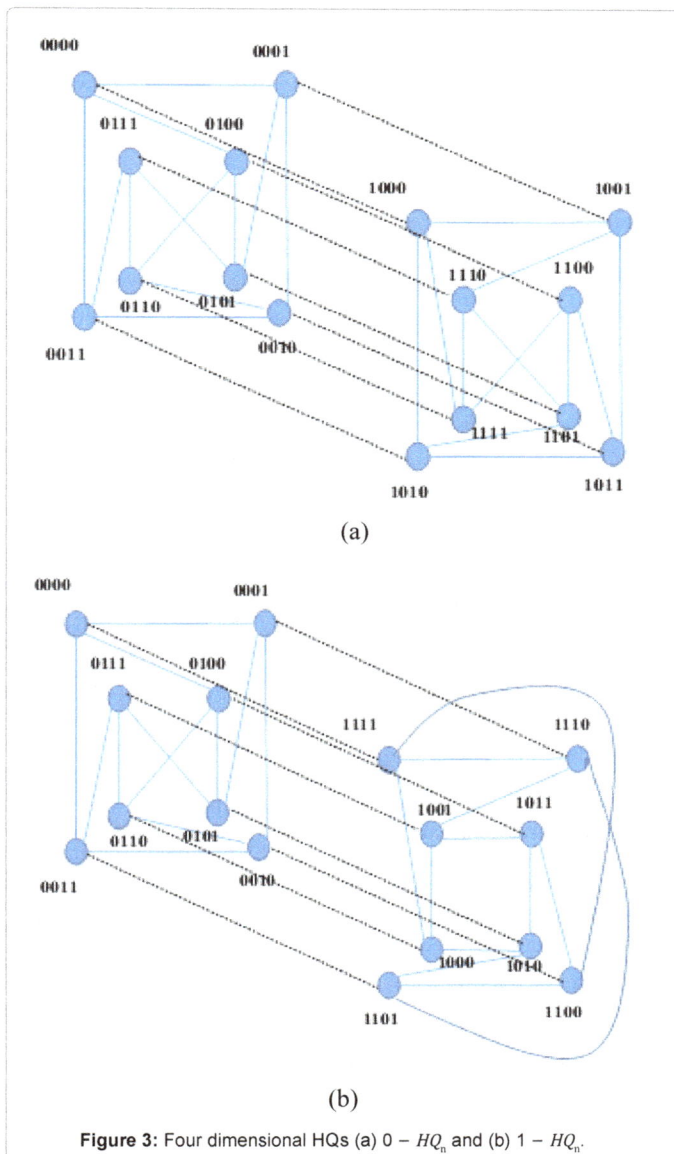

(a)

(b)

**Figure 3:** Four dimensional HQs (a) $0 - HQ_n$ and (b) $1 - HQ_n$.

failure rate estimation through the adjacency link matrix provides the necessary background information. Initially, the set of neighbor nodes is initialized with zero. The availability of elements (channels) for the transmission node during the run time defines the cardinality. The algorithm to compute the matched edge is expressed as follows:

| Cardinality-based Matched Edge set prediction |
|---|
| For each secondary node (N$_i$) //i=1, 2, …n |
| Set the neighbor set as null Ni={0} |
| For each cardinal value (N$_j$) //j=1,2, …, Nn |
| Send(Hello, Sj) //Sj–Neighbor of Si |
| If Si receives (RREP, Sj) |
| Ni=Ni∪{Sj} |
| End For |
| End For |
| For k=1, 2, …, Nn |
| For l=1, 2, …, Ch // Ch-Channels |
| Sense (ch$_k$, l) |
| If (ch$_k$, l==True) |
| Add the channel to ALM |
| End If |
| End For |
| End For |

For each node, the number of secondary nodes on the path (P) is extracted. For each secondary node, the routing request (RREQ) and the reply (RREP) are communicated in between the source and destination nodes as per AODV. Then, the cardinality values for the particular node are estimated. The channels available at the time are selected for transmission.

## Performance Analysis

This section presents the performance analysis of the proposed HPHC-based fault-free path prediction for an efficient WSN communication. We utilize the Network Simulator-2 (NS-2) and MATLAB to model the fault model and the cardinality-based fault-free path estimation respectively. Besides, the performance of proposed HPHC is compared with the existing works such as FDWSN, PFDWSN and FNCM [23] regarding the various parameters such as throughput, faulty node detection accuracy, Packet Delivery ratio (PDR) and end-to-end delay. The message overhead, end-to-end delay and the PDR values are estimated with the variations of faulty nodes and no fault condition. The simulation configuration parameters for proposed work implementation are listed in Table 1.

### Packet delivery ratio

The measure of the sum of packets received by the destination to the sum of packets generated is referred as packet delivery ratio.

$$PDR = \frac{\text{Sum of packets received by the destination}}{\text{Sum of packets generated in the source}} \quad (8)$$

This section investigates the effect of proposed HPHC on PDR values with respect to minimum and maximum simulation periods.

Figure 4 graphically illustrates the PDR variations with respect to the simulation period values. For minimum simulation period (10 ms), the PDR value of HPHC is 90.58% for the absence of fault and 58.9% for 50 faulty nodes. Similarly, the PDR values are 94.96 and 66.60% for maximum simulation period (100 ms). The utilization Hamiltonian formulation and hypercube network maintained the PDR value in stable level.

Figure 5 shows the PDR variations corresponding to the linear increase of number of nodes from 25 to 200. In existing methods, the FDWSN offers 89.89 and 91.12% PDR which are more compared to the existing PFDWSN for 25 and 200 nodes respectively. But, the cube structure extension increases the PDR values to 92.715% for maximum nodes (200). The comparative analysis of proposed HPHC with the PFDWSN shows that the HPHC offers 1.72% improvement due to the cube extension.

### End-to-end delay

The end-to-end delay of the packet is the time it takes to reach the destination and it depends on the number of components as follows:

$$delay_{E-E} = N_L(delay_{TX} + delay_{prop} + delay_{process} + delay_{queue}) \quad (9)$$

| Parameters | Values |
|---|---|
| Number of nodes | 200 |
| Topology size | 1000 × 1000 m$^2$ |
| Data packet size | 800 bits |
| Data aggregation energy | 5 nJ/bit/signal |
| BS location | x=150, y=160 |
| Duration of round | 20 s |
| Initial energy | 0.5 J |
| Sensing range | 10 m |
| Frame/second | 5 frame |
| Threshold transmission range | 200 m |
| Control packet size | 80 bits |
| TTL control packet | 3 |
| Stand-by node energy consumption | 15 mW |
| Round time | 20 s |
| MAC header | 68 bits |

**Table 1:** Simulation parameters.

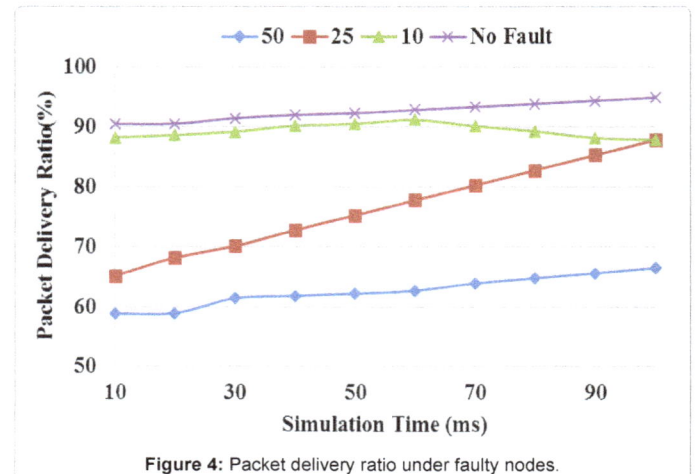

**Figure 4:** Packet delivery ratio under faulty nodes.

Where, $N_L$–Number of links;

delay$_{TX}$–Transmission delay (BW/L);

delay$_{prop\_}$Propagation delay;

delay$_{process}$–Processing delay;

delay$_{queue}$–Queuing delay.

The parameters that affect the end-to-end delay are bandwidth and packet length (bits). Besides, the parameters that affect the propagation delay are congestion level, physical link length and speed respectively. Hence, the preservation of link is the major requirement to reduce the end-to-end delay.

Figure 6 graphically illustrates the end-to-end delay variations with the linear increase of number of nodes from 25 to 200 nodes. For 25 nodes, the end-to-end delay value for PFDWSN is minimum like 25 ms and it is 37.02 ms for 200 nodes. The delay values for proposed HPHC are 25.18 and 28.28 ms for 25 and 200 nodes respectively. The comparative analysis between the proposed HPHC with the PFDWSN shows the 23.61% reduction in end-to-end delay values.

Figure 7 shows the end-to-end delay variations with respect to the absence and presence of faulty nodes respectively. For minimum simulation period (10 ms), the end-to-end delay value of HPHC is 0.0263 ms for the absence of fault and 0.0496 ms for 50 faulty nodes. Similarly, the delay values are 0.06096 and 0.10275 ms for maximum simulation period (100 ms). The utilization Hamiltonian formulation and hypercube network linearly increases the time consumption.

## Faulty node detection rate

The faulty node detection rate is defined as the ratio of the number of faulty sensor nodes detected to the total number of faulty nodes available in the network. Figure 8 graphically presents the faulty node detection rate variations with respect to the minimum and maximum number of nodes like 10 and 100 nodes. For minimum number of nodes (10), the faulty node detection rate for proposed HPHC is 1 and it is linearly decreased to 0.51 ms for 100 nodes. The existing FNCM offers 0.2655 and 0.12 ms for 10 and 100 nodes respectively. The comparative analysis between the proposed HPHC with the existing FNCM shows that the HPHC offers 73.45 and 76.47% reduction in end-to-end delay values respectively compared to the existing FNCM method.

## Throughput

The number of data packets sent over the total simulation period refers throughput. The mathematical formulation for throughput is expressed as:

$$\text{Throughput} = \frac{\text{Number of data packets sent (bits)}}{\text{Time period (secs)}} \quad (10)$$

In this section, the percentage of throughput is investigated corresponding to the number of nodes variation. Figure 9 shows the graphical variations of throughput with respect to the variation in simulation period from 10 to 100 ms. The percentage throughput values corresponding to HPHC are 96.63 and 98.62% for 10 and 100 ms respectively. Similarly, the percentage values corresponding to FDWSN are 93.61 and 93.14% respectively. The comparative analysis between the proposed HPHC with the existing FDWSN shows that

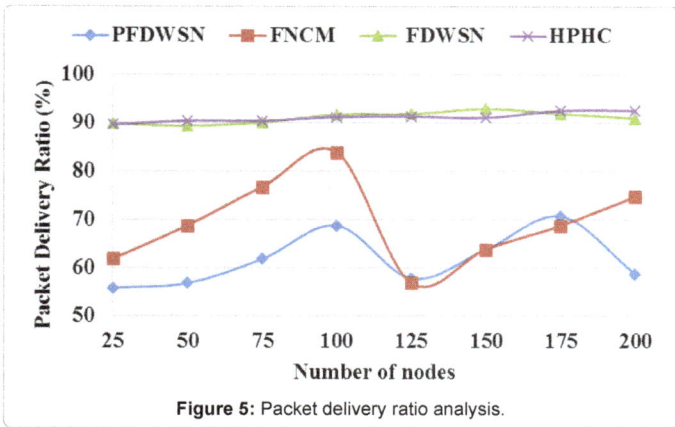

**Figure 5:** Packet delivery ratio analysis.

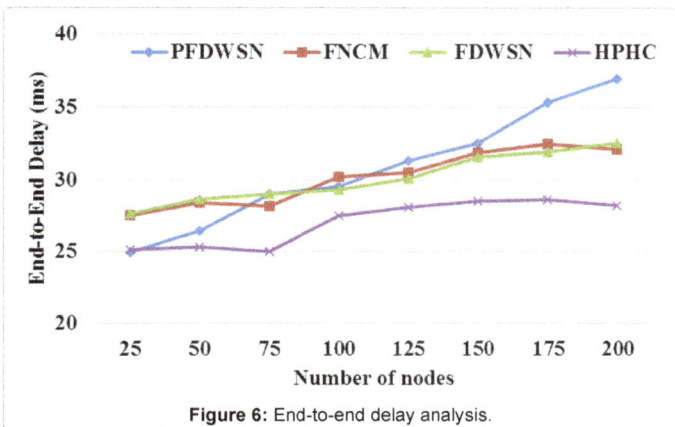

**Figure 6:** End-to-end delay analysis.

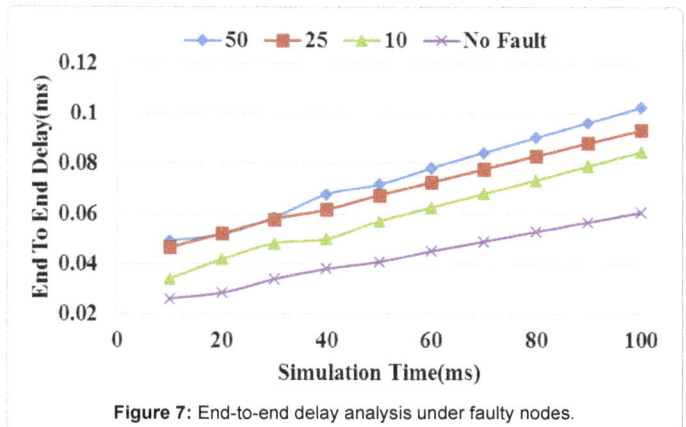

**Figure 7:** End-to-end delay analysis under faulty nodes.

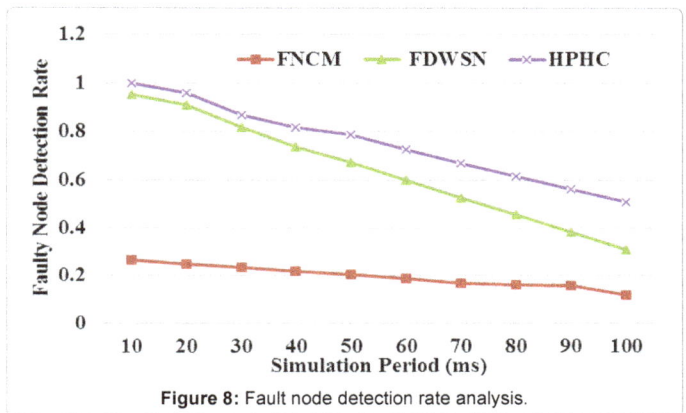

**Figure 8:** Fault node detection rate analysis.

the HPHC offers 3.13 and 5.56% improvement with respect to the simulation time values.

## Average message overhead

Under the delay constraint, the ratio of the size of control packets to the total number of data packets successfully transmitted to the destination refers routing overhead. The analysis of control packet flow for minimum and maximum simulation periods of 10 and 100 ms.

The average message overhead for minimum (10) and maximum simulation period values (100 ms) are 0.57 and 1 under the absence of fault nodes.

The link preservation and the quality factor update through Hamiltonian cube formulation increase the overhead values to 2.3659 and 3.2189 respectively which are the acceptable levels. The cube extension-based link preservation in proposed work efficiently overcomes the fault scenario.

## Conclusion and Future Work

Due to the battery-operated sensors, the recharge and replace mechanisms, the WSN suffered from the energy conservation and minimum network lifetime. The identification of fault nodes on the transmission path played the major role in energy conservation. With the dense deployment of sensor nodes, the failures in node and link are high that disrupted the entire communication. This paper proposed the suitable alternative fault-free path prediction model to perform the communication among the nodes. Initially, the sensor nodes are deployed in the WSN environment. Once the initialization of source and destination nodes are over, the path between them is predicted through the Hamiltonian path prediction model. During the failure scenario, this paper estimated the node and link parameters such as RSSI, queue size, response time, and bandwidth are individually estimated and group them into the QF. Based on the QF, the proposed work predicts the fault-free link to alleviate the unnecessary transmissions to the fault node and reduced the energy consumption. The comparison between the proposed HPHC network with the existing fault detection mechanisms regarding the performance measures such as Packet Delivery ratio (PDR), fault node detection rate, throughput and end-to-end delay assured the effectiveness of HPHC in WSN communication. The major focus of this paper to identify the fault-free path after the faults occurred (Figure 10). The occurrence of faults lies in two ways such as self and external. The prevention of external faults and the spreading of internal faults to other node will be considered as the future work to improve the communication performance.

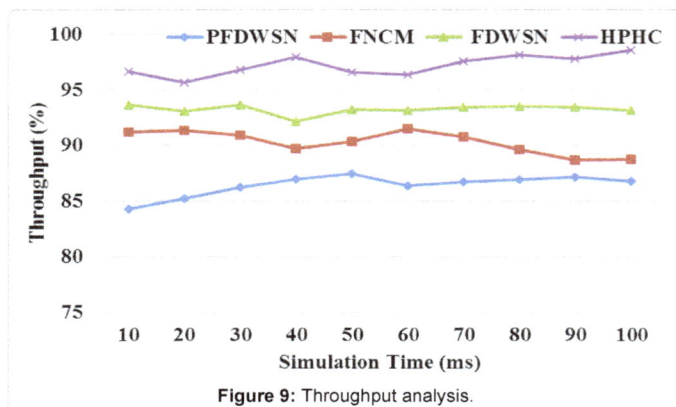

**Figure 10:** Control packet overflow analysis.

## References

1. Halim T, Islam MR (2012) A study on the security issues in WSN. Int J Comput Appl T 53.

2. Sha K, Gehlot J, Greve R (2013) Multipath routing techniques in wireless sensor networks: A survey. Wireless Pers Commun pp: 1-23.

3. Sathish S, Ramesh L, Kumar SG (2014) A survey on node recovery from a failure in wireless sensor networks. Int J Adv Res Comp Sci Technol 2: 158-161.

4. Pasqualetti F, Bicchi A, Bullo F (2012) Consensus computation in unreliable networks: A system theoretic approach. IEEE Transactions on Automatic Control 57: 90-104.

5. Sharanapriya RP (2014) Design and implementation of dead nodes recovery algorithm to improve the life time of a wireless sensor network. Int J Sci Res 3: 1419-1422.

6. Hashemi SE, Motameni H, Ghaleh MR, Esmaeili S (2013) Clustering and routing wireless sensor network based on the parameters of distance, density, energy and traffic with the help of fuzzy logic. Int J Comput Sci Issues 10: 9-14.

7. Bianchin G, Cenedese A, Luvisotto M, Michieletto G (2015) Distributed fault detection in sensor networks via clustering and consensus: In Decision and Control (CDC). IEEE 54th Annual Conference pp: 3828-3833.

8. Re GL, Milazzo F, Ortolani M (2012) A distributed Bayesian approach to fault detection in sensor networks via Global Communications Conference (GLOBECOM). IEEE pp: 634-639.

9. Jain AGS (2014) Routing Techniques in wireless sensor networks. Int J Comput Appl 94: 15-20.

10. Chouikhi S, El Korbi I, Ghamri-Doudane Y, Saidane LA (2015) A survey on fault tolerance in small and large scale wireless sensor networks. Comput Commun 69: 22-37.

11. Bao F, Chen R, Chang M, Cho JH (2012) Hierarchical trust management for wireless sensor networks and its applications to trust-based routing and intrusion detection. IEEE Transactions on Network and Service Management 9: 169-183.

12. Duche RN, Sarwade NP (2012) Sensor node failure or malfunctioning detection in wireless sensor network. ACEEE Int J Commun 3: 57-61.

13. Duche RN, Sarwade NP (2014) Sensor node failure detection based on round trip delay and paths in WSNs. IEEE Sensors J 14: 455-464.

14. Lee MH, Choi YH (2008) Fault detection of wireless sensor networks. Comput Commun 31: 3469-3475.

15. Akbari A, Dana A, Khademzadeh A, Beikmahdavi N (2011) Fault detection and recovery in wireless sensor network using clustering. IJWMN 3: 130-138.

16. Paradkar V, Chandel GS, Patidar K (2015) Fault Node Discovery and Efficient Route Repairing Algorithm for Wireless Sensor Network. IJCSIT 6: 1710-1715.

17. Mahapatro A, Khilar PM (2013) Energy-efficient distributed approach for clustering-based fault detection and diagnosis in image sensor networks. IET Wireless Sensor Systems 3: 26-36.

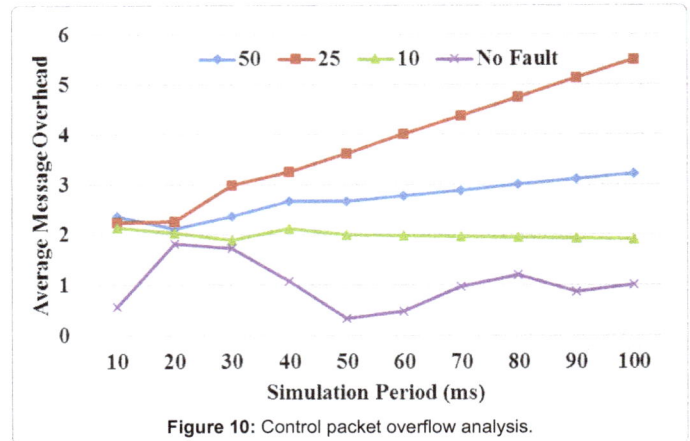

**Figure 9:** Throughput analysis.

18. Ould-Ahmed-Vall E, Ferri BH, Riley GF (2012) Distributed fault-tolerance for event detection using heterogeneous wireless sensor networks. IEEE Transactions on Mobile Computing 11: 1994-2007.

19. Lo C, Lynch JP, Liu M (2013) Distributed reference-free fault detection method for autonomous wireless sensor networks. IEEE Sensors J 13: 2009-2019.

20. Deniz F, Bagci H, Korpeoglu I, Yazıcı A (2016) An adaptive, energy-aware and distributed fault-tolerant topology-control algorithm for heterogeneous wireless sensor networks. Ad Hoc Networks 44: 104-117.

21. Ding Y, Hu Y, Hao K, Cheng L (2015) MPSICA: An intelligent routing recovery scheme for heterogeneous wireless sensor networks. Inf Sci 308: 49-60.

22. Lau Bc, Ma EW, Chow TW (2014) Probabilistic fault detector for wireless sensor network. Expert Systems with Applications 41: 3703-3711.

23. Chanak P, Banerjee I (2016) Fuzzy rule-based faulty node classification and management scheme for large scale wireless sensor networks. Expert Systems with Applications 45: 307-321.

24. Hsieh SY, Chang NW (2006) Hamiltonian path embedding and pancyclicity on the Mobius cube with faulty nodes and faulty edges. IEEE Transactions on Computers 55: 854-863.

25. Cheng CW, Hsieh SY (2016) Edge-fault-tolerant pancyclicity and bipancyclicity of Cartesian product graphs with faulty edges. J Comput Syst Sci 82: 767-781.

26. Lai PL (2012) A systematic algorithm for identifying faults on hypercube-like networks under the comparison model. IEEE Transactions on Reliability 61: 452-459.

# 20

# Design of a Selective Filter-Antenna with Low Insertion Loss and High Suppression Stopband for WiMAX Applications

**Alkhafaji MK\*, Sahbudin RKZ, Ismail AB and Hashim SJ**

*Department of Computer and Communication System Engineering, Center of Excellence of Wireless and Photonics Network (WIPNET), UPM, 43400-Serdang, Malaysia*

### Abstract

This paper presents a selective quasi–elliptic bandpass filter-antenna. The presented filter-antenna has a low insertion loss in the passband and relatively high stopband rejection. This structure consists of a quasi–elliptic bandpass filter direct coupled with patch antenna. The bandpass filter consists of four ($\lambda/4$) spiral square resonators. It has operates between (3.25–3.6) GHz so it is suitable for WiMAX applications. A CST Microwave Studio Suite software has used to simulate the filter-antenna circuit. The simulated results of the patch antenna and the results of the filter-antenna appears a good matching between the two circuits.

**Keywords:** Quasi–elliptic; Bandpass filter; Filter-antenna; Insertion loss; Spiral square resonators; CST Microwave studio suite; Patch antenna

## Introduction

The large and rapid progress of wireless communication, systems emerged as the urgent need to reduce the size, weight, and cost of the receiving and transmitting circuits. It has become one of the important requirements in the recent years. To obtain these aims, several attempts could be design at the same time in a single circuit module. In order to reduce the overall circuit size, a predesigned bandpass filter with appropriate configuration is inserting directly into the feed location of a patch antenna. The bandpass filter can be integrate completely with the antenna for a required bandwidth [1]. A bandpass filter is composed of resonators having the identical resonant frequency as the antenna. This case leads to interference between the return loss and the antenna gain responses, especially at the band edges. It is usually that the impedance bandwidth of the antenna is different from that of the bandpass filter [2]. Typically, input/output ports design of a bandpass filter is usually takes as 50 Ω terminations. Although, the antenna input impedance may be not perfectly match to a 50 Ω at the band edges. The regression due to mismatch thus takes place. Both components have in general arranged at the highly front–end of communication system. Integration of the antenna and the bandpass filter has been considered for boosting the total performance and decreasing the circuit size [3]. In the recent years, many academic scholars began to study the co–design process for the bandpass filters and the antennas. By optimizing the impedance at the interfaces of the filter and the antenna, the impedance bandwidth has enhanced [4]. The filter-antennas, which designed by using the bandpass filter synthesis method, the shaped antenna and the rectangle patch antenna has replaced with the last resonator, and the load impedance of a bandpass filter [5]. The dimensions of the shaped antenna and the rectangle patch antenna in order of half–wavelength. To get an easy mobility, for modern wireless communication systems, the transmitting and receiving components should have minimum dimensions. The integration of the bandpass filter to the antenna in a single module contributes to make the dimensions of the transmitters and receivers minimum [6]. Usually, in RF/Microwave front–end systems, a bandpass filter is cascaded to the antenna to reject the spurious signals that received or transmitted by the antenna [7]. Different research groups have examined the capability to use dielectric resonator (DR) antenna simultaneously for filtering, packaging, and oscillating purposes. However, there is no work has been done to integrate a filter and an antenna using a single resonator [8].

Bandpass filter provides a low insertion loss within the passband and a high attenuation in the stopband, and they can be realized by cascading lowpass filters and highpass filters, where the layouts are efficient and uncomplicated [9]. Edge coupled line filters become compact in size when their resonators folded, therefore the spurious pass zone becomes closer to the essential band and the selectivity of the filter becomes more flat [10]. The design process and purpose of the filter-antenna are differing from those of the traditional antenna. The filter-antenna circuit is not just another impedance matching technique, but also a forming for a filter and antenna gain and input return loss [11].

This paper presents an integration between a quasi–elliptic bandpass filter and the patch antenna to create a filter-antenna. This circuit has good specifications in terms of the matching between the two devices and a relatively low insertion loss in the pass band and has a high suppression on a spurious harmonics in the stop band. The filter-antenna circuit presented here is suitable for WiMAX applications, because the circuit operates within the frequency range (3.25–3.6) GHz. The selectivity of this circuit considered good in general.

This paper presents a filter-antenna design using ($\lambda/4$) spiral square resonator as the basic unit. Figure 1 depicts the layout of the basic unit resonator (Table 1).

## Quasi–Elliptic Bandpass Filter Design and Results

This filter has in-between solution for Chebyshev and Elliptic-function type filters. The transfer function for a quasi–elliptic filter is Amaya et al. [12]:

**\*Corresponding author:** Alkhafaji MK, Faculty of Engineering, Department of Computer and Communication System Engineering, Center of Excellence of Wireless and Photonics Network (WIPNET), UPM, 43400-Serdang, Malaysia E-mail: mkks67@yahoo.com

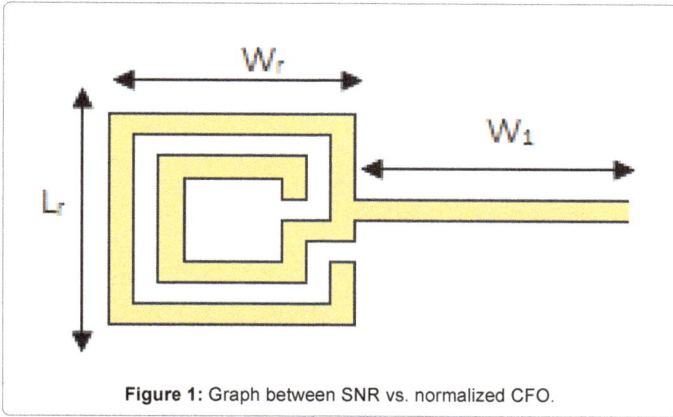

**Figure 1:** Graph between SNR vs. normalized CFO.

| Wr | W1 | W2 | W3 | W4 | Wm | Lr | Ls |
|------|-------|-----|-----|------|------|------|-----|
| 6 | 11.20 | 3.5 | 0.6 | 1.2 | 2.4 | 6 | 1 |
| L1 | L2 | L3 | L4 | L5 | L6 | L7 | - |
| 2.4 | 1.8 | 0.6 | 6.4 | 9.55 | 8.85 | 3.6 | - |

**Table 1:** Quasi-elliptic bandpass filter dimensions (mm).

$$|s_{21}| = \frac{1}{1+ \in^2 F_N^2(\Omega)} \quad (1)$$

Where $\Omega$ is the frequency variable, which is normalized to the passband cutoff frequency of a lowpass prototype filter, and $\in$, is a ripple constant related to a given return loss LR described by:

$$\in = \frac{1}{\sqrt{10^{-(L_R/10)-1}}} \quad (2)$$

In addition, $F_N(\Omega)$ for the selective filter has stated as a function of the pair of attenuation poles so that $\Omega = \Omega_s (\Omega_s > 1)$ match to their frequency response locations. The locations of the pair of attenuation poles of the bandpass filter namely $\omega_{s1}$ and $\omega_{s2}$, are given by

$$\omega_{s1} = \omega_0 \frac{-\Omega_s FBW + \sqrt{(\Omega_s FBW)^2 + 4}}{2} \quad (3)$$

$$\omega_{s2} = \omega_0 \frac{\Omega_s FBW + \sqrt{(\Omega_s FBW)^2 + 4}}{2} \quad (4)$$

Before integrating filter and antenna, a four-spiral resonator quasi-elliptic bandpass filter has designed. The bandpass filter operates at $f_c$=3.33 GHz with Fractional Bandwidth (FBW) of 10%. The filter designed using the standard filter synthesis technique. In order to obtain the design parameters, RT/Duroid 5880 substrate had been used ($\varepsilon_r$=2.2), and h=0.786 mm. Figure 2 shows the layout of the bandpass filter,

The response of the bandpass filter represented by S11 (dB) and S21 (dB) parameters has illustrated in Figure 3. The curves for various values of the distance between two-neighbor resonators (Ls) show a good performance in passband and stopband. Figure 4 shows the bandpass filter Voltage/current matrix coefficient of the Z-impedance.

## Antenna Design and Results

Figure 5 illustrates the geometry of the patch antenna design. As shown in the figure, the radiating component is a rectangular patch

antenna with top T–shape band notch, and two bottom L-shape notches. These top and bottom notches have found to improve the patch antenna bandwidth as shown in Figure 6 which is illustrated the reflection coefficient parameter S11 (dB) of the patch antenna, and as it is seen that the patch antenna has a center frequency located at 3.42 GHz. The patch antenna fed by a 50 Ω microstrip transmission line. CST Microwave Studio Suite has employed to perform the design and optimization process. The design parameters of the patch antenna have stated in Table 2. Figure 7 shows the Voltage Standing Wave ratio (VSWR) of the patch antenna.

## Filter-antenna Design and Results

Figure 7 shows the geometry of the quasi–elliptic filter-antenna structure. The microstrip line feeding of the patch antenna has replaced by a quasi–elliptic filter and the performance of this arrangement have observed and compared with the performance of the patch antenna fed by the microstrip transmission line. Figure 8 shows the reflection coefficient S11 (dB) of the filter-antenna for various values of the distance Ls. Table 3 shows the dimensions of the filter-antenna bottom view (mm).

Figure 9 shows the optimized performance of the filter-antenna, it illustrated the reflection coefficient S11 (dB) and gain (dB). It can be

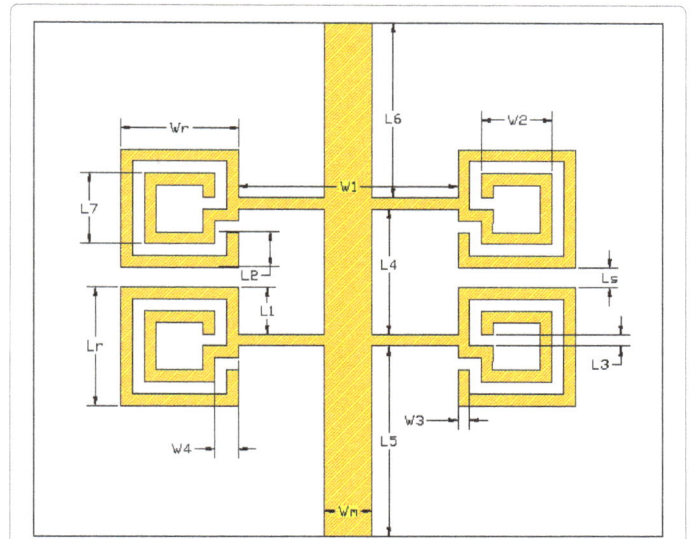

**Figure 2:** Direct couple four spiral resonators of the quasi-elliptic bandpass filter design.

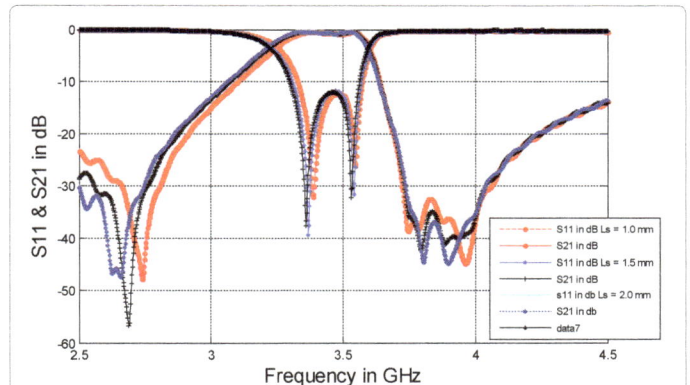

**Figure 3:** VSWR of the proposed quasi–elliptic bandpass filter.

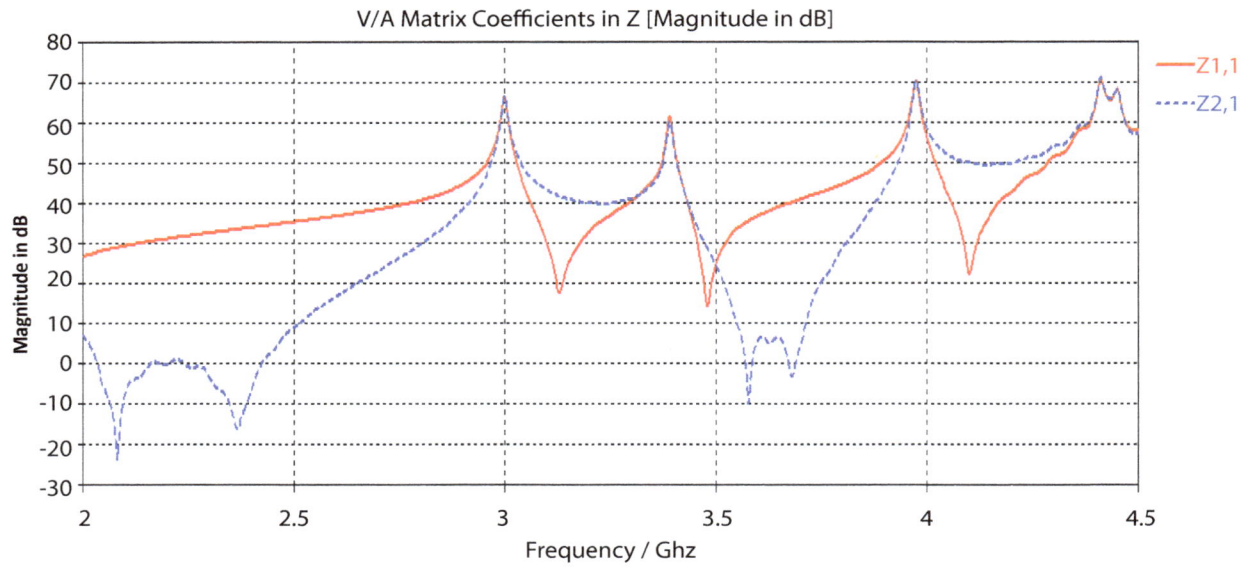

**Figure 4:** Voltage/Current matrix coefficients for the Z-impedance.

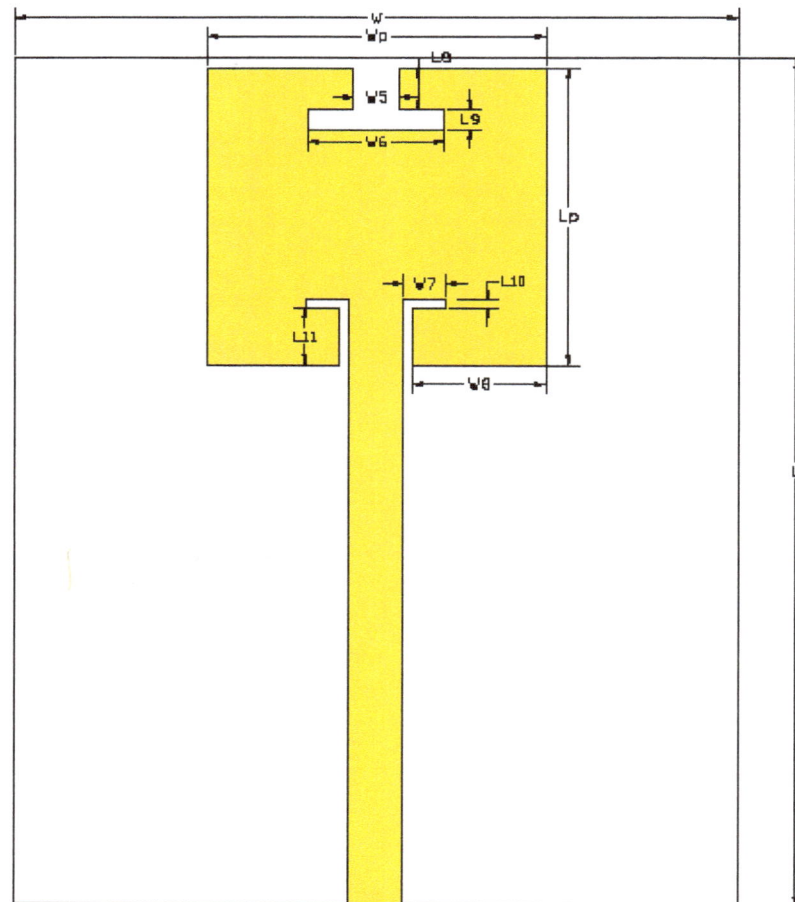

**Figure 5:** The geometry of the patch antenna.

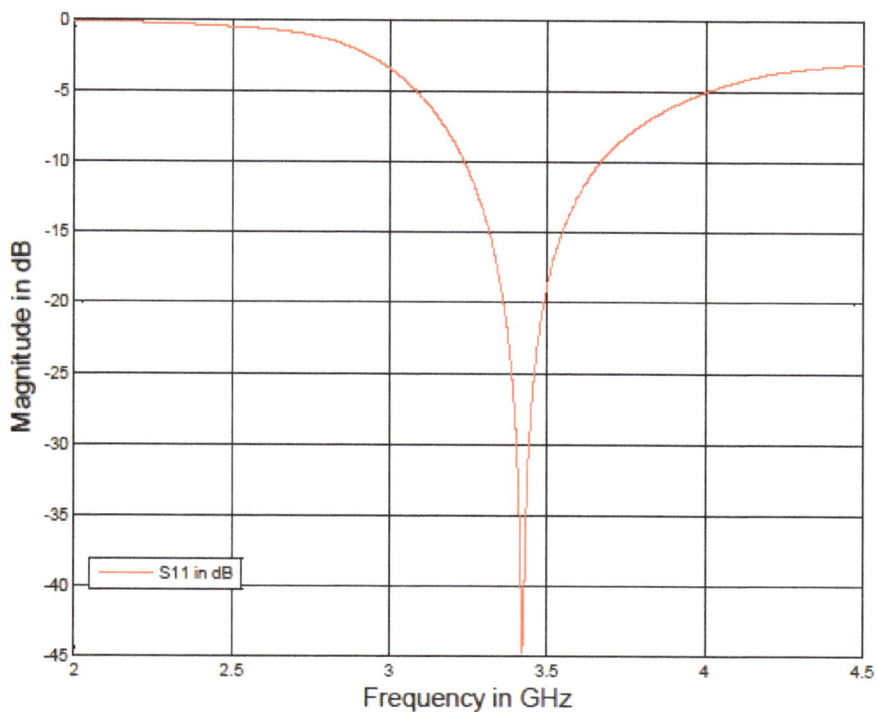

**Figure 6**: Return loss of the proposed patch antenna.

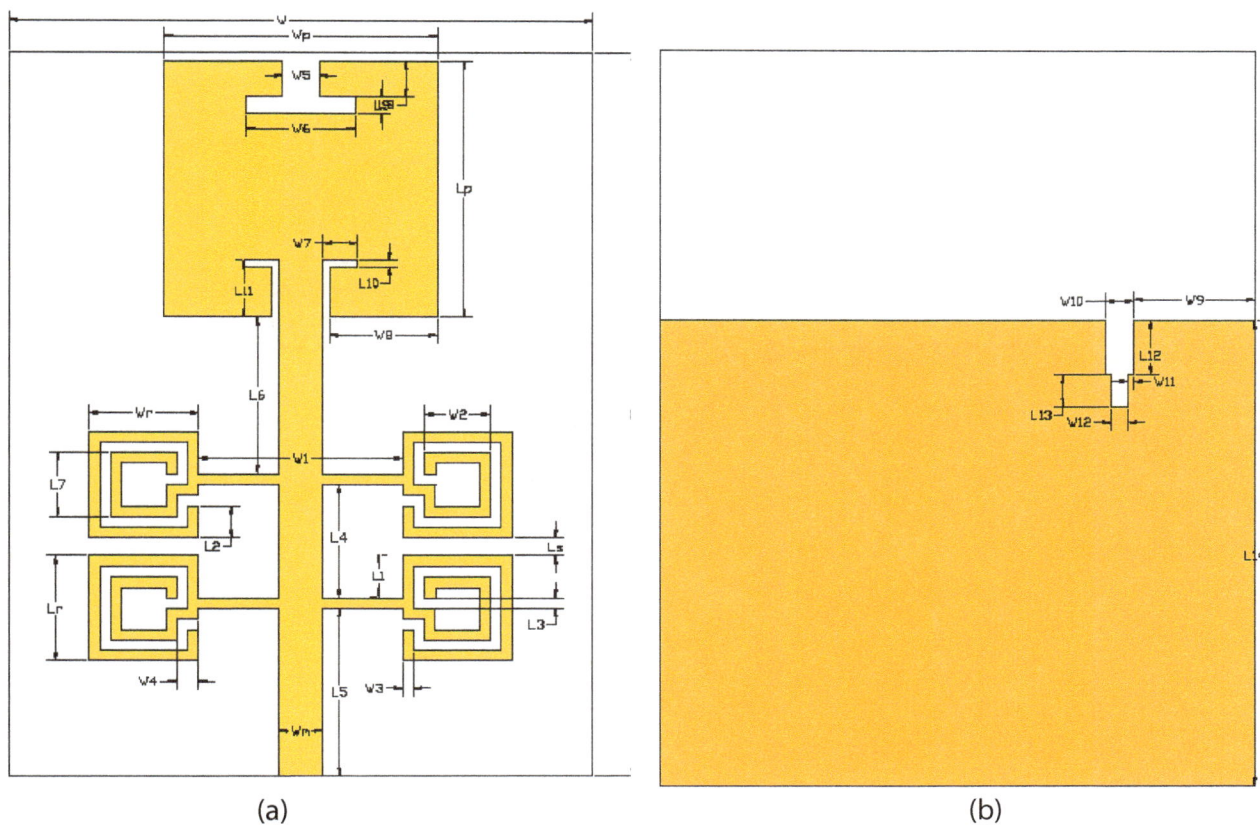

(a)

(b)

**Figure 7**: The geometry of the proposed filter-antenna (a) top view (b) bottom view.

seen from that the filter-antenna has a bandwidth equal to 350 MHz (from 3.25MHz–3.6MHz), and a center frequency located at 3.425 GHz with the peak gain of about 4.9 dBi. This is clearly indicates the high matching between the filter and the antenna. In addition, a radiation null located at 3.75GHz, and the frequency skirt selectivity becomes better.

Figure 10 shows the Voltage Standing Wave Ratio (VSWR) of the filter-antenna and that of the patch antenna.

Figure 11 shows the simulated farfield gain and directivity of the proposed filter-antenna at the center frequency.

## Conclusion

A selective quasi–elliptic bandpass filter-antenna with low insertion loss and high suppression stop band has discussed in this paper. The design has accomplished by first establishing and analyzing the basic unit ($\lambda/4$) resonator and the complete four-resonator quasi–elliptic bandpass filter. The patch antenna with T–shape band notch from the top and two L–shape notches from the bottom had used as a radiating component. In this design the feeding microstrip transmission line of the patch antenna, has replaced by the quasi–elliptic bandpass filter to give filtering specifications. The

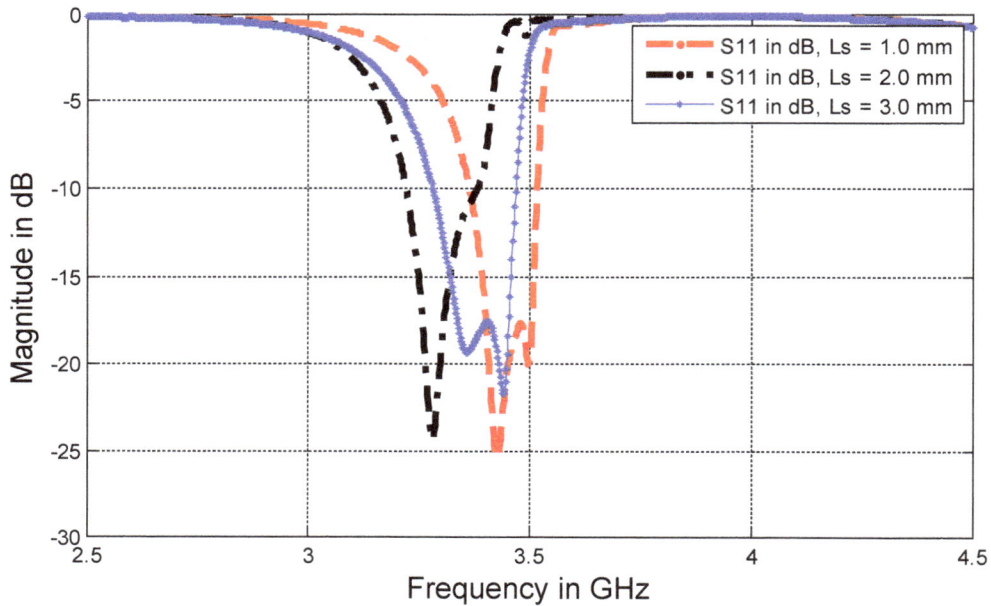

**Figure 8**: Return loss of the proposed filter-antenna for various values of Ls.

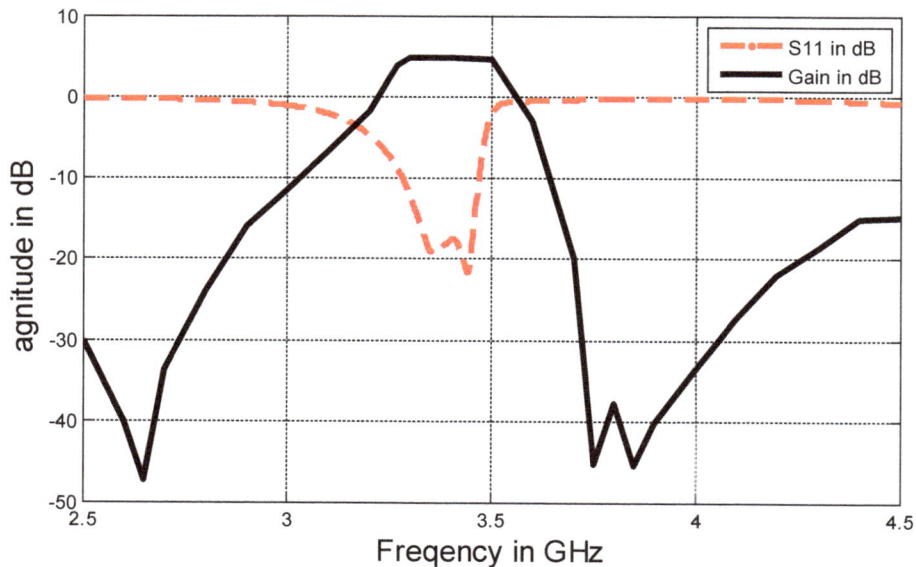

**Figure 9**: Filter-antenna performance (S11–parameter and gain).

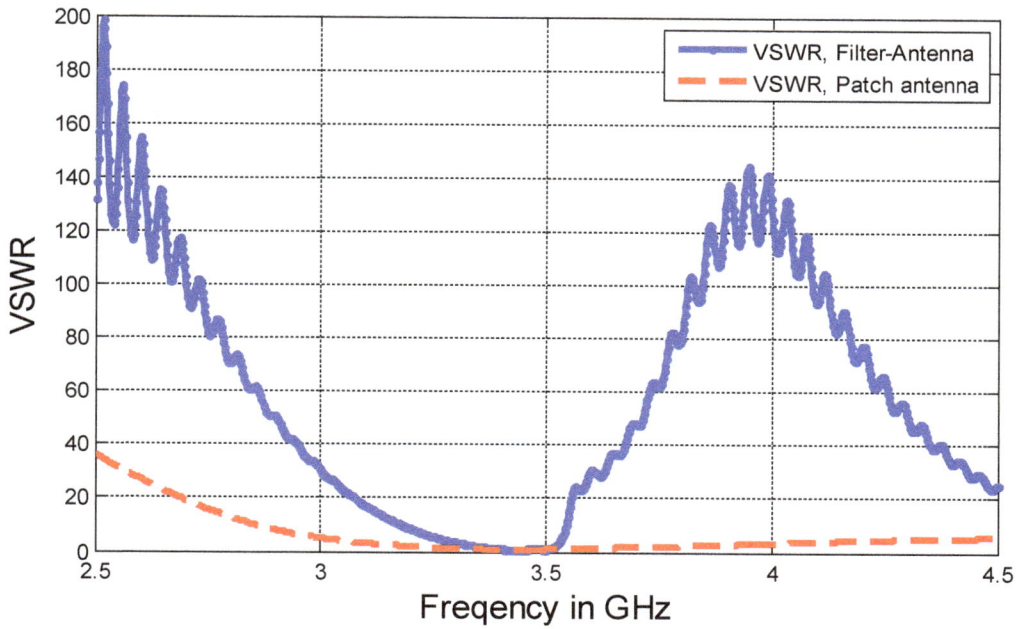

**Figure 10**: VSWR of the filter-antenna and the patch antenna.

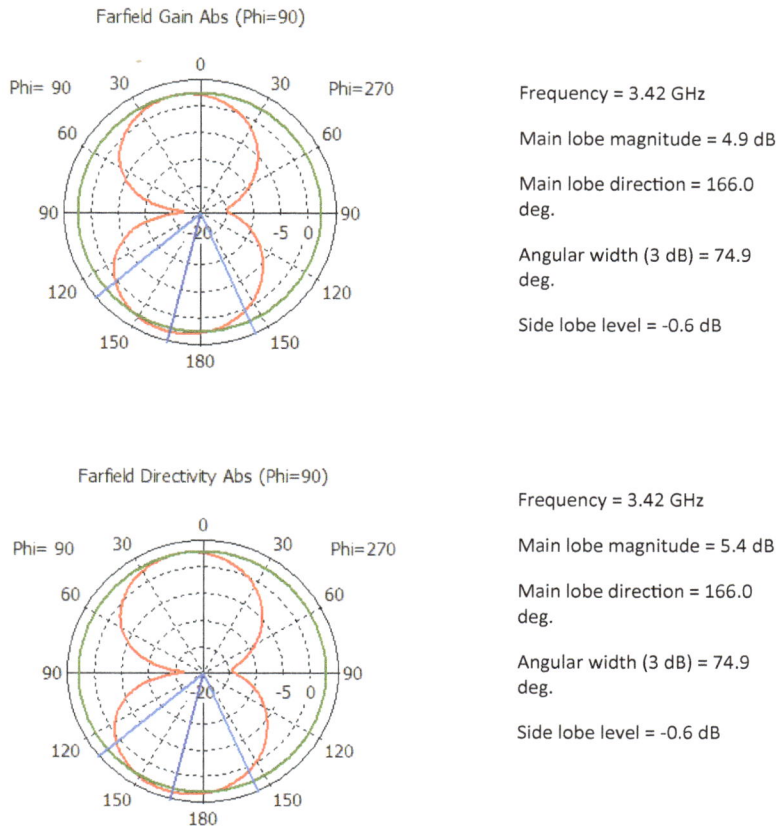

Farfield Gain Abs (Phi=90)

Frequency = 3.42 GHz

Main lobe magnitude = 4.9 dB

Main lobe direction = 166.0 deg.

Angular width (3 dB) = 74.9 deg.

Side lobe level = -0.6 dB

Farfield Directivity Abs (Phi=90)

Frequency = 3.42 GHz

Main lobe magnitude = 5.4 dB

Main lobe direction = 166.0 deg.

Angular width (3 dB) = 74.9 deg.

Side lobe level = -0.6 dB

**Figure 11**: Simulated farfield gain and directivity of the proposed filter-antenna at 3.42 GHz.

| Wp | Lp | W5 | W6 | W7 | W8 | L8 | L9 | L10 | L11 |
|----|----|----|----|----|----|----|----|-----|-----|
| 15 | 14.5 | 2 | 6 | 1.9 | 5.9 | 2 | 1 | 0.4 | 2.8 |

**Table 2**: Design parameters of the proposed antenna (mm).

| W9 | W10 | W11 | W12 | L12 | L13 | L14 |
|----|-----|-----|-----|-----|-----|-----|
| 6.5 | 1.5 | 0.3 | 0.9 | 3 | 1.8 | 26 |

**Table 3**: Filter-antenna bottom view dimensions (mm).

proposed structure provides good skirt selectivity and low loss in the pass band with high reject in the stop band.

## References

1. Chuang CT, Chung SJ (2009) New printed filtering antenna with selectivity enhancement. In Microwave Conference, 2009. EuMC 2009. European pp: 747-750.

2. Yusuf Y, Gong X (2010) A new class of 3-D filter/antenna integration with high quality factor and high efficiency. In Microwave Symposium Digest (MTT). IEEE MTT-S International pp: 892-895.

3. Lin CK, Chung SJ (2011) A filtering microstrip antenna array. IEEE Transactions on Microwave Theory and Techniques 59: 2856-2863.

4. Chen X, Zhao F, Yan L, Zhang W (2013) A compact filtering antenna with flat gain response within the passband. IEEE Antennas and Wireless Propagation Letters 12: 857-860.

5. Zuo J, Chen X, Han G, Li L, Zhang W (2009) An integrated approach to RF antenna-filter co-design. IEEE Antennas and Wireless Propagation Letters 8: 141-144.

6. Kufa M, Raida Z, Mateu J (2014) Equivalent circuits of planar filtering antennas fed by apertures. 2014 20th International Conference on Microwaves, Radar, and Wireless Communication (MIKON) pp: 1-4.

7. Lin CK, Chung SJ (2009) A compact edge-fed filtering microstrip antenna with 0.2 dB equal-ripple response. Microwave Conference, 2009. EuMC 2009. European pp: 378-380.

8. Sung Y (2010) Microstrip resonator doubling as a filter and as an antenna. IEEE Antennas and Wireless Propagation Letters 9: 467-470.

9. Tu WH (2010) Sharp-rejection broadband microstrip bandpass filters using loaded open-loop resonator. Microwave Symposium Digest (MTT). IEEE MTT-S International pp: 1-1.

10. Partal HP (2010) Cross coupled wiggly line hairpin filters with high selectivity and spurious suppression. Microwave Conference (EuMC), 2010 European pp: 1265-1268.

11. Lin CK, Chung SJ (2011) A compact filtering microstrip antenna with quasi-elliptic broadside antenna gain response. IEEE Antennas and Wireless Propagation Letters 10: 381-384.

12. Amaya R, Momciu A, Haroun I (2013) High-performance, compact quasi-elliptic bandpass filters for V-band high data rate radios. IEEE Transactions on Components, Packaging and Manufacturing Technology 3: 411-416.

# 21

# Effectiveness of a Novel Power Control Algorithm in Heart Rate Monitoring of a Mobile Adult: Energy Efficiency Comparison with Fixed Power Transmission

**Debraj Basu\*, Gourab Sen Gupta, Giovanni Moretti and Xiang Gui**

*School of Engineering and Advanced Technology, Massey University, New Zealand*

## Abstract

In this paper, experiments are conducted to evaluate the efficacy of a novel adaptive power control algorithm in terms of energy efficiency in heart rate monitoring scenario of a mobile adult in a typical home environment. As part of health care, persons with heart related problems are required to be monitored by logging for example, their heart rate on a regular basis to check for any anomaly. At the same time, it is expected that the person in question should be able to move freely within the given facility. The wireless sensors that are attached to the person send periodic data to the central base station. Since the person is mobile, the distance between the transmitting sensor and the base station changes with time. Since the signal path-loss is primarily dependent on distance and the number and type of obstructions between the transmitter and the receiver, it may be wise to use transmission power control to modulate the transmit power. Using power control, the sensor can adjust the level that is sufficient to send the data through the wireless channel without wasting energy. Conservation of energy is critical in wireless sensor network scenarios because they are powered by batteries which have limited lifetime. A critical application like the heart rate monitoring sensor is expected to operate for a reasonable amount of time before the battery dies. The novel adaptive power control algorithm uses intelligent modulation methods to ramp up or ramp down the transmission power level as and when required. By this method, the operational lifetime of the wireless sensor can be extended. As part of the experimental methodology for this paper, two subjects of different age groups have been used. Experimental results show that there is at least a 12% increase in the energy savings using the proposed algorithm.

**Keywords:** Mobile wireless sensors; Heart rate monitoring; Energy consumption optimization; Adaptive power control

## Introduction

The proliferation of low power wireless sensor networks and their discreet presence have introduced a new paradigm in data collection and analysis of target parameters in both indoor and outdoor environments. Especially in healthcare, body wearable and implantable sensors are used to continuously monitor the vital physiological parameters of patients in hospitals and elderly at home who enjoy independent living [1]. From such "living records", medical practitioners can draw useful inferences about the health and well-being of an individual. This can be used for self-awareness and analysis to assist in making behaviour changes, and to share with caregivers for early detection of any ailment and allow appropriate intervention. At the same time such procedures are effective and economic ways of monitoring age-related illnesses [2]. An overview of a simple WSN application in healthcare is shown in Figure 1 [3,4]. The wearable and implanted sensors collect vital health parameters like the pulse rate, EEG, blood insulin level, etc. and transmit to the access point. The gateway that is shown in Figure 1 can be with the person or with the access point.

One unique challenge with body wearable sensors is that these sensors are mobile. Therefore, the communication layer must adapt quickly with the changing environment so that the transmission power can be re-calibrated for reliable transmission [5].

### Need to control power for wireless mobile sensor nodes

When a sensor is mobile, it means that its communicating distance from the base station is changing with time. The energy loss is primarily dependent on distance. Unwanted obstructions can also lead to signal degradation due to absorption. Beside there are effects of fading and multipath propagation of radio signal in indoor wireless communication [6].

For acceptable performance, the average received $E_b/N_0$ at the base station should be above a threshold value. If that value is maintained on average, then it means that the wireless sensor is performing as expected. Since most of these mobile wireless sensors are battery powered, they have limited energy resources. Therefore, if a mobile node is near to a base station and the received $E_b/N_0$ far exceeds the required threshold, then there is waste of energy. Similarly, when the same sensor node is far from the base station, the node is required to pump in more power than the present value. It may also happen that due to an obstruction between the transmitting node and the base station, the node is transmitting at a power level that is enough to deliver packets. If that obstruction is removed or the node moves to a position in which the obstruction is cleared, then it should adjust its transmitting power down.

Previous work [7] has provided an empirical analysis of the impact of power control for mobile sensor network. The focus of this research paper is on residential health monitoring, in-hospital patient monitoring and sports monitoring that require mobile sensing. Cellular networks deal with mobility by using different types of hand-

---

**\*Corresponding author:** Debraj Basu, School of Engineering and Advanced Technology, Massey University, New Zealand
E-mail: debrajece.ciem@gmail.com

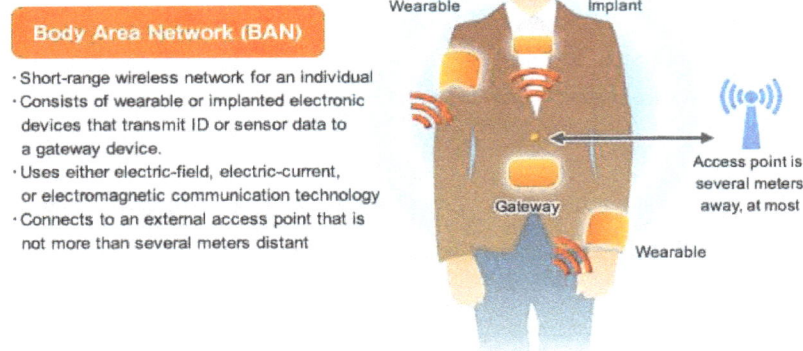

**Figure 1:** Body area network with sensors connect with the access point via a gateway [3-4].

shake mechanism. However, the emphasis is on the energy constrained sensor nodes. This paper suggested that received signal strength indication (RSSI) data may not be sufficient to evaluate link quality when sensor nodes are mobile. They proposed an active probing scheme for sensor applications that send data periodically and those which are triggered by an event (i.e., event driven). In active probing scheme, the mobile node counts the number of consecutive packets that are successfully transmitted at the current power level. If that is more than a predefined threshold, then the power level is decremented by one level. However for any un-acknowledged packet, the transmission power level is incremented by one level, until the maximum power level is reached. This paper has also modified this approach by using the link quality indication (LQI) values that are provided by CC2420 transceiver modules. In order to make good use of the LQI values of the acknowledgement packets, it allowed the radio to transmit several packets at each of the power levels. It finds the optimal power level region where consistent LQI values higher than 100 is observed. This optimal value is used as the benchmark to set the new transmission power level.

In GSM, a power control algorithm is employed to achieve desired signal strength for faithful communication between the mobile station (MS) and the base transceiver station (BTS). Power control also reduces interference and improves cell capacity. During a connection between the BTS and the MS in a cell, the MS measures the channel's RF link quality after every 480 milliseconds [8]. In this way an acceptable link quality is maintained which can also improve the battery lifetime of the mobile device. However, the research work that is presented here has aimed at saving energy by cutting down on the cost of sensing the RF channel *before* actual transmission. This is because the wireless sensor devices are battery-powered with capacity in the order of 250-300 mAh [9] and have far less capacity than the batteries used by mobile devices (~1500 -3500 mAh) [10-12].

There are a few non-RSSI based power control algorithm that uses matrices like the packet delivery ratio (PDR) or the packet reception rate (PRR) to estimate link quality rather than RSSI or LQI. Among them Practical-TPC [13] and ART [14] are worth mentioning.

P-ATPC is a receiver oriented protocol that is considered robust in dynamic wireless environments and uses packet reception rate (PRR) values to compute the transmission power that should be used by the

sender in the next attempt. The receiver monitors all incoming packet and counts the successes and failures of the packet transmission within the current sampling window. After the sampling window period is over, the P-TPC protocol computes the next transmission power level and sends to the transmitter. This new power level will be used during the next sampling window. P-ATPC has two main components. One component (fast online model identification FID) estimates the model between the PRR and the transmission power. It initialises or reconfigures the second component proportional-integral with anti-Windup (PI-AW). The PI-AW computes the transmission power level based on the difference between the current PRR and the application specific PRR requirement. P-ATPC runs two feedback loops. The inner loop involves the PI-AW that adapts the transmission power based on the PRR measurement. The outer loop involves FID to adjust the parameters based on the updated power model that defines the relationship between the PRR and transmission power. P-ATPC also initializes the power model parameters before the feedback loops kick in. In this initial phase, each link is set to transmit a sequence of probe packets using highest to lowest power level to build the transmission power model.

ART (Adaptive and Robust Topology control) protocol has been designed for complex and dynamic radio environments. It adapts the transmission power in response to variation in link quality or degree of contention. Empirical studies presented in this paper have shown that the PRR and the signal strength can vary over time. This is primarily due to effect of movements of objects and people in between the transmitter and the receiver during the busy hour of the day and corresponding fading of signal. Analysis of the paper has suggested that RSSI and LQI may not be good or the most reliable indicators of link quality, especially in dynamic indoor radio environment. ART changes the transmission power of a link based on the observed PRR. It has an initialization phase when the ART protocol monitors all the outgoing packets for its successful or failed transmission within a sliding window of predefined size. It compares the number of failures within that window with a minimum and a maximum threshold failure. Based on the comparison, it does anyone of the following:

1) Remains in the same power level or

2) Increases the transmission power level or

3) Reduces the transmission power level and enters a trial to compare the new failure rate count with the predefined threshold.

In these type of non-RSSI based adaptive protocols, there is a running overhead cost as the link quality is sampled after a given interval of time. The sampling period will itself depends on the dynamics of the link quality.

The novel non-RSSI based channel estimation and output power control algorithm is proposed in our earlier papers [15,16]. It does not use RSSI/LQI data as side information for channel link quality estimation. The basis of this lightweight adaptive algorithm is the states where each state represents one cycle of packet transmission. The details of the adaptive power control algorithm are explained in the next section.

## Basics of an Adaptive Power Control Algorithm

The packet success rate performance of a mobile sensor node depends primarily on the distance between the base station and the sensor, obstacles in the communication path, and fading due to movement of objects in between. In particular, when a node operates on the fringes of the communicable distance, the power amplifier pumps in maximum power so that it is discernible at the receiving station. It uses its allocated retry limit to achieve the threshold packet success rate. On the other hand, the radio conditions can be most favourable so that successful data communication is possible at the lowest available output power. At intermediate stages of communication distances, the transmitter may use higher power levels. In a non-RSSI based power control approach, the key is to keep track of the success or failure of packets at a particular transmission power. The outcome of the last packet transmission gives an indication as to what is the expected outcome when a new packet transmission starts. This adaptive power control can be most successfully applied when a node transmits quite frequently. Nodes that transmit once every hour or a day do not come under the purview of this research. The reason is that the transmitter does not send probe packets for channel estimation neither it can avail RSSI information of the last data transmission. Therefore, the last power level is an indicator of the channel condition. Now, this channel condition can be transient or semi-permanent. Transient condition can occur because of momentary dip in the signal level due to fading or due to change of distance or any obstruction in between the transmitter and receiver.

In case the channel change is transient, the adaptive algorithm must drop-off fast to the lower state. When the change in the link quality or the channel condition is not transient, then the adaptive algorithm

| State | 1 | 2 | 3 | 4 |
|---|---|---|---|---|
| Available Power Levels | Minimum (M) | | | |
| | Low (L) | Low (L) | | |
| | High (H) | High (H) | High (H) | |
| | Maximum (X) | Maximum (X) | Maximum (X) | Maximum (X) |
| Number of retries | 3 | 2 | 1 | 3 |

**Table 1:** States, power levels, and retry limits.

| | | Next State | | | |
|---|---|---|---|---|---|
| | | 1 (MLHX) | 2 (LHX) | 3 (HX) | 4 (X) |
| Current State | 1 (MLHX) | Succeed at level M | Succeed at level L | Succeed at level H | Failed or Succeed at level X |
| | 2 (LHX) | Not applicable | Not applicable | Succeed at level H | Failed or Succeed at level X |
| | 3 (HX) | No transition | Not applicable | Not applicable | Failed or Succeed at level X |
| | 4 (X) | No transition | No transition | Not applicable | Not applicable |

**Table 2:** State transition matrix when state levels go up.

must continue to transmit at a higher power level to ensure that packets are delivered successfully. Even in this case, when the distance between the transmitting node and the base station improves or the obstruction is removed, then the transmitter should be able to back off to a lower state.

## Non- RSSI/LQI based channel estimation and power control algorithms for energy efficiency

The basis of this lightweight adaptive algorithm is the states where each state represents one cycle of packet transmission. In each state there are output power levels in increasing order which can be used by the transmitter. State transition occurs depending on the power level at which the transmission is successful or failed. State 4 uses only the maximum power level and is allowed to transmit 4 times. There is no direct transition from state 4 to state 1 or 2. Similar conditions hold true when transiting from 3. The most energy efficient state is 1. The more it stays in state 1, the more it saves energy. State 4 is where the maximum energy may be used to transmit the packet. The adaptive algorithm is designed in such a way that it takes into account of performances in each state. It also has a unique drop-off algorithm that allows it to drop down to a lower state when deemed necessary. It is guided by the drop-off factor R. In this paper, R values of 0.01, 0.05, 0.1, 0.5 and 1 are used. Higher value of R means higher rate of drop-off. Figure 2 shows the state transition diagram of the adaptive power control algorithm. State transition occurs depending on the power level at which the transmission is successful or has failed.

The objective of the adaptive power control algorithm is to respond to the packet error rate and move to a new state with different retry limits. The adaptive algorithm is designed in such a way that it takes into account the performance in each state. Each state has a different retry limit. Increasing state number indicates poorer channel quality. The proposed adaptive algorithm does not allow retransmission in the same power level except when it is in state 4 and transmitting at 0 dBm. When the system is in state 4, it is considered the worst channel condition and three retries are allowed. The retry limit of state 1 is three. However, the retry limit of states 2 and 3 have been set at 2 and 1. The asymmetry is because the increase in the retry limit in states 2 and 3 can increase the current consumption while only marginally improving the packet success rate.

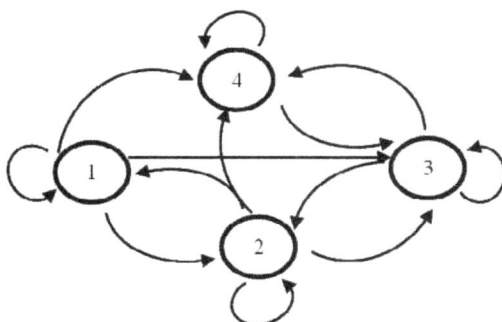

**Figure 2:** State transition diagram of the adaptive algorithm.

|  |  | Next State | | | |
|---|---|---|---|---|---|
|  |  | 1 (MLHX) | 2 (LHX) | 3 (HX) | 4 (X) |
| **Current State** | **1 (MLHX)** | Success at state M | Not applicable | Not applicable | Not applicable |
|  | **2 (LHX)** | Probabilistic model that depends on the number of successes in level L | Probabilistic model that depends on the number of successes in level L | Not applicable | Not applicable |
|  | **3 (HX)** | No transition | Probabilistic model that depends on the number of successes in level H | Probabilistic model that depends on the number of successes in level H | Not applicable |
|  | **4 (X)** | Not applicable | Not applicable | Probabilistic model that depends on the number of successes in level X | Probabilistic model that depends on the number of successes in level X |

**Table 3:** State transition matrix when state levels go down. Here, $P_{drop-off}$ =probability of drop-off; S=the number of successes in that power level of the higher state; R=drop-off factor.

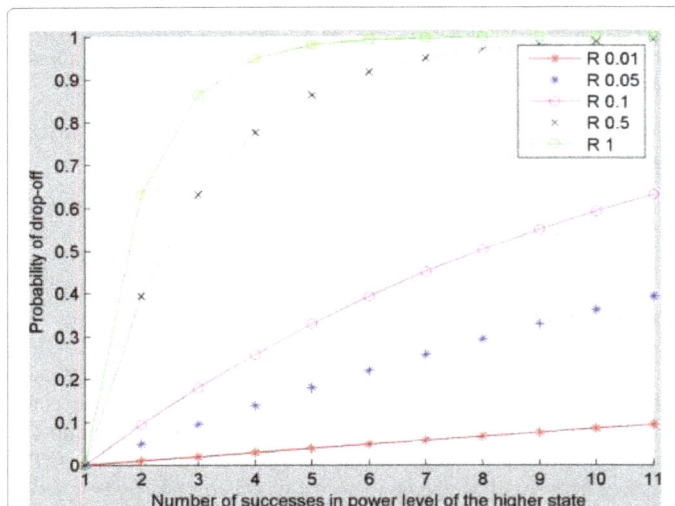

**Figure 3:** The curves behave differently depending on the value of R. A low R value indicates slow back off while a high R indicates fast back off. When the number of successes is 0, the probability of transition is 0. This drop-off algorithm takes into account of all the previous successes indicating that it also uses past history while dropping-off.

Table 1 shows the available power levels based on the states. Transmission starts at the lowest available power level of that particular state. The transmitter can be in any one of the states during the start of transmission of a packet. There are two separate algorithms that determine the state transitions, one from a lower state to higher state and the other from a higher to lower states. The logic to transit to lower states also includes situations when it remains in the same state or transit to a lower state.

Table 2 describes the state transition matrix when state level goes up. All the state transition decisions depend on the success or failure of the packet being transmitted to the destination hub.

Table 3 describes the state transition logic when state level goes down. The primary objective of the adaptive algorithm is to save energy by transmitting at a power level that is enough to send the packet

successfully through the channel. For example, when the system is in state 4, it is transmitting at the maximum power. With time, the channel condition can improve and packet can be successfully transmitted at a lower power level. If the system drops down to state 3, the transmission starts at a lower power level. This drop-off from a higher state to a lower state is determined by a drop-off algorithm which is probabilistic in nature.

In the proposed adaptive algorithm, the drop-off or the back-off process is dependent on the number of successes (S) in the higher power level and a drop-off factor (R). By default, the drop-off factor is 1. The probability of the system to drop-off to a lower power level is represented by Equation (1).

$$P_{drop-off} = 1 - ^{(-RS)} \tag{1}$$

Here, $P_{drop-off}$ =probability of drop-off; S=the number of successes in that power level of the higher state; R=drop-off factor.

The plots in Figure 3 show the state transition probability based on different values of R. When there is a state change, the value of S is reset to 0. Overall, the value of R indicates how fast the system will fall from a higher state to a lower state. When there is no success, the probability of state transition is 0, meaning that there will be no state transition. At the same time, when the number of successes is too high, it converges to 0.

Back-off algorithms are extensively used in data communication (both wired and wireless) by MAC protocols to resolve contention among transmitting nodes to acquire channel access. In a MAC protocol, the back-off algorithm chooses a random value from the range [0, CW], where CW is the contention window size. The contention window is usually represented in terms of time slots.

The number of time slots to delay before the nth retransmission attempt is chosen as a uniformly distributed random integer r in the range 0<r<2k.

Where k=min (n, 10), 10 is the maximum number of retries allowed.

The $n^{th}$ retransmission attempt also means that there have been n collisions. For example, after the first collision, it has to retransmit. Based on the back-off algorithm, the sender will choose between 0

| Device: Receiver (Hub) | nRF24L01p with PA and LNA |
|---|---|
| Arduino Mega Development board | Microcontroller board based on the ATmega2560 [20] |
| Transmission mode peak current | 115 mA |
| Reception mode peak current | 45 mA |
| PA gain | 20 dB |
| LNA gain | 10 dB |

**Table 4:** Features of nRF24L01p receiver [19].

| Device : Transmitter | nRF24L01p |
|---|---|
| Arduino Mega Development board | Microcontroller board based on the AT mega 2560 |
| Transmission @ 0 dBm output power (MIN) | Current drawn: 11.3 mA |
| Transmission @ -6 dBm output power (LOW) | Current drawn: 9 mA |
| Transmission @ -12 dBm output power (HIGH) | Current drawn: 7.5 mA |
| Transmission @ -18 dBm output power (MAX) | Current drawn: 7 mA |

**Table 5:** Features of nRF24L01p transmitter [21].

and one time slot for the retransmission. After the second collision, the sender will wait anywhere from 0 to three time slots (inclusive). After the third collision, the senders will wait anywhere from 0 to seven time slots (inclusive), and so forth. As the number of retransmission attempts increases, the number of possibilities for delay increases exponentially [17,18].

Similarly, an exponential operator is used in this novel adaptive algorithm to decide to switch from a higher state to a lower state. The drop-off algorithm is dynamic as it re-evaluates at every successful transmission. It gets reset to 0 when it leaves the state and jumps to a lower state and starts a new packet transmission at a lower power level [17,18].

## Experimental Methodology

### Choice of hardware

The adaptive power control algorithm has a unique channel estimation method without RSSI side information. The hardware used in the research includes the nRF24L01p transceiver module that acts as a transmitting sensor. For the receiver at the hub, another nRF24L01p transceiver module is used that has an additional PA and LNA. The receiver has a maximum output power level of 20 dBm. The reason to choose a high power transmitter at the hub is to make the path between the hub and the sensor practically error free. The primary features of the receiver and the transmitter receiver are presented in Tables 4 and 5 respectively [19-21].

The transceiver can transmit at four power levels: -18 dBm, -12 dBm, -6 dBm and 0 dBm. In general a wireless transceiver has different modes of operation. All the software programming was done in C in the open-source Arduino (version 1.0.5-r2) software (IDE) [22]. The programs or sketches in Arduino are used to interface with the nRF24L01p modules to do the necessary changes.

With regards to the heart rate data, a set of heart rate data are preloaded in the transmitter module and made to transmit after every 5 seconds. These data has been borrowed from PhysioNet which offers free web access to large collections of recorded physiologic signals [23,24]. Here we are testing the energy efficiency of the protocol, so we have used heart rate data from PhysioNet while the subjects provide the sensor mobility.

### Location and subject

The experimental setup is a typical house with the base station powered by Mains while the transmitting sensor is piggybacked on the subjects. Two subjects of different age groups ([30-35] years and [65-70] years) are chosen to observe the effect of age on the adaptive protocol. The subjects are allowed to roam freely inside the house and follow their routine activities. In this paper, subject 1 refers to the person in age group 65-70 and subject 2 refers to age group 30-35. Data were collected for a period of approximately five hours on different days of the week, starting from afternoon till late evening. In the 1st set of experiments, the mean distance between the subject and the base station was approximately 5 meters. During the 2nd set of experiments, the mean distance was changed to roughly 10 meters. The different distances are expected to have an effect on the evaluation parameters which are presented in subsection 3.3.

### Evaluation parameters

The evaluation parameters are

• Average cost per successful transmission

• Expected success rate or protocol efficiency [25]

One of the parameters for the optimization is the energy consumed per useful bit transmitted over a wireless link [26,27]. Similarly in this paper, the cost per successful transmission has been considered.

$$C_{s\_avg} = \frac{C_T}{P_s - P_L} \qquad (2)$$

Where

$C_{S\_avg}$=average energy cost per successful transmission

$C_T$=total cost of transmission

$P_L$=number of lost packets

$P_S$=Number of packets to send

| Fixed power transmission | | | |
|---|---|---|---|
| Output power | PSR % | Avg. Cost per successful transmission mJ | Protocol Efficiency % |
| -18 dBm | 96.99 | 0.03489 | 87.08 |
| -12 dBm | 99.61 | 0.0333 | 97.42 |
| -6 dBm | 99.86 | 0.03944 | 98.67 |
| 0 dBm | 99.86 | 0.04945 | 98.81 |
| Non-RSSI based adaptive power control | | | |
| Drop-off factor R | PSR % | Avg. Cost per successful transmission mJ | Protocol Efficiency % |
| 0.01 | 99.75 | 0.0438 | 93.58 |
| 0.05 | 99.8 | 0.03272 | 96.49 |
| 0.1 | 99.78 | 0.03217 | 96.68 |
| 0.5 | 99.82 | 0.03155 | 96.94 |
| 1 | 99.86 | 0.03137 | 97.25 |

Table 6: Average cost, PSR and protocol efficiency with subject 1 and mean distance equals to 5 meters.

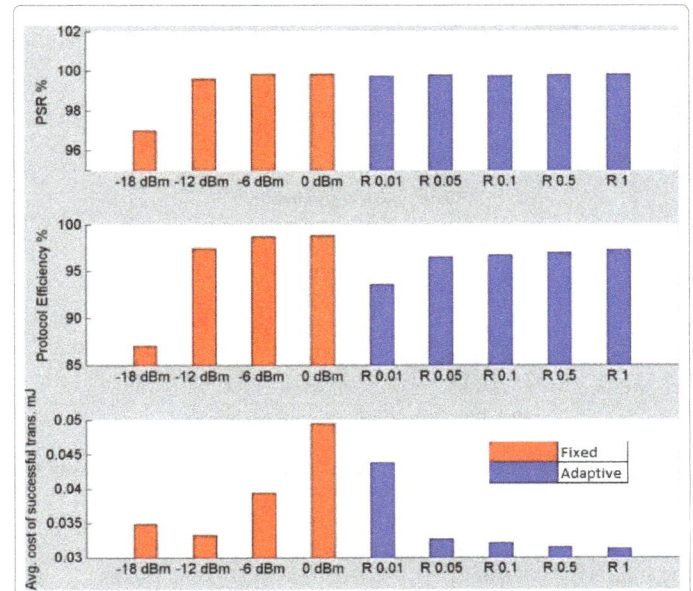

Figure 4: Subject 1: Comparison of the minimum cost and the corresponding PSR and protocol efficiencies due to different transmission strategy for subject 1 shows that the adaptive protocol can save upto 6% energy when R=1 as compared to fixed transmission at -12 dBm.

All cost values are measured in mJoules. The total cost of transmission includes the expenditure for the first transmission attempt of a packet and the subsequent retries if the first attempt fails. The total packet to send count does not include the retry packets. Therefore the denominator in equation 3 is only the count of successfully transmitted packets.

The expected success rate or efficiency is defined as the expected number of successes and takes into account the average number of retries [3]. It can also be defined as the expected number of successes per 100 transmissions. Mathematically,

$$Succ_{rate} = \frac{P_s - P_L}{P_s + \text{Re}t_T} \qquad (3)$$

$$Succ_{rate} = \frac{P_s - P_L}{P_s + \text{Re}t_T} \qquad (4)$$

Where

$Succ_{rate}$=expected success rate

$Ret_T$=total number of retries

Here $P_s - P_L$=total number of successes ($P_{succ}$). If both the numerator and denominator are divided by Ps, then in percentage term,

$$Succ_{rate}(\%) = \frac{PSR}{P_s + \text{Re}t_{avg}} \qquad (5)$$

where $Ret_{avg}$=average number of retries per packet and is defined as

$$\text{Re}t_{avg} = \frac{\text{Re}t_T}{P_s} \qquad (6)$$

Here,

$$PSR = \frac{P_{succ}}{P_s} 100 \qquad (7)$$

This parameter indicates the total number of transmissions (on average) to achieve a given packet success rate (PSR).

## Experimental Results and Analysis

The results and analysis of the experiments are presented in this section.

| Fixed power transmission | | | |
|---|---|---|---|
| Output power | PSR % | Avg. Cost per successful transmission mJ | Protocol Efficiency % |
| -18 dBm | 98.40 | 0.03325 | 91.28 |
| -12 dBm | 99.95 | 0.03257 | 99.55 |
| -6 dBm | 99.95 | 0.03915 | 99.38 |
| 0 dBm | 99.95 | 0.04918 | 99.34 |
| Non-RSSI based adaptive power control | | | |
| Drop-off factor R | PSR % | Avg. Cost per successful transmission mJ | Protocol Efficiency % |
| 0.01 | 99.95 | 0.03334 | 98.08 |
| 0.05 | 99.95 | 0.03073 | 99.06 |
| 0.1 | 99.95 | 0.03054 | 99.38 |
| 0.5 | 99.9 | 0.03063 | 99.14 |
| 1 | 100 | 0.03058 | 99.1 |

**Table 7:** Average cost, PSR and protocol efficiency with subject 2 and mean distance equals to 5 meters.

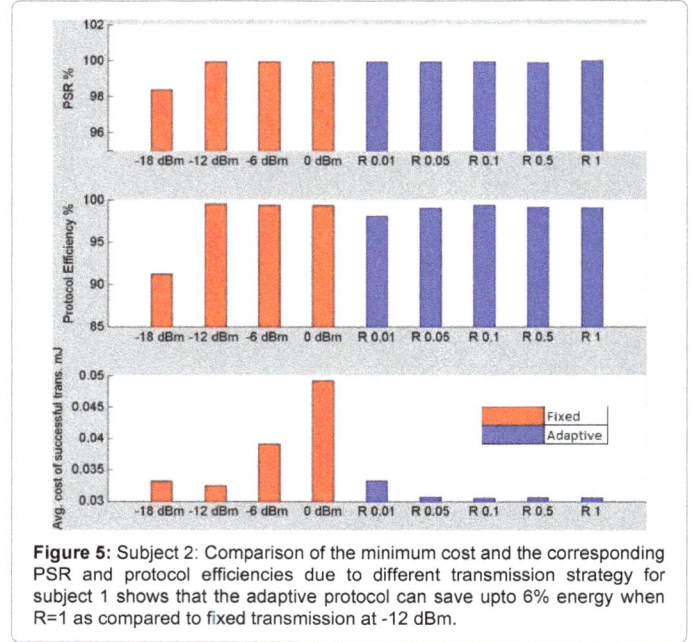

**Figure 5:** Subject 2: Comparison of the minimum cost and the corresponding PSR and protocol efficiencies due to different transmission strategy for subject 1 shows that the adaptive protocol can save upto 6% energy when R=1 as compared to fixed transmission at -12 dBm.

| Fixed power transmission | | | |
|---|---|---|---|
| Output power | PSR % | Avg. Cost per successful transmission mJ | Protocol Efficiency % |
| -18 dBm | 90.45 | 0.04547 | 67.27 |
| -12 dBm | 98.79 | 0.03458 | 93.94 |
| -6 dBm | 99.73 | 0.0395 | 98.52 |
| 0 dBm | 99.86 | 0.04944 | 98.82 |
| Non-RSSI based adaptive power control | | | |
| Drop-off factor R | PSR % | Avg. Cost per successful transmission mJ | Protocol Efficiency % |
| 0.01 | 99.79 | 0.04455 | 93.22 |
| 0.05 | 99.87 | 0.03134 | 98.66 |
| 0.1 | 99.9 | 0.0308 | 99.12 |
| 0.5 | 99.92 | 0.03079 | 98.87 |
| 1 | 99.9 | 0.03069 | 99.09 |

**Table 8:** Average cost, PSR and protocol efficiency with subject 1 and mean distance equals to 10 meters.

| Fixed power transmission | | | |
|---|---|---|---|
| Output power | PSR % | Avg. Cost per successful transmission mJ | Protocol Efficiency % |
| -18 dBm | 80.18 | 0.0654 | 47.08 |
| -12 dBm | 95.40 | 0.04015 | 81.23 |
| -6 dBm | 98.00 | 0.04275 | 91.2 |
| 0 dBm | 98.92 | 0.05186 | 94.29 |
| Non-RSSI based adaptive power control | | | |
| Drop-off factor R | PSR % | Avg. Cost per successful transmission mJ | Protocol Efficiency % |
| 0.01 | 98.43 | 0.05102 | 86.64 |
| 0.05 | 98.6 | 0.03475 | 93.51 |
| 0.1 | 98.75 | 0.03427 | 93.47 |
| 0.5 | 98.79 | 0.03354 | 93.91 |

**Table 9:** Average cost, PSR and protocol efficiency with subject 2 and mean distance equals to 10 meters.

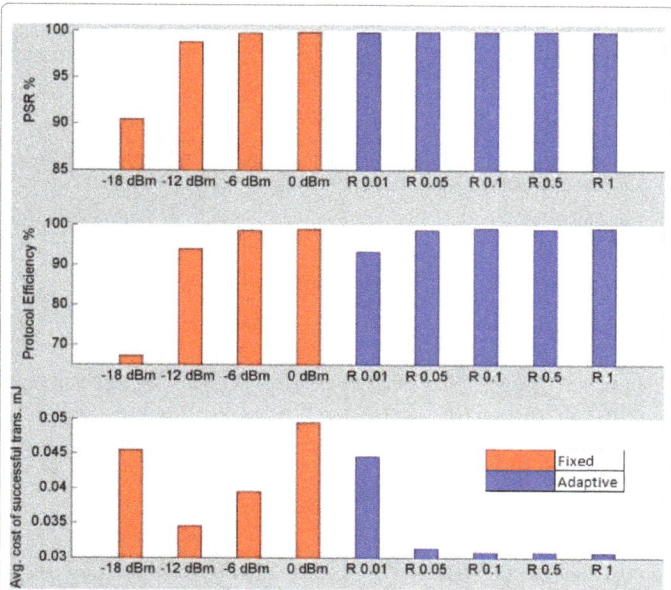

**Figure 6:** Subject 1: Comparison of the minimum cost and the corresponding PSR and protocol efficiencies due to different transmission strategy for subject 1 shows that the adaptive protocol can save upto 13% energy when R=1 as compared to fixed transmission at -12 dBm.

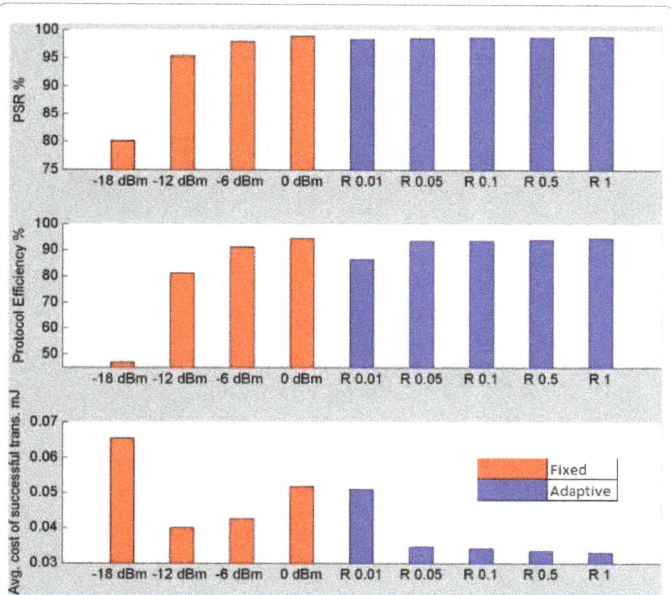

**Figure 7:** Subject 2: Comparison of the minimum cost and the corresponding PSR and protocol efficiencies due to different transmission strategy for subject 1 shows that the adaptive protocol can save upto 21% energy when R=1 as compared to fixed transmission at -12 dBm.

## Scenario 1: Mean distance is 5 meters

Table 6 presented the average values of three runs of the experiments with subject 1 when the distance between the base station and the transmitting mobile node is varying with mean distance approximately equal to 5 meters.

Figure 4 compares the compares the performance parameters of each of the transmission strategies based on the PSR.

Based on Figure 4 it can be observed that the PSR in all the cases are ~100%. The protocol efficiency values are all above 95%. Since the mean distance is very small (~ 5 meters), the adaptive protocol could only marginally perform better than the fixed power transmission. It is still able to save 6% energy. A low value of R means that the adaptive system will back-off at a slow pace. This is the reason that the minimum cost is achieved at R=1 because it has the ability to back-off fast to a lower state and transmit at a lower power as compared to when R is set at 0.01.

Not much change is observed when the experimental data for subject 2 is analysed. Table 7 and Figure 5 present the experimental results in details.

It can be observed from the Figures 4 and 5 that since the mean distance between the subjects and the base station are very small, power control may not be able to save significant amount of energy.

### Scenario 2: Mean distance is 10 meters

In this 2nd set of experiments, the mean distance is changed to approximately 10 meters. The results are tabulated in Tables 8 and 9 and comparison are presented in Figures 6 and 7 respectively for subjects 1 and 2.

The minimum amount of energy per successful transmission is consumed in fixed transmission mode when the power level is set at -12 dBm. Figure 6 shows that if the drop-off factor is set at 1, the adaptive protocol can save 13% energy as compared to fixed transmission at -12 dBm.

Table 9 and Figure 7 present the analysis and graphical comparison of the evaluation parameters when data from subject 2 is used with a mean distance of 10 meters.

The minimum amount of energy per successful transmission is consumed in fixed transmission mode when the power level is set at -12 dBm. Figure 7 shows that if the drop-off factor is set at 1, the adaptive protocol can save 21% energy as compared to fixed transmission at -12 dBm.

Overall, it can be observed that the use of adaptive protocol in typical home environment for sensor monitoring purposes can save energy and extend the operational lifetime before batteries are replaced. The younger adult as subject 2 of the experiment is expected to move faster as compared to the elder adult, denoted by subject 1. This variability in the mobility is not reflected in the 1st set of results as the distance is small. As the mean distance changed, the experimental results show that the savings for subject 2 is more than subject 1. This is because subject 2 is more active than subject 1 and the adaptive protocol finds enough space to modulate the power level, thereby saving more energy.

## Conclusion

The results in this paper demonstrate the advantages and limitations of using power control under different channel conditions to achieve energy efficiency. When the link quality is good (mean distance ~5 meters and very few obstructions in between the transmitter and the receiver), the adaptive power control algorithm is able to save energy marginally as compared to fixed power transmission. This is because there was no scope of output power manoeuvring to achieve energy efficiency. When the mean distance is roughly doubled to 10 meters, the adaptive power control approach has proved to be energy-saving as compared to fixed power. It is able to save upto 21% energy. It can be observed from Table 5 that the output power levels scale poorly with the corresponding current rating. A drop in output power by 63 times only halves the current consumption approximately. Considering this values, the energy savings by the adaptive protocol is significant. More experiments will be conducted as part of future research scopes in other types of radio environments.

## References

1. Darwish A, Hassanien AE (2011) Wearable and Implantable Wireless Sensor Network Solutions for Healthcare Monitoring. Sensors (Basel) 11: 5561-5595.

2. Ko JG, Lu C, Srivastava MB, John A (2010) Wireless Sensor Networks for Healthcare. Proceedings of the IEEE 98: 1947-1960.

3. Renesas (2015) Renesas Solutions for Wireless Sensor Networks? Part 2: Body Area Networks.

4. Alemdar H, Ersoy C (2010) Wireless sensor networks for healthcare: A survey. Computer Networks 54: 2688-2710.

5. Shnayder V, Chen B, Lorincz K, FulfordJones TRF, Welsh M (2005) Sensor Networks for Medical Care. Technical Report TR-08-05. Division of Engineering and Applied Sciences, Harvard University.

6. Trappaport (2002) Wireless Communications- principles and practice. Prentice Hall PTR.

7. Ko J, Terzis A (2010) Power Control for Mobile Sensor Networks: An Experimental Approach. SECON. pp: 1-9.

8. Eberspächer J, Vögel HJ, Bettstetter C, Hartmann C (2009) GSM - Architecture, Protocols and Services. Wiley & Sons.

9. http://data.energizer.com/PDFs/cr2320.pdf

10. EB-B600BUBEST (2015) Samsung Corporation, Korea.

11. Gsmarena (2015) Apple iphone 5-4910. Accessed on: November 2015.

12. Vandervell A (2015) Nokia Lumia 1520: Battery Life and Verdict.

13. Fu Y, Sha M, Hackmann G, Lu C (2012)Practical Control of Transmission Power for Wireless Sensor Networks. In IEEE International Conference on Network Protocols pp: 1-10.

14. Hackmann G, Chipara O, Lu C (2006) Robust Topology Control for Indoor Wireless Sensor Networks. In ACM conference on Embedded network sensor systems pp: 57-70.

15. Basu D, Sen Gupta G, Moretti G, Gui X (2015) Protocol for improved energy efficiency in wireless sensor networks to support mobile robots. In Proc. International Conference on Automation, Robotics and Applications (ICARA). pp: 230-237.

16. Basu D, Sen G, Moretti G, Gui X (2015) Performance comparison of a novel adaptive protocol with the fixed power transmission in wireless sensor networks. Journal of Sensor and Actuator Networks, Multidisciplinary Digital Publishing Institute (MDPI AG). pp: 274-292.

17. Tanenbaum AS (2002) Computer Networks. Chapter 4. 4th edn. Prentice Hall PTR.

18. IEEE Standard for Information technology (2015) Telecommunications and information exchange between systems- Local and metropolitan area networks- Specific requirements. Part 3: Carrier sense multiple access with Collision Detection (CSMA/CD). Access Method and Ph. IEEE Computer Society.

19. Elecfreaks (2015) 2.4G Wireless nRF24L01p with PA and LNA. Accessed on: November 2015.

20. Arduino (2015) Arduino Mega 2560. Accessed on: November 2015.

21. Nordicsemi (2015) nRF24L01+ SingleChip 2.4GHz Transceiver Product Specification v1.0. Accessed on: November 2015.

22. Arduino (2015) Arduino Software: arduino.cc. Accessed on: November 2015.

23. Physionet (2015) Physionet: The research resource for complex physiologic signals. Accessed on: November 2015.

24. Physionet (2015) Moody GBm, RR Intervals, Heart Rate, and HRV Howto.

25. Zhang D, Ya-nan Z, Zhao CP, Wen-bo Dai (2012) A new constructing approach for a weighted topology of wireless sensor networks based on local-world theory for the Internet of Things (IOT). Computers and Mathematics with Applications. pp: 1044-1055.

26. Sheu JP, Hsieh KY, Cheng YK (2009) Distributed Transmission Power Control Algorithm for Wireless Sensor Networks. Journal of Information Science and Engineering 25: 1447-1463.

27. Schmidt D, Berning M, When N (2009) Error correction in single-hop wireless sensor networks - A case study. In: Conference and Exhibition of Design, Automation and Test in Europe. pp: 1530-1591.

# 22

# Degrading the Effect of Channel Impairment on STBC-OFDM System in Rayleigh Fading Channel

**Sinha H[1]\*, Meshram MR[2] and Sinha GR[1]**

[1]*Shri Shankaracharya Technical Campus Bhilai, Chhattisgarh, India*
[2]*Department of Electronics and Telecommunication Engineering, Government Engineering College, Jagdalpur, Chhattisgarh, India*

### Abstract

This paper indicates the space time block code provides transmit diversity in wireless fading channel during the STBC-OFDM system. OFDM is enforced in broadband wireless access systems as the way to beat wireless channel impairments and to enhance and breadth potency. This research is to analyze the performance of non STBC-OFDM system and STBC-OFDM system. The performance has been evaluated for construction modulation (QAM) technique for orthogonal frequency division multiplexing (OFDM) systems using multiple transmit diversity antenna system in the channel. Conjointly the equation of SNR and BER has been derived for 16-QAM for multiple transmit and receive antenna for STBC-OFDM system by considering the have an effect on of Inter-symbol interference (ISI), phase noise and temporal arrangement noise in Raleigh fading channel.

**Keywords:** STBC; OFDM; Phase noise

## Introduction

Physical limitations of the wireless medium produce a technical challenge for reliable wireless communication techniques that improve spectral potency and to attain high information rates with low information measure. To overcome varied channel impairments cherish signal attenuation, interference, section noise, temporal order interference have made a colossal contribution to the expansion of wireless communications [1]. Multiple-input multiple-output (MIMO) primarily based communication systems area unit capable accomplishing these objectives, the OFDM system carries the message information on orthogonal subcarriers for parallel transmission; it's wide for its information measure potency used combating the distortion caused by the frequency selective channel or equivalently, the inter-symbol interference in the multi-path attenuation channel [2]. However, the advantage of the OFDM is effectively helpful only when the orthogonality is maintained between sub carriers because of orthogonal characteristics additional information can be transmitted at a precise quantity of band breadth as compare to the opposite technique. Other way to increase diversity gain by mistreatment area time blocks writing STBC technique [3]. The combination of the OFDM with STC with multi path attenuation channel accomplish high spectral potency and high data rate STBC-OFDM will increase the antenna diversity gain [3-5]. This paper derived analytically SNR and BER performance in multiple antenna system. just in case the orthogonality isn't sufficiently invulnerable by any means, its performance could also be degraded thanks to inter-symbol interference (ISI) and inter-channel interference (ICI), temporal order jitter [6]. Obtaining the results of temporal order interference, inter-symbol interference (ISI) problems in OFDM systems. Fading describes the variation of the native channel h (t) due to the varied phases and amplitudes of scatterers. Once the freelance scatter $n_s$ is massive and every one scattered contributions area unit noncoherent (not LOS) and of roughly equal energy, then by the central limit theorem h (t) could be a complicated Gaussian variable with zero mean and independent quadrature components. In this paper to analysis non-STBC OFDM system and other part shows the analysis of STBC-OFDM. At the end of the research analysis to determine the equation for Signal to Noise Ratio (SNR) and Bit Error Rate (BER) have been derived analytically using four transmitting antennas and one receiving antenna.

## Orthogonal Frequency Division Multiplexing (OFDM)

Orthogonal Frequency Division Multiplexing (OFDM) might be a competent technique to perform multicarrier modulation with mainly utilization of bandwidth and high performance characteristics profile against multipath fading channel. At the instant time Orthogonal Frequency Division Multiplexing is broad used for its bandwidth efficiency property credit to its orthogonal characteristic additional information are often transmitted at the certain amount of bandwidth as compare to the other systems. The affects Inter Symbol Interference (ISI) is in addition very low compare to the different type of multiplexing techniques. OFDM has been adopted and enforced in wire and wireless communication system. Unfortunately OFDM is extremely sensitive to the synchronization errors like Carrier frequency offset (CFO), Inter Symbol interference (ISI), and timing jitter [7]. Orthogonality take into account the time-limited advanced exponential signals that represent the various subcarriers at fk=k/Tsym within the OFDM signal, wherever $0 \leq t \leq Tsym$. These signals clearly define to be orthogonal if the integral of the products for its common (fundamental) periodic zero, the Equation (1) given below,

$$\frac{1}{Tsym}\int_0^{Tsym} e^{j2\pi fkt}e^{-j2\pi fi}dt = \frac{1}{Tsym}\int_0^{Tsym} e^{j2\pi \frac{k}{Tsym}t}e^{-j2\pi \frac{k}{Tsym}t}dt$$

$$= \frac{1}{Tsym}\int_0^{Tsym} e^{j2\pi \frac{(k-t)}{Tsym}t}$$

$$= \begin{cases} 1, \forall \text{ int } egerk = i \\ 0, otherwise \end{cases}$$

(1)

**\*Corresponding author:** Sinha H, PhD Scholar, Shri Shankaracharya Technical Campus, Bhilai, Chhattisgarh, India, E-mail: Sinha.hemlata552@gmail.com

Taking the discrete samples with the sampling instances at t ¼ nTs ¼ nTsym=N, n ¼ 0; 1; 2; ---; N_1, Equation (2) can be written in the discrete time domain as

$$\sum_{n=0}^{N-1} e^{j2\pi\frac{k}{Tsym}\cdot nT} e^{-j2\pi\frac{k}{T}nTs} = \frac{1}{N}\sum_{n=0}^{N-1} e^{j2\pi\frac{k}{Tsym}\cdot\frac{nT}{N}} e^{-j2\pi\frac{f}{Tsym}\cdot\frac{nTsym}{N}}$$

$$= \frac{1}{N}\sum_{n=0}^{N-1} e^{j2\pi\frac{(k-1)}{N}n} \tag{2}$$

$$= \begin{cases} 1, \forall \text{ int } egerk = i \\ 0, otherwise \end{cases}$$

The above orthogonality equation is an essential condition for the OFDM signal to be ICI-free. OFDM is extremely sensitive to the synchronization errors like Carrier frequency offset (CFO), timing jitter. The Carrier frequency offset CFO arise primarily due to Doppler Effect, the result is caused by the CFO decrease the signal amplitude and makes interference between the carriers [5]. Timing error would occur either once the clock signal is not properly recovered or once sampling circuit imperfect. Propagation delay of the ICI causes timing error. Other term is AWGN is introduced within the channel through that information is transmitted. The aim of this paper is to analysis the signal to noise ratio (SNR) and by changing the CFO in timing jitter [7]. Analysis has been done and the graphs have been plotted with the assistance of MATLAB Program and other graph also for Bit Error Rate (BER) in keeping with totally different SNR.

### Phase noise

Phase noise is rapidly short term, time-varying and random fluctuations that effects on the phase of a signal waveform caused by time domain instabilities occurs in early stages of receiver part, especially in demodulation stage. Presence of phase noise is increased errors for overall system. Therefore this term must be eliminated in order to enhance the error performance.

### Timing jitter

In the sampling circuit at the receiver additional error may occur in the determination of the best sampling phase. This means that the sampling instants are non-ideal and is given by- tn = nT + ξn ; Where ξn is the timing jitter of the nth sampling instant normalized by the symbol period T. Elective measurement of phase noise, but in the time-domain associated with actuations in the times of zero-crossings of a phase error.

### CFO

In order to suppress the ICI and thereby reduce SNR degradation, the residual CFO must be sufficiently small. For example, when using the 64QAM constellation, it is better to keep the residual CFO below O. OI/s to ensure that SNR < 0.3 dB for moderate SNR. The absolute value of CFO is fε, is either an integer multiple or a fraction of Δf. Now if the fε is normalized to the sub carrier spacing Δf then normalized CFO of the channel is expressed as Where δ is an integer and |ε| ≤ 0.5 If the CFO occurs then the symbol transmitted on a certain sub carrier k, will shift to another sub carrier ks = k + δ.

The transmitted OFDM signal for the month symbol is given by the N point complex modulation sequence is given by Equation (3),

$$y_m(n) = \sum_{k=0}^{N-1} Xm(k) e^{j\frac{2\pi}{N}nk} \tag{3}$$

Where n ranges from 0 to N+Ng-1 After passing through a Rayleigh fading channel and LO, the received signal impaired by AWGN and PN can be modeled as;

Where,

$$y_m = \left[\sum_{k=0}^{N-1} Xm(k) Hm(k) e^{j\frac{2\pi}{N}n(k+\epsilon)}\right] e^{i\varphi_m(n)} + W_m(n) \tag{4}$$

Or

$$y_m = S_m(n) e^{i\varphi_m(n)} + W_m(n) \tag{5}$$

$$S_m(n) = \sum_{k=0}^{N-1} Xm(k) Hm(k) e^{j\frac{2\pi}{N}n(k+\epsilon)} \tag{6}$$

Here Hm(k) is the transfer function of the Rayleigh fading channel at the frequency of the $k^{th}$ carrier and Wm(n) is the complex envelope of the AWGN with zero mean and variance $\sigma^2$. SNR Equation (7) is given below,

$$SNR(\varepsilon, \sigma_u^2) \geq \frac{\gamma\{\sin c^2(\pi\varepsilon)\}}{\left[1 + \left[0.5947(\sin\pi\varepsilon)^2 + \left\{\frac{\sigma_u^2}{2N}\sin c^2(\pi\varepsilon)\sum_{r=1}^{N-1}\frac{1}{\sin^2\left(\frac{\pi\gamma}{N}\right)}\right\}\right]\right]}$$

$$; |\varepsilon| \leq 0.5 \tag{7}$$

This Signal to Interference plus Noise Ratio or SINR is in terms of CFO and the variance of $\sigma^2$. This is without timing jitter (ξ). The equation (8) for SNR is including the timing jitter (ξ) is shown below,

$$SNR(\varepsilon, \sigma_u^2, \xi) \geq \frac{\gamma(1-\xi)\{\sin c^2(\pi\varepsilon)\}}{1 + \gamma(1-\xi)\left[0.5947(\sin\pi\varepsilon)^2 + \left\{\frac{\sigma_u^2}{2N}\sin c^2(\pi\varepsilon)\sum_{r=1}^{N-1}\frac{1}{\sin^2\left(\frac{\pi\gamma}{N}\right)}\right\}\right] + \gamma\xi}$$

$$; |\varepsilon| \leq 0.5 |\xi| \leq 1 \tag{8}$$

Now the Bit Error Rate can be determined with the help of E0/N0 the Equation (9) below,

$$BER = 0.5 \, ^*erfc\left(\sqrt{(SNR)}\right) \tag{9}$$

Orthogonal frequency division multiplexing (OFDM) transforms a frequency selective channel into huge set of individual frequency non-selective narrowband channels that is fitted to a multiple-input multiple-output (MIMO) structure that needs a frequency non-selective characteristic at every channel once the transmission rate is high enough to form the entire channel frequency selective. Therefore, a MIMO system using OFDM, denoted MIMO-OFDM, is ready to realize high spectral efficiency. However, the adoption of multiple antenna parts at the transmitter for spatial transmission outcomes a superposition of multiple transmitted signals at the receiver weighted by their corresponding multipath channels and makes the reception tougher. This imposes a true challenge on the way to design a sensible system which will supply a real spectral efficiency improvement. If the channel is frequency selective, the received signals are distorted by ISI that makes the detection of transmitted signals troublesome. OFDM has emerged jointly of best ways that to get rid of such ISI [8,9].

### Space time block code

MIMO-STBC structure concatenated with OFDM is modeled then the flexibility of OFDM system to vary the frequency selective channel to flat fading channel is delineated there is lots of technical purpose which should be consider in implementation of real MIMO system. Another problems is that the assumption of synchronization of the receive signal at totally different receive of antenna. This problem is very essential whereas doing decoding in the within the receiver. Rich scattering environment is another assumption that is sometimes created in MIMO-OFDM system. Therefore applying of MIMO-OFDM system in outside wireless application raise the technical difficulty that is have to be compelled to be self-addressed. However, the adoption of multiple antenna components at the transmitter for special transmission leads to a superposition of multiple transmitted signals at the receiver weighted by their corresponding multipath channels and makes the reception more difficult. This imposes a true challenge on a way to design a reasonable system which will supply a real spectral efficiency improvement. If the channel is frequency selective, the received signals are distorted by ISI that makes the detection of transmitted signals tough. OFDM has emerged in performance of best ways that to get free of such ISI. Orthogonal frequency division multiplexing (OFDM) because it mentioned in transforms a frequency selective channel into an huge set of individual frequency non-selective narrowband channels, that is fitted to a multiple-input multiple-output (MIMO) structure that needs a frequency non-selective characteristic at every channel once the transmission rate is high enough to form the full channel frequency selective. Therefore, a MIMO system using OFDM, denoted MIMO-OFDM, is in a position to attain high spectral efficiency. In this paper MIMO-STBC structure concatenated with OFDM is modeled then the ability of OFDM system to change the frequency selective channel to flat fading channel is portrayed [10,11]. Attenuation in a multipath wireless environment makes it extremely difficult for the receiver to determine the transmitted signal unless the receiver is provided with some form of diversity i.e., some less-attenuated replica of the transmitted signal is provided to the receiver. In some applications, the only practical means of achieving diversity is deployment of antenna array at the transmitter and/or receiver end. As the current trend of communication systems demands highly power-efficient and bandwidth-efficient schemes, techniques that provide such desirable properties are considered very valuable in next generation wireless systems. Making use of multiple antennas increases the capacity of the system with the associated higher data rates than single antenna systems. Space-Time coding is power-efficient and bandwidth-efficient methods of communication over a fading channel by using multiple transmit antennas systems.

In this paper the performances have been shown in non-STBC OFDM system and with STBC OFDM. The performance of both the system has been derived in equation of Signal to Noise Ratio (SNR) and Bit Error Rate (BER) for 2:1 transmission system. This paper extended the work by deriving the equations for 4:1 transmission system. After that performances have been shown by plotting various graphs. Results also show the effect of the Inter Carrier Interference (ICI), Channel Estimator Error, and the Additive White Gaussian Noise (AWGN). Mapping is used to calculate the BER and equations for the SNR and BER have been derived analytically.

## System model for STBC-OFDM

We consider an OFDM system with transmit diversity, in which the total system bandwidth is divided into N equally spaced and orthogonal sub-carriers. We investigate the system with four transmission antennas and one receiving antenna. During the first time instant, the four symbols [X0 X1 X2 X3] are transmitted from four antennas simultaneously, with X0, X1, X2 and X3 transmitted from all four antennas. In the second time slot [-X1* X0* -X3* X2], third time slot [-X2* -X3* X0* X*] and fourth time slot [X3 -X2 -X1 X0] are

transmitted. This encoding of the transmitted symbol sequence from the transmit antennas is given by then encoding matrix Equation (10),

$$\begin{bmatrix} H0 & H1 & H2 & H3 \\ -H1* & H0* & -H3* & H2* \\ -H2* & -H3* & H0* & H2* \\ H3 & -H2 & -H1 & H0 \end{bmatrix} \tag{10}$$

For each transmit antenna, a block of N complex-valued data symbols {X(k)} for k=0 to N-1 are grouped and converted into a parallel set to form the input to the OFDM modulator, where k is the sub carrier index and N is the number of sub carriers. The modulator consists of an Inverse Fast Fourier transform (IFFT) block. The output of the IFFT at each transmitter is the complex baseband modulated OFDM symbol in discrete time domain and is given by equation (11),

$$x(n) = \frac{1}{\sqrt{N}} \sum_{k=0}^{N-1} X(k) e^{j\frac{2\pi}{N}nk}; 0 \le n \le N-1 \tag{11}$$

The channel is modeled by a tapped delay line with channel coefficients that are assumed to be slowly varying such that they are almost constant over the two transmission instants. The channel frequency response for the $k^{th}$ subcarrier is Where h(p) is the complex channel gain of the $p^{th}$ multipath component.

$$H(k) = \sum_{k=0}^{L-1} h(p) e^{j\frac{2\pi}{N}pk} \tag{12}$$

Where h(p) is the complex channel gain of the $p^{th}$ multipath component. The time-domain received signals at the first and second transmission instances at the input to the FFT block are respectively given by Equations (13)-(16)

$$y^0(n) = \left( h0(n) \Diamond x0(n) + h1(n) \Diamond x1(n) + h2(n) \Diamond x2(n) + h3(n) \Diamond x3(n) + w(n)0 \right) e^{j\theta(n)} \tag{13}$$

$$y^1(n) = \left( -h0(n) \Diamond x1^*(n) + h1(n) \Diamond x0^*(n) - h2(n) \Diamond x3^*(n) + h3(n) \Diamond x2^*(n) + w(n)1 \right) e^{j\theta(n)} \tag{14}$$

$$y^2(n) = \left( -h0(n) \Diamond x2^*(n) - h1(n) \Diamond x3^*(n) + h2(n) \Diamond x0^*(n) + h3(n) \Diamond x2^*(n) + w(n)2 \right) e^{j\theta(n)} \tag{15}$$

$$y^3(n) = \left( h0(n) \Diamond x3(n) - h1(n) \Diamond x2(n) - h2(n) \Diamond x1(n) + h3(n) \Diamond x0(n) + w(n)3 \right) e^{j\theta(n)} \tag{16}$$

Where $\Diamond$ represents linear convolution, subscripts indicate antenna index, and superscripts indicate transmission instant. The complex Gaussian random variable w(n) represents the Additive White Gaussian Noise (AWGN) term with $\zeta$ w 2= E[|w(n))|2], and $\theta(n)$ is the phase noise. Detection with Imperfect Channel Estimation: In the presence of imperfect channel estimation, we assume a channel estimation model such that the channel estimates H of the true channel H is given equation (17)

$$\begin{bmatrix} H0 & H1 & H2 & H3 \\ -H1* & H0* & -H3* & H2* \\ -H2* & -H3* & H0* & H2* \\ H3 & -H2 & -H1 & H0 \end{bmatrix} =$$

$$\begin{bmatrix} H0+\varepsilon0 & H1+\varepsilon1 & H2+\varepsilon2 & H3+\varepsilon3 \\ -H1*+\varepsilon1* & H0*+\varepsilon0 & -H3*-\varepsilon3* & H2*+\varepsilon2* \\ -H2*-\varepsilon2* & -H3*-\varepsilon0 & H0*+\varepsilon0* & H2*+\varepsilon0* \\ H3+\varepsilon3 & -H2-\varepsilon2 & -H1-\varepsilon1 & H0+\varepsilon0 \end{bmatrix} \tag{17}$$

Where $\varepsilon0$, $\varepsilon1$, $\varepsilon2$ and $\varepsilon3$ are the errors in the channel estimate from the first, second, third and fourth transmit antennas respectively, and are modeled as independent zero-mean complex Gaussian random variables with variances $2\sigma\varepsilon_0^2, 2\sigma\varepsilon_1^2, 2\sigma\varepsilon_2^2, 2\sigma\varepsilon_3^2$ respectively.

Variance: As the noise signal has both positive and negative amplitude, it is squared and then the mean has been taken, which is variance. We consider the variance of noise for calculation. The variance of the noise W, after some mathematical manipulations, is given by $\sigma_w^2 = E[|W|^2]$.

Calculation of SNR and BER: The bit error rate for the case of 16QAM modulation using Gray code mapping for (b1b2b3b4). It is important to note that although the presentation is only for 16QAM, the following analysis is valid for all square QAM constellations. The conditional BER for bit b1, condition on H0, H1, H2, and H3 is given by Equation (18),

$$P_1\left(b_1 \mid H_0, H_1, H_2, H_3\right) = \frac{1}{2} * \left[Q\left(\sqrt{\frac{\left(2*\frac{E_g}{5}\right)\left(H0^2+H1^2+H2^2+H1^2\right)}{6\alpha^2+6\beta^2+6\omega^2}}\right) + Q\left(\sqrt{\frac{\left(2*\frac{E_g}{5}\right)H0^2+H1^2+H2^2+H3^2}{6\alpha^2+6\beta^2+6\omega^2}}\right)\right] \quad (18)$$

and for bit b3 is given by Equation (19)

$$P_3\left(b3 \mid H_0, H_1, H_2, H_1\right) = \frac{1}{2}\left[Q\left(\sqrt{\frac{9*\left(2*\frac{E_g}{5}\right)\left(H0^2+H1^2+H2^2+H3^2\right)}{6\alpha^2+6\beta^2+6\omega^2}}\right) Q\left(\sqrt{\frac{\left(2*\frac{E_g}{5}\right)H0^2+H1^2+H2^2+H3^2}{6\alpha^2+6\beta^2+6\omega^2}}\right) Q\left(\sqrt{\frac{\left(2*\frac{E_g}{5}\right)H0^2+H1^2+H2^2+H3^2}{6\alpha^2+6\beta^2+6\omega^2}}\right) + \left(\sqrt{\frac{25*\left(2*\frac{E_g}{5}\right)H0^2+H1^2+H2^2+H3^2}{6\alpha^2+6\beta^2+6\omega^2}}\right)\right] \quad (19)$$

From which the SNR $\gamma$ is given by Equation (20),

$$\gamma = \frac{\left(2*\frac{E_g}{5}\right)\left(H0^2+H1^2+H2^2+H3^2\right)}{6\alpha^2+6\beta^2+6\omega^2} \quad (20)$$

Probability density function given by Equation (21),

$$P(\gamma) = \frac{1}{2*6\gamma^2}\exp^{-\frac{1}{2*6\gamma^2}} \quad (21)$$

Due to the symmetry of square M-QAM constellations, the BER for the in-phase and quadrature bits are equal such that Pe (b1) = Pe (b2) and Pe (b3)= Pe (b4 ). Therefore the average BER is obtained by averaging the conditional BER of b1 and b3 over the PDF of the SNR $\gamma$. The average BER is therefore given by equation (22),

$$P_e = \frac{1}{2}*\int_0^\infty \left[Pe\left(b1 \mid H0, H1, H2, H3\right) + Pe\left(b3 \mid H0, H1, H2, H3\right)\right]P(\gamma)d\gamma \quad (22)$$

## Results

The Figure 1 shows the graph between SNR vs. normalized CFO where x-axis denotes the normalized CFO and y-axis denotes the SNR (dB). It is observed that when CFO is zero then SNR is high. And after that SNR is exponentially decreasing with the increasing of CFO. Graph shows the four plots for different value of variance of the phase noise. Figure 2 shows the graph between SNR vs. timing jitter. It is observed that if the delay or the timing jitter increases then the signal strength decreased. For this the SNR is also decreasing. There are plotting shows different graphs for different values of phase noise and CFO. Figure 3 shows the SNR vs. BER graph at different value of noise. From the graph it has been seen that for v=0.3, we get highest SNR curve. And for this highest SNR value the BER is decreasing fast compared to the other value of phase value v and corresponding SNR. Figure 4 show the graph of SNR vs. BER for 4:1 transmission system along with 2:1 and 1:1 system. It is clearly seen that 4:1 graph is closer to the two axes than the others. It

means that the BER is decreasing fast when the number of antennas increases. It shows that transmission power decreases when number of antennas increase. Another analysis can be drained from the graph shown is that or fixed value of transmission power the noise term can be reduced by increasing the number of antennas.

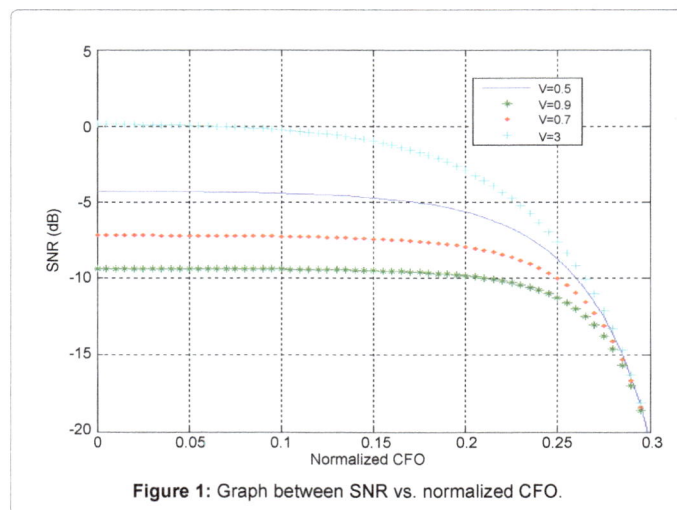

**Figure 1:** Graph between SNR vs. normalized CFO.

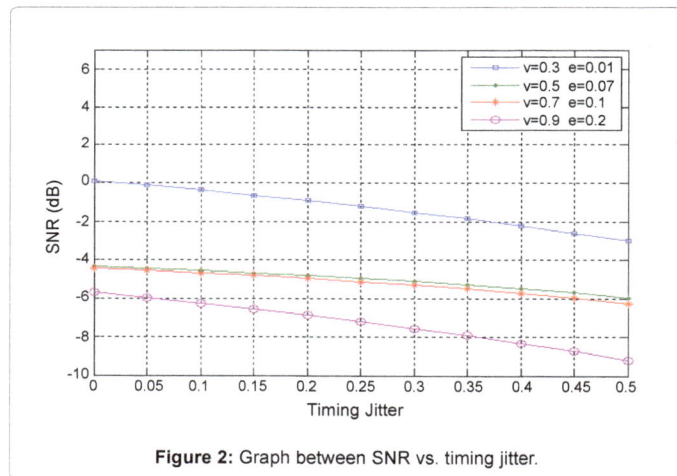

**Figure 2:** Graph between SNR vs. timing jitter.

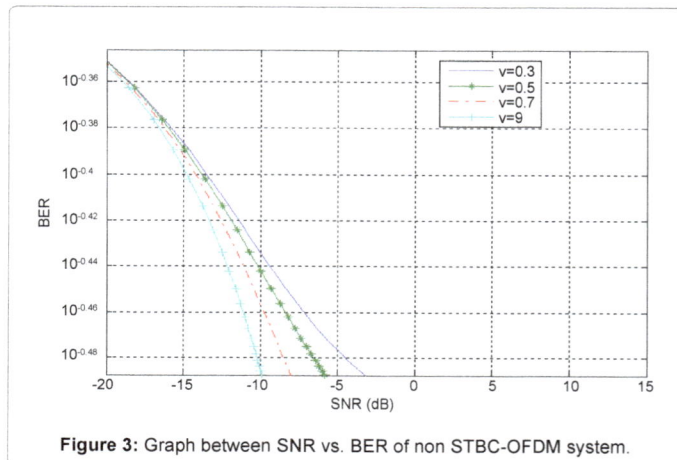

**Figure 3:** Graph between SNR vs. BER of non STBC-OFDM system.

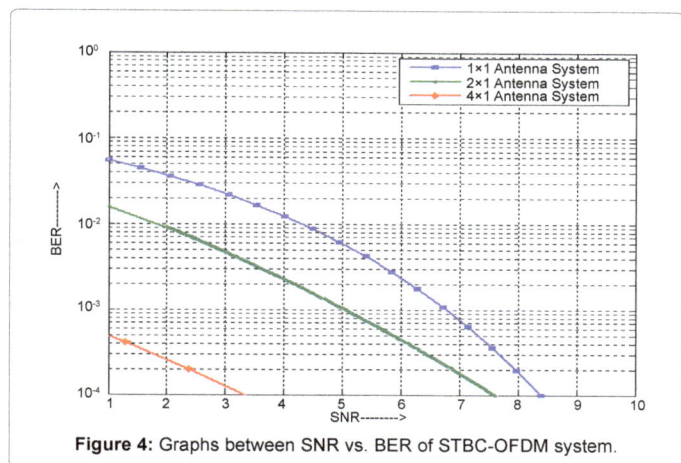

**Figure 4:** Graphs between SNR vs. BER of STBC-OFDM system.

## Conclusion

In this paper the equations for SNR and BER using 2:1 and 4:1 transmission system are derived. The result obtained with good accuracy. The effects of phase noise, timing jitter and carrier interference on the system are analyzed. The SNR vs. BER curve shows that increasing the diversity gain improve the performance of the system.

## References

1. Agarwal D, Tarokh V, Naguib A, Seshadri N (1998) Space–time coded OFDM for high data rate wireless Communication over wideband channels. In Proc IEEE VTC 3: 2232-2236.

2. Bolcskei H, Gesbert D, Paulraj AJ (2002) On the capacity of OFDM-based spatial multiplexing systems. IEEE Trans Commun 50: 225-234.

3. Alamouti SM (1998) A simple transmit diversity technique for wireless communications. IEEE J Select Areas Commun 16: 1451-1458.

4. Tarokh V, Jafarkhani H, Calderbank AR (1999) Space-Time Block Codes from Orthogonal Designs. IEEE Transac Inform Theo 45: 1456-1467.

5. Armstrong J (1999) Analysis of New and Existing Methods of Reducing Intercarrier Interference Due to Carrier Frequency Offset in OFDM. IEEE Trans Commun 47: 365-369.

6. Zogakis TN, Cioffi JM (1996) The effect of timing jitter on the performance of a discrete multitone system. IEEE Trans Commun 44: 799-808.

7. Wu S, Bar-Ness Y (2004) OFDM Systems in the Presence of Phase Noise: Consequences and Solutions. IEEE Trans Commun 52: 1988-1996.

8. Zhou E, Zhang X, Zhao H, Wang W (2005) Synchronization algorithms for MIMO OFDM systems. In Proc IEEE Wirless Commun Netw Conference 1: 18-22.

9. Priotti P (2004) Frequency synchronization of MIMO OFDM systems with frequency selective weighting. In Proc IEEE Vehic Technol Conference 2: 1114-1118.

10. Jalloh M, Das P (2008) Performance Analysis of STBC-OFDM Transmit Diversity with Phase Noise and Imperfect Channel Estimation pp:1-5.

11. Zelst AV, Schenk TCW (2004) Implementation of a MIMO OFDM-based wireless LAN system. IEEE Transac Sig Proces 52: 483-494.

# 23

# Path Loss Prediction of Wireless Mobile Communication for Urban Areas of Imo State, South-East Region of Nigeria at 910 MHz

**Nnamani Kelvin N\* and Alumona TL**

*Electronics and Computer Engineering, Nnamdi Azikiwe University, Awka, Anambra State, Nigeria*

## Abstract

This paper provides an extension of path loss prediction in urban city of Imo State of Nigeria with a measured set of propagation at 910 MHz band. This paper work discusses and implements Okumura, Hata, cost-231, walfisch-Ikegami model, Sagami-Kuboi Model even though Hata and cost-231 Hata Models are extensively used in path loss analysis for GSM and CDMA systems comparison with the set results developed. It is of utmost importance that this paper work guides network designers in mobile cellular propagation and inculcates an accurate method of designing, deploying and managing of their network for proper attenuation.

**Keywords:** Path loss; Propagation models; Hata model; Urban area; PCHIP (Piecewise cubic Hermite Interpolating polynomial); WCDMA (Wireless code division multiple access systems); GSM (Global System for mobile communications); WIMAX (Worldwide interoperability for microware access)

## Introduction

Path loss is the degradation in received power of an electromagnetic signal when it propagates through space. Path loss is due to several effects such as free space path loss, refraction, diffraction, reflection, coupling and cable loss, and absorption. Path loss depends on several factors such as type of propagations, environments, distance between the transmitter and receiver, height and location of antennas. Also, the signal from the transmitting antenna may take multiple paths (multipath) to reach the receiving side, which results in either increase or decrease of received signal level depending on the constructive or destructive interference of the multipath waves [1].

Path loss is highly inevitable in evaluating networks quality and capacity as regards efficient and reliable coverage areas in the growth of mobile communication [2]. This article centres its results on the experimental and statistical analysis at GSM frequency of 910 MHz using Okumara model which is most widely used propagation models are used extensively in network planning, particularly for conducting feasibility studies and during initial deployment. They are also very useful for performing interference studies as the deployment proceeds numerous experiments have been carried out in urban city of Imo State, for checking the applicability of suitable path loss models in mobile communications. This research aims at enhancing the quality of wireless service in Imo State of Nigeria by carrying out site specific measurements and developing an acceptance path loss model for the state.

## Existing Models

### Free space propagation model

The wave is not reflected or absorbed in free space propagation model. The ideal propagation radiates in all directions from transmitting source and propagating to an infinite distance with no degradation. Attenuation occurs due to spreading of power over greater areas. Power flux at the transmitter can be calculated using equation [3].

$$P_d = P_t / 4\pi d^2 \tag{1}$$

Where $P_d$ is the power density at a distance, d from an isotropic source, in watts/square meter.

$P_t$ is the transmitted power, in watts

d is the distance in meters, from the source.

The power is spread over an ever-expanding sphere if radiating elements generates a fixed power. As the sphere expands, the energy will be spread more thinly. The power received can be calculated from the antenna if a receiver antenna is placed in power flux density at a point of a given distance from the radiation. To calculate the effective antenna aperture and received power, the formulas are shown in equations below.

The amount of power captured by the antenna at the required distance, d, depends on the effective aperture of the antenna and the power flux density at the receiving element. These are mainly three factors by which the actual power received depends upon by the antenna:

(a) The aperture of receiving antenna (b) The power flux density (c) and the wavelength of received signal.

For isotropic antenna, effective area is given by

$$A_e = \lambda^2 / 4\pi . \tag{2}$$

Power received is gien by.

$$P_r = P_d \, X \, A_e = \frac{P_t \, X \, \lambda^2}{(4\pi d)^2} . \tag{3}$$

Pathloss is,

$$L_p = \text{power transmitted} \left(P_t\right) - \text{power received} \left(P_r\right). \tag{4}$$

Therefore, $L_p = P_t \text{-} P_r$

---

**\*Corresponding author:** Nnamani Kelvin N, Electronics and Computer Engineering, Nnamdi Azikiwe University, Awka, Anambra State, Nigeria
E-mail: Tisaman62@yahoo.com

Substituting equation 3 and 4, we get

$$L_p \text{ (dB)} = 20\log_{10}(4p) + 20\log_{10}(d) - 20\log_{10}(\lambda) \qquad (5)$$

Then substituting ($\lambda$ (in Km) = 0.3/f (in MHz), and rationalizing the equation produces the generic free space pathless formular,

$$L_p \text{ (dB)} = 32.5 + 20\log_{10}(d) + 20\log_{10}(f). \qquad (6)$$

## Plane earth propagation model

The effects of propagation model on ground are not considered for the free space propagation mode. Some of the power will be reflected due to the presence of ground and then received by the receiver when a radio wave propagates over ground. The free space propagation model is modified and referred to as the "plane earth" propagation model by determining the effects of the reflected power. Thus, these model suites better for the true characteristics of radio wave propagation over ground. This model computes the received signal to be the sum of a direct signal which reflected from a smooth, flat earth. The relevant input parameters include; the length of the path, the antenna heights, the operating frequency and the reflection coefficient of the earth. The coefficient will vary according to the type of terrain either water, wet ground, desert etc. [4].

For this, path loss equation is given by

$$L_{pe} = 40\log_{10}(d) - 20\log_{10}(h1) - 20\log_{10}(h2). \qquad (7)$$

Here 'd' is the path length in meter, h, and h2 are the antenna heights at the base station and the mobile, respectively.

Furthermore, if the mobile height changes (as it will in practice) then the predicted pathloss will also be changed.

## Empirical Propagation Models

Empirical propagation models will be discussed in this section, amongst them are Okumara and Hata models.

Okumura and Hata are among the two empirical propagation models. The two basic propagation models are free space loss and plane earth loss would be requiring detailed knowledge of the location and constitutive parameters of building, terrain features, every tree and terrain feature in the area to be covered. It is too complex to be practical and would be providing an unnecessary amount of detail therefore appropriate way of accounting for these complex effects is by an empirical model. There are many empirical prediction models like, cost 231, Hata model, Okumura-Hata model, sakagami-kuboi model, cost 231, walfisch-Ikegami model [5-7].

## Okumura propagation mode

In mobile communications, the terrain between the transmitter and the receiver plays a very important role in determining the signal strength at the receiver. Okumura model is one of the popular models, especially used for urban areas. It is generally applied for frequencies in the range of 150 MHz-1920 MHz, for a distance separation ranging from I km to 100 km, and for antenna heights from 30 m to 1000 m.

The pathloss is given as:

$$L_p \text{ (dB)} = L_F + A_{mu}(f,d) - G(h_{te}) - G(h_{re}) - G(\text{AREA}) \qquad (8)$$

Where $L_p$ is the median value of the propagation Path loss.

$L_F$ is the free space propagation loss

$A_{mu}$ is the median attenuation relative to the free space.

$G(h_{te})$ is the base station antenna height gain factor

$G(h_{re})$ is the mobile antenna height gain factor

$G(\text{AREA})$ is the gain due to the type of the environment.

## Hata's propagation model

The Hata model is used for frequency range of 150 MHz to 1500 MHz. The median pathloss from the Hata is given as:

$$L_p \text{ (urban) (dB)} = 69.55 + 26.16\log_{10}f_c$$
$$-13.82\log_{10}h_{te} - \propto (h_{re}) + (44.9 - 6.55\log_{10}h_{te})\log_{10}d \qquad (9)$$

where, $f_c$ is frequency from 150 MHz to 1500 MHz

$h_{te}$ is effective transmitter antenna height of the base station ranging from 30 m to 200 m.

$h_{re}$ is effective receiver antenna height (mobile) ranging from 1 m to 10 m.

d is distance between the transmitter and the receiver in km.

$\propto(h_{re})$ is correction factor for effective mobile antenna.

The mobile antenna correction factor for a small to medium city is obtained as:

$$\propto (h_{re}) = (1.1\log_{10}f_c - 0.7) h_{re} - (1.56\log_{10}f_c - 0.8) \qquad (10)$$

The mobile antenna correction factor for a large city is given as

$$\propto (h_{re}) = 8.29 (\log_{10}1.5 h_{re})^2 - 1.1 \text{ dB for } f_c \leq 300 \text{ MHz}. \qquad (11)$$

$$\propto (h_{re}) = 3.2 (\log_{10} 11.75h_{re})^2 - 4.97\text{dB for } f_c \geq 300\text{MHz}. \qquad (12)$$

The pathloss for a suburban area from the Hata model is given by the following equation

$$L_p\text{(dB)} = L_p\text{(urban)} - 2\left\{\log_{10}\left(\frac{f_c}{28}\right)\right\}^2 - 5.4. \qquad (13)$$

The pathloss for the open rural areas is obtained from the following equation:

$$L_p \text{ (dB)} = L_p\text{(urban)} - 4.78 (\log_{10}f_c)^2 + 18.33\log_{10}f_c - 40.94 \qquad (14)$$

The Hata model predicts the mean signal pathloss for transmitter receiver separation of more than 1km. Therefore, it is very much suited for large cell mobile communications, but not for personal communication systems (PCS, radius < I km)

## Data Collection Procedures

The event of this research was taken on a hot Sunny day about three times on August, 2014 with average temperature of 27°C in urban city Imo State of Nigeria.

The test was carried out in urban areas of Imo State such as Okigwe Road, Tetlow Road, Ikenegbu layout, wetheral road and government house roundabout with necessary analysis taken at frequency range of 910 MHz using a net monitor application of NOKIA 3310 CDMA. The net monitor is software compatible with some Nokia phones, with the capability of giving information on a BTS over the air interface. Thus, the signal strength information sent over the air interface between the BTS and the mobile station were read. The base station antenna lengths of 42 m operated by GSM base station over a transmitted power of 25 W mounted on a tower.

The measured data of pathloss (in dB) against their corresponding receive-transmit separation distance median values over the period of the investigation are presented as in Tables 1 and 2 grouped as sites A and B respectively.

The BTS were selected to cover the urban (group A) and sub urban (group B) terrain in Imo State, South-East region of Nigeria. The terrain group A consisted of sites located near dense vegetation, highly populated areas with nucleated settlements such as linear and nodal settlements as seen in Edinburgh, shopping complex along wetheral road, cherubim junction and Imo State government house round about while group B composed mainly of down stairs fitted with communication gadgets such as Imo State library, Nicon insurance, mobis mall along ikenegbu layout.

## Results and Analysis

Applying linear regression formulae,

$$e(n) = \sum_{i=1}^{K} \{L_p(di)\acute{L}_p(do)\}^2 \qquad (15)$$

Where $L_p(di)$ is the measured path loss at distance di and $\acute{L}_p(di)$ is the estimated path loss using equations below [8]:

$$P_L(dB) = P_L(do) + 10\,n\log_{10}\left(\frac{di}{do}\right) \text{ and replacing it in}$$

equation above, yields

$$e(n) = \sum_{i=1}^{K}\left\{L_p(di) - L_p(do) - 10n\log_{10}\left(\frac{di}{do}\right)\right\}^2 . \qquad (16)$$

Appling differentiation in equation (16) with respect to n, and equating $\frac{\delta E(n)}{\delta n}$ to zero,

$$\sum_{i=1}^{K}\left\{\{L_p - L_p(do)\}\sum_{i=1}^{K}10n\log_{10}\left(\frac{di}{do}\right)\right\} = 0$$

$$\sum_{i=1}^{K}\{L_p - L_p(do)\} = \sum_{i=1}^{K}\left\{10n\log_{10}\left(\frac{di}{do}\right)\right\}$$

Then, 'n' given by

$$n = \frac{\sum_{i=1}^{K}\{L_p(di) - L_p(do)\}}{\sum_{i=1}^{K}10\log\left(\frac{di}{do}\right)} . \qquad (17)$$

The combined path loss model for shadowed Imo urban environment is expressed as

$$L_p(d) = L_p(d) + 10n\log_{10}\left(\frac{d}{d_o}\right) + S . \qquad (18)$$

Where S is the shadow fading variation about the linear relationship and has a r.m.s value that reduces the Error as seen below in the given equation.

$$\sqrt{\sum_{i=1}^{K}\frac{(Pm - pr)^2}{N}} \qquad (19)$$

In Figure 1 below, ORIGINLAB program was used to plot the regression analysis of path loss against distance. The significance of this using least square method, shows that the intercept or maximum path loss is about 80dB with acceptable range of 10dB. In the analysis, the linear plot of the path loss yields a correlation coefficient of 0.898 which is less than unity and this shows strong positive relationship between median path loss (dB) and its equivalent distance. The coefficient of determination, $r^2 = 0.80717$. This value of 0.80717 can be expressed in percentage as 80.8% and it can interpreted to mean that 80.8% of the variation in the path loss is propagated over distance. The slope of the graph is 26.37255 with error of 3.19874. The intercept is 78.91176 with error of 3.27774.

In Figure 2, a linear regression model way fitted using a scatter plot of the experimental data on path loss against distance. This reveals a first order polynomial trend. The method of least squares method was employed to estimate the correction coefficient of 0.9881 which is close to Unity and thus shows a strong positive path loss propagation over a distance. Also, the coefficient of determination of this regression suggested that about 98% variation in path loss over a propagated distance is attained. Also, the maximum number of measured data points was about 129dB with range of 10dB which is the acceptable range i.e. 4 to 20dB.

| Distance (km) | Median Rx (dBm) | Median PL (dB) |
|---|---|---|
| 0.10 | -54 | 88 |
| 0.20 | -58 | 90 |
| 0.30 | -61 | 94 |
| 0.40 | -65 | 96 |
| 0.50 | -70 | 84 |
| 0.60 | -74 | 86 |
| 0.70 | -71 | 82 |
| 0.80 | -76 | 98 |
| 0.90 | -80 | 100 |
| 1.00 | -82 | 106 |
| 1.10 | -86 | 110 |
| 1.20 | -83 | 107 |
| 1.30 | -90 | 113 |
| 1.40 | -92 | 118 |
| 1.50 | -94 | 120 |
| 1.60 | -91 | 123 |
| 1.70 | -93 | 130 |

Table 1: Measured median path loss for group a sites.

| Distance (km) | Median Rx (dBm) | Median PL (dB) |
|---|---|---|
| 0.10 | -48 | 81 |
| 0.20 | -51 | 85 |
| 0.30 | -56 | 87 |
| 0.40 | -57 | 83 |
| 0.50 | -53 | 88 |
| 0.60 | -58 | 95 |
| 0.70 | -63 | 98 |
| 0.80 | -67 | 102 |
| 0.90 | -72 | 107 |
| 1.00 | -75 | 109 |
| 1.10 | -77 | 113 |
| 1.20 | -80 | 110 |
| 1.30 | -76 | 117 |
| 1.40 | -84 | 120 |
| 1.50 | -88 | 122 |
| 1.60 | -90 | 126 |
| 1.70 | -93 | 129 |

Table 2: Measure median path loss for group B sites.

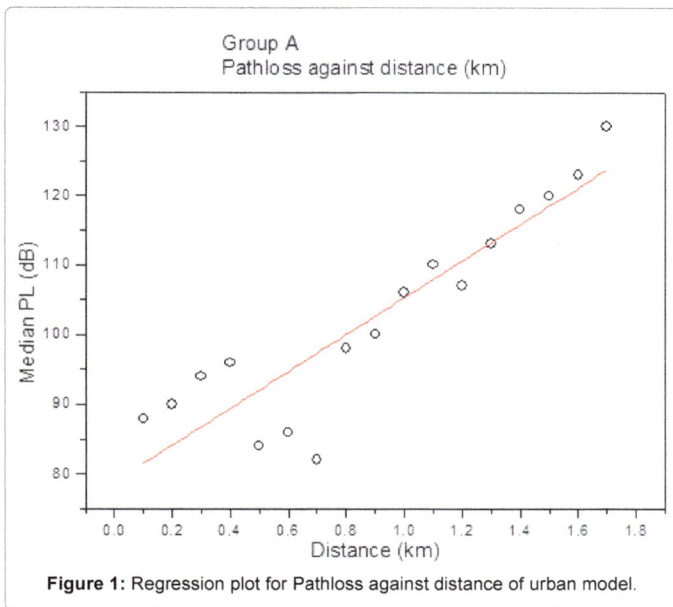

**Figure 1:** Regression plot for Pathloss against distance of urban model.

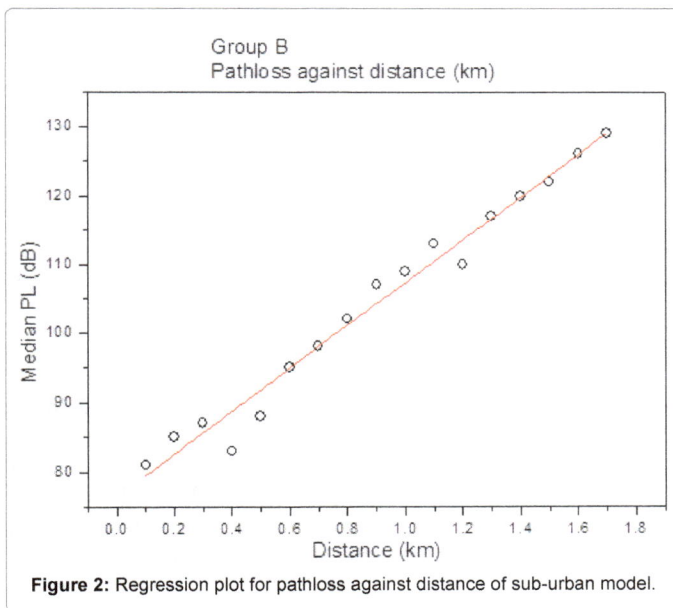

**Figure 2:** Regression plot for pathloss against distance of sub-urban model.

## Conclusion

This work was aimed at predicting the mean signal strength of Owerri. However, most propagation models aim to predict the median path loss. Today's predictions models differ in their applicability over different environmental and terrain conditions. There are many predictions methods based on deterministic processes through the availability of improved data values, but still the Okumura- Hata models is most commonly used empirical propagation model. That is because of the ITU-R recommendation for its proven reliability and its simplicity.

### References

1. Ubom EA, Idigo VE, Azubogu ACO, Ohaneme CO, Alumona TL (2011) Path loss characterization of Wireless propagation for South-south region of Nigeria. International Journal of computer Theory and Engineering 3: 360-364.

2. Nwalozie Gerald C, Ufoaroh SU, Ezeagwu CO, Ejiofor AC (2014) Path loss Prediction for GSM Mobile networks for urban Region of Aba, South-East Nigeria. International Journal of computer Science and Mobile computing 3: 267-281.

3. Shalangwa DA, Jerome G (2010) Path loss Propagation Model for Gombe Town, Adamawa State, Nigeria. International Journal of Computer Science and Network Security 18.

4. Adebayo TL, Edeko FO (2006) Characterized propagation path loss at 1.8GHz band for Benin. Research Journal for Applied Sciences 1: 92-96.

5. Hata M (1980) Empirical formular for propagation loss in land mobile radio services. IEEE trans-veh Tech 29: 317-325.

6. Azubogu ACO (2010) Empirical statistical propagation path loss model for suburban environment of Nigeria at 800MHZ band. The IUP Journal of Science and Technology.

7. Nadir Z (2012) Empirical Path loss characterization for Oman. IEEE Computing, communications & applications conference (IEEE ComcomAp 2012). University of Science and Technology, Hong Kong, China.

8. Erceg V, Greenstein LJ, Tjandra SY, Parkoff SR, Gupta A, et al. (1999) An Empirically based path loss model for wireless channels in suburban Environments. IEEE Journal on selected areas in Communications 17: 1205-1211.

# 24

# Bifurcation Behavior of a Capacitive Micro-Beam Suspended between Two Conductive Plates

**Azizi A[1]\*, Mobki H[2] and Rezazadeh G[3]**

[1]*Department of Engineering, German University of Technology, Halban, Oman*
[2]*Department of Mechanical Engineering, University of Tabriz, Iran*
[3]*Department of Mechanical Engineering, Urmia University, Iran*

## Abstract

In this paper, bifurcation and pull-in phenomena of a capacitive micro switch suspended between two stationary plates have been studied. The governing dynamic equation of the switch has been attained using Euler Bernoulli beam theorem. Due to the nonlinearity of the electrostatic force, the analytical solution for the derived equation is not available. So the governing differential equation has been solved using combined Galerkin weighted residual and Step-By-Step Linearization Methods (SSLM).

To obtain the fixed points and study the local and global bifurcational behavior of the switch, a mass-spring model has been utilized and adjusted so that to have similar static/dynamic characteristics with those of Euler-Bernoulli beam model (in the first mode). Using 1-DOF model, mathematical and physical equilibrium points of the switch have been obtained for three different cases. It is shown that the pull-in phenomenon in the present micro-switch can be occurred due to a pitchfork or transcritical bifurcations as well as saddle node bifurcation which are transpired in the classical micro-switches. And for some cases primary and secondary pull-in phenomena are observed where the first one is due to a transcritical bifurcation and the second one is due to a saddle node bifurcation. In addition the dynamic response of the switch to a step DC voltage has also been studied and the results show that in contrast to the classical micro-switches, the ratio of the dynamic pull-in to the static one depends on the gaps and voltages ratio where for the classical one is approximately a constant value.

**Keywords:** Electrostatic actuation; Micro-switch; Nonlinear dynamic; Transcritical bifurcation; Saddle node bifurcation

## Introduction

"Bifurcation, a French word introduced into nonlinear dynamics by Poincare, is used to indicate a qualitative change in the features of a system, such as the number and type of solutions, under the variation of one or more parameters on which the considered system depends" [1]. Electro statically actuated Micro Electro Mechanical (MEM) and Nano Electro Mechanical (NEM) systems form a broad class of devices, which bifurcational behavior can be observed due to the nonlinearity of the electrostatic force.

Nowadays, because of the advantages of the electrostatic actuators, such as favorable scaling property, low driving power, large deflection capacity, relative ease of fabrication, and others, have led to their being more widely applied for electrostatic-actuator applications in MEM systems. The MEM switch is one of the most important devices in such systems. The structural elements that are used in MEM devices are typically simple elements including micro-beams, plates, and membranes. Electro statically actuated micro-beams (e.g., cantilever and fixed–fixed micro-beams) are used in many MEM devices such as capacitive MEM switches and resonant sensors. Manufacturing and design of these devices are, to some extent, in a more mature stage than those of some other MEM devices [2].

One of the most significant issues in the capacitive micro and nano switches is the pull-in instability. In these devices, a movable beam/plate is suspended over a stationary plate, and a potential difference is applied between them. As the micro-structure is balanced between two forces, namely, electrostatic (attractive) and mechanical (elastic restoring) forces, both of these forces are increased when the applied voltage increases. When the voltage reaches a critical value, pull-in instability occurs. Pull-in is a situation at which the elastic restoring force can no longer balance the electrostatic attractive force. Further increasing the voltage will cause the structure to have a sudden displacement jump, causing structural collapse and failure. Pull-in instability is a snap-through like behavior and it is a saddle-node bifurcation type of instability [3].

The pull-in phenomenon usually occurs in many micro-machine devices which require bi-stability for their operation, such as in the MEM and NEM switches [4,5]. Many studies have been developed in the analysis of instability of the MEM structures due to their nonlinearity [6,7]. Zhang and Zhao [8] studied the Pull-in instability of a MEM switch under electrostatic actuation. Taghizadeh and Mobki [9] analyzed the pull-in phenomenon of a torsional micro-mirror. In nano scale, Dequesnes et al., [10,11] and Hosseini et al., [12] investigated the static and dynamic stability of a carbon nano tube. Mobki et al., [13] studied the static and dynamic pull-in phenomenon of a capacitive nano-beam, considering length scale-parameter.

In the case of bifurcational behavior of the MEM and NEM systems, some works have been done. Lin and Zhao [14-16] studied bifurcation and pull-in phenomenon of NEM actuators with considering van der Waals (vdW) and Casimir forces. They showed that the pull-in voltage causes a saddle node type bifurcation in these devices. Many researchers

---

**\*Corresponding author:** Azizi A, Department of Engineering, German University of Technology, Oman, E-mail: Aydin.azizi@gutech.edu.om

have shown that the saddle node bifurcation, can be transpired in the electro statically MEM/NEM devices [3,9,13-18].

In spite of many research works accomplished on the stability of MEM structures, there is not enough comprehensive work explaining the stability of these switches from the type of bifurcation point of view.

In the present study, different types of bifurcation are investigated for a capacitive micro-beam suspended between two conductive plates. It should be mentioned that this specific sort of micro-switch allows us to produce different bifurcation types, by varying of some parameters. The micro-beam (micro-switch) is actuated by electrostatic forces induced by DC polarization voltages. Also, to study the bifurcation of the beam, the nonlinear equation of the dynamic motion of the Euler Bernoulli beam using a one term Galerkin weighted residual method is converted to a lumped mass-spring model. Compared to the distributed model, the mass spring has lower accuracy in studying the static and dynamic behavior of the micro-beams. In order to overcome this shortcoming, in this work, the accuracy of the model has been improved using corrective coefficients. By solving the static deflection equation, the fixed points of the micro-beam are obtained for different conditions, resulting in saddle-node, pitchfork, and transcritical bifurcations. In order to study the stability of the fixed points, motion trajectories are produced in phase portraits. Moreover, dynamic response of the system to a step DC voltage also is investigated

## Model Description

Figure 1a shows the schematic view of a fixed-fixed beam with length L, which is suspended between two stationary conductor plates. The beam distances from the bottom and top plates are $G_1$ and $G_2$, respectively. The illustrated beam is attracted toward the lower and upper plates by applying the voltages of $V_1$ and $V_2$, respectively. By different combination of the applied voltages and gaps, different bending states, and hence different bifurcation types, can be obtained.

In the present study, the micro-beam is assumed to be an isotropic material with modulus of elasticity E, density ρ, width b, thickness h, cross section area A and moment of inertia I.

As mentioned before, in order to simplify the analysis of the bifurcation behavior of the micro-beam, the illustrated model in Figure 1b can be simulated by an equivalent mass-spring model, which is shown in Figure 2.

## Mathematical Modeling

The static deflection equation of the presented micro-switch may be computed using Euler Bernoulli beam theory. Of course the mechanical behavior the nano and micro-beams is size dependent, but as the main propose of the present paper is to study the qualitative bifurcational behavior of a micro-beam therefore the micro-beam is modeled based on classical theories. The electrostatic forces per unit length of the beam applied to the lower and upper plates can be obtained [19] as in Eqs. (1) and (2) respectively:

$$q_{elec}(V_1, \hat{w}) = \frac{b\varepsilon_0 V_1^2}{2(G_1 - \hat{w})^2} \tag{1}$$

$$q_{elec}(V_2, \hat{w}') = \frac{b\varepsilon_0 V_2^2}{2(G_2 - \hat{w}')^2} \tag{2}$$

Where, $\hat{W}(x)$ and $\hat{W}'(x)$ are the flexural deflection of the micro-beam with respect to the lower and upper plates, respectively. Also, $\varepsilon_0 = 8.854 \times 10^{-12}\, C^2\, N^{-1}\, m^{-2}$ is the permittivity of the vacuum within the gaps? Considering that, the deflection of the micro-beam with respect to the bottom plate equals to the negative value regarding to the upper plate, as:

$$\hat{W} = \hat{W}' \tag{3}$$

So the governing equation of the motion of the micro-beam, using Euler Bernoulli beam theorem, is:

$$EI\frac{\partial^4 \hat{w}}{\partial \hat{x}^4} + \rho A\frac{\partial^2 \hat{w}}{\partial t^2} = \frac{b\varepsilon_0}{2}\left(\frac{V_1^2}{(G_1 - \hat{w})^2} - \frac{V_2^2}{(G_2 + \hat{w})^2}\right) \tag{4}$$

Where right hand terms of this equation are resultant electrostatic forces per unit length of the beam ($q_{elec}(V_1, V_2, \hat{W})$). In Eq. (4) first and second right hand terms indicate the imposed forces from lower and upper plates, respectively.

In order to compose a lumped mass-spring model for the micro-switch, Eq. (4) is converted to Eq. (5) [4]:

$$m\frac{\partial^2 \hat{y}}{\partial t^2} + K\hat{y} = \frac{b\varepsilon_0 L}{2}\left(\frac{V_1^2}{(G_1 - \hat{y})^2} - \frac{V_2^2}{(G_2 + \hat{y})^2}\right) \tag{5}$$

where m=ρAL is the mass of the micro-beam, and K is the spring constant for the beam and defined as the ratio of the applied uniform force 'q' to the maximum beam deflection '$Y_{max}$'. Thus, the spring constant depends on the cross-section/shape of the beam as well as on the boundary conditions. Considering the mass-spring model, the

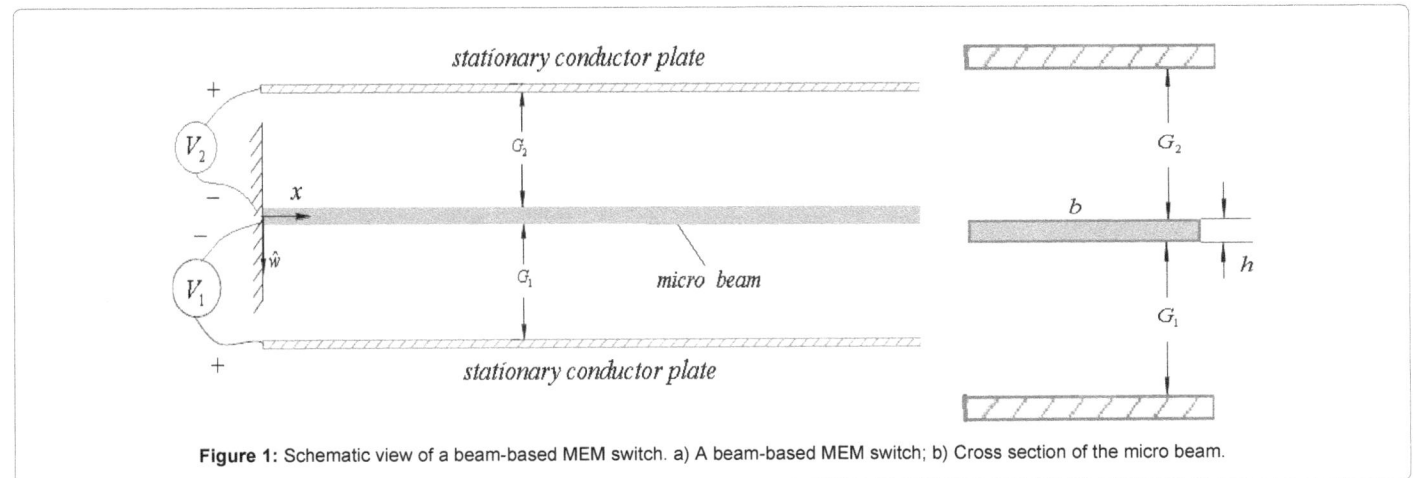

**Figure 1:** Schematic view of a beam-based MEM switch. a) A beam-based MEM switch; b) Cross section of the micro beam.

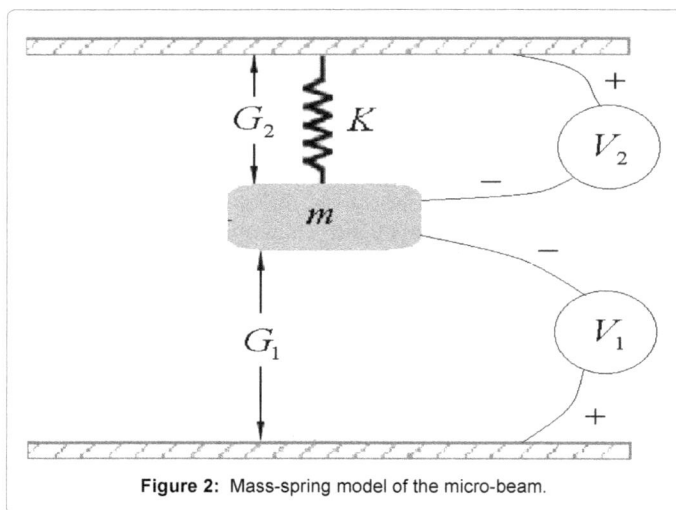

**Figure 2:** Mass-spring model of the micro-beam.

spring constant for cantilever and fixed-fixed beams are $8EI/L^3$ and $384EI/L^3$, respectively [4]. In order to increase the accuracy of the mass-spring model, and to adjust it with the distributed model as well, corrective coefficients of $a_0$ and $b_0$ are applied as below:

$$a_0 m \frac{\partial^2 \hat{y}}{\partial t^2} + K\hat{y} = \frac{b_0 b \varepsilon_0 L}{2} \left[ \frac{V_1^2}{(G_1 - \hat{y})^2} - \frac{V_2^2}{(G_2 + \hat{y})^2} \right] \tag{6}$$

Where $a_0 m$ is the equivalent mass of the micro-beam. The parameters $a_0$ and $b_0$ are determined; so that the first natural frequencies and the static pull-in voltages obtained using both mass-spring and the distributed models, become equal.

For convenience Eqs. (4) and (6) can be rewritten in a non-dimensional form by defining the following parameters:

$$w = \frac{\hat{w}}{G_1}, \ y = \frac{\hat{y}}{G_1}, \ x = \frac{\hat{x}}{L}, \ \tau = \frac{t}{t^*}, \ s = \frac{G_2}{G_1}, \ p = \frac{V_2}{V_1} \tag{7}$$

Where $\tau$ is the dimensionless time, and $t^* = \sqrt{\frac{a_0 m}{K}}$ for the mass-spring, and $t^* = \sqrt{\frac{\rho A L^4}{EI}}$ for the Euler-Bernoulli beam models. Therefore Eqs. (4) and (6) may be written as:

$$\frac{\partial^4 w}{\partial x^4} + \frac{\partial^2 w}{\partial \tau^2} = \alpha V_1^2 \left[ \frac{1}{(1-w)^2} - \frac{p^2}{(s+w)^2} \right] \tag{8}$$

$$\frac{d^2 y}{d\tau^2} + y = \alpha' V_1^2 \left[ \frac{1}{(1-y)^2} - \frac{p^2}{(s+y)^2} \right] \tag{9}$$

Where $\alpha$ and $\alpha'$ are the non-dimensional parameters of the electrostatic forces in the distributed and mass-spring models, respectively. These parameters are:

$$\alpha = \frac{\varepsilon_0 b L^4}{2EIG_1^3}, \ \alpha' = \frac{b_0 \varepsilon_0 b L}{2KG_1^3}, \ (\alpha'_{\text{cantilever}} = \frac{b_0 \alpha}{8} \ \alpha'_{\text{fixed-fixed}} = \frac{b_0 \alpha}{384}) \tag{10}$$

## Numerical Approach

Due to the presence of the nonlinear terms in Eq. (8), the analytical solution methods may not be employed to obtain the pull-in voltage '$V_{\text{pull-in}}$' of the micro-beam. Hence, the SSLM method together with Galerkin based reduced integration method are implemented to solve this equation. By using SSLM, the smooth and continuous behavior of the beam, as well as the magnitude of the nonlinear forces, can be approximated in every iteration step [13].

## Static deflection

The use of static SSLM calls for smooth forces application. In the case of electrostatic forces, the voltages can be gradually increased from zero to the final value, so it satisfies the quasi-equilibrium condition.

Denoting superscript 'i' as the counting step, and $w^i$ being the non-dimensional displacement of the micro-structure, subjected to $V_1^i$ and $V_2^i$; and increasing the applied DC voltage in each step, the dimensionless static deflection at (i+1)th step can be obtained as:

$$V_1^{i+1} = V_1^i + dV_1 \ \& \ V_2^{i+1} = V_2^i + dV_2 \ \Rightarrow \ w^{i+1} = w^i + \delta w = w^i + \varphi^i \tag{11}$$

By considering small values for $dV_1$ and $dV_2$, the variable will be small enough; so that we can approximate the excitation function with the first two term of its Taylor series expansion in each step. As a result, Eq. (8) for the ith step and quasi-static case will be:

$$\frac{d^4 w^i}{dx^4} = q_{elect}(V_1^i, V_2^i, w^i) \tag{12}$$

and for (i+1)th step:

$$\frac{d^4 w^{i+1}}{dx^4} = q_{elect}(V_1^{i+1}, V_2^{i+2}, w^{i+1}) \tag{13}$$

Substituting $W^{i+1}$, $V_1^{i+1}$ and $V_2^{i+1}$ from Eq. (11) into Eq. (13) and using Taylor expansion, results in:

$$\frac{d^4 w^i}{dx^4} + \frac{d^4 \phi^i}{dx^4} = q_{elec}(V_1^i, V_2^i, w^i) + (\frac{\partial q_{elec}}{\partial w}\big|_{w^i})\phi^i + \frac{\partial q_{elec}}{\partial V_1}dV_1 + \frac{\partial q_{elec}}{\partial V_2}dV_2 \tag{14}$$

With subtracting Eq. (12) from Eq. (14) one can obtain:

$$\frac{d^4 \phi^i}{dx^4} - (\frac{\partial q_{elec}}{\partial w}\big|_{w^i})\phi^i = +\frac{\partial q_{elec}}{\partial V_1}dV_1 + \frac{\partial q_{elec}}{\partial V_2}dV_2 \tag{15}$$

This is a linear ordinary differential equation which represents the variation of the deflection along the micro-beam. This linear differential equation may be solved using Galerkin method in which $\varphi(x)$ can be expressed as:

$$\phi(x) = \sum_{j=1}^{\infty} q_j \psi_j(x) \tag{16}$$

Where $\psi_j(x)$ is the jth shape function satisfying the boundary conditions of the micro-beam. The primary variable in step i "$\varphi^i(x)$" is approximated by truncating the summation series to the finite number, N:

$$\phi_N(x) \cong \sum_{j=1}^{N} q_j \psi_j(x) \tag{17}$$

By substituting Eq. (17) into Eq. (15) and multiplying them by $\psi_r$ (x) as a weight function in the Galerkin method, and integrating the outcomes from x=0 to 1, a set of algebraic equations will be generated. Solving these set of equations in each step, the deflection at any given point, under applied voltage can be determined. The pull-in voltage of the micro-beam is subsequently obtained in the last step when the instability occurs.

## Dynamic analysis

For obtaining the response of the system excited by the time dependent voltage, dynamic analysis of the micro-beam has also been performed. Applying a minor modification on Eq. (4), by assuming the generalized deflections are function of time i.e.,

$$\hat{w}(x, t) = \sum_{j=1}^{N} q_j(t) \psi_j(x) \tag{18}$$

and using the Galerkin approximation method, the equation of dynamic response will be:

$$[M][\ddot{q}]+[K][q]=[F] \qquad (19)$$

Where,

$$[M]_{rj} = \rho A \int_0^L \psi_j(x)\psi_r(x)dx \,,$$

$$[K]_{rj} = EI \int_0^L \frac{d^4\psi_j}{dx^4}\psi_r dx \,,$$

$$[F]_r = \int_0^L \frac{b\varepsilon_0}{2}\left(\frac{V_1^2}{(G_1-\hat{w})^2} - \frac{V_2^2}{(G_2+\hat{w})^2}\right)\psi_r(x)dx \qquad (20)$$

are the effective mass, spring and actuating force matrices, respectively q(t) can be obtained from above set of ordinary differential equations (Eq. (19)) using an integration scheme.

## Results and Discussion

### Validation of the numerical method

The convergence of the numerical method and validation of the results may be investigated by comparing them with those given by Mobki et al. [19], and Osterberg [20]. The considered case studies in Osterberg; Medio and Lines [20,21] are fixed-fixed and cantilever micro-switches with width of 50 μm, thickness of 3 μm, Young's modulus of 169 GPa, $G_1=1$ μm and $V_2=0$V.

### Bifurcation analysis

The fixed points for a fixed-fixed micro-switch with L=400 μm, $G_1=3$ μm, b=4 μm, h=2 μm and E= 169 Gpa has been obtained. Based on Eq. (9), physical fixed points for the micro-switch exist for –s<y<1, but mathematically, these points may also exist in the range of 1<y or y<-s. At the fixed points, the micro-beam's velocity is zero, hence considering Eq. (9), equilibrium points are obtained by solving the following equation [21]:

$$f(\alpha', s, p, y) = y(1-y)^2(s+y)^2 - \alpha'V_1^2[(s+y)^2 - p^2(1-y)^2] = 0 \qquad (21)$$

The order of polynomial in Eq. (21) is five with respect to y, having maximum five real roots. In order to check the stability in the vicinity of an equilibrium point(y=$y_i$), the following Jacobian matrix is used [15].

$$J = \begin{bmatrix} 0 & 1 \\ \alpha'V_1^2[\dfrac{2}{(1-y_i)^3} + \dfrac{2p^2}{(s+y_i)^3}]-1 & 0 \end{bmatrix} \qquad (22)$$

Where $Y_i$ is a fixed point. Eigen vlaues of the Jacobian satisfy $\lambda^2 - \alpha'V_1^2[\frac{2}{(1-y_i)^3} + \frac{2p^2}{(s+y_i)^3}]+1=0$. For $\lambda^2<0$, it has two pure imaginary roots, meaning that the fixed point ($y_i$, $V_1$) is a center point but when $\lambda^2>0$, indicating two real eigenvalues with opposite signs; which means the corresponding equilibrium point ($y_i$, $V_1$) is an unstable saddle point. Using this method, the stability in the vicinity of each equilibrium point can be identified.

Bifurcation analysis for the present micro-switch is accomplished for three different cases as following.

**Geometrical symmetric micro-switch with balanced electrostatic force (P=S=1):** Figure 3 depicts bifurcation diagram of a micro-beam versus applied voltage $V_1$ as a control parameter in the case of p=s=1. Based on the afore mentioned procedure, stability of each branch in Figure 3 can be distinguished (in this paper dashed and continues curves represent unstable and stable branches for bifurcation diagram, respectively).

In Figure 3 by increasing the control parameter $V_1$, three physical fixed points get closer together and for $V_1$=8.97V, i.e., pull-in voltage, they coalesce and change to one unstable saddle node. This condition represents subcritical pitchfork bifurcation [1], for the presented MEM switch where p=s=1.

Figures 4-7 present motion trajectories of the presented micro-switch for different values of the applied voltage $V_1$ with different initial conditions. As shown in Figures 4-6, in each case, there is a physical region of periodic set of center point which is bounded with heteroclinic orbit (black bold curve). Figures 4-7 show that with increasing the applied voltage, the physical region of the periodic set is contracted and when the applied voltage is equal to the pull-in voltage, this physical region of the periodic set vanishes, rendering the system unstable for any initial condition (In this paper, continues and dashed curves represent periodic and unstable orbits for phase diagrams respectively).

**Geometrical non-symmetric micro-switch with balanced electrostatic force (p=s ≠ ζ; ζ ≠ 1):** Bifurcation diagram of a micro-switch in the case of p=s=2 is shown in Figure 8. In the corresponding bifurcation diagram, a transcritical bifurcation occurs due to an exchange of stability between the trivial (y=0) and nontrivial branches in $V_1$=10.33V (primary pull-in). For this bifurcation no fixed points appear or disappear, only their stability properties change [22].

Furthermore, by increasing the controlling parameter $V_1$, the distance, between two nontrivial physical fixed points, is decreased and for $V_1$=11.04V they meet together in a saddle node bifurcation (secondary pull-in). For this case there is no physically stable branch, after pull-in voltage. But as shown in this figure, for the applied voltage10.33<$V_1$<11.04, there is a physically stable branch after pull-in voltage. This phenomenon can be observed in the capacitive micro-beam suspended between two conductive plates in the condition of p=s ≠1.

Figures 9 and 10 shows bifurcation diagrams for the micro-switch in the case of p=s=3 and p=s=4 respectively. As shown in these figures, the stable branches appear after primary pull-in voltage. For the case of p=s=3 and p=s=4, the mentioned branches can be observed in the ranges of 11.09<V1<13.02 and 11.49<V1<14.75 respectively. With comparison of Figures 8-10 with each other, it may be noted that with increasing of the parameter p, the mentioned stable branch is extended, for the case of p=s and G2>G1.

Figures 11-16 present motion trajectories of a micro-switch with p=s=2 (Figure 8) for different values of the applied voltage$V_1$, with different initial conditions. As it is seen in Figures 12 and 13, there are physical region of periodic set of center point and a region of repulsion of unstable saddle node in each diagram. It must be noted that the substrate position acts as a singular point and velocity of the system near this singular point tends to infinity. The region of periodic set of the physical center point is bounded by a bold closed orbit (homoclinic orbit). Depending on the location of the initial condition, the system can be stable or unstable. Similar to the previous case, Figures 11-13 shows that with increasing the applied voltage, the physical region of the periodic set is contracted and when the applied voltage equals to the primary pull-in voltage (Figure 14), there is no region of periodic set and the system becomes unstable for any initial conditions.

Figures 15a and 15b shows phase diagrams of the micro-switch for the case of 10.33<$V_1$<11.04. As shown in these figures, physical region of the periodic set exists in $V_1$=10.8V and disappears in $V_1$=11.04V (Figures 16a and 16b). This condition represents the saddle-node bifurcation in $V_1$=11.04V (secondary pull-in), which is shown in Figure 8.

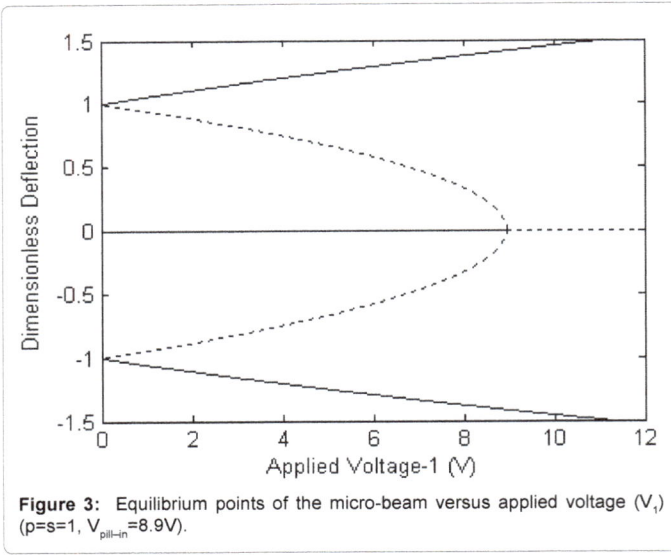

**Figure 3:** Equilibrium points of the micro-beam versus applied voltage ($V_1$) (p=s=1, $V_{pill-in}$=8.9V).

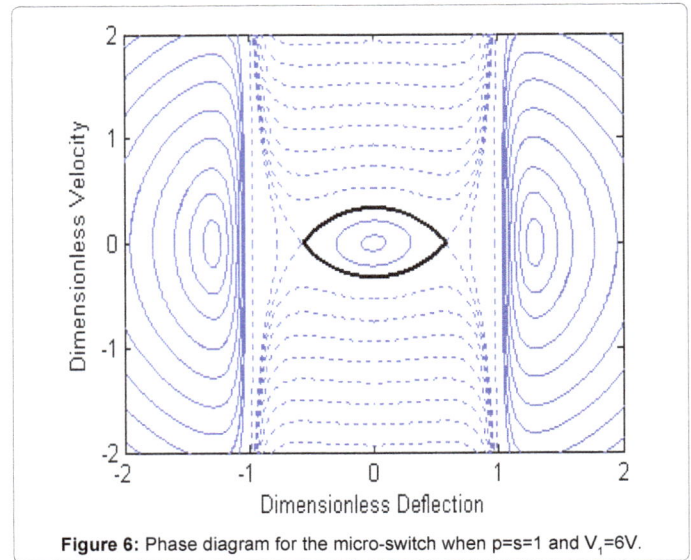

**Figure 6:** Phase diagram for the micro-switch when p=s=1 and $V_1$=6V.

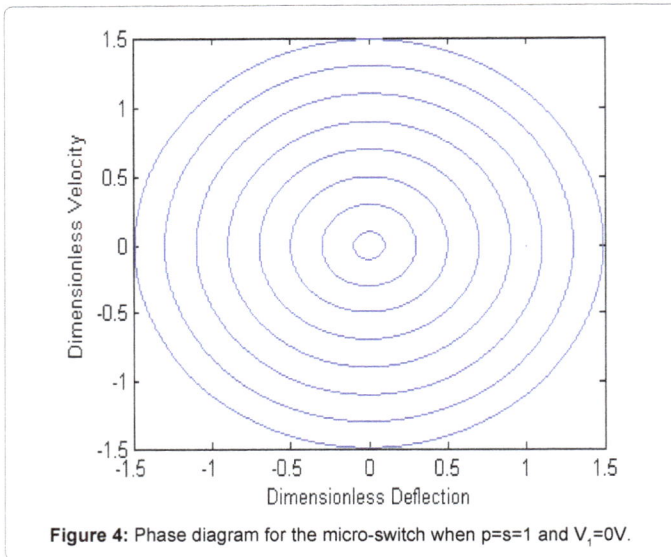

**Figure 4:** Phase diagram for the micro-switch when p=s=1 and $V_1$=0V.

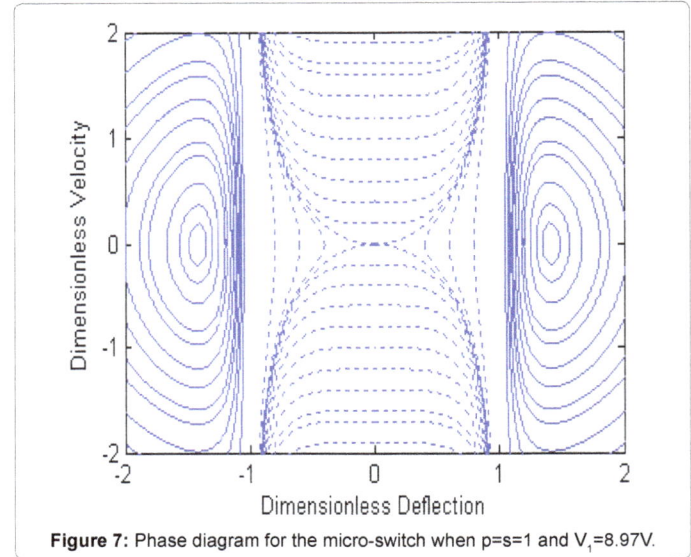

**Figure 7:** Phase diagram for the micro-switch when p=s=1 and $V_1$=8.97V.

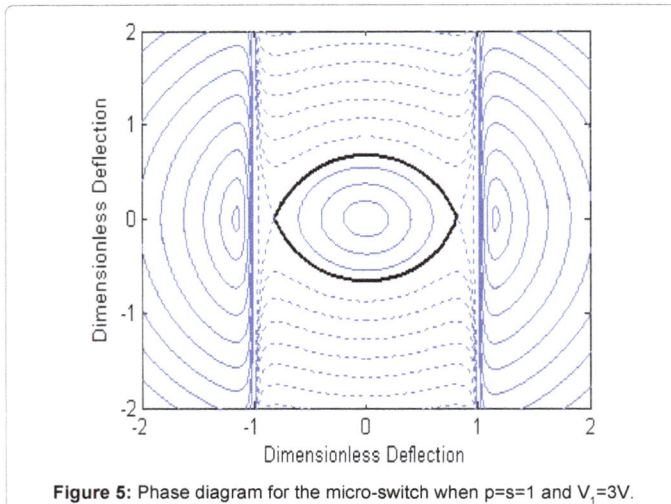

**Figure 5:** Phase diagram for the micro-switch when p=s=1 and $V_1$=3V.

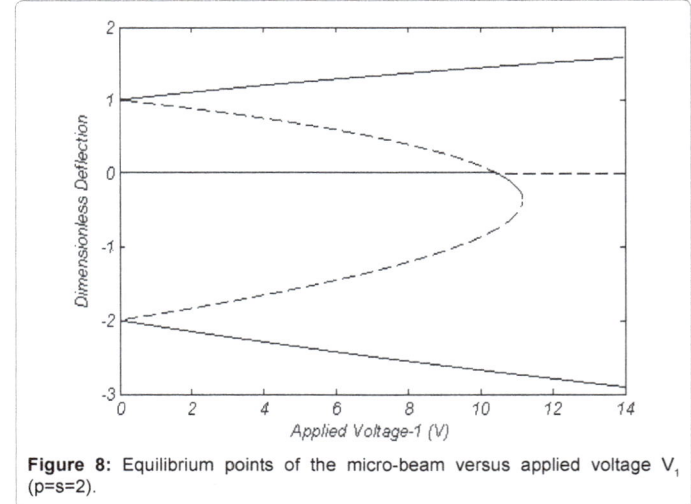

**Figure 8:** Equilibrium points of the micro-beam versus applied voltage $V_1$ (p=s=2).

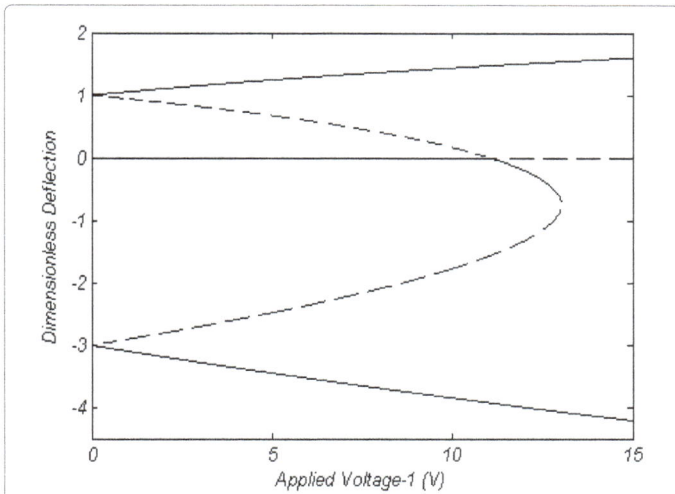

**Figure 9:** Equilibrium points of the micro-switch versus applied voltage $V_1$ (p=s=3).

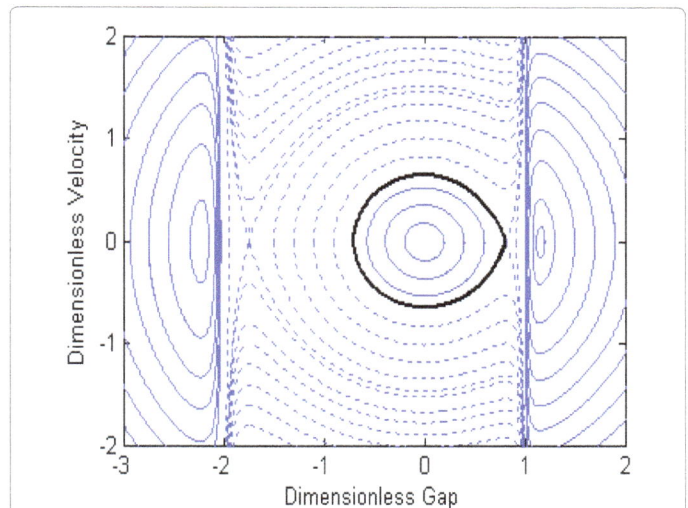

**Figure 12:** Phase diagram for the micro-switch when p=s=2 and $V_1$=3.

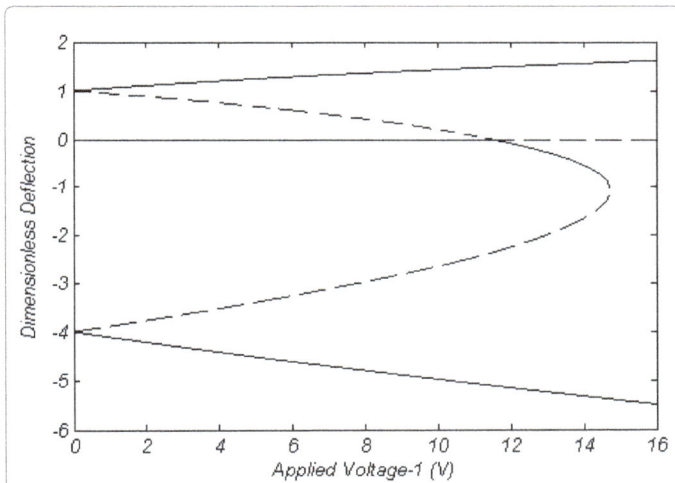

**Figure 10:** Equilibrium points of the micro-switch versus applied voltage $V_1$ (p=s=4).

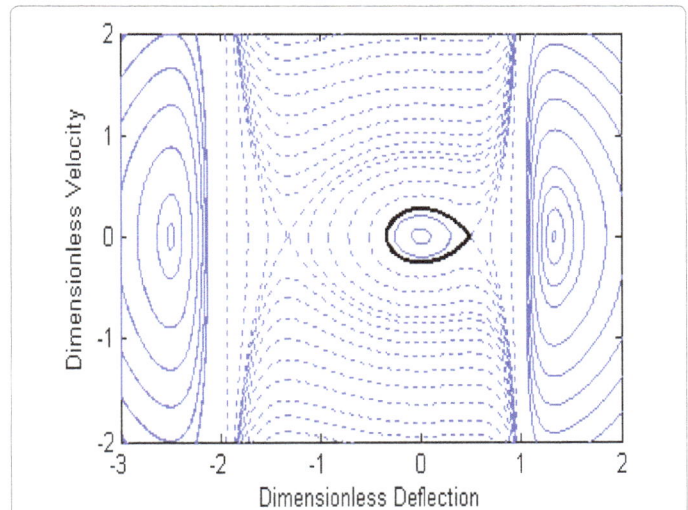

**Figure 13:** Phase diagram for the micro-switch when p=s=2 and $V_1$=7V.

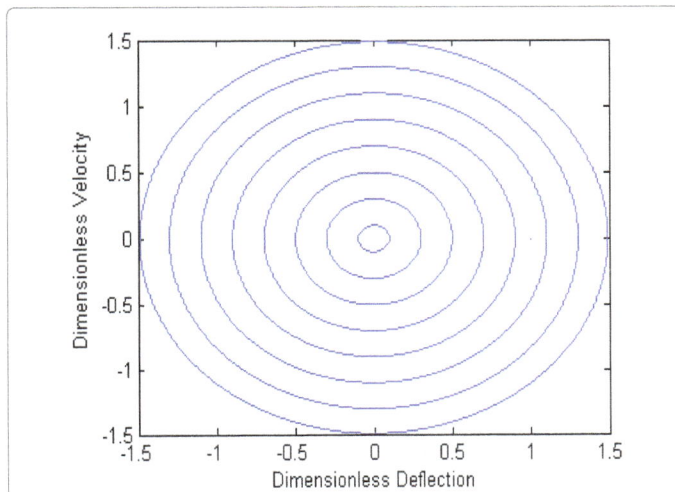

**Figure 11:** Phase diagram for the micro-switch when p=s=2 and $V_1$=0V.

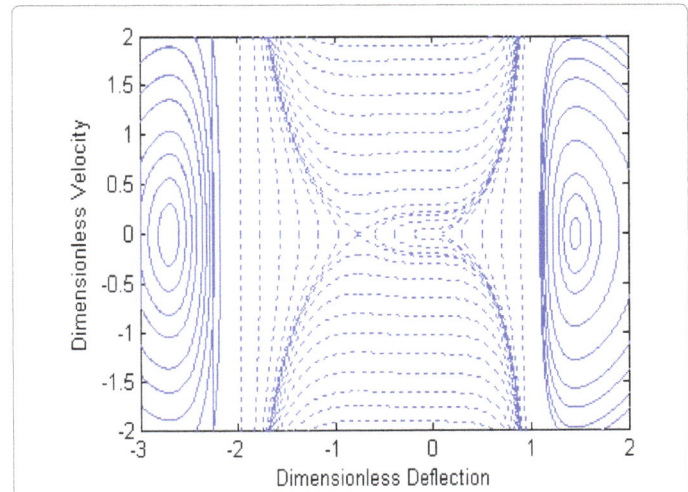

**Figure 14:** Phase diagram for the micro-switch when p=s=2 and $V_1$=10.33V.

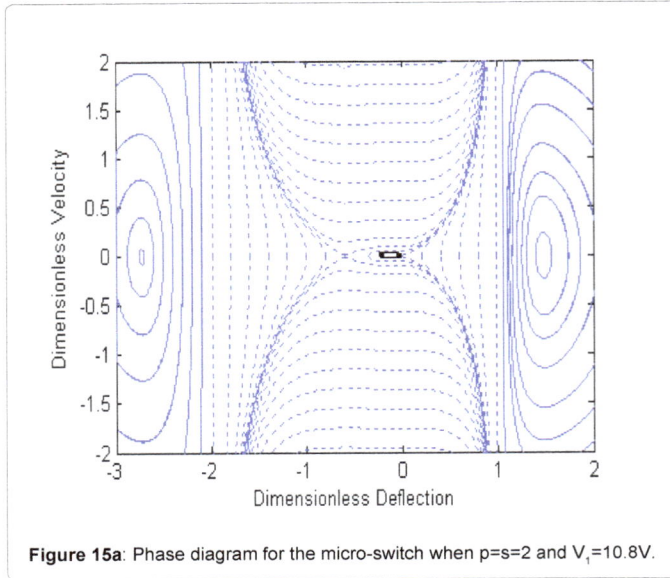

**Figure 15a**: Phase diagram for the micro-switch when p=s=2 and V₁=10.8V.

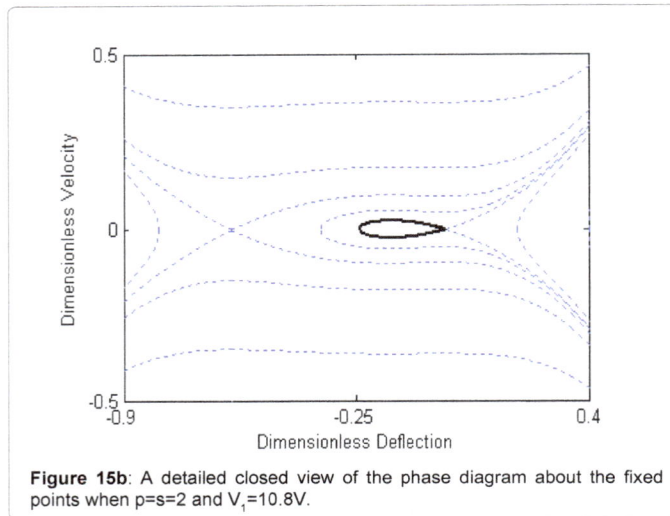

**Figure 15b**: A detailed closed view of the phase diagram about the fixed points when p=s=2 and V₁=10.8V.

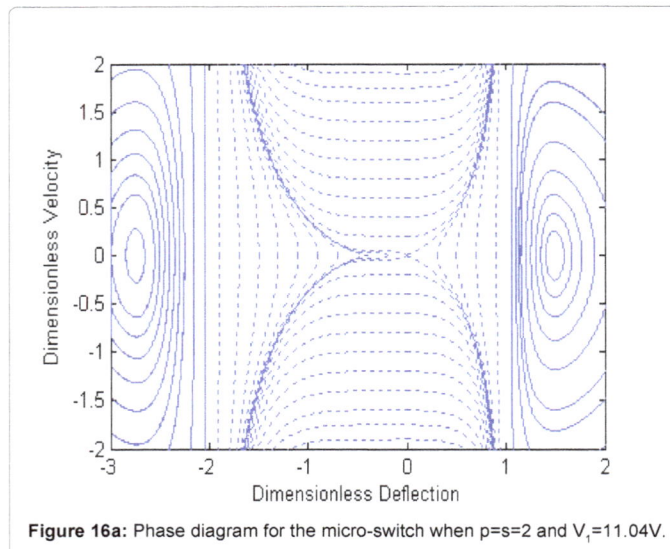

**Figure 16a**: Phase diagram for the micro-switch when p=s=2 and V₁=11.04V.

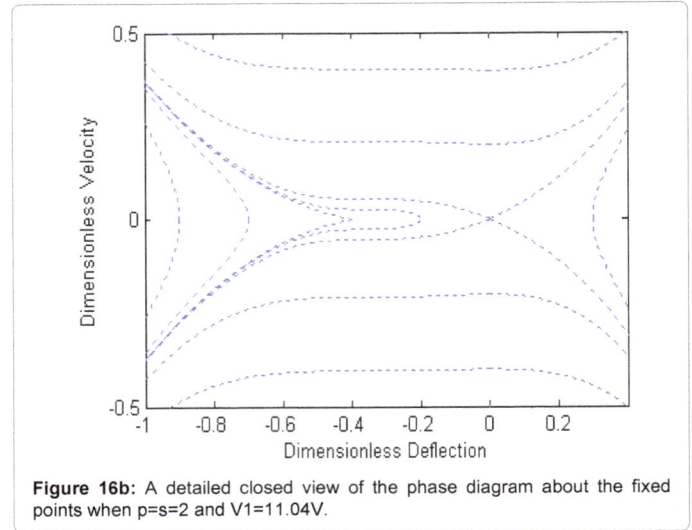

**Figure 16b**: A detailed closed view of the phase diagram about the fixed points when p=s=2 and V1=11.04V.

It must be noted that in the classic micro-switches the pull-in phenomenon is occurred due a saddle node bifurcation, whereas in the presented micro-switch in the symmetric case (p=s=1) the pull-in phenomenon is occurred owing to a pitch fork bifurcation and in the non-symmetric cases p=s≠1 this phenomenon is first happened primarily due to a transcritical bifurcation (can be called as primary pull-in phenomenon) and secondarily due to a saddle node bifurcation (can be called as secondary pull-in phenomenon).

In addition, it can be said that the first case p=s=1 can be obtained from the second case decreasing the value of the voltages and gap ratios (ς). With decreasing the value of (ς) the transcritical and saddle node bifurcation points approaches together and at the case when these ratios are equal to 1 (ς=1) only one saddle node bifurcation point is observed.

**Geometrical non-symmetric micro-switch with un-balanced electrostatic force (p ≠ s ≠ 1):** Figure 17 shows bifurcation diagram of the micro-switch in the case of $p \neq s$ (p=1, s=2). As shown in this figure, by increasing the control parameter V₁, two physical fixed points get close to each other and in V₁=7.21V, (pull-in voltage), they meet together in a saddle node bifurcation point.

Figures 18-21 shows phase diagram of the micro-switch for different values of the applied voltage $V_1$, with different initial conditions. As shown in Figures 18-20 there is a physical region of the periodic set of the center point in each diagram. The region is bounded by a bold closed orbit (homoclinic orbit). Depending on the location of the initial condition in the phase diagram, the system can be stable or unstable. Figures 18-21 show that with increasing the applied voltage, the physical region of the periodic set is contracted until the applied voltage reaches to the pull-in voltage; the region of the periodic set vanishes making the system unstable for any initial condition.

## Dynamic response

The static bifurcation was investigated in previous subsection, and in this part we study the dynamic response of the micro-switch subjected to a step DC voltage. The solution of dynamic response is based upon Galerkin reduced order method. The physical characteristics of the studied fixed-fixed beam are L=600 μm, h=2 μm, G₁=2 μm, b=4 μm, E=169 Gpa. The static pull-in voltage for this micro-switch in the case of p=0 is 10.54 V whereas for the dynamic case is 9.62 V, which is 91.3%

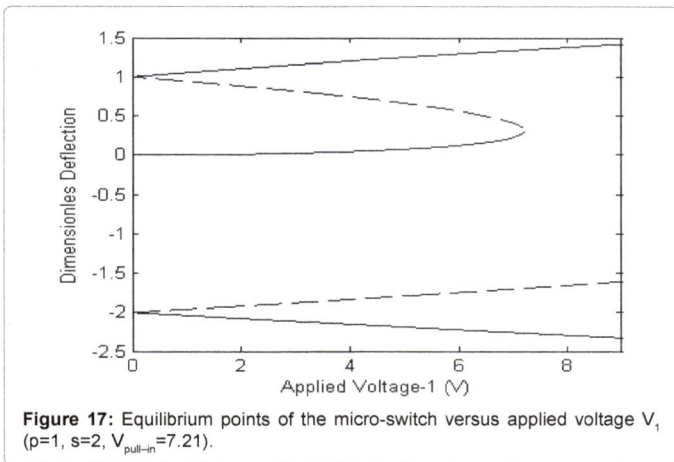

**Figure 17:** Equilibrium points of the micro-switch versus applied voltage $V_1$ ($p=1$, $s=2$, $V_{pull-in}=7.21$).

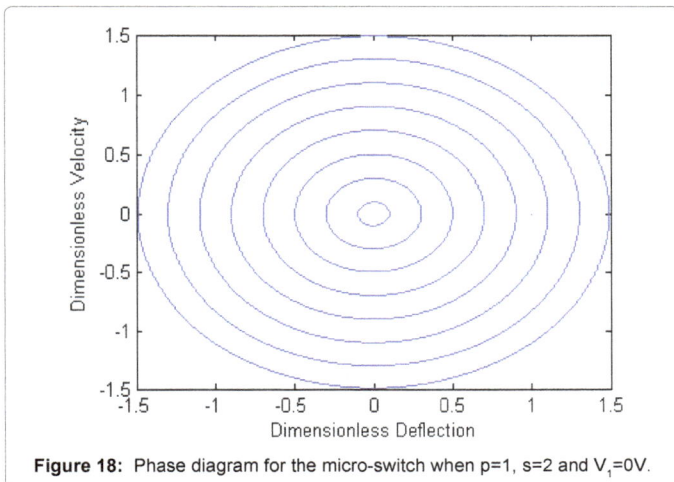

**Figure 18:** Phase diagram for the micro-switch when $p=1$, $s=2$ and $V_1=0V$.

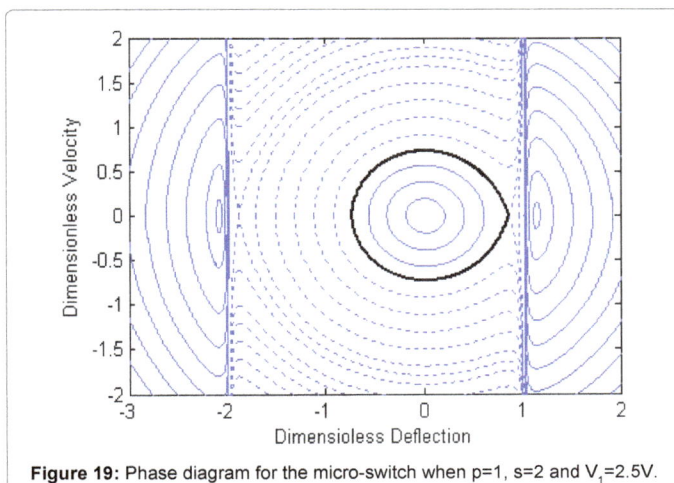

**Figure 19:** Phase diagram for the micro-switch when $p=1$, $s=2$ and $V_1=2.5V$.

leads to instability in the mentioned beam and the dynamic pull- in phenomenon occurs.

For micro switches, with $p \neq 0$, the dynamic pull-in condition may be different from reported results [13,23]. In micro-switches with $p \neq 0$, the dynamic pull-in voltage may be equal or less than the static one. This condition occurs because of the application of two electrostatic forces with opposite directions. In addition to applied voltage, the parameter s can be effective on the dynamic response of the micro-switch.

Figure 23 shows the magnitude of dynamic pull- in voltages versus s ($1<s<10$) for $p=0.5, 0.75, 1$. As shown in this figure, with increasing of parameter s, for any value of p, the dynamic pull-in voltage approaches to 9.62, which agrees with the result [13,23]. On the other hand, with decreasing of the s, the dynamic pull-in voltage increases. As shown in this figure for $p=1$ and $s=1.61$ the dynamic pull-in voltage reaches to the static one. This condition for $p=0.75$ occurs when $s=1.19$. But as shown in this figure, for $p=0.5$ dynamic pull-in voltage is less than the static one for every magnitude of the s. for this case, maximum value of dynamic pull-in voltage is 10.07 V, which is occurred in $s=1$. It must be noted that the dynamic pull-in voltage cannot be more than the static one. In exact word, based on reported results of refs [13,23] and obtained results of this paper, it can be said that dynamic pull-in voltage is in the range of $0.9V_{s-pull-in} \leq V_{d-pull-in} \leq 1V_{s-pull-in}$, which $V_{s-pull-in}$ and $V_{d-pull-in}$ represent static and dynamic pull-in voltages, respectively.

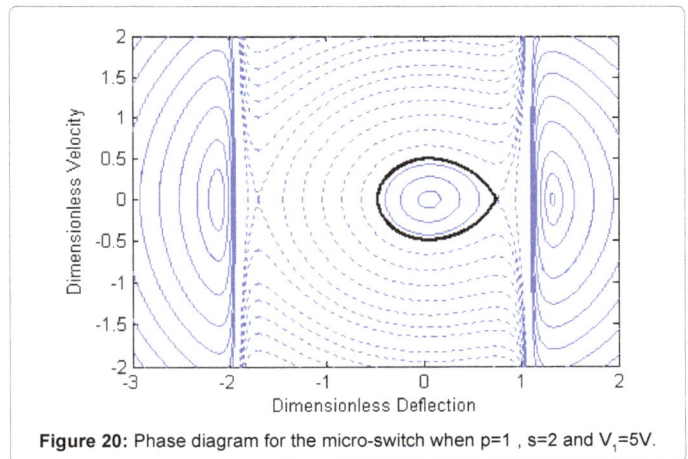

**Figure 20:** Phase diagram for the micro-switch when $p=1$, $s=2$ and $V_1=5V$.

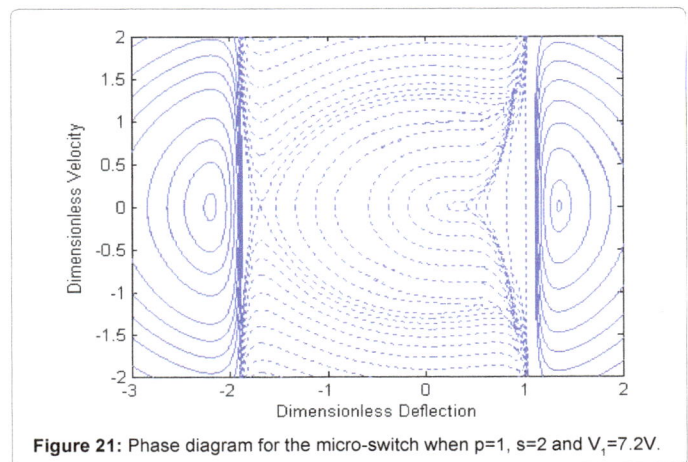

**Figure 21:** Phase diagram for the micro-switch when $p=1$, $s=2$ and $V_1=7.2V$.

of the static case. The obtained percentage is in good agreement with the reported results in Mobki et al.; Seydel [13,23]. Figure 22 shows the time history of the center deflection of the micro-beam subjected to 9.61V and 9.62V applied DC voltages, respectively. As shown in these figures, the beam is in its stable condition under 9.61V and starts to vibrate after application of this voltage, however, 9.62V applied voltage

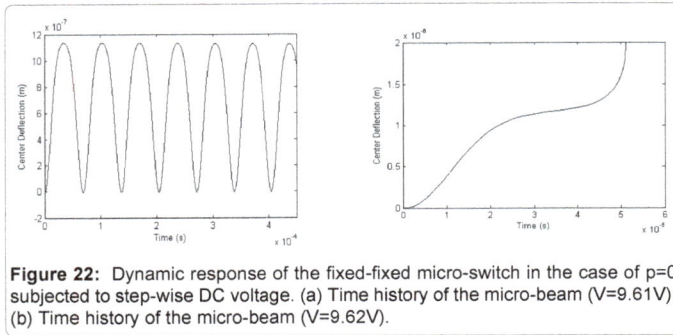

**Figure 22:** Dynamic response of the fixed-fixed micro-switch in the case of p=0 subjected to step-wise DC voltage. (a) Time history of the micro-beam (V=9.61V); (b) Time history of the micro-beam (V=9.62V).

So for p=1 in range of s ≤ 1 and p=0.75 in range of s ≤ 1.19, dynamic pull-in voltage is the same with static pull-in voltage.

The time history and phase portrait of the micro-beam with s=2 and p=1 and various step DC voltages is shown in Figure 24. As shown in Figure 24a the response of the micro-beam to DC voltage lower than 10.34V is periodic but for higher than this value, the response is non-periodic (unstable). This condition represents pull-in condition. As shown in this figure, with increasing of applied voltage from pull-in value, pull-in time is decreased. Figure 24b shows a metamorphosis of how a periodic orbit approaches to homoclinic orbit at dynamic pull-in voltage (V=10.34V). Indeed, the periodic orbit is ended at dynamic pull-in voltage where a homoclinic orbit is formed. In other words, when the applied voltage approaches the dynamic pull-in voltage, the periods of the closed orbits tend to infinity. It can be said that, the homoclinic bifurcation happened, when the periodic orbit collides with a saddle point at dynamic pull-in voltage. It must be noted that the scenario of instability in the case of applying step DC voltage is different from its static application. As Figure 17 shows, when applied DC voltage approaches the static pull-in voltage, the system tends to an unstable equilibrium position by undergoing to a saddle node bifurcation. A saddle node bifurcation, which is seen in the static application of the DC voltage, is a locally stationary bifurcation. This kind of bifurcation can be analyzed based on locally defined eigenvalues. In addition to local bifurcations, periodic orbits encounter phenomena that cannot be analyzed based on locally defined eigenvalues. Such phenomena are global bifurcations [24]. Furthermore for DC step excitation voltages lower than 9.2 V, the response is linear and the trajectories in the phase plane have symmetric forms. Increasing the voltage of the step excitation the trajectories in the phase plane shows a symmetric breaking for voltages between 9.2 and 10.33 V.

## Conclusion

The governing equation to analyze the dynamic motion of the micro-beam suspended between two conductive plates and subjected to electrostatic forces was presented. Due to the nonlinearity of this equation, it was solved using SSLM and Galerkin weighted residual method. The bifurcation behavior of the micro-switch under various conditions and excited voltages and stationary electrodes distances from the micro-beam were obtained using a modified mass-spring model. By solving the equation of the static deflection, fixed points of the micro-switch was determined, for three different cases of p=s=1, p=s ≠ 1 and p ≠ s ≠ 1. It was shown that, the pull-in phenomena were occurred by undergoing to a pitchfork and a saddle node bifurcation in the cases of p=s=1 and p ≠ s respectively. For these cases it was shown that, there are five fixed points in range of $0<V<V_{pull-in}$, and also it was shown that this number decreases to three fixed points for $V \succ V_{pull-in}$. Also was shown that for the electro-statically balanced case p=s one

of the fixed points is a trivial solution. For the case when p=s≠1 it was shown that primary and secondary pull-in phenomena are observed. In this case the primary pull-in phenomenon is due to a transcritical bifurcation and the secondary one is due to a saddle node bifurcation. Furthermore it was shown that, a stable center point exists between these unstable fixed points. The length of the stable branch between these unstable fixed points is increased with simultaneously increasing

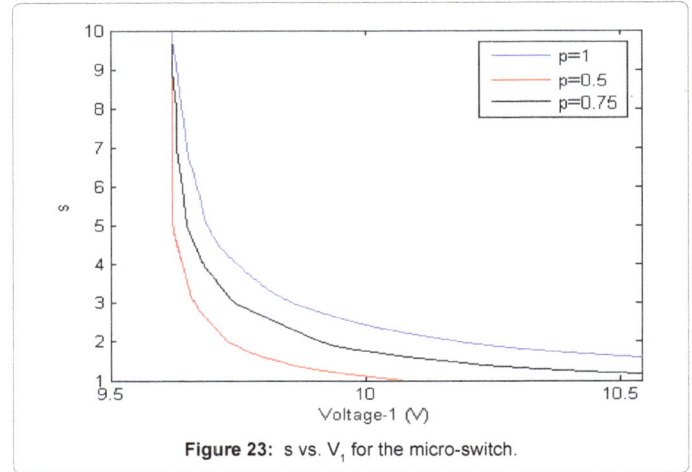

**Figure 23:** s vs. $V_1$ for the micro-switch.

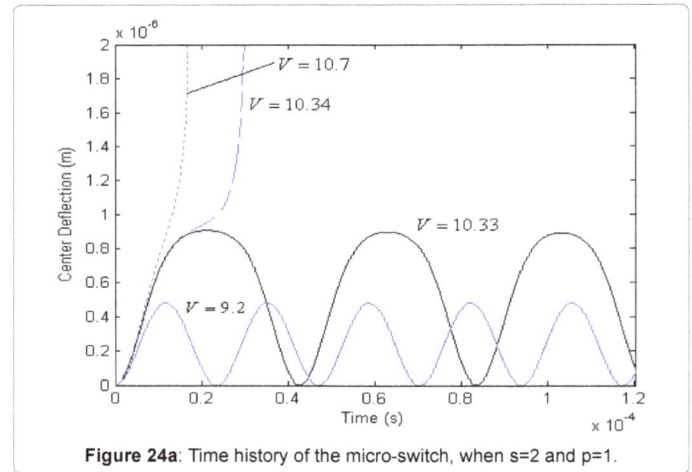

**Figure 24a:** Time history of the micro-switch, when s=2 and p=1.

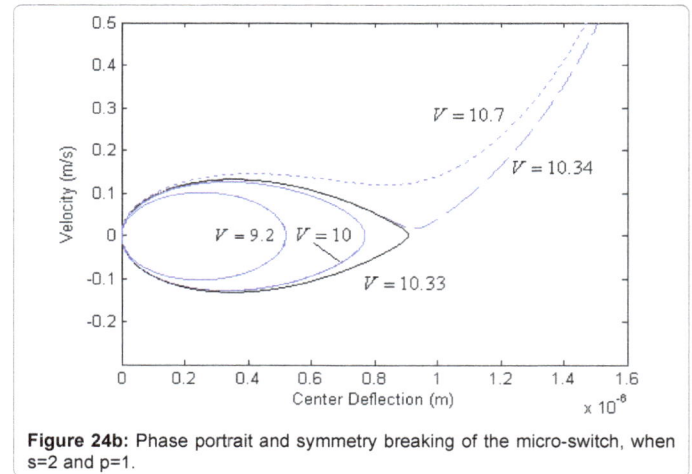

**Figure 24b:** Phase portrait and symmetry breaking of the micro-switch, when s=2 and p=1.

the voltages and gap ratios (ς). In another word the case of p=s=1 is a special case of the second one when these ratios are equal to 1 and in this case two bifurcation points join together and only one saddle node bifurcation is observed.

In the case of applying step DC voltage, the results show that, contrary to the classic micro-switch, for presented micro-switch, the ratio of the dynamic pull-in voltage to the static one is not a constant value and located in the range of 0.9 to 1. In another word when the system is in the electro-statically balanced case (p=s) and exist the trivial solution, as there is no inertial forces in the system, the static and dynamic pull-in values are equal.

## References

1. Nayfeh AH, Balachandran B (2004) Applied Nonlinear Dynamics.

2. Rezazadeh G (2007) A comprehensive model to study nonlinear behavior of multilayered micro beam switches. Microsyst Technol 14: 135-141.

3. Zhang Y, Zhao YP (2006) Numerical and analytical study on the pull-in instability of micro-structure under electrostatic loading. J Sens Actuators A Phys 127: 366-367.

4. Lin WH, Zhao YP (2003) Dynamic behavior of Nanoscale Electrostatic actuators. Chinese Phys Lett 20: 2070-2073.

5. Arani GA, Ghaffari M, Jalivand A, Kolahchi R (2013) Nonlinear nonlocal pull-in instability of boron nitride nanoswitches. Acta Mech 224: 3005-3019.

6. Jia LX, Yang J, Kitipornchai S (2011) Pull-in instability of geometrically nonlinear micro-switches under electrostatic and Casimir forces. Acta Mech 218: 161-174.

7. Mobki H, Rezazadeh G, Sadeghi M, Vakili-Tahami F, Seyyed-Fakhrabadi MM (2013) A comprehensive study of stability in an electro-statically actuated micro-beam. Int J Nonlinear Mech 48: 78-85.

8. Zhang Y, Zhao YP (2006) Numerical and analytical study on the pull-in instability of micro-structure under electrostatic loading. Sensors and Actuators A 127: 366-380.

9. Taghizadeh M, Mobki H (2014) Bifurcation analysis of torsional micromirror actuated by electrostatic forces. Arch Mech 66: 95-111.

10. Dequesnes M, Tang Z, Aluru NR (2004) Static and Dynamic Analysis of Carbon Nanotube-Based Switches. J Eng Mater-T Asme 126: 230-237.

11. Dequesnes M, Rotkin SV, Aluru NR (2002) Calculation of pull-in voltages for carbon-nanotube-based nanoelectromechanical switches. Nanotechnology 13: 120-131.

12. Hosseini M, Sadeghi-Goughari M, Atashipour SA, Eftekhari M (2014) Vibration analysis of single-walled carbon nanotubes conveying nanoflow embedded in a viscoelastic medium using modified nonlocal beam model. Arch Mech 66.

13. Mobki H, Sadeghi MH, Rezazadeh G, Fathalilou M, Keyvani-janbahan AA (2014) Nonlinear behavior of a nano-scale beam considering length scale-parameter. Appl Math Model 38: 881-1895.

14. Lin WH, Zhao YP (2005) Casimir effect on the pull-in parameters of nanometer switches. Microsyst Technol 11: 80-85.

15. Lin WH, Zhao YP (2005) Nonlinear behavior for nanoscale electrostatic actuators with Casimir force. Chaos Soliton Fract 23: 1777-1785.

16. Lin WH, Zhao YP (2007) Stability and bifurcation behavior of electrostatic torsional NEMS varactor influence by dispersion forces. J Phys D Appl Phys 40: 1649-1654.

17. Shabani R, Sharafkhani N, Tariverdilo S, Rezazadeh G (2013) Dynamic analysis of an electrostatically actuated circular micro-plate interacting with compressible fluid. Acta Mech 224: 2025-2035.

18. Sharafkhani N, Rezazadeh G, Shabani R (2012) Study of mechanical behavior of circular FGM micro-plates under nonlinear electrostatic and mechanical shock loadings. Acta Mech 223: 579-591.

19. Mobki H, Rashvand K, Afrang S, Sadegh MH, Rezazadeh G (2014) Design, Simulation and Bifurcation Analysis of a Novel Micromachined Tunable Capacitor with Extended Tunability. T Can Soc Mech Eng 38: 15-29.

20. Osterberg P (1995) Electrostatically actuated microelectromechanical test structures for material property measurement. Ph.D. thesis, MIT, Cambridge, USA.

21. Medio A, Lines M (2003) Nonlinear Dynamics A Primer. Cambridge University Press, England.

22. Ananthasuresh GK, Gupta RK, Senturia SD (1996) An approach to macromodeling of MEMS for nonlinear dynamic simulation. In: Proceedings of the ASME International Conference of Mechanical Engineering Congress and Exposition (MEMS). Atlanta, GA, pp: 401-407.

23. Seydel R (2010) Practical Bifurcation and Stability Analysis. In: Verlag S (ed.) NY, USA.

24. Kuznetsov YA (1997) Elements of Applied Bifurcation Theory (2nd edn.). In: Verlag S (ed.) NY, USA.

# 25

# Optimization of Resource Allocation in OFDM Communication System for Different Modulation Technique using FRBS and PSO

**Farhana Mustafa\* and Padma Lohiya**

*Department of E&TC, D.Y. Patil College of Engineering, Akurdi, Pune, India*

### Abstract

OFDM is technique that is chosen for high data rate communication and is important for 4th generation communication system. Resources such as power, bandwidth are limited, thus intelligent allocation of these resources to users are crucial for delivering the best possible quality of services. Fuzzy Rule Based System (FRBS) and Particle Swarm Optimization (PSO) algorithm are used for optimization of code rate, modulation and power. FRBS is used for adapting code rate and modulation size while PSO is used for power allocation.

**Keywords**: OFDM; PSO; FRBS; Optimization; Resource allocation

## Introduction

Orthogonal Frequency Division Multiplexing (OFDM), offers a considerable high spectral efficiency multipath delay spread tolerance, immunity to frequency selective fading channels and power efficiency [1,2]. As a result, OFDM has been chosen for high data rate communication and is used for 4th generation technology. In Orthogonal Frequency Division Multiplexing (OFDM) technique, a single very high data rate stream is divided into several low data rate streams using Inverse Fast Fourier Transform (IFFT). Then these streams are modulated over different orthogonal subcarriers. This is to divide one large frequency selective channel into a number of frequency non-selective sub-channels. Moreover, addition of appropriate cyclic prefix (CP) and interleaver makes the system almost inter-symbol-interference (ISI) free. It has been widely deployed in many wireless communication standard such as based mobile worldwide interoperability for microwave access (mobile WIMAX), 3GPP long term evolution (LTE) based on OFDM access technology. In OFDM every sub channel experiences a different channel condition so the use of same modulation and code rate may not be suitable for all subcarriers. Also, flat power is not beneficial since sub-channels may need different power. This situation demands adaptive resource allocations for an optimum utilization.

## Prior Work

In the optimal power allocation and user selection solution was derived based on Lagrange dual decomposition proposed by Wong et al. [3] for maximizing the system energy efficiency. A low complexity algorithm for proportional resource allocation in OFDMA system was proposed in ref. [4], where linear method and root finding algorithm were used to allocate power and data rates to users. A gradient based solution was proposed by Rajendrasingh et al. [5], for downlink OFDM wireless systems and a 96.6% utility was achieved. A Genetic Algorithm based adaptive resource allocation scheme was proposed by Reddy [6] to increase the user data rate where water-filling principle was used as a fitness function. The water filling theorem is based on a continuous relationship between the allocated power and the achievable capacity. OFDM Systems Resource Allocation using Multi-Objective Particle Swarm Optimization. Another paper with adaptive resource allocation based on modified GA and particle swarm optimization (PSO) for multiuser OFDM system was propose by Kennedy and Eberhart [7]. In this paper it has shown that MOPSO power optimization is better than 3GPP LTE and NSGA II Algorithm. An optimization problem for power constraints and use of GA algorithm to maximize the

sum capacity of OFDM system with the total power constraint was investigated in ref. [8-11]. Also it was shown that GA is better than conventional methods. A scheme for resource allocation in downlink MIMO OFDMA with proportional fairness where dominant Eigen channels obtained from MIMO state matrix are used to formulate the scheme with low complexity in ref. [8], scheme provides much better capacity gain than static allocation method. A PSO based Adaptive multi carrier cooperative communication technique which utilizes the subcarrier in deep fade using a relay node in order to improve the bandwidth efficiency [9] where centralized and distributed versions of PSO were investigated. Atta-ur-Rahman et al. in ref. [10,11] , used GA and Water-filling principle in conjunction with FRBS for adaptive coding, modulation and power in OFDM systems, where GA was used to adapt the power.

The paper is organized as follows: section III deals with the Multi Modulation OFDM system where QAM modulation is taken in consideration with M=4, 16, 32, 64, 128, 256, system description is given in section III. FRBS and PSO aspects are discussed for FRBS rule are define in section IV. Section V describes the simulation and results for OFDM system. VI concludes the paper.

## Multi-Modulation OFDM System

The system model considered is OFDM equivalent baseband model with N number of subcarriers. It is assumed that complete channel state information (CSI) is known at receiver. The frequency domain representation of system is given by

$$r_k = h_k \sqrt{P_k} X_k + Z_k; k = 1, 2, 3 \dots N \tag{1}$$

where amplitude, transmit symbol and the Gaussian noise of sub carrier k=1, 2,..., N respectively. The overall transmit power of the system and the noise distribution is complex Gaussian with zero mean and unit variance. It is assumed that signal transmitted on the k[th] subcarrier is propagated over Rayleigh at fade channel and each

---

**\*Corresponding author:** Farhana Mustafa, Department of E&TC, D.Y. Patil College of Engineering, Akurdi, Pune, India
E-mail: farhanabhatt23@gmail.com

subcarrier faces a different amount of fading independent of each other. This can be given mathematically

The proposed adaptation model is given in Figure 1.

## Coded modulation

Performance of standard modulation and codes being used in IEEE 802.11n1g/b are analyzed in terms of bit error rate (BER) and SNR. Calculation of coding scheme, modulation scheme and channel is estimated. The code rate are taken from the set C

$$C = \{\frac{1}{4}, \frac{1}{3}, \frac{1}{2}, \frac{2}{3}, \frac{3}{4}\} \tag{2}$$

Modulation symbol are taken from

$$M = \{2, 4, 8, 16, 32, 64, 128\} \tag{3}$$

Total number of MCPS can be given by

$$P = C \times M \tag{4}$$

## Fuzzy rule Base System and Practicle Swarm Optimization

### Fuzzy rule base system

To maximize the data rate FRBS is used for optimum selection of code modulation pair (CMP) per subcarrier based upon received SNR and QoS. The steps involved in creation of FRBS are described below

**Data acquisition**: The information about SNR and BER obtained from Coded Modulation can be expressed as "for a given SNR and specific QOS which modulation code pair can be used.

**Rule formulation**: Rules for every pair are obtained by the appropriate fuzzy set used.

**Elimination of conflicting rule**: This is used for eliminating conflicting rules.eg If there are two different pairs with same throughput like [2,1/2] and[4,1/4], both have same throughput i.e.1×1/2=0.5. Thus [2,1/2] is chosen since it have less modulation/ demodulation, coding/ decoding cost.

**Completion of Look up Table**: If complete numbers of IO pairs are not present, then those parts are filled by heuristic or expert knowledge. Example a modulation code pairs is suggested by rule for a certain SNR and QOS. Then that rule can also be used for slightly above SNR and poor QOS (Table 1).

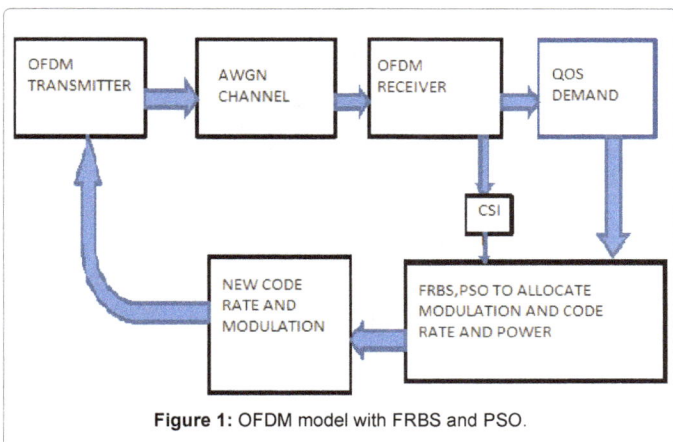

**Figure 1:** OFDM model with FRBS and PSO.

| SR. NO. | PARAMETERS | VALUES |
|---|---|---|
| 1 | NO. OF SUBCARRIERS | 52 |
| 2 | CODE RATE | ½ , ¼ |
| 3 | MODULATION | 16, 32, 64, 128,256 |
| 4 | CHANNEL | AWGN |
| 5 | BER | 10E-2,10E-3,10E-4,10E-5,10E-6, 10E-7 |

**Table 1:** Parameters.

For instance [64, 1/2] is suggested for 20dB SNR and SER=$10^{-2}$ then this pair can be used for 21 to 25dB SNR at $10^{-1}$ SER

**Fuzzy rule base creation:** The input output pair for design of FRBS are of the form

$$(x_1^s, x_2^s, y_3^s); s = 1, 2, 3, \ldots S \tag{5}$$

where x1s represents received SNR, $x_2^s$ represents BER (QOS) and y3s represents the output MCP.

**Fuzzy set:** Input for Fuzzy inference system is given as SNR and BER or minus log BER.

$$MLBER = -\log(BER) \tag{6}$$

$$BER = 10^{-q} \tag{7}$$

$$MLBER = -\log(10^{-q}) = q \tag{8}$$

where q is 0 to 10, there will be one output as MCP. Where BER is $\log_2(\frac{1}{M})$ of SER

**Membership function:** Membership function used in FIS (fuzzy inference system) is triangular. Triangular membership function is simple to implement as well as calculation of arithmetic operation is easier than Bell, Sigmoidal ,Gaussian.

In FIS system AND is used for MIN and OR as MAX.

**Rule base:** 7 sets of SNR and 10 sets MLBER are taken. Total number of rules taken is 70 that will be used in FIS system.

**De-fuzzifier:** Standard Center Average Defuzzifier (CAD) is used for defuzzification. CAD is to perform a linear combination over the computed weights at the fuzzy inference engine and then modify this combination by novelization. CAD provide continuity and homogeneity and has less computational complexity.

## Particle Swarm Optimization

Particle Swarm Optimization is a stochastic optimization technique developed by Eberhart and Kennedy inspired by the social behavior of flocks of bird. Each particle is represented by a position and velocity vector. Let Dimensions of position and velocity vectors are defined by the number of decision variables in optimization problem. Soft PSO has been utilized for finding the optimum power vector for all the sub carriers depending upon the channel conditions and their QOS demand. Each sub-channel have different channel condition so different channel should have different power allocation depending upon the channel condition. Power allocation will be done with the help of PSO. Different power allocation is done for different users and thus the optimization of Power is done for different users in OFDM system.

Let $\overline{x}_i(t)$ represents the position particle $p_i$ at time t, then it is given as

$$\overline{x}_i(t) = \overline{x}_i(t-1) + \overline{v}_i(t) \tag{9}$$

The position of $p_i$ is then changed by adding a velocity $\overline{v}_i(t)$.

Each particle know its best position (p-best) and global best (g-best).Thus the particle will tend to attain its g-best at final iteration. g-best will give the optimum allocated power

$$\overline{v}_i(t) = \overline{v}_i(t-1) + C_1 r_1 (\overline{x}_{pbest} - \overline{x}(t)) + C_2 r_2 (\overline{x}_{gbest} - \overline{x}(t)) \text{ where } \quad C_1$$

and $C_2$ are constants and is normally equal to 2.0. $r_1$ and $r_2$ are random variables.

## Simulation Results

Simulation is performed to have optimal power allocation.

### Simulation parameters

### Results

OFDM system with different code rate and modulation are simulated. Input of this graph is given to Fuzzy inference system and PSO for optimization of code rate and modulation Figure 2.

In Figure 3 Symbol Error Rate vs Signal to Noise Ratio are calibrated for different modulation QAM techniques such as 4, 16, 32, 64, and 128 with code rate ½, where EbN0 is taken from 0:33. This represents SER decrease exponentially with respect to EbN0 for different modulation.

Figure 4 represent SER vs SNR plot with respect to each modulation such as 16, 32,64,128 with code rate as ¼ . SER decrease exponentially with respect to EbN0 for different modulation.

Figure 5 shows the impact on throughput for different values of SNR and QoS demand after incorporating the constraint. In this diagram a higher numbered CMP reflects a high throughput.

Using FRBS in OFDM system it will give optimum modulation and code rate represented by Figure 5.

Power allocation with respect to noise interface is shown in Figure 6. This figure indicates that the 3rd user has less interference thus less power will be allocated.

## Conclusion

In this paper FRBS and Particle swarm algorithm are used for optimization of code rate and modulation. FRBS are used for

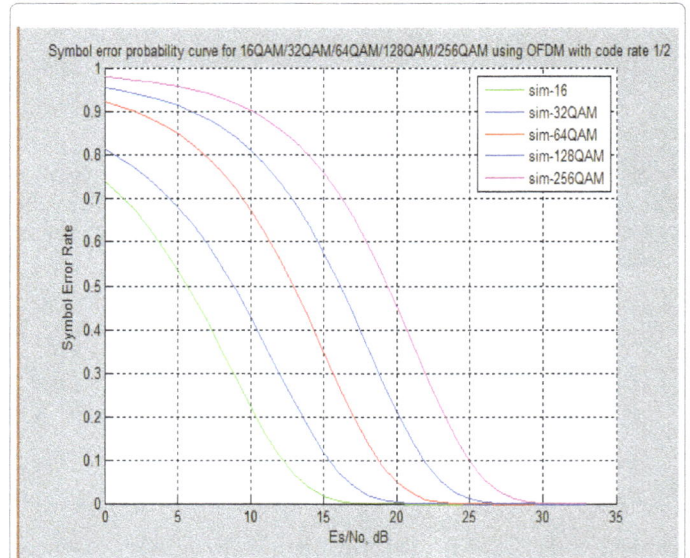

**Figure 3:** SER comparison of different QAM with code rate 1/2 code.

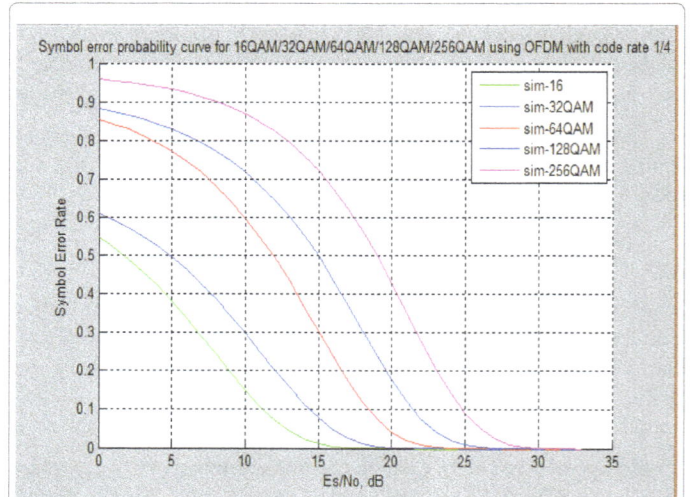

**Figure 4:** SER comparison of different QAM with code rate1/4 code.

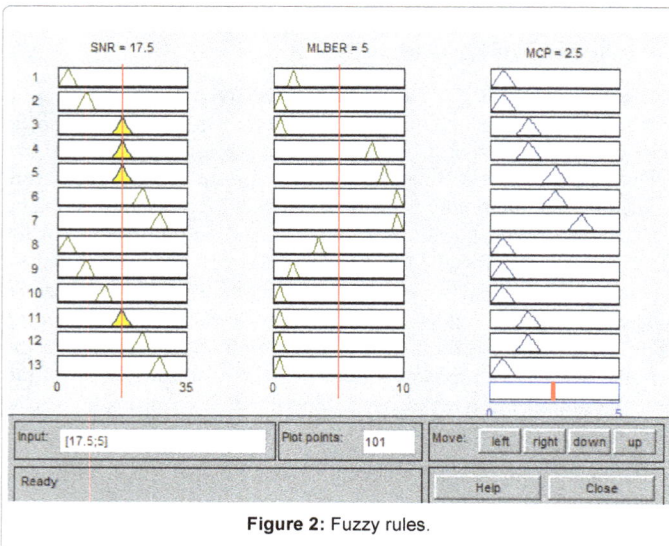

**Figure 2:** Fuzzy rules.

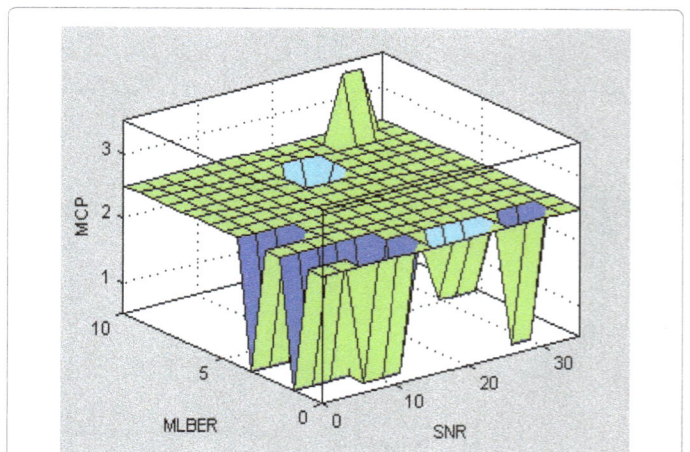

**Figure 5:** Rule surface.

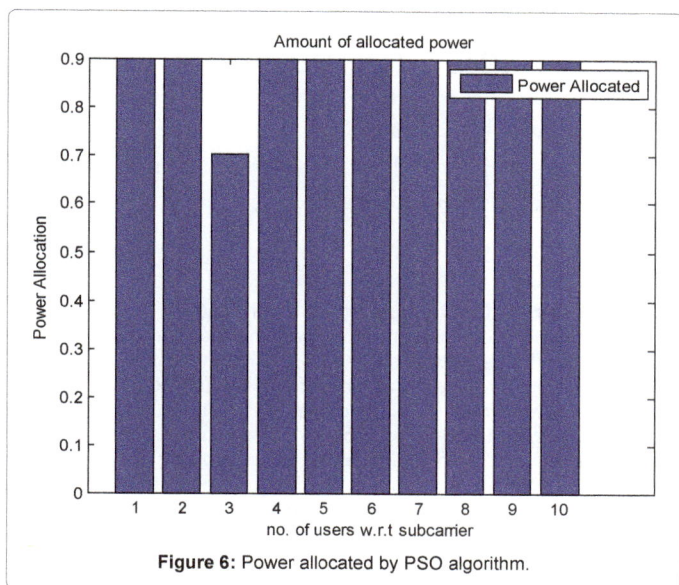

**Figure 6:** Power allocated by PSO algorithm.

optimization of code rate and modulation. Higher numbered CMP reflects a higher throughput. PSO is used for allocation of power for each sub-channel depending upon the channel condition. After using FRBS in OFDM system it will give optimal modulation and Code rate. Thus y using FRBS and PSO optimal power allocation can be done with specified modulation techniques for specified sub channel.

### References

1. Wu Y, Zou WY (1995) Orthogonal frequency division multiplexing: A multi-carrier modulation scheme, IEEE Trans Consumer Electonics 41: 392-399.

2. Derrick Wing Kwan Ng, Ernest SLo, Robert Schober (2010) Energy-Efficient Resource Allocation in Multiuser OFDM Systems with Wireless Information and Power Transfer, University at Erlangen-Nurnberg, Germany.

3. Wong IE, Zukang Shen, Evans BL, Andrews LG (2004) A low complexity algorithm for proportional resource allocation in OFDMA systems, Dept of Electr & Comput Eng, Texas Univ, Austin, TX, USA.

4. Reddy YB, Gajendar N, Taylor, Portia (2007) Computationally Efficient Resource Allocation in OFDM Systems: Genetic Algorithm Approach , Dept of Math & Comput. Sci, Grambling State Univ, LA, p 36-41.

5. Rajendrasingh A, Rughooputh Harry CS (2012) OFDM Systems Resource Allocation using Multi-Objective Particle Swarm Optimization.International Journal of Computer Networks & Communications 4.

6. Atta-ur-Rahman, Qureshi IM, Malik AN, (2012) A Fuzzy Rule Base Assisted Adaptive Coding and Modulation Scheme for OFDM Systems.J Basic Appl Sci Res 2: 4843-4853.

7. Kennedy J, Eberhart RC (1995) Particle Swarm Optimization, Proceedings of IEEE Conference on Neural Networks 4: 1942-1948.

8. Gheitanchi S, Ali F, Stipidis E (2007) Particle Swarm Optimization for Resource Allocation in OFDMA. Proc International Conference on digital Signal Processing 383-386.

9. Atta-ur-Rahman, Qureshi IM, Muzaffar MZ (2011) Adaptive Coding and Modulation for OFDM Systems using Product Codes and Fuzzy Rule Base System. International Journal of Computer Applications (IJCA) 35: 41-48.

10. Atta-ur-Rahman, Qureshi TM, Malik AN (2012) Adaptive Resource Allocation in OFDM Systems using GA and Fuzzy Rule Base System.World Applied Sciences Journal 18: 836-844.

11. Atta-ur-Rahman (2013) Optimum Resource Allocation in OFDM Systems using FRBS and Particle Swarm Optimization. Barani Institute of Information Technology, Rawalpindi, Pakistan Institute of Signals, Systems and Soft-computing (ISSS), Islamabad, Pakistan, 175-181.

# 26

# Compared with a-Fe$_2$O$_3$ and ZnxFe$_3$-XO$_4$ Thin Films Grown by Chemical Spray Pyrolysis

**Saritaş S[1], Turgut E[2], Kundakci M[1], Gürbulak B[1]\* and Yildirim M[1]**

[1]Department of Physics, Ataturk University, 25250, Erzurum, Turkey
[2]Department of Electrical and Energy, Aşkale Vocational School of Higher Education, Ataturk University, 25250, Erzurum, Turkey

### Abstract

This work describes hematite (a-Fe$_2$O$_3$) and ZnxFe$_3$-XO$_4$ thin films prepared by Chemical Spray Pyrolysis (CSP) method. CSP method allows an optimal control of stoichiometry and impurity incorporation, hematite films modified with Zn$^{2+}$ was also prepared. Moreover, the most attracting characteristics of the hematite are its stability in neutral and basic solutions, abundance and band gap energy (2.0–2.2 eV) which permits it to absorb approximately 40% of the incident solar spectrum on earth. Nevertheless, the performance of hematite electrodes for water oxidation is restricted by their poor charge transport properties. Hematite has low conductivity and low charge-carrier mobility. In addition, the photoexcited electron–hole pairs have short life time (~10$^{-12}$ s), which makes the hole diffusion length to be also short (2–4 nm). The charge transport properties of hematite can be improved by dopping. We demonstrated to increase the conductivity of hematite by dopping it with metal cations with 2+ charges which improved the photocatalytic properties. Doping with metal cations with 2+ charges has also brought good photoelectrochemical results. So we iron oxide and Zn-doped iron oxide compounds have been investigated.

The structural, optical and magnetic properties of a-Fe$_2$O$_3$ and ZnxFe$_3$-xO$_4$ compounds have been extensively investigated. XRD, XPS, Raman, FE-SEM and AFM techniques have been used for structural analysis; Absorption technique has been used for optical properties; Hall and Vibrating Sample Magnetometer (VSM) techniques have been used for magnetic properties.

**Keywords:** Hematite; Zn-doped iron oxide; Photoelectrochemical (PEC); Thin film

## Introduction

Photocatalysis has been attracting much research interest because of its wide applications in renewable energy and environmental restoration; however materials limitations have significantly hindered their efficiency. Researchers to find different techniques to use solar energy as an alternative for future energy needs. The objective of our research is to improve the efficiencies of PEC cells by identifying and engineering corrosion-resistant semiconductors that exhibit the optimal conduction and valence band edge alignment for PEC applications [1,2]. There are many materials that are found to show good photocatalytic activity in the presence of Ultraviolet (UV) and visible light. The most common method to directly convert solar energy into electric energy is Photovoltaic (PV), and this process utilizes semiconductors which generate electron–hole pairs upon illumination with visible light, thereby producing electric power in solar cells. However, the utility of photovoltaic cells is limited by poor conversion efficiency. To overcome these problems, researchers have tried to find suitable methods to produce hydrogen (H$_2$) from photocatalysis of H$_2$O using sunlight, which can be used in fuel cells for power generation. However, the applications of these materials are limited to the UV portion of sunlight. a-Fe$_2$O$_3$ has an advantage over the other conventional materials like TiO$_2$, ZnO, WO$_3$ etc. in using solar energy for photocatalytic applications due to its lower band gap ~2.2 eV value. As a result of which Fe$_2$O$_3$ is capable of absorbing a large portion of the visible solar spectrum (absorbance edge ~600 nm). Also its good chemical stability in aqueous medium, low cost, abundance and nontoxic nature makes it a promising material for photocatalytic water treatment and water splitting applications [3].

However, the photocatalytic performance of a-Fe$_2$O$_3$ is limited by certain factors such as high recombination rate of electrons and holes, low diffusion lengths of holes (2–4 nm). And poor conductivity, which led to both low efficiencies and a larger requisite over potential for photo- assisted water oxidation [4-8]. Many attempts have been made by researchers to overcome these anomalies of a-Fe$_2$O$_3$ such as lowering the recombination rate by forming nanostructures, enhancement in conductivity by doping with suitable metals and improving the charge transfer ability [9,10].

Photocatalytic ability in materials is one of the most interesting research topics due to its usefulness in various fields such as H$_2$ generation [11-13], artificial photosynthesis [14,15], waste water treatment [16-18], removal of toxic gases from air [19-21].

Fossil fuels have been the most consumed energy by the World during the last 40 years. Indeed, fossil fuels provided approximately 87% of global energy consumption in 2013. Using this kind of energy will continue to provoke the emission of greenhouse gases (e.g. CO$_2$) that pollute and damage our environment. Therefore, optimizing the technology of clean and renewable energies is urgent in order to diminish the use of fossil fuels, and, thus, it will permit the preservation of our environment for the next generations [22-27].

Spintronics is another application field of technology for Iron Oxide, whereas conventional electronic devices ignore the spin property and rely strictly on the transport of the electrical charge of electrons. Spintronics is an emergent nano technology which deals with spin dependent properties of an electron instead of or in addition to its charge dependent properties. Adding the spin degree of

**\*Corresponding author:** Gürbulak B, Department of Physics, Ataturk University, 25250, Erzurum, Turkey, E-mail: gurbulak@atauni.edu.tr

freedom provides new effects, new capabilities and new functionalities. Spintronic devices offer the possibility of enhanced functionality, higher speed, and reduced power consumption. High-volume information-processing and communications devices are at present based on semiconductor devices, whereas information-storage devices rely on multilayers of magnetic metals and insulators. Spin transistors would allow control of the spin current in the same manner that conventional transistors can switch charge currents, which was first spin device proposed for metal-oxide geometry [28-30].

## Experimental Details

CSP is one of the solution based coating technique to produce metallic and semiconductor thin or thick films. Apart from the many other thin film fabrication methods, this technique is quite simple and comparatively cost effective. Dense, porous or multi-layered films in any composition can be fabricated using this versatile method. Temperature control unit, substrate heater, deposition solution and atomizer are components of the CSP setup. Different type of atomizers such as air blast, ultrasonic or electrostatic can be employed depending on properties of liquid and operating conditions to obtain coatings with desired properties. Atomized droplets of deposition solution spread over the surface of the substrates with respect to the temperature of substrate, volume and momentum of the droplets. The interaction of droplets with substrate surface, aerosol transport, and evaporation of solvent and decomposition of precursor are consecutive or simultaneous processes of this processing technique and these processes are accompanied by decomposition temperature. Therefore, temperature is the main parameter of CSP and significantly affects the microstructural, optical and electrical properties of the resultant thin film. Moreover, air flow rate, nozzle distance and viscosity of deposition solution are the other controllable processing parameter to produce high quality coatings. The salts given in Table 1 were prepared as 0.1 molar solutions in deionized water. The substrate was sprayed with argon gas onto a substrate heated to 320°C at a distance of 30 cm [30-34].

## Results and Discussion

The structural, optical and magnetic properties of a-$Fe_2O_3$ and ZnxFe_3-xO_4 compounds have been extensively investigated. XRD, XPS, Raman, FE-SEM and AFM techniques have been used for structural analysis; Absorption technique has been used for optical properties; Hall and Vibrating Sample Magnetometer (VSM) techniques have been used for magnetic properties.

The XRD diffraction pattern of the iron oxide structure growing on the glass substrate is given in Figure 1 and it has been determined that the structure has a tetragonal structure (Table 2). The lattice constants a=b=8.33 Å, c=24.99 Å. Four evident peaks are observed, of which characteristic hematite peak is observed at 32.30 degrees.

In Figure 1, XRD diffraction patterns of $Fe_2O_3$ and ZnxFe_3-xO_4 thin films are given. As can be seen from this figure, when compared with the ZnxFe_3-xO_4 structure of the $Fe_2O_3$ structure, the XRD peaks are narrower and more intense, while the peaks of the newly formed structure are observed and the resulting structure is polycrystalline (Table 3).

In Figure 2 the value of the energy of the band gap is calculated 2.16 eV, 2.14 eV with the fit drawn on the energy graph against the $(\alpha h\upsilon)^2$ $(cm^{-1} eV^2)$ of the $Fe_2O_3$, ZnxFe_3-xO_4 thin films grown by CSP technique, respectively. As $Fe_2O_3$ thin filminin gives absorption at smaller wave lengths, it shifts at larger wave length as a result of doping (Table 4). This may mean that the energy band gap is causing the contraction of the band gab due to the fact that the imperfections in the structure constitute the possibility of transition at the band edge [35,36].

Figure 3 shows the Raman shift of the stretching vibration mode of $Fe_2O_3$ films are seen. There are Raman active states of the hematite phase that these peaks are relatively narrow and severe. In the ZnxFe_3-xO_4 compound, the peak of the Raman shift peaks belonging to the hematite phase falls. In addition, the peaks showing the ramping changes of the stretching vibration mode of the ZnxFe_3-xO_4 film is due to the presence of multiple phases due to polycrystalline crystal structure [31].

XPS is used for the analysis of the elemental and chemical state information of the investigated surfaces and can be made approximately 10 nm from the surface. It is able to detect the ions and ligand energies attached to the chemical ligands of the sample studied.

As shown in Figure 4a; Tables 5 and 6 the binding energy of the 2p3/2, 2p1/2 orbitals for the $Fe^{3+}$ ($Fe_2O_3$) ion are 711 and 724 eV, respectively. The 1s orbital binding energy of the $O^{2-}$ ion is 531 eV. The peak intensities of the connecting electrons are very close to one another and the number of non-bonding electrons is small, which can be seen as the reason for the insulating properties [37].

We can also say that the conductivity is low because the atomic percentage of the oxygen atom is 65.96% (Table 6) and the oxygen vacancies causing the conductivity are low.

As can be seen in Figure 4b; Tables 5 and 6 the graph showing the binding energies of the $Fe^{3+}$, $Zn^{2+}$ and $O^{2-}$ ions of the ZnxFe_3-xO_4 composition is almost identical to the atomic oxygen content of the $Fe_2O_3$ compound, There has been no drop in severity. It is believed that

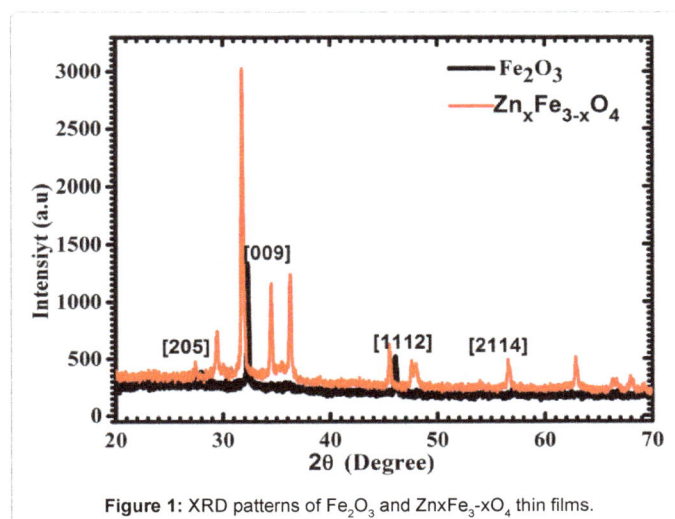

**Figure 1:** XRD patterns of $Fe_2O_3$ and ZnxFe_3-xO_4 thin films.

| Film | Used Chemical Salt | Solution Molar Ratio | Substrate Temperature (°C) | Carrier Gas | Grown Time (min) |
|---|---|---|---|---|---|
| $Fe_2O_3$ | $FeCl_3.6H_2O+FeCl_2.4H_2O+NaOH$ | 02:00.2 | 320 | Argon | 35 |
| ZnxFe_3-x O_4 | $FeCl_3.6H_2O+FeCl_2.4H_2O+NaOH+Zn(NO_3)_2 \cdot 6H_2O$ | 1:2:0.25:0.1:0.01 | 320 | Argon | 35 |

**Table 1:** Experimental details of the $Fe_2O_3$ and ZnxFe_3-xO_4 thin films grown by chemical spray pyrolysis technique.

| 2θ° | (hkl) | FWHM | Lattice Constant | d (A) | Crystal System | Chemical Formula | Reference Code |
|---|---|---|---|---|---|---|---|
| 28,14 | 205 | 0.071 | a=b=8,33 c=24,99 | 3,21 | Tetragonal | $Fe_2O_3$ | 00-015-0615 |
| 32,30 | 9 | 0.437 | a=b=8,33 c=24,99 | 2,79 | Tetragonal | $Fe_2O_3$ | 00-015-0615 |
| 46,02 | 1112 | 0.174 | a=b=8,33 c=24,99 | 1,94 | Tetragonal | $Fe_2O_3$ | 00-015-0615 |
| 57,02 | 2114 | 0.001 | a=b=8,33 c=24,99 | 1,60 | Tetragonal | $Fe_2O_3$ | 00-015-0615 |

**Table 2:** Structural properties obtained from XRD patterns of $Fe_2O_3$ thin film.

| 2θ° | (hkl) | FWHM | d (A) | Crystal system | Chemical formula | Lattice constant | Reference Code |
|---|---|---|---|---|---|---|---|
| 27,45 | 15 | 0,150 | 3,24 | Hexagonal | $Fe_2O_3$ | a=5,560 b=5,560 c=22,550 | 01-076-1821 |
| 29,44 | 7 | 0,135 | 3,03 | Hexagonal | $Fe_2O_3$ | a=5,560 b=5,560 c=22,550 | 01-076-1821 |
| 31,77 | 220 | 0,171 | 2,81 | Cubic | $ZnFe_2O_4$ | a=b=c=8,30 | 01-073-1963 |
| 34,50 | 311 | 0,001 | 2,59 | Cubic | $ZnFe_2O_4$ | a=b=c=8,44 | 01-086-0507 |
| 36,28 | 222 | 0,191 | 2,47 | Cubic | $ZnFe_2O_4$ | a=b=c=8,30 | 01-073-1963 |
| 45,49 | 249 | 0,100 | 1,99 | Hexagonal | $ZnFe_2O_4$ | a=12,80 b=12,80 c=57,26 | 00-045-1186 |
| 47,60 | 331 | 0,001 | 1,91 | Cubic | $ZnFe_2O_4$ | a=b=c=8,30 | 00-016-0653 |
| 47,97 | 331 | 0,262 | 1,89 | Cubic | $ZnFe_2O_4$ | a=b=c=8,30 | 00-016-0653 |
| 56,57 | 511 | 0,148 | 1,62 | Cubic | $ZnFe_2O_4$ | a=b=c=8,30 | 00-016-0653 |
| 62,86 | 440 | 0,001 | 1.47 | Cubic | $ZnFe_2O_4$ | a=b=c=8,35 | 01-073-1963 |
| 66,54 | 531 | 0,090 | 1,40 | Cubic | $ZnFe_2O_4$ | a=b=c=8,35 | 01-073-1963 |
| 68,02 | 22-Mar | 0,090 | 1,37 | Hexagonal | $Fe_2O_3$ | a=5,560 b=5,560 c=22,550 | 01-076-1821 |

**Table 3:** Structural properties obtained from XRD patterns of $Zn_xFe_{3-x}O_4$ thin film.

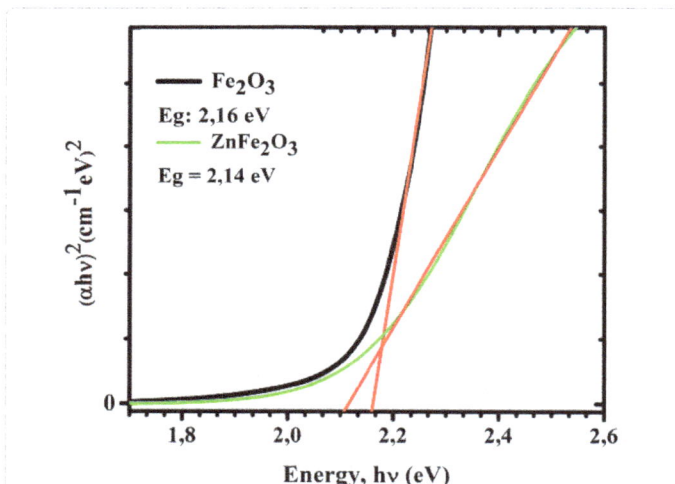

**Figure 2:** Plot of $(\alpha h \upsilon)^2 (cm^{-1} eV^2)$ vs. photon energy hu of $Fe_2O_3$ and $Zn_xFe_{3-x}O_4$ thin films.

this may be due to the fact that some of the Zn may have been linked to O, such that the Zn is linked to the $Fe_2O_3$ compound.

It is known that Zn is alloyed with Fe, and as it is known, when an alloy occurs, it often occurs that an element has a defect in the structure of the other element. It can be said that the decrease of Fe in the $Zn_xFe_3$-

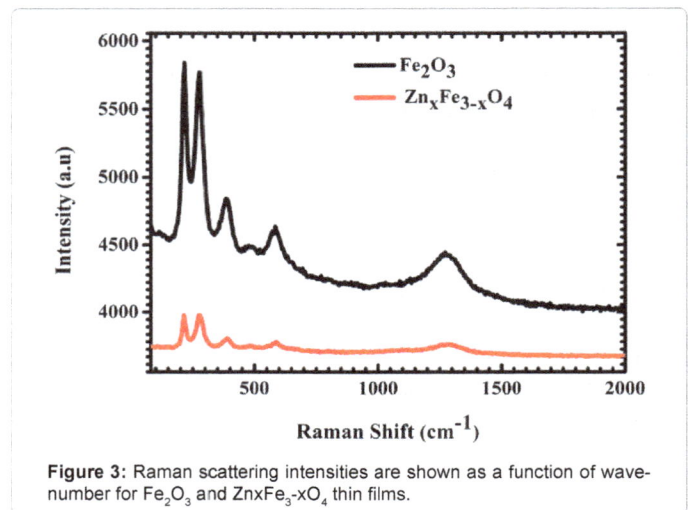

**Figure 3:** Raman scattering intensities are shown as a function of wavenumber for $Fe_2O_3$ and $Zn_xFe_{3-x}O_4$ thin films.

| Elements | Raman Shift (cm) | Mod |
|---|---|---|
| $Fe_2O_3$ | 216;277; 216;277 | Hematit; [(A1g) 225], [(Eg) 249],[(Eg) 295],[(Eg) 302] |
| $Zn_xFe_3$-$xO_4$ | 212;274;384; 212;274;384 | Zinc ferrit [F2g(2) 355] [ F2g(3) 451] [26-28] |

**Table 4:** Raman shift and modes of $Fe_2O_3$ ve $Zn_xFe_{3-x}O_4$ thin films.

$xO_4$ compound in $Fe_2O_3$ compound is due to the Zn incorporated in the structure. This also supports XRD results. For the $Fe^{3+}$ ($Fe_2O_3$) ion, the binding energy of the 2p 3/2, 2p1/2 orbitals is 711.65 eV and 724.3 eV, respectively. The 1s orbital binding energy of the $O^{2-}$ ion is 530 eV. The binding energy for 2p3/2 and 2p1/2 orbitals for $Zn^{2+}$ ion is 1024.23 and 1047.98 eV, respectively. If we think that the reduction in the amount of binding oxygen can be interpreted as oxygen vacancies in the structure, the cause of the increase in conductance will become apparent.

In Figure 5a, the pure $Fe_2O_3$ compound was given a 200 nm scaled FE-SEM image at about 171,000 magnifications taken at 6.6 mm working distance with an inlens detector. As can be seen from this figure, it can be said that the surface/volume ratio in which a stacked leaf-like image exists, it is a widely used material for gas sensor application [29]. Also films surface cover with $OH^{-1}$ groups.

Figure 5b shows FE-SEM images of the $Zn_xFe_3$-$xO_4$ compound at 400,000 magnifications. It is possible to say that the composition is homogeneously dispersed on the surface and that there is a nano porous structure. However, it was found that the decrease of the nanoporous structure in the $Zn_xFe_3$-$xO_4$ compound films grown by adding the solution of zinc nitrate solution prepared for the $Fe_2O_3$ thin film was confirmed that the Zn element was mixed more into the structure and the pores were closed. This nano porous structure can be used as a suitable substrate for forming many nano rod structures.

Figure 6 shows two-dimensional and three-dimensional (5 × 5 μm) AFM images of $Fe_2O_3$ thin filmin. As it can be seen, there are locally pebbles, and circular-like clusters are arranged regularly. The roughness value is about 28 nm with a maximum height of 51 nm and a maximum depth of 79 nm. The average roughness value Rq\RMS value is 33 nm, which is almost consistent with the linear roughness value.

In Figure 7, the two-dimensional and three-dimensional (5 × 5 μm) AFM images obtained for the $Zn_xFe_3$-$xO_4$ film showed that the particles in the structure showed a sharper image. There are pits and hills almost

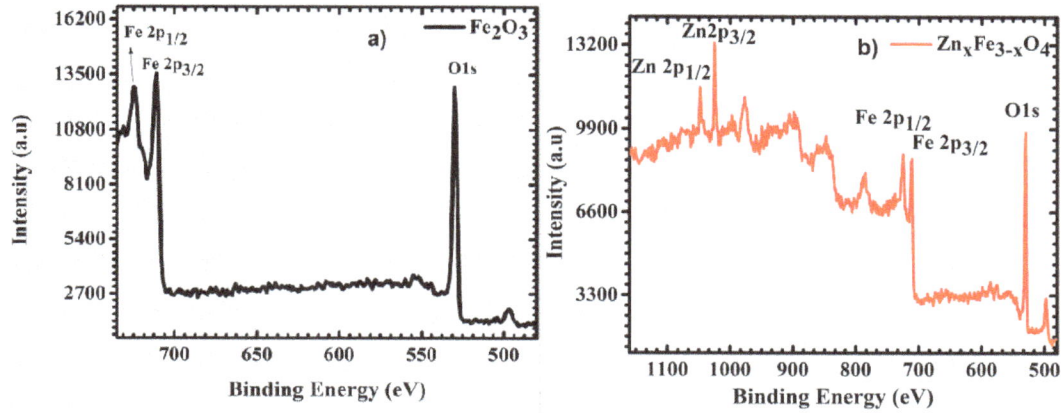

**Figure 4:** XPS analysis of a) $Fe_2O_3$ and b) $ZnxFe_3$-$xO_4$ thin films.

| Elements | Experimental Ion Binding Energy (eV) | | Literature Ion Bonding energy (eV) | | |
|---|---|---|---|---|---|
| | $2p_{1\backslash2}$ | $2p_{3\backslash2}$ | İyon | $2p_{1\backslash2}$ | $2p_{3\backslash2}$ |
| Fe | 724,30 | 711,65 | $Fe^{3+}$ ($Fe_2O_3$) | 724,30 | 710,70 |
| Zn | 1047,98 | 1024,23 | $Zn^{2+}$ (ZnO) | 1044,7 | 1021,70 |

**Table 5:** XPS measurement results of the binding energy according to orbitals of Fe3+, Zn2+ ions.

| Compound | Orbital | Intensity | % Atomic |
|---|---|---|---|
| $Fe_2O_3$ | O 1s | 12969 | 65.96 |
| | Fe 2p3\2 | 13470 | 34.04 |
| $ZnxFe_3$-$xO_4$ | O 1s | 9847 | 67.39 |
| | Fe 2p 3\2 | 8718 | 27.93 |
| | Zn 2p 3/2 | 13377 | 4.67 |

**Table 6:** The atomic percentage of the elements in the structure, orbital, peak intensity, bound by ions in the compounds.

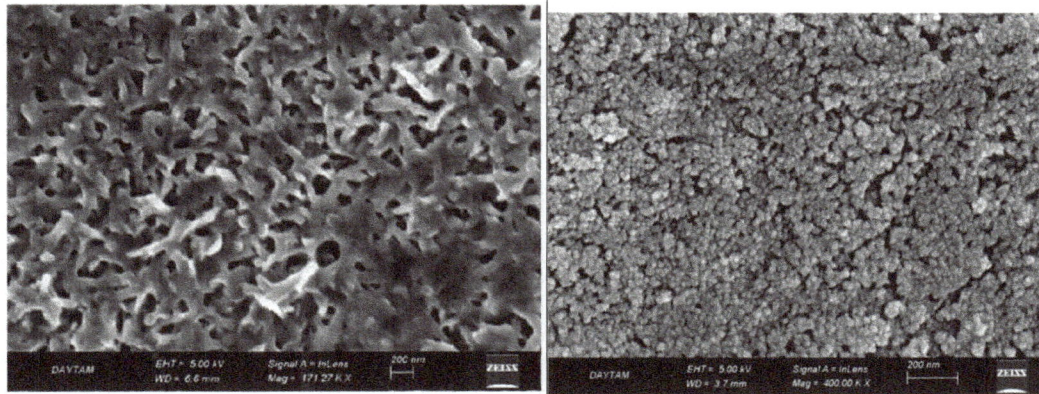

**Figure 5:** FE-SEM images of (a) $Fe_2O_3$ and (b) $ZnxFe_3$-$xO_4$.

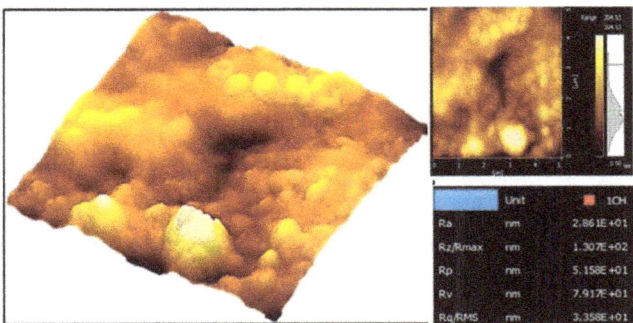

**Figure 6:** AFM images of the $Fe_2O_3$ thin film (Average height is 270.1 nm).

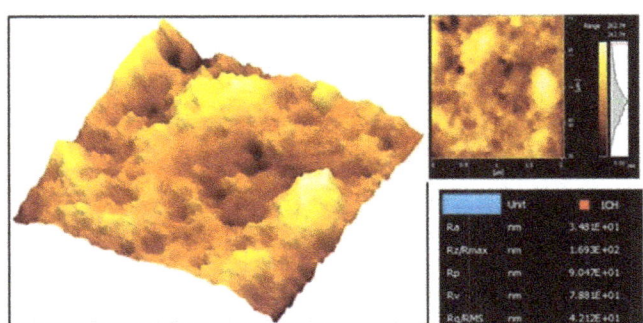

**Figure 7:** AFM images of the $ZnxFe_3$-$xO_4$ thin film (Average height is 266.6 nm).

everywhere resembling craters. The roughness value is about 34 nm, with a maximum height of 90 nm and a maximum depth of 78 nm. The average roughness value Rq\RMS value is 41 nm, which is almost in line with the linear roughness value. The variability of the colors tone indicates that the height difference in the topography is great. It is possible to say that the surface consists mostly of hills and pits.

Figure 8 shows that the sharp morphology of the ZnxFe$_3$-xO$_4$ thin-film AFM image in the three- dimensional (5 × 5 μm) AFM image of ZnxFe$_3$-xO$_4$ with a magnification of 5 min resulted in a large structure of sharp-pointed surface area and that the existing morphology stood out as shown in Figure 8. Here, a large surface area appears to be formed and the roughness value is 2.2 nm. It can be considered that such less rough ZnxFe$_3$-xO$_4$ thin films are an ideal material for gas sensor application [30].

Results of Hall measurement of p-type Fe$_2$O$_3$ thin film are considered, it is expected that the carriers in the valence band of the semiconductor materials may pass to the acceptor levels by thermal excitation and contribute to the conductivity (Figure 9a). In the enlarged film the carrier density is reduced and the resistivity is increased. We can say that the holes in the valence band of the p- type semiconducting material are compensated by the donor type defects and impurities and therefore the density of the hole carrier is lowered (Figure 9b). In this case, the decrease in Hall mobility due to the increased temperature is consistent with the literature (Figure 10a). Phonon scattering of the carriers becomes predominant as temperature increases the vibration amplitude of the cage ions. This may result in an average free path reduction.

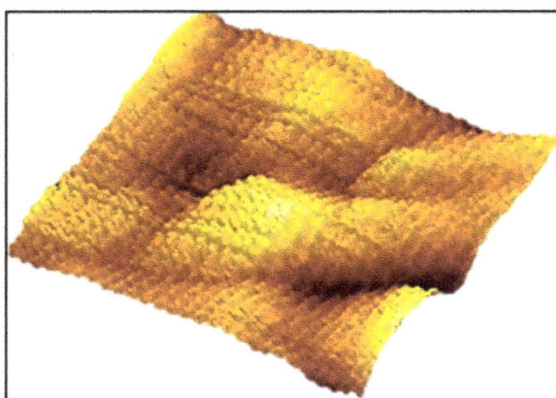

**Figure 8:** AFM images of the ZnxFe$_3$-xO$_4$ thin film (5 min grown, average height is 17.5 nm).

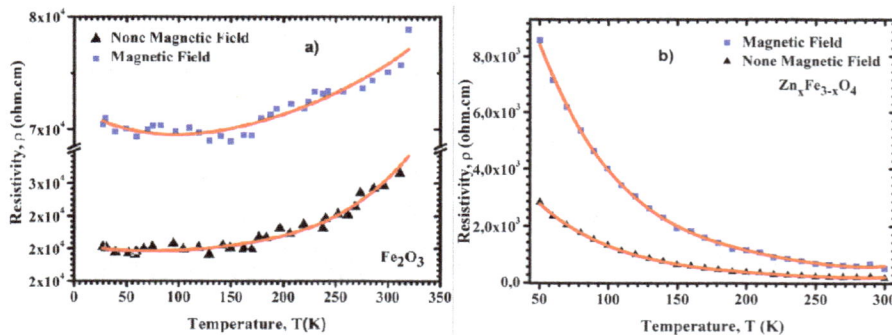

**Figure 9:** Resistivity versus temperature for (a) a-Fe$_2$O$_3$ and (b) ZnxFe$_3$-xO$_4$.

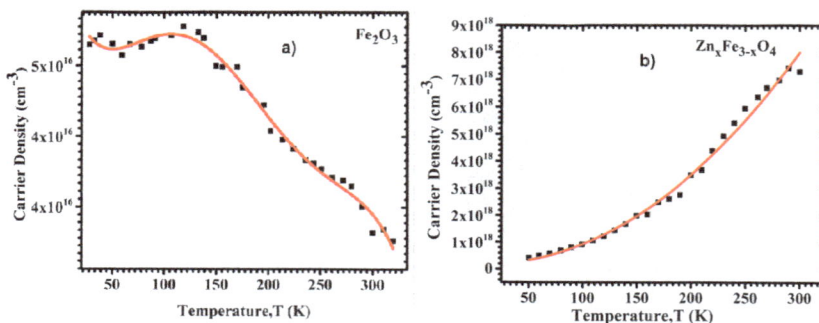

**Figure 10:** Carrier density versus temperature for (a) a-Fe$_2$O$_3$ and (b) ZnxFe$_3$-xO$_4$.

As can be seen in Figure 10b it is seen that the p-type $ZnxFe_3-xO_4$ semiconducting compound exhibits a semiconductor character in the result of the increase in the temperature (50-300 K). Due to the increased temperature the resistivity decreases. In addition, the magnetic field increases the resistivity of the material and shows a positive magneto resistance effect. We can say that the results obtained are in the expected direction and agree with each other.

Figure 10b shows the increase in carrier density due to the increasing temperature of p-type $ZnxFe_3-xO_4$ thin film. The increase in the number of whole carriers in the valence band with p-type semiconductor-induced thermal excitation can be seen under normal conditions.

Figure 11a and 11b shows the decrease in Hall mobility due to the increased temperature of $ZnxFe_3-xO_4$ thin films. As a result, the carrier density has increased by two orders of magnitude.

In Figure 12a, the magnetic hysteresis curve of the $Fe_2O_3$ thin film is observed to be relatively narrow. The saturation magnetic torque value is $4.4 \times 1.10-5$ emu, which corresponds to a value of 15.78 Oe. In addition, the coercive force is -0.66 Oe and the remanence magnetic moment is 0.74. It has an emu value of $10^{-5}$. In these values, it has been determined that $Fe_2O_3$ has a soft magnetic property. Soft magnetic materials are used in devices that are exposed to alternative magnetic fields and therefore must have low energy losses. They are commonly used in transformers, electric motors, generators, dynamo and switch circuits.

In Figure 12b, the magnetic hysteresis curve of $ZnxFe_3-xO_4$ thin film seems to be relatively wide. The saturation magnetic moment value is $3.94.10^{-5}$ emu, which corresponds to 1.72 kOe. The coercive force is

-2.98 kOe and the remanence magnetic moment is 3.31. It has an emu value of $10^{-5}$. When the hysteresis curve is taken into consideration, it can be said that the material exhibits hard magnetism and is difficult to demagnetize. One of the most important application areas of hard magnetic materials is motors. In addition, hard magnets are preferred in wireless drills, screwdrivers, automobile windshield wipers, water sprayers, contact circuits, ventilation systems, recorders, clocks. Other applications that benefit from hard magnets include speakers, headphones, and computer hardware in the audio system.

## Conclusion

Undoped and modified with $Zn^{2+}$ hematite films were synthesized by the CSP method at 320°C temperature. Raman shifts (216 $cm^{-1}$, 277 $cm^{-1}$, 383 $cm^{-1}$, 584 $cm^{-1}$, 1272 $cm^{-1}$) of the stretching vibration mode of $Fe_2O_3$ films are seen. There are Raman active states of the hematite phase that these peaks are relatively narrow and severe. In the $ZnxFe_3-xO_4$ compound, the peak of the Raman shift peaks belonging to the hematite phase falls. In addition, the peaks showing the ramping changes of the stretching vibration mode of the $ZnxFe_3-xO_4$ film is due to the presence of multiple phases due to polycrystalline crystal structure and this is consistent with XRD results.

It was demonstrated that the $Zn^{2+}$ doped influenced on the photocatalytic performance of films. The $Zn^{2+}$ doped hematite film exhibited a better photocatalytic performance than undoped hematite film. Hematite has low conductivity and low charge-carrier mobility. In addition, the photoexcited electron–hole pairs have short life time ($\sim 10^{-12}$ s), which makes the hole diffusion length to be also short (2–4 nm). The charge transport properties of hematite can be improved by dopping. The better performance of $Zn^{2+}$ doped film was attributed to

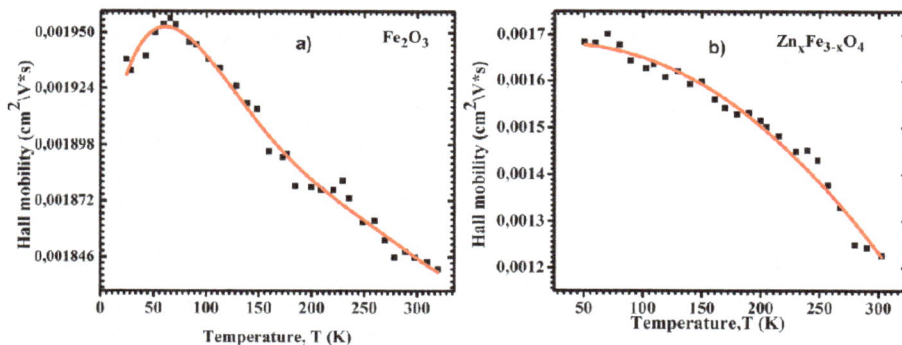

**Figure 11:** Hall Mobility versus temperature for (a) a-$Fe_2O_3$ and (b) $ZnxFe_3-xO_4$.

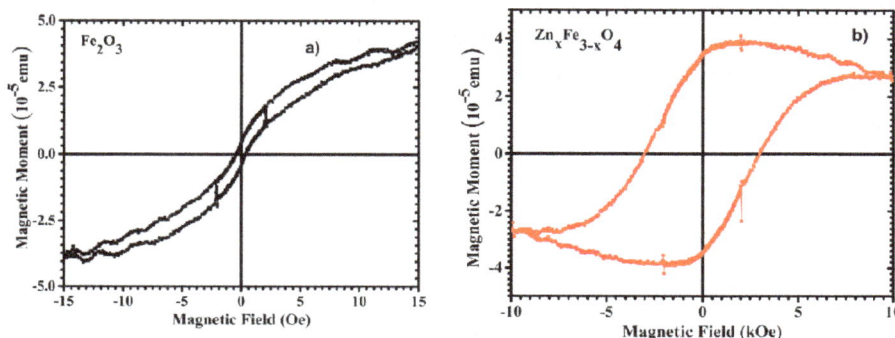

**Figure 12:** M-H curves for (a) $Fe_2O_3$ and (b) $ZnxFe_3-xO_4$ thin films.

their higher carrier density that improved their conductivity. Because the carrier concentration of ZnxFe3-xO4 is $7.2 \times 10^{18}$ and the carrier concentration of $Fe_2O_3$ is $4.2 \times 10^{16}$. So the carrier concentration of ZnxFe3-xO4 is nearly $10^2$ times greater. When we consider the magnetization situation, the hardest magnetization property is $ZnxFe_3$-$xO_4$ film that the magnetic hysteresis curve of $ZnxFe_3$-$xO_4$ thin film seems to be relatively wide. The saturation magnetic moment value is $3.94.10^{-5}$ emu, which corresponds to 1.72 kOe, the softest magnetization feature is $Fe_2O_3$ film that the magnetic hysteresis curve of the $Fe_2O_3$ thin film is observed to be relatively narrow. The saturation magnetic torque value is $4.41.10^{-5}$ emu, which corresponds to a value of 15.78 Oe. It can be said here that Zn, which has no magnetic property, causes pinning which make defects in the structure difficult to move the domains.

Also when we evaluate the films grown using CSP technique in terms of their applications; $Fe_2O_3$ and $ZnxFe_3$-$xO_4$ films are suitable for spintronic applications. Spintronics is an emergent nano technology which deals with spin dependent properties of an electron instead of or in addition to its charge dependent properties. We can see in the results of VSM that the $Fe_2O_3$ compound can control its magnetic properties by dopping with Zn. As a result controlling magnetic properties is very important for spin transistor applications.

## References

1. Hoseinzadeh S, Ghasemiasl R, Bahari A, Ramezani AH (2017) The injection of Ag nanoparticles on surface of WO3 thin film: Enhancedelectrochromic coloration efficiency and switching response. Journal of Materials Science: Materials in Electronics (In Publish).

2. Hoseinzadeh S, Ghasemiasl R, Bahari A, Ramezani AH (2017) n-type WO3 semiconductor as a cathode electrochromic material for ECD devices. Journal of Materials Science: Materials in Electronics (In Publish).

3. Mishra M, Chun DM (2015) α-Fe2O3 as a photocatalytic material: A review. Applied Catalysis A: General 498: 126-141.

4. Sartoretti CJ, Ulmann M, Alexander BD, Augustynski J, Weidenkaff A (2003) Photoelectrochemical oxidation of water at transparent ferric oxide film electrodes. Chem Phys Lett 376: 194-200.

5. Kennedy HJ, Frese WKJ (1978) Photo oxidation of water at α-Fe2O3 electrodes. J Electrochem Soc 125: 709-714.

6. Morin JF (1951) Electrical properties of α-Fe2O3 and α-Fe2O3 containing titanium. Phys Rev 83: 1005-1010.

7. Morin JF (1954) Electrical properties of α-Fe2O3. Phys Rev 93: 1195-1199.

8. Dare-Edwards PM, Goodenough BJ, Hamnett A, Trevellick RP (1983) Electrochemistry and photoelectrochemistry of iron(III) oxide. J Chem Soc Faraday Trans 79: 2027-2041.

9. Sivula K, Formal LF, Grätzel M (2011) Solar Water Splitting: Progress Using Hematite (α- Fe2O3) Photoelectrodes. Chem Sus Chem 4: 432-449.

10. Katz JM, Riha SC, Jeong NC, Martinson ABF, Farha OK, et al. (2012) Toward solar fuels: Water splitting with sunlight and "rust"? Coord Chem Rev 256: 2521-2529.

11. Barroso M, Mesa CA, Pendlebury SR, Cowan AJ, Hisatomi T, et al. (2012) Dynamics of photogenerated holes in surface modified α-Fe2O3 photoanodes for solar water splitting. PNAS 109: 15640-15645.

12. Ni M, Leung MKH, Leung DYC, Sumathy K (2007) A review and recent developments in photocatalytic water-splitting using TiO2 for hydrogen production. Renew Sustain Energy Rev 11: 401-425.

13. Maeda K, Domen K (2007) New Non-Oxide Photocatalysts Designed for Overall Water Splitting under Visible Light. J Phys Chem C 111: 7851-7861.

14. Concepcion JJ, House RL, Papanikolas JM, Meyer TJ (2012) Chemical approaches to artificial photosynthesis. PNAS 109: 15560-15564.

15. Bora DK, Hu Y, Thiess S, Erat S, Feng X, et al. (2013) Between photocatalysis and photosynthesis: Synchrotron spectroscopy methods on molecules and materials for solar hydrogen generation. J Electron Spectrosc 190: 93-105.

16. Sun S, Wang W, Zeng S, Shang M, Zhang L (2010) Preparation of ordered mesoporous Ag/WO3 and its highly efficient degradation of acetaldehyde under visible-light irradiation. J Hazard Mater 178: 427-433.

17. Arabatzis IM, Stergiopoulos T, Andreeva D, Kitova S, Neophytides SG, et al. (2003) Characterization and photocatalytic activity of Au/TiO2 thin films for azo-dye degradation. J Catal 220: 127-135.

18. Sonawane RS, Dongare MK (2006) Sol–gel synthesis of Au/TiO2 thin films for photocatalytic degradation of phenol in sunlight. J Mol Catal A: Chem 243: 68-76.

19. Lin L, Chai Y, Zhao B, Wei W, He D, et al. (2013) Photocatalytic oxidation for degradation of VOCs. Open J Inorg Chem 3: 14-25.

20. Mo J, Zhang Y, Xu Q, Joaquin Lamson J, Zhao R (2009) Photocatalytic purification of volatile organic compounds in indoor air: A literature review. Atmos Environ 43: 2229-2246.

21. Wang W, Soulis J, Yang YJ, Biswas P (2014) Comparison of CO2 photoreduction systems: A review". Aerosol Air Qual Res 14: 533-549.

22. Gasparov LV, Tanner DB, Romero DB, Berger H, Margaritondo G, et al. (2000) Infrared and Raman studies of the verwey transition in magnetite. Phys Rev B: Condens Matter Mater Phys 62: 7939-7944.

23. Chamritski I, Burns G (2005) Infrared-and Raman-active phonons of magnetite, maghemite, and hematite: a computer simulation and spectroscopic study. J Phys Chem B 109: 4965-4968.

24. Bersani D, Lottici PP, Montenero A (1999) Micro-Raman investigation of iron oxide films and powders produced by sol–gel syntheses. J Raman Spectrosc 30: 355-360.

25. Jubb AM, Allen HC (2010) Vibrational spectroscopic characterization of hematite, maghemite, and magnetite thin films produced by vapor deposition applied materials and interface. ACS Appl Mat Inter 2: 2804-2812.

26. Singh JP, Srivastava RC, Agrawal HM, Kumar R (2011) Micro-Raman investigation of nanosized zinc ferrite: effect of crystallite size and fluence of irradiation. J RAMAN Spectrosc 42: 1510-1517.

27. De Faria DLA, Silva SV, De Oliveira MT (1997) Raman microspectroscopy of some iron oxides and oxyhydroxides. J Raman Spectrosc 28: 873-878.

28. Rivero M, Del CA, Mayoral A, Mazario E, Sanchez-Marcos J, et al. (2016) Synthesis and structural characterization of ZnxFe3-xO4 ferrite nanoparticles obtained by an electrochemical method. RSC Adv 6: 40067-40076.

29. Pawar NK, Kajale DD, Patil GE, Wagh VG, Gaikwad VB, et al. (2012) Nanostructured Fe2O3 thick film as an ethanol sensor. International Journal on Smart Sensing and Intelligent Systems 5: 441-457.

30. Yuan HL, Liu E, Yin YL, Zhang W, Wong PKJ, et al. (2016) Enhancement of magnetic moment in ZnxFe3-xO4 thin films with dilute Zn substitution. Appl Phys Lett 108: 232403.

31. Bellido-Aguilar DA, Tofanello A, Souza FL, Furini LN, Constantino CJL (2016) Effect of thermal treatment on solid–solid interface of hematite thin film synthesized by spin-coating deposition solution. Thin Solid Films 604: 28-39.

32. I.E. AGENCY (2013) Key world energy statistics.

33. BP Company (2014) BP statistical review of world energy.

34. Solangi KH, Islam MR, Saidur R, Rahim NA, Fayaz H (2011) A review on global solar energy policy. Renew Sust Energ Rev 15: 2149-2163.

35. Van De Krol R, Gratzel M (2012) Photoelectrochemical hydrogen production. Springer Science.

36. Wang Z, Roberts RR, Naterer GF, Gabriel KS (2012) Comparison of thermochemical, electrolytic, photoelectrolytic and photochemical solar-to-hydrogen production Technologies. Int J Hydrogen Energy 37: 16287-16301.

37. Dincer I (2012) Green methods for hydrogen production. Int J Hydrogen Energy 37: 1954-1971.

# 27

# Remote Patient Monitoring System Based Coap in Wireless Sensor Networks

**Pandesswaran C[1], Surender S[2] and Karthik KV[3]***

*St. Joseph's College of Engineering, Chennai, India*

## Abstract

Patient monitoring is a process of collecting medical parameters of a patient present at a remote location. To deal with this particular issue in a hospital or medical assistance centre, this paper proposes an alternative vision that includes support through internet based on wireless networks and also compatible with existing infrastructure. Most of the existing paper deals with monitoring through sophisticated hardware in a private network and each have its own overheads of switching to a new system model. This paper deals with the enhanced Constrained Application Protocol (CoAP) using multi hop flat topology, which makes the patients being monitored by a central system. It also provides secure communication among the patient nodes by using the public key algorithm. We aim to minimize as well improve implementation method more domestically and secure communication between client and server.

**Keywords:** Patient monitors; Constrained Application Protocol (CoAP); Internet of things; Sensor networks

## Introduction

A wireless sensor network (WSN) of spatially distributed autonomous sensors to monitor physical or environmental conditions, such as temperature, sound, pressure, etc. and to cooperatively pass their data through the network to a central coordinator. The more modern networks are bi-directional, also enabling control of sensor activity. The development of wireless sensor networks was motivated by military applications such as battlefield surveillance; today such networks are used in many industrial and consumer applications, such as industrial process monitoring and control, smart parking system, and so on. The Internet of Things is the network of physical objects accessed through the internet, as defined by technology analysts and visionaries. This patient monitoring system (Figure 1) is an integral part of health care that deals with medication and treatment to inaccessible areas. The current situation indicates the presence of bio telemetry where the hardware involved is complex and expensive. The proposed method is more domesticated where a particular patient can be monitored through mobile phone or computers by a concerned person or a doctor as long as each of them have access to internet and stay connected. It indicates the advantage of Implementing CoAP over existing technical resources. To understand the proposed model, consider a server-client model where it consists of one client and multiple servers. The client end serves as the monitoring mode where each and every server hosts a patient data indigenously. The server end hosts a page on the web that has a two way communication with the server only. The page contains data that are medical parameters in our case and also a set of options or buttons to serve the purpose of the application. The care takers or concerned persons who are relatives and friends of server (patient) are given access to view the parameters. The information palette can be designed or modified in their choice or knowledge. The hospital end (clients) will have more detailed information and can even control or adjust the equipment manually or through an automated algorithm by the connected actuators with the internet.

The goal of this work is to:

- Create a wireless topology with one client and multiple servers.

- Establish link between each server with client.

- Servers can be accessed via Copper Web Browser plugin through which sensor data can be monitored.

- Each server is provided with separate URL; Client should be able to log onto the URL of particular server and enter his/her patient number and acquire necessary details from his platform.

- Sensors attached to the patient's body provides the information about the parameters in healthcare.

- As per the request from the client end, information is displayed.

- When implemented, the request by doctor, then information can also be displayed in his/her mobile devices, thus reducing the time spent for manual diagnosis.

The contributions in this paper are: A simulative evaluation of the proposed approach. The remainder of this paper is organized as follows. The next section discusses works related to CoAP (Figure 2) implementation in health care. Section 2 describes the proposed work of the paper and section 4 shows the evaluation methodology that is protocol to be used along with the simulated environment. The results are presented and discussed in Section 5. Section 6 concludes the paper.

## Related Work

Internet of things made complicated functions very simple, leading to formation of more astute environment and smart appliances making a better and safe place to live in. Some previous works was based on an easy web interface that provides information about temperature, humidity and led status of sensors [1]. Other works is done on various applications like car parking which was implemented through COOJA simulator to offer parking facilities will get cars off the street and into parking spaces sooner thus contributing to congestion control in highly congested urban areas [2]. In order to develop a low power efficient

***Corresponding author:** Karthik KV, St. Joseph's College of Engineering, Chennai, Tamil Nadu, India, E-mail: vijaykarthikindia@gmail.com

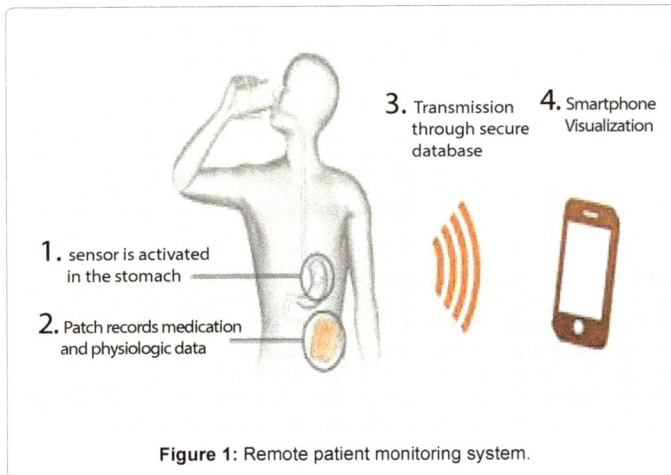

**Figure 1:** Remote patient monitoring system.

CoAP system works have been carried out as implementation for CONTIKI OS that leverages a generic radio duty cycling mechanism to achieve high energy efficiency [3]. The encryption technique between the client and server nodes provides secured data transfer between them [4].

In this paper [5], the implementation works in wireless sensor networks and not in mobile networks. The data transmitted and received during the process lacks security. So, Data can be accessed by irrelevant people. This method lacks privacy. This paper also gives an idea to establish communication between different motes. In this paper [6], the process works both in mobile networks as well as wireless sensor networks but lacks secure data transmission and communication between different servers or motes.

In this paper [2], the method can be implemented in mobile networks as well as wireless sensor networks but lacks data security and remote monitoring services. This paper establishes communication between various motes and it is implemented in CONTIKI OS. In this paper [3], the process is implemented without data security and remote monitoring services. From these above papers, the demerits of data insecurity and communication between motes can be rectified in order to implement the secured remote monitoring process.

This is done by improving the monitoring process to be carried in mobile networks and data security between client and server is established to provide privacy from others. The mobile monitoring is established by including CoAP protocol in browsers using COPPER plugin. The data security is established by using secure public key encryption technique to provide secured data transmission between different motes.

The already existing protocols such as REST, 6LoWPAN, and CoAP are implemented in the LINUX based CONTIKI environment with extended data security options that can readily be downloaded to commercially supported Contiki motes. These proposed features highlights the future possibility in a remote patient monitoring system. The achieved results showcases the easier implementation and simulations without involving specific hardware requirements [7]. They not only monitor the patient parameters but also provide effective indications and diagnostic solutions to the situation present

This work is done on Implementation of CoAP and its Application in Transport Logistics highlighting the use of the CoAP protocol for the retrieval of sensor data during land or sea transportation. Thus these early works are in interest for the betterment of environment.

## Overview of CoAP

### Constrained Application Protocol (CoAP)

It is a software protocol intended to be used in very simple electronics devices that allows them to communicate interactively over the Internet. It is particularly targeted for small low power sensors, switches, valves and similar components that need to be controlled or supervised remotely, through standard Internet networks [8]. Therefore, efficiency is very important. CoAP can run on most devices that support UDP or a UDP analogue.

Message types involved in CoAP are Confirmable, Non confirmable, Acknowledgement, Reset messages. Confirmable requires Acknowledgement whereas Non Confirmable doesn't require Acknowledgment. Reset message indicates missing of few contexts.

- GET: The GET method helps in retrieving information from that of server.

- POST: The POST method requests the server to provide data for the user.

- PUT: The PUT method helps in updating or creating the information in the corresponding URI.

- DELETE: The DELETE method helps in erasing the information stored.

The usage of CoAP in patient monitoring application is to form a client-server relationship, whereas Doctors use the browser as client and the server program runs in the GUI for updating the browser dynamically. Thus the Method definitions are used efficiently for remote monitoring application

## Proposed Architecture for Remote Patient Monitoring Application

### System architecture

As per Figure 3, Patient is monitored by the sensors mounted on them. Servers will host the information on the network. The server collects the data from the sensors. On the other side, Doctors are acting as clients. If doctors request server for data, the CoAP protocol helps to connect server with client to display the current status and check for variation of sensor values [9]. If there is any variation, Doctors can give medication indicating emergency situation for the patient. This medication is done in browser using CoAP Protocol. The data of patient is stored in some data storage device for future purposes.

In multi-hop flat wireless topology, there are one or more intermediate nodes along the path that receive and forward packets via wireless links. During transmission and reception in multi-hop networks, nodes communicate with each other using wireless channels and do not have the need for common infrastructure or centralized control.

The above Figure 4 explains the Multi-hop topology with Border router, Client and server.

**Server:** A Restful server shows how to use the REST layer to develop server-side applications.

**Border router:** Border Router keeps radio turned on. Enabling of it helps in connection between that of client as well as server to that of CoAP web Address. Border-Router [10] has the same stack and fits into mote memory.

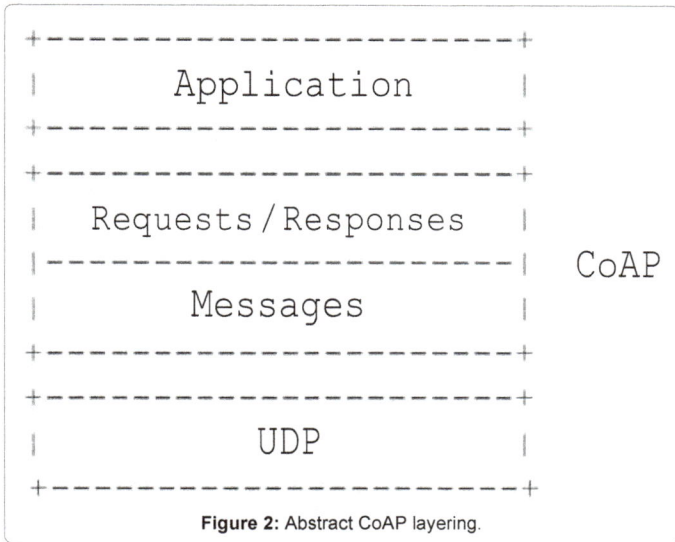

**Figure 2:** Abstract CoAP layering.

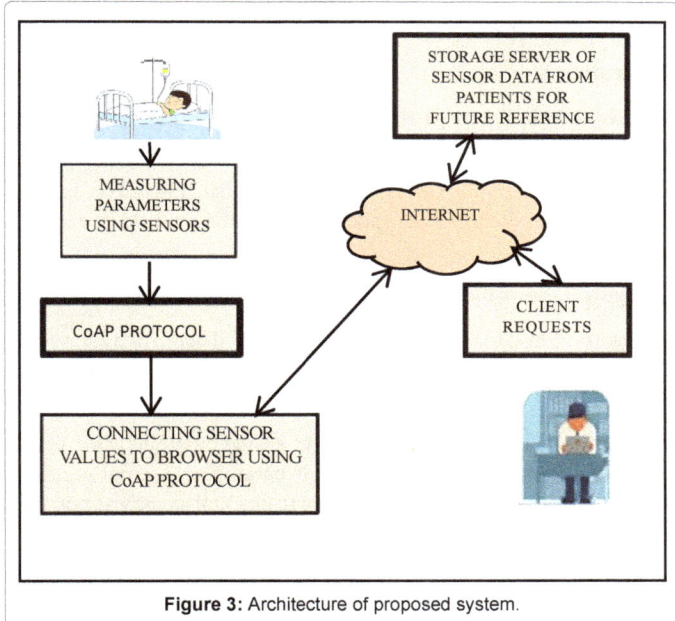

**Figure 3:** Architecture of proposed system.

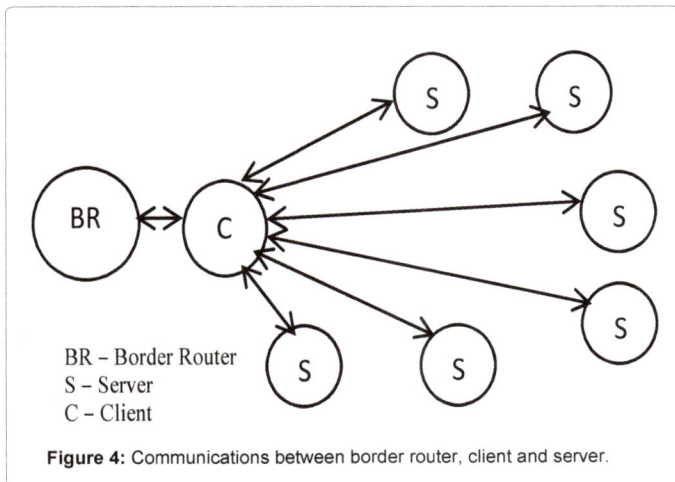

BR – Border Router
S – Server
C – Client

**Figure 4:** Communications between border router, client and server.

**Client:** A CoAP client cycles through four resources on event detection such as GET, PUT, PUSH, and DELETE. It gets connected to the server via multihop topology.

The communication between server and clients are of multi hop fashion. The data from the motes are sent to that of Copper Web Browser which can be viewed by the user.

### Flow chart of client operation in browser

From Figure 5, Flowchart explains the methodology to carry out this particular application. The address of the server is typed in the URL using a copper plugin which runs on top of the http. The hosted servers responds to the requests only after permission of access is granted.

The hosted webpage displays the status of the patient along with the sensor values that has been under operation. The obtained values can also be stored on the cloud or offline for future references. The collected values are compared with the standard or normal diagnostic values to compute an action. Then, a particular or a group of parameters can be modified to stimulate an action in order to bring the patient condition to normal or to the require state.

### Data security

The conversation between sensors nodes and client can be eavesdropped by the adversary. The adversary can be aware of the conversation between the sensors and can forge the data. Security is the main pre concern to socialize this network for common usage. The goal of security services in WSNs is to protect the information and resources from attacks.

The security in sensor network will be employed by public.

Key cryptography because it is easy to distribute keys in public key cryptography than symmetric key cryptography.

This is because of the random deployment of the sensor nodes in the network.

Thus we purpose, Public key algorithm using Elliptic Curve Cryptography algorithm for preventing replay attack in sensor network as well as for data confidentiality and authentication between patient nodes and client.

### Simulation Output

Constrained Application Protocol is simulated via COOJA simulator. COOJA simulator is a java based simulator and it is also called as "cross level simulator".

The simulation can be done via the terminal as follows,

- Open terminal to connect the client with server using the command, $ Make connect-router-cooja.

- This command is included in the same border router environment to bridge the connection using CoAP protocol.

Figure 6 will give the bridging terminal window between client and server.

Figure 7 gives the idea of connection exhibiting between different servers of patients with client of doctor being at another mote. The motes arrangement can be viewed in COOJA simulation window.

Figure 8 gives the final output can be viewed in Browser with a Mozilla Firefox add-on called Copper. Patient is a resource added to the left side window of the browser in the address bar **coap://**

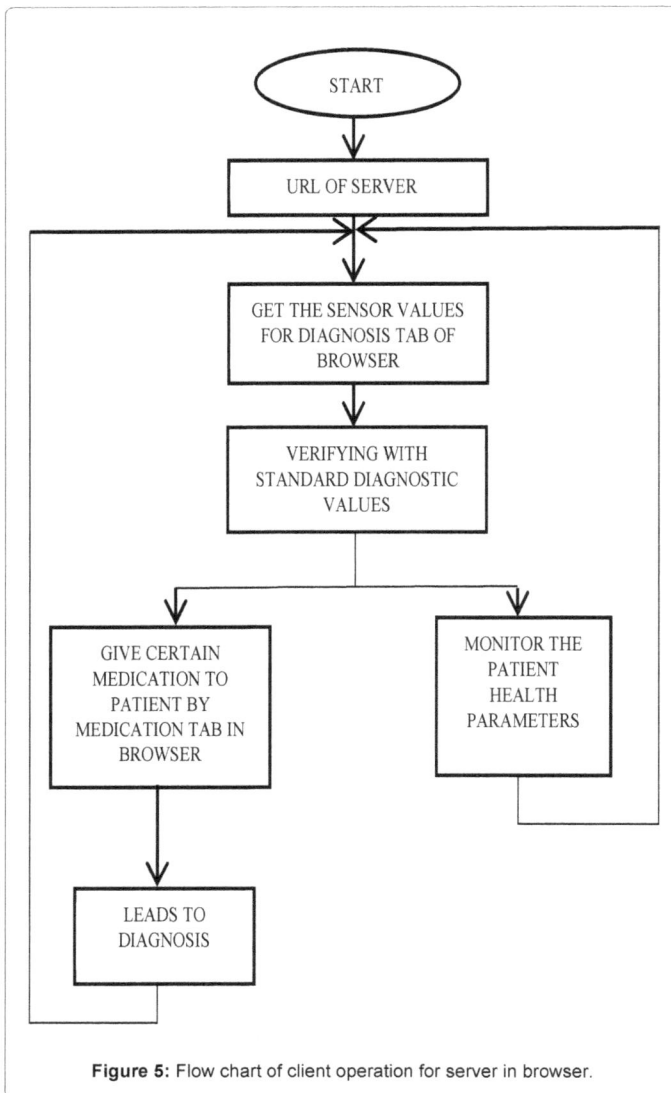

**Figure 5:** Flow chart of client operation for server in browser.

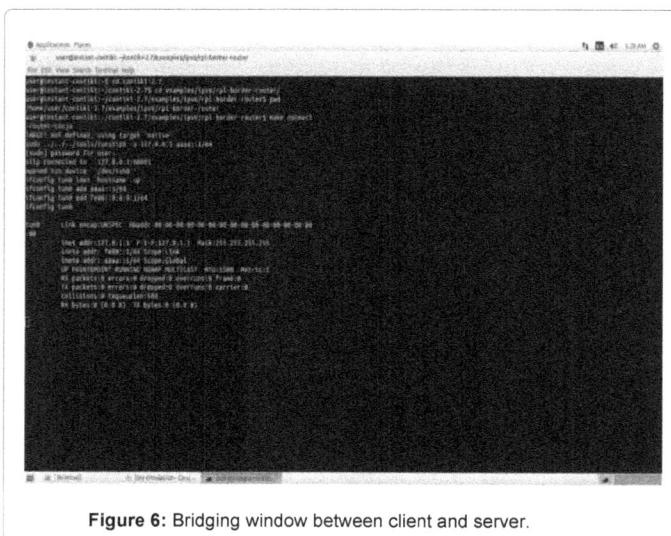

**Figure 6:** Bridging window between client and server.

[**aaaa::212:7402:2:202**]/connects client to the particular server [11].

This shows the process done in Mozilla Firefox browser:

- Discovery–calls the resources included in the server.
- Ping–Tab helps to check the connectivity of the server and client.
- GET, POST, PUT, DELETE are the Method definitions which gets in to effect only after Pinging.

From Figure 9, it shows the sub-resources added to the browser in order to perform varied operation in server end.

### Diagnosis

The diagnosis function is performed, when client wants to read the sensors data from server. In this, client transmits command to the server to read the data from sensors.

On the other hand, server gathers the data from sensors and transmits it back to client. The transmission of command from client to server and reception of data from server to client is done using CoAP protocol. Depending on the data received, clients or doctors recommend medication to the patients.

### Medication

Used to indicate the patient to go to nearest hospital by actuator.

This medication helps patient to identify some changes in their body conditions and corresponding remedial measures can be implemented by doctors in hospitals based on the sensor data received from the servers.

Figure 9 shows the browser screen indicating the Sensor Values,

Figure 10 indicates any change in standard diagnostic values leads medication to be done in order to prevent patient from detrimental effect of his/her health.

Figure 11 indicates change in patient's body by the actuator signal, this is done to aware the patient that is parameters are not accurate. This is the indication to patient to move to hospital as early as possible.

Figure 12 shows that if every parameters are within range then the browser will indicate to the client that the patient is normal and further investigation of that particular patient is not required.

From these above simulation results, we can ensure that Patient's diagnosis evaluation will go to its pinnacle point.

### Conclusion and Future Work

We presented our low-power CoAP implementation for Contiki that leverages a generic radio duty cycling mechanism to achieve a high energy efficiency on a COOJA simulator. We experimentally evaluated our implementation in a multihop network and showed that the use of a duty cycle results in a low power consumption, at the cost of a higher latency. Our protocol-independent REST Engine provides an abstraction to create RESTful Web services. In future work, we plan to evaluate the possibilities and limitations of the RESTful approach towards the low cost microcontrollers such as Arduino family, MSP Launchpads and other ARM processors through construction of custom motes and bring in more features of contiki like how the IP-based IoT performs in terms of latency, reliability, and battery-lifetime.

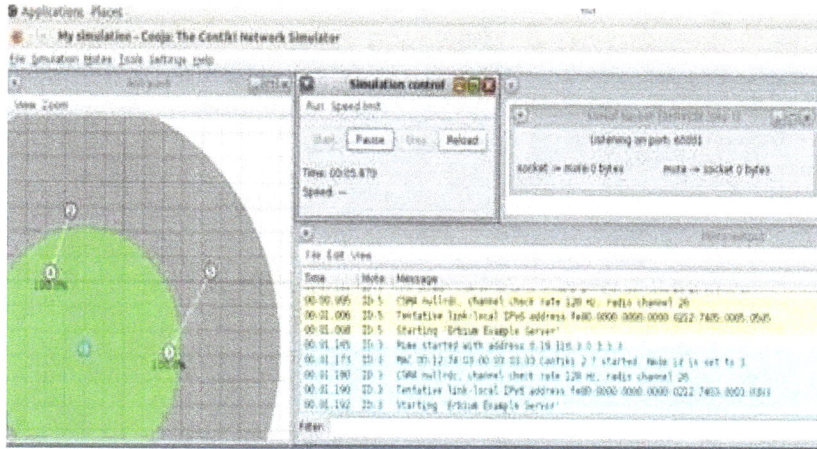

**Figure 7:** Motes connection representing server and client.

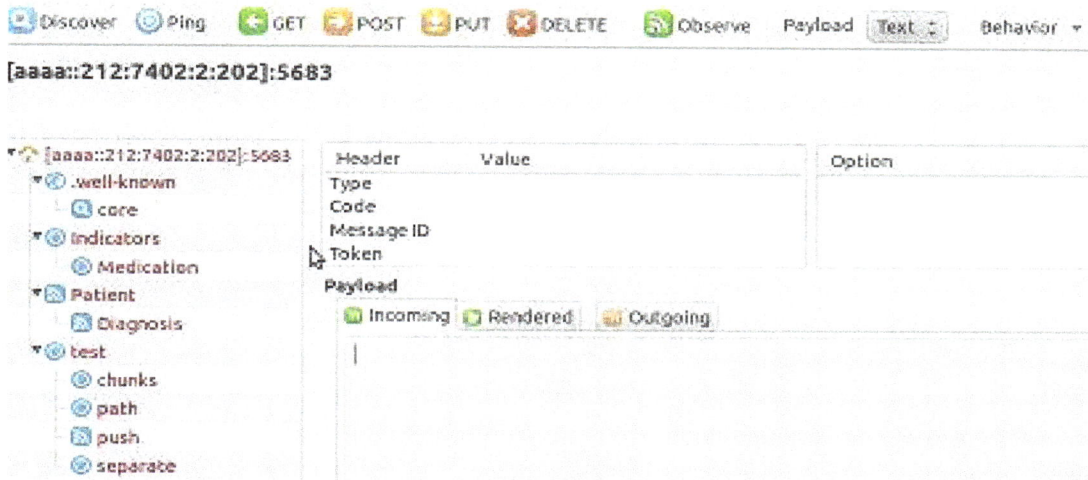

**Figure 8:** CoAP interface in browser.

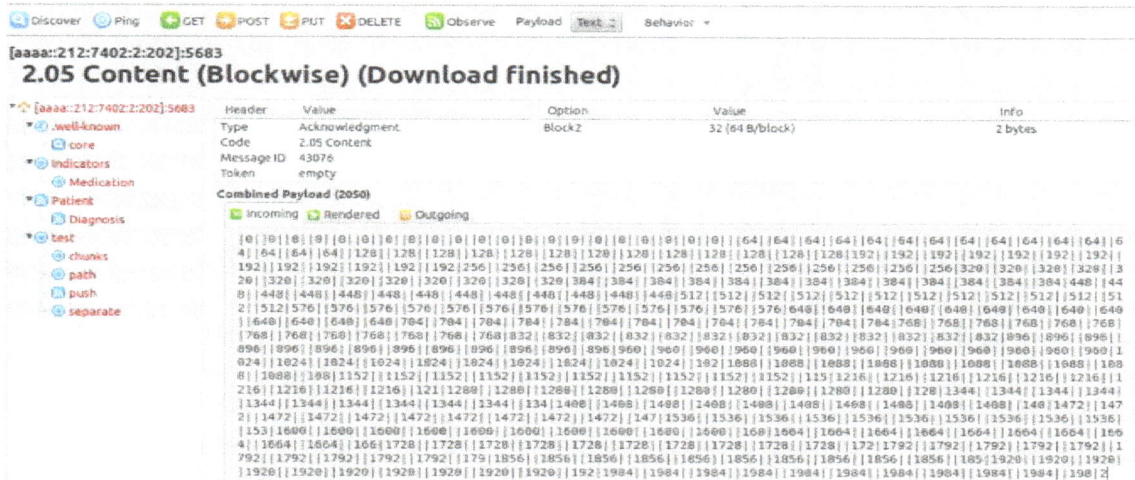

**Figure 9:** Browser screen indicating the sensor values.

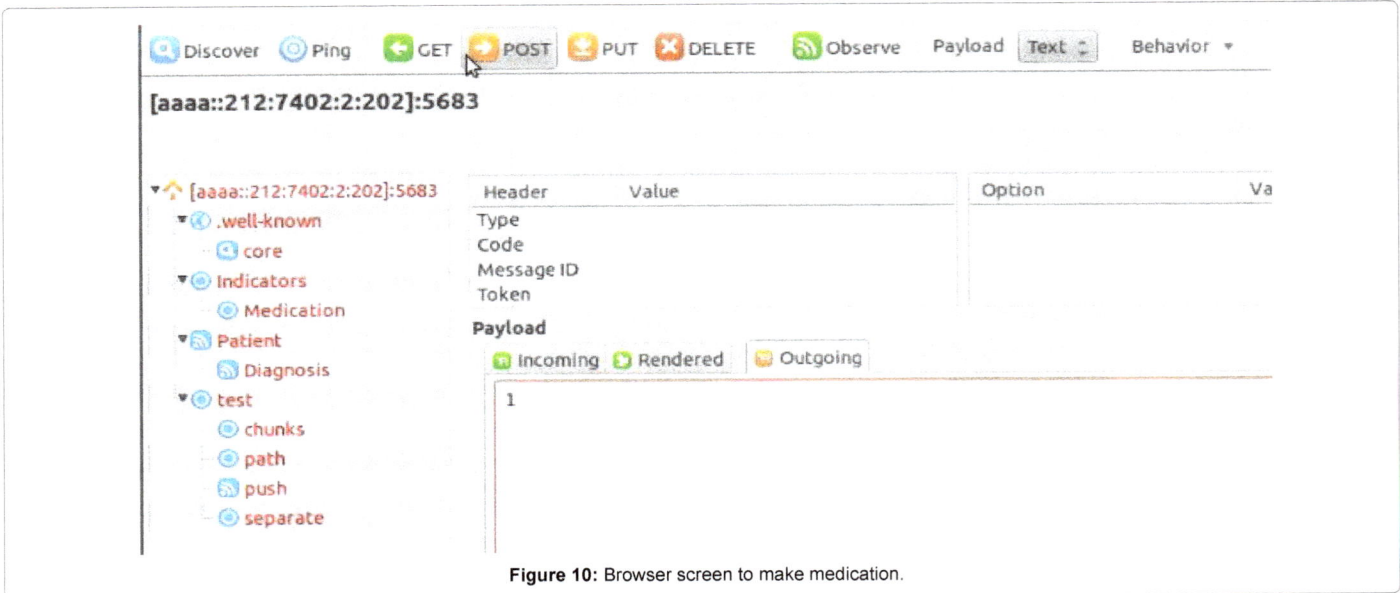

**Figure 10:** Browser screen to make medication.

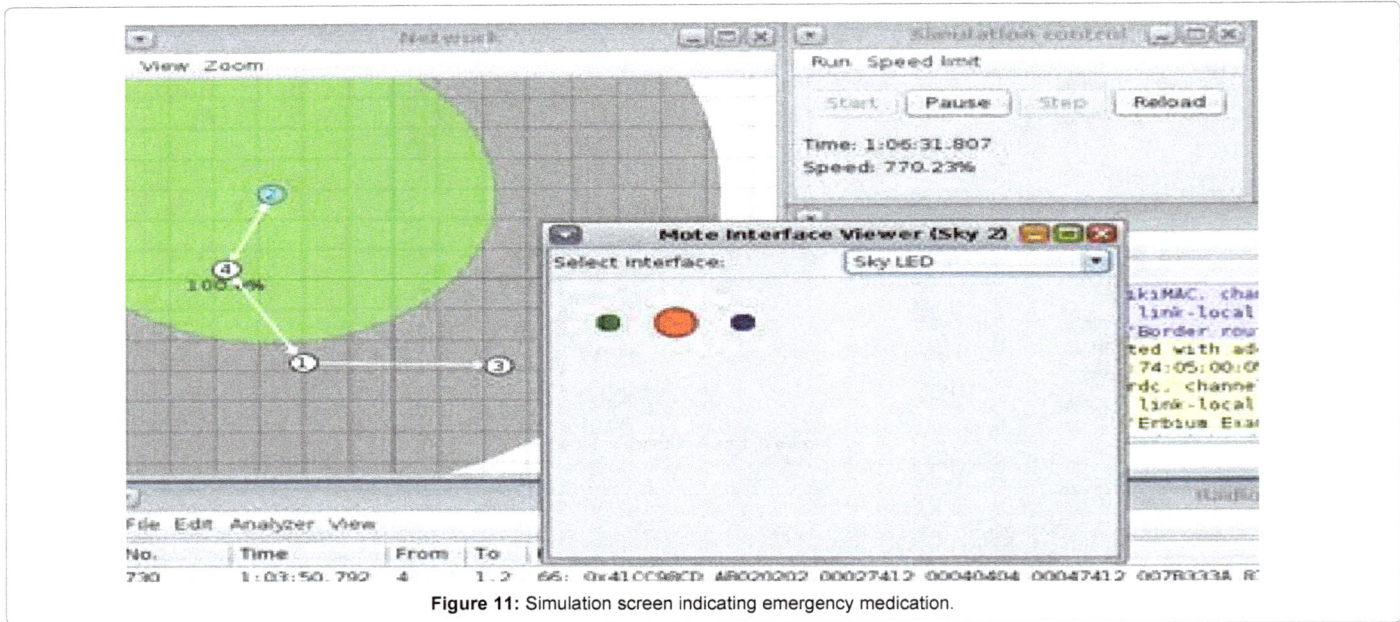

**Figure 11:** Simulation screen indicating emergency medication.

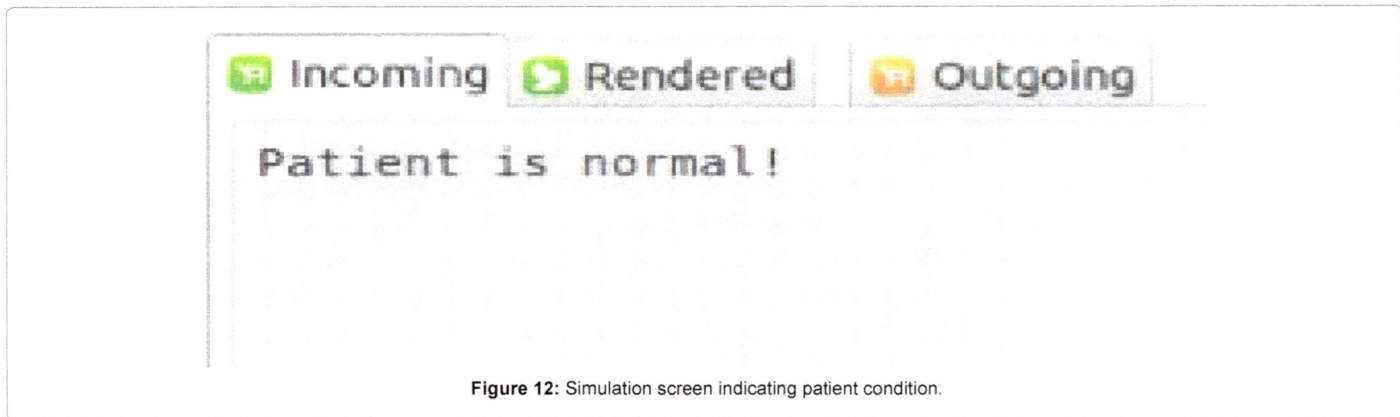

**Figure 12:** Simulation screen indicating patient condition.

## References

1. Raptopoulou E (2014) "CoAP-enabled Sensors for the Internet-of-Things" Department of Applied Informatics & Multimedia, Technological Educational Institute of Crete.

2. Aarthi R, Renold AR (2014) Coap Based Acute Parking Lot Monitoring System Using Sensor Networks. ICTACT J Commun Technol 5.

3. Kovatsch M (2013) CoAP for the web of things: from tiny resource-constrained devices to the web browser", Proceedings of ACM conference on Pervasive and ubiquitous computing adjunct publication. ACM pp: 1495-1504.

4. Villaverde BC, Pesch D, Alberola RDP, Fedor S, Boubekeur M (2012) Constrained application protocol for low power embedded networks: A survey. IEEE Xplore pp: 702-707.

5. Khattak HA, Ruta M, Sciascio E (2014) CoAP-based healthcare sensor networks: A survey. IEEE Xplore pp: 499-503.

6. Catarinucci L, Donno DD, Palano L (2015) An IoT-Aware Architecture for Smart Healthcare Systems. IEEE Internet of Things Journal 2: 515-526.

7. Karimi K, Atkinson G (2013) What the Internet of Things (IoT) Needs to Become a Reality. White Paper, FreeScale and Arm.

8. Dimcic T, Krco S, Gligoric N (2012) CoAP (Constrained Application Protocol) implementation in M2M Environmental Monitoring System. E-society Journal 21.

9. Chander RPV, Elias S, Shivashankar S, Manoj P (2012) A REST based design for Web of Things in smart environments. International Conference on Parallel Distributed and Grid Computing pp: 337-342.

10. Shelby Z, Hartke K, Bormann C, Frank B (2013) Constrained Application Protocol (CoAP). Corecoap pp: 1-118.

11. Sunkari V (2015) Framework for Providing Security and Energy Saving Using Elliptic Curve Cryptography in Wireless Sensor Networks. IJCSIT 6: 3604-3608.

# 28

# A Clinical Trial of Translation of Evidence Based Interventions to Mobile Tablets and Illness Specific Internet Sites

Carol E Smith[1,2], Ubolrat Piamjariyakul[1]*, Marilyn Werkowitch[1], Donna Macan Yadrich[1], Noreen Thompson[1], Dedrick Hooper[3] and Eve-Lynn Nelson[3-5]

[1]School of Nursing, University of Kansas Medical Center, USA
[2]School of Preventive Medicine and Public Health, University of Kansas Medical Center, USA
[3]Center for Telemedicine and Telehealth, University of Kansas Medical Center, USA
[4]Pediatrics and Telemedicine, University of Kansas Medical Center, USA
[5]Institute for Community Engagement, University of Kansas Medical Center, USA

## Abstract

This article describes a method to translate an evidence based health care intervention to the mobile environment. This translation assisted patient participants to: avoid life threatening infections; monitor emotions and fatigue; keep involved in healthy activities. The mobile technology also decreased costs by reducing for example travel to visit health care providers. Testing of this translation method and its use by comparison groups of patients adds to the knowledge base for assessing technology for its impact on health outcome measures. The challenges and workflow of designing materials for the mobile format are described.

Transitioning clinical trial verified interventions, previously provided in person to patients, onto tablet and internet platforms is an important process that must be evaluated. In this study, our evidence based guide's intravenous (IV) homeCare interventions (IVhomeCare) were delivered via Apple iPad mini™ tablet audiovisual instruction / discussion sessions and on a website. Each iPad audiovisual session (n = 41), included three to five families, a mental health specialist, and healthcare professionals. Patients and their family caregivers readily learned to use the wireless mobile tablets, and the IVhomeCare interventions, as described here, were successfully translated onto these mobile technology platforms. Using Likert scale responses on a questionnaire (1 = not helpful and 5 = very helpful) participants indicated that they gained problem solving skills for home care through iPad group discussion (M = 4.60, SD = 0.60). The firewall protected videoconferencing in real time with multiple healthcare professionals effectively allowed health history taking and visual inspection of the patient's IV insertion site for signs of infection. Supportive interactions with peer families on videoconferencing were documented during discussions. Discussion topics included low moods, fatigue, infection worry, how to maintain independence, and need for support from others with their same lifelong IV experiences. The visual family interactions, discussions with professionals, and the iPad internet links were highly rated. Mobile distance care delivery can result in saved time and money for both healthcare professionals and families.

**Keywords:** Mobile technology; Intravenous catheter care; Group sessions; Translating to iPad and internet platforms

## Background and Significance

Patients depending on daily intravenous (IV) catheter home care have a complicated medical regimen requiring significant time, energy, and resources [1]. These families have ongoing fears about risks of infection, hospitalization, and deteriorating health and function, as well as worries about finances and coverage to meet health care needs [2-4]. It is often challenging for patients and family caregivers to understand and adhere to the twice daily, multi-step IV therapy procedures [5,6]. Stresses associated with the regimen and the day-to-day demands of employment and family life place pressure on family members who may neglect their own physical and mental health. Notably, we have developed and validated a series of interventions that have reduce patients' IV infections, depression and fatigue and improve their families access to information and illustrations of homecare problem solving and healthy living activities [7-10]. The interventions include: infection, depression, and fatigue prevention; problem solving IVhomeCare with health professionals; and maintaining family health [11,12]. These and other interventions have previously been delivered in person, by telephone, and over internet [13-15].

## Related Work

In previously validated interventions, patients and family caregivers managing complex and lifelong daily IV homecare needs have responded well to having access to visual instructions [16,17]. Support from peers, reinforcement of healthy living activities, and illustrated step-by-step guides improved safety and increased adherence to complex care at home [18]. Annual re-hospitalizations of patients on IVs result from complications that could be reduced using our research based interventions [19]. Specific IV homecare interventions (IVhomeCare) can be used daily by patients in their homes with professionals providing directions and clarifying IV procedures [20]. Professionals can also reinforce adherence to daily IV aseptic care routines that can reduce catheter infections and complications [21]. We have previously published, and other research has replicated our results indicating that our IVhomeCare interventions for infection prevention and mood and emotion monitoring have resulted in reducing incidents of IV catheter infections and decreasing clinically significant depression [22-24]. Also our intervention for problem solving partnerships with healthcare

*Corresponding author: Ubolrat Piamjariyakul, School of Nursing, University of Kansas Medical Center, USA, E-mail: upiamjariyakul@kumc.edu

professionals and for short daily restorative naps result in greater patient problem-solving skills, reduced daytime fatigue, and improved quality of life [20]. Overall, these interventions have been associated with family success stories and lower healthcare costs [25]. To build on these outcomes and enhance interactive communication, we translated these in-person interventions to mobile technology platforms (example: tablets) with asynchronous Internet-based components available 24/7.

A number of advantages have been suggested for increased interest in mobile technologies in health care, including fewer families having a landline telephone connection, increasing use of mobile devices rather than desk or laptop computers, near ubiquitous wireless coverage with standardized policies for data security, availability even when traveling and ease of use of mobile devices, and improved clarity and detail of visual images and video [26]. Our patient survey results overwhelmingly indicated interest in mobile technologies rather than landline phones or desktop computers [16,27]. The Pew Research Center survey found that 72% of adults in the U.S. reported they seek information, care, or support from a healthcare professional via the Internet [28,29]. Thus, based on this patient and family input and with our focus on further extending our family-centered approach and meeting patient needs in the home, we translated our previously validated interventions onto mobile technology (i.e., Apple iPad mini™ tablets). The figure and tables herein describe the translated interventions and illustrate the work flow related to uploading and delivering the important information via mobile devise.

Using tablets equipped with secure encryption-based videoconferencing, patients and their family members / caregivers reliably connected are from their homes to healthcare professionals in their offices [30,31]. Further, the videoconference discussion sessions allowed the families and the professionals to interact with one another in real-time [32]. These telehealth technologies can be used to engage families in evidence-based interventions for IV catheter care, thus reducing care burden in home settings [33-35]. Our IVhomeCare interventions both provide peer support and address families who lack access to healthcare professional guidance and information about serious but preventable IV problems at home [36].

## Objectives

The objectives of this study were to: (1) Translate IVhomeCare and healthy living interventions and deliver these via synchronous videoconferencing family group sessions through mobile tablets. (2) Offer IVhomeCare information via asynchronous Internet available 24/7. (3) Summarize the patient and family evaluations of the videoconferencing sessions with health professionals and other patients.

## Methods

The university medical center's Institutional Review Board (IRB) approved the study procedures, following all IRB data management policies and Health Insurance Portability and Accountability Act (HIPAA) regulations for the iPad group intervention sessions and data collection. All participating patients and caregivers provided informed consent. Our clinical trial compared IV patients randomly assigned to either (1) the iPad comparison attention placebo group (without iPad appointments or our Internet site access) or (2) the experimental group who had iPad appointments and who were also provided with a translated mobile access interventions via iPad. Following the study, the comparison placebo group patients / family members were given the Internet access and one clinic appointment with professionals

to evaluate. Patients and family caregivers who attended the group discussion sessions via the iPad mini were invited to evaluate the program.

The methods for translating the evidence-based IVhomeCare intervention components included: (1) delivering synchronous group videoconferencing sessions via the tablets; and (2) uploading the asynchronous IVhomeCare interventions (written information, forms, illustrations and graphics) to the mobile device Internet links. These two intervention approaches provided information and professional guidance to promote and support independent self-management for complex IV infusion care at home and to provide evidence-based information available 24/7. The audiovisual group discussion among peers and professionals was conducted on encrypted and firewall-protected videoconference software to ensure privacy and confidentiality. The technology used to deliver the intervention components is a wireless mobile tablet (iPad mini) with unlimited 4G data plan, which was loaned to each family at no cost.

All tablets were put in hard cover stands that not only protected the tablet, but also allowed the tablets to be placed on tables or other surfaces rather than users holding them. This eliminated the iPad pixel blur during the video sessions, which sometimes results in unclear video. The tablet, power supply, and hard cover stand were all shipped to the user in a Fed Ex tablet box, along with a how-to use guide and shipping instructions on how to return the tablet to the System Coordinator at no cost to the user.

Our procedures were developed to give the end user (study participant) a secure, high quality, and user-friendly videoconferencing experience. While security and high quality are critical components, the ease-of-use is equally important to ensure user satisfaction and increase success of the project.

### Delivering the synchronous videoconferencing sessions

Tablets offer visual patient monitoring by healthcare professionals, allowing early detection to avoid IV sepsis. Such visual monitoring can prevent the frequent emergency room visits or hospitalizations that may result when early signs of problems are missed with non-visual telephone calls. Mobile connections have the added advantage of not exposing patients to infection risks from hospitals or clinic visits; as this group of patients is especially vulnerable to bloodstream infections [37].

Each imaged iPad tablet with the transferred intervention materials was listed in our study record for tracking purposes. Recorded log data included the assigned tablet name, serial number (SN), Mobile Equipment Identifier (MEID), Integrated Circuit Card Number (ICCID), and Cellular Data Number. The iCloud username and password were also recorded in case the tablet needed to be disabled, if lost or stolen. Upon return, the Systems Coordinator noted the study log of its successful return as well as assessed the tablet, power cord, and hardcover stand for any damage. The coordinator also disinfected the tablet using the cleaning protocols established by Apple. After completing updates, the Systems Coordinator recorded any applications installed to tablets by the user into the study log and then deleted all data by selecting "Erase all Content and Settings" on the tablet.

### Communications technology

Our Network engineer and system coordinator developed an approach for setting up each iPad prior to giving to the end user (our study participant patient) and once the iPad was returned from the patient the medical center and apple directions (Figure 1). The Polycom™ RealPresence software was pre-loaded on tablet devices using all best practices for HIPAA-compliant, encrypted audio-visual delivery in order to securely connect to the videoconferencing sessions omg. With the participants' full knowledge and consent, the sessions were videotaped for later review. Families were not required to have previous experience using the iPad mini. An iPad user help guide was uploaded onto each iPad mini, and a printed copy was also shipped to homes along with the iPad. In addition, our telehealth systems coordinator was available by phone to support families' iPad usage.

Our project allowed users to connect via videoconference to a fully encrypted Multipoint Conferencing Unit (MCU). This was accomplished by allowing an iPad to become an endpoint that is internal to the University of Kansas Medical Center (KUMC) network. This network resides behind a firewall for privacy. The Polycom RealPresence mobile application, also called the Polycom app, was available at no cost via the Apple app store, and was loaded on to the iPads. The KUMC Network Engineer created a unique username and password for each individual tablet. The flowchart in Table 1 details each of these activities followed for managing the iPad technology that allowed the tablets access to the KUMC network via its external Video Border Proxy (VBP). The VBP was purchased for the study to provide encryption to video devices external to the KUMC network. The VBP was given a URL registration point to which external units registered. It was only when the Polycom app was registered to the VBP that the conference was encrypted. Therefore, the Systems Coordinator set the Polycom app to auto-register every time it was opened, saving the users (our study participant patients) the burden of registering each time. The VBP also allowed the user access to the application's "Directory," which the technician used to store the dial-in information in order to connect to the MCU. The user would simply open the Polycom app, access the Directory, and call the only entry available (Figure 1). During the study our coordinator was contacted by one-third of the patients, all of whom wanted practice in connecting and using the iPad within a week of receiving it. The remaining patients connected per our graphic instructions.

**Scheduling and establishing the connection:** The iPad mobile group sessions consisted of pre-scheduled iPad videoconferencing appointments with patients and their family members and healthcare professionals. Prior to the session, the coordinator completed test connections with each family resolve any connection problems or difficulties with adequate lighting. On the day of the session, the IT expert connected all families with the health professional interventionists. All participants appeared on screen in small separate picture tiles so that all were able to view each other and the facilitators on the tablet screen.

**iPad session training and procedures:** Before beginning the intervention, the professionals were trained as facilitators to: (1) maintain the communication to ensure participation and confidentiality; (2) focus the discussion to identify patient-centered problems and concerns; (3) enhance problem-solving partnerships with health professionals; (4) encourage families to share their experiences and offer one another support; (5) clarify misinformation shared by families; and (6) encourage families to discuss questions and concerns with their primary healthcare providers.

**Mobile group sessions:** Sessions were facilitated by an interventionist with extensive IV homecare experience and a mental health clinical specialist. This specialist both facilitated the discussion and addressed the depression often found among these homes bound

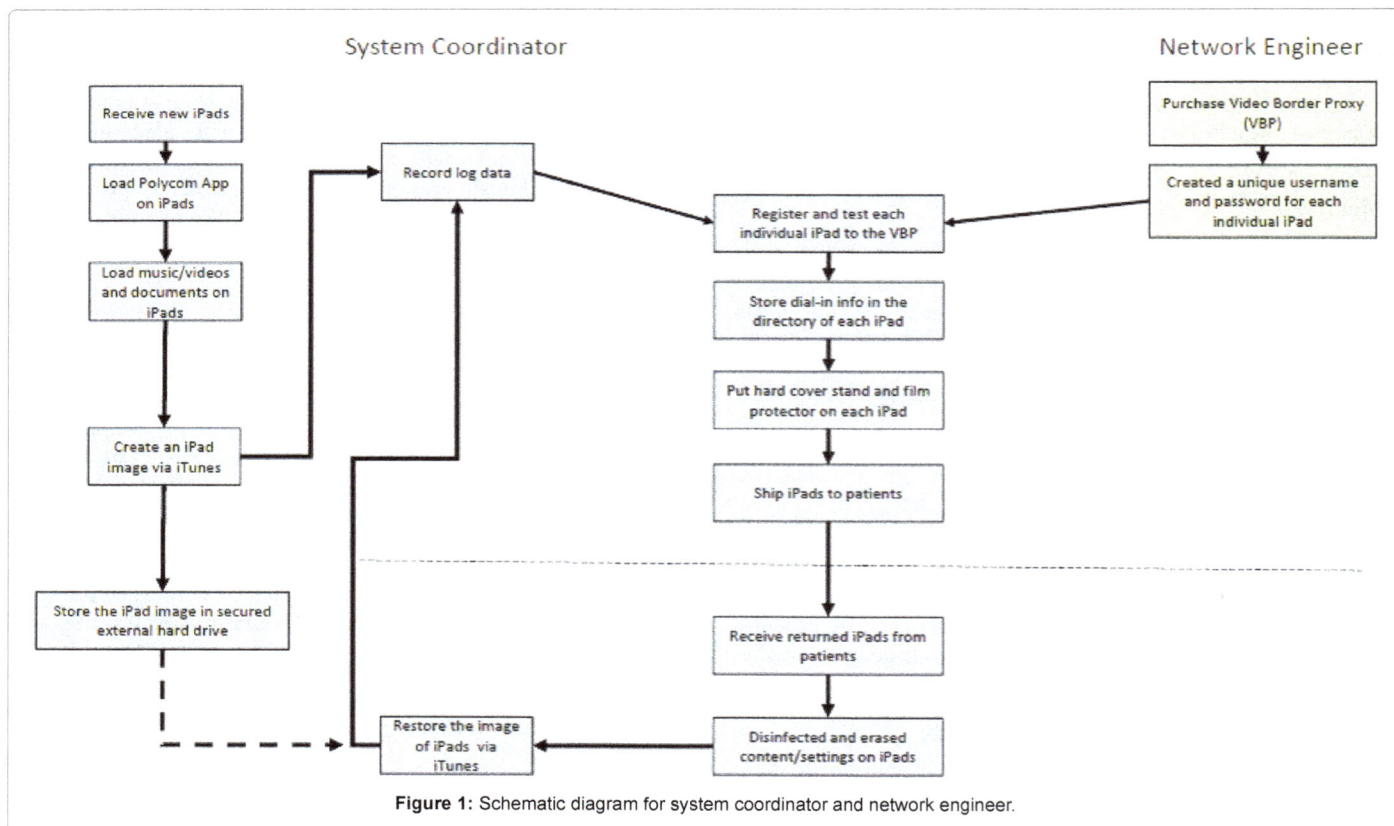

**Figure 1:** Schematic diagram for system coordinator and network engineer.

| Flow chart for uploading and delivering the translated intervention content via iPad |
|---|
| KUMC Network Engineer created a unique username and password for each individual tablet. This allowed the tablet to access the KUMC network via its external Video Border Proxy (VBP). The VBP was purchased for the study in order to provide encryption to video devices external to the KUMC network. |
| After received new iPads, System Coordinator used two iPads to create templates for the two research groups: "intervention" and "control". All educational documents were created in HTML format and placed on KUMC web server. Music files and training videos were loaded on iPads. On the iPad homepage, these HTML documents were grouped and named "HPN Resources" in a folder, which allowed study participants to access these documents. |
| System Coordinator also loaded the Polycom App on iPads, via a free Real Presence app from the Apple App store, and recorded log data:<br>1. Assigned tablet name<br>2. Serial Number (SN)<br>3. Mobile Equipment Identifier (MEID)<br>4. Integrated Circuit Card Number (ICCID)<br>5. Cellular Data Number<br>6. iCloud username and password |
| The two iPads were loaded with different resources for "intervention" and "control" groups. Next they were created with different iPad images via iTunes. Both image files were stored in the secure external hard drive for restoring iPads later. |
| System Coordinator restored the images via iTunes on new iPads for new study participants. |
| System Coordinator registered and tested each individual iPad to the VBP. He set the app to auto-register every time the Polycom application was opened, thus saving the user the burden of registering each time. The VBP also allowed the user access to the application's "Directory," which the coordinator used to store the dial in information in order to connect to the MCU (Multipoint Conferencing Unit). The user could then easily open the app, access the Directory, and call the only entry. |
| System Coordinator put hard cover stand and film protector on iPads and shipped to users / study participants. |
| Upon return, the System Coordinator noted the study log, assessed the iPad for any damage. System Coordinator disinfected and erased all contents and settings on iPads. Restored images again and sent to the next user / study participant. |

**Table 1:** Flow chart of detailed intervention activities.

IV patients. The mobile group session began with introductions, establishing a welcoming group environment focused on mutual respect. The facilitators then delivered self-management information for the IVhomeCare intervention and facilitated the group discussion. Sessions were scheduled for up to 90 minutes, with the average session lasting 56 minutes. The multiple patients and family members / caregivers engaged in discussions, shared experiences, and offered peer support to each other.

The IVhomeCare intervention included two scheduled mobile group sessions with multidisciplinary healthcare professionals in their office and three to five families in their own homes. A family included the patient and at least one family member or other caregiver. The number of participants in these sessions ranged from four to nine depending on the numbers of family members. In each mobile group session, the facilitators: (1) introduced the participants and themselves and reviewed related HIPAA regulations asking group members to keep information shared confidential; (2) explained session ground rules (similar to other support groups), including using first name only and sharing only what one wants others to know; (3) reminded individuals that the facilitators and other group members are not providing medical advice nor endorsing any specific IV products; and (4) encouraged use of the iPad mini Internet links to IVhomeCare resources on our website [38].

Session content focused on our empirically supported IVhomeCare interventions and the asynchronous components of these interventions: Infection prevention monitoring intervention; Mood and emotion monitoring intervention; Problem-solving partnership intervention; Fatigue monitoring and restorative nap intervention; Family healthy living activities intervention; and Daily self-monitoring and early reporting using checklist. Details of their IV homecare concerns and use of the iPad to link to health professionals and to the IVhomeCare intervention website information closed the discussion.

The facilitators provided guidance if the group reached an impasse and corrected misinformation. The facilitators strove to refrain from lecturing, and they promoted nonjudgmental information sharing. As a follow-up to these interventions, online automated text message prompts were sent via an encrypted email account (restrictively used for this study) to each family member through the iPad mini. The prompts were used to reinforce intentions to use the IVhomeCare interventions.

Because families managing lifelong IVs at home rarely meet others in similar situations, peer support was an essential component of the group sessions. Social isolation is common in this population due to patients frequently being homebound and having limited time to socialize given their daily IV treatments [39]. Our iPad mobile sessions with multiple families decreased social isolation by allowing families to share their stories and support one another around home IV care issues [40]. In addition, a majority of adults reported that it was beneficial to interact with others who have the same healthcare concerns. Such interaction with peer patient groups and their families' help patients learn "how to cope with a health issue or get quick relief." However, they turned to healthcare professionals when they needed specific advice for complex homecare.

## Translating the asynchronous IVhomeCare interventions to the iPad tablets

New opportunities and challenges arise when translating previously validated interventions to the mobile environment, necessitating attention to maximize the benefit of the new technology [41,42]. Table 2 summarizes the IVhomeCare interventions and describes considerations for translating them for mobile delivery, including access to Internet platforms.

**Internet resources:** The iPad mini tablets were pre-loaded with previously validated IVhomeCare Resources website links, to provide the benefit of Internet access at any time that meets patient and family needs. The IVhomeCare intervention website provides: evidence-based symptom monitoring algorithms; IVhomeCare illustrations and step-by-step support for independent homecare management; and problem-solving partnership guides with healthcare professionals [43].

Families readily learned to use the iPad mini's interactive touch-screen features that provide a fast and easy way for families to access the IVhomeCare interventions [44]. The IVhomeCare Resources icon on the iPad mini home-screen connects directly to the IVhomeCare intervention website links, and the individual intervention webpages are available 24/7 for quick reference of a specific topic (Table 3). The IVhomeCare Resources icon also connects to digital versions of

| Intervention Name | Description of IV Home Care Intervention Content | Advantages / Challenges to Translating the IV Home Care Information into the Mobile Environment |
|---|---|---|
| Internet Links to IV home Care Resources | • Scientifically based algorithms guide families through specific management and reporting of the 42 most common IV homecare problems.<br>• Internet website links with online access provide information in one place and allow continuously updated information, compared with static paper versions:<br>o Oley Foundation Lifeline Newsletter for alerts and information about IV home care.<br>o Caregiving advocacy resources, such as National Caregiving Alliance and other caregiver advocacy.<br>o Bilingual and other website links.<br>• Guide on how to determine quality of information. | • Internet web links easily pre-loaded on the iPad and materials readily accessible to patients and their family members.<br><br>• Support group participants reinforced use of the Oley Foundation resources, noting which resources were particularly beneficial in their own lives and sharing experience posting questions on the Oley Foundation social media resources. |
| Infection Prevention Monitoring | • Audiovisual scenarios illustrate:<br>o Hand washing techniques.<br>o Maintenance of equipment and storage area.<br>o Care of the IV catheter insertion site.<br>• A music jingle that times hand washing for 1 minute (per CDC guidelines).<br>• Self-monitoring checklist guides patients to follow a daily routine for self-monitoring IV care (Table 4).<br>• Interactive digital game guides identification of infection symptoms. | • Mobile device provides ease of access and repeated viewing, even when the patient is outside the home setting.<br>• Pre-loaded photographs show possible arrangements for the large volume of storage of sterile equipment and supplies, a major issue.<br>• Music guiding hand washing is available on the touch-screen.<br>• Future updates will allow the procedure checklist to be in a "fill-in form" format, providing automatic alerts on steps repeatedly missed.<br>• Support group participants supported one another in applying the scenarios to their daily life. |
| Mood and Emotion Monitoring | • Journal writing about low moods (Table 5).<br>• Audiovisual scenes illustrate:<br>o Mood-elevating activities and diaries.<br>o Self-monitoring checklist that encourages monitoring mood by:<br>  ▪ A 10-item scale for rating daily mood / emotional reactions [11] (Table 4).<br>  ▪ Engaging in mood elevating activity (Table 4).<br>• Interactive digital game guides confidence building for positive outlook. | • New options are available for online journal writing in order to easily access the journal, avoiding risks of paper documentation (e.g., loss or damage to the paper version, less privacy).<br>• Online tools allow increased options for patients to track their progress graphically and note changes over time.<br>• Future online checklist options will include searchable forms and associated alerts to the patient / family / provider should depression warning signs arise.<br>• Support group participants provided real-life examples of mood-elevating activities, ranging from knitting to participating in car shows.<br>• Support group participants provided positive reinforcement to other participants as well as normalizing experiences, all of which are associated with mood enhancement. |
| Problem-Solving Partnership | • Audiovisual scenes illustrating problem-solving with healthcare professionals.<br>• Scientifically based algorithms with step-by-step guides for solutions to the most common IV problems.<br>• Prompts for establishing an IV home care routine.<br>• Prompts for working with healthcare providers and insurers to address problems. | • Mobile access facilitates use of problem-solving resources, particularly at times when the patient / family are under additional stress and can benefit from such strategies.<br>• The digital space for saving screens is greatly enhanced on the tablet format.<br>• Support group participants gave real-life examples that illustrated the problem-solving steps, such as how to request delivery of needed infection control supplies to an upper-level apartment when one is at work. |
| Fatigue Monitoring and Restorative Nap | • Fatigue monitoring sheets are provided.<br>• Science-based information indicating short naps improve attention to detail.<br>• A music-guided restorative daytime nap. | • Fatigue monitoring forms are more readily accessible with mobile tools.<br>• Tablets allow various types of music (soft rock, classical, etc.) and ease of use with the ability to take the mobile device anywhere in one's home or outside the home when on vacation, etc.<br>• Support group participants shared times that they had tried the restorative nap and problem solved barriers to such naps together. |
| Family Healthy Living Activities | • Illustration of health promotion and healthy living activities:<br>o Simple daily walking exercise.<br>o Lower sugar and salt intake and healthy eating. | • The synchronous session allowed the facilitators to reinforce that information on healthy activities varies with patients and family caregivers need clarification. Also, realistic limits on suggested activities can be discussed, improving intentions for participation. |
| Group Discussions and Peer Support | • Exemplar peer and family stories shared through the web resources.<br>• Support gained from other families experiences about managing IV home care.<br>• Group discussions conducted after each IV Home Care intervention session. | • Increased group discussion and peer support through the mobile videoconferencing sessions, with social support and encouragement or preventing feelings of isolation when homebound. |

**Table 2:** Comparison of traditional IV home care in-person content and translation of this content into interventions for the mobile environment.

the IVhomeCare intervention information about: (1) monitoring for infection prevention; (2) monitoring mood and emotions; (3) supporting problem-solving partnerships with healthcare professionals; (4) monitoring fatigue and providing a restorative nap guide to reduce daytime fatigue; and (5) sharing healthy living activities including daily exercise and physician recommended sugar / salt intake. This information is described in more detail below.

The IVhomeCare website incorporates the National Institutes of Health (NIH) Web Literacy and National Disabilities Act standards including large-font text, simple graphics, and easy-to-locate, straightforward navigation symbols [46,47]. This website provides IVhomeCare guidance and illustrated instructions that encourage the patient and family members to use the information in daily IVhomeCare routines. In addition, the IVhomeCare Resources includes a list of

| |
|---|
| · **Infection Prevention.** Video scenarios depict hand washing techniques using our prescribed timed music jingle and checklist guides for infection control. The interactive digital game Improving Sepsis Recognition and Management through a Mobile Education Game, a case-based interactive learning using a set of typical infection cases that follow evidence-based treatment algorithms. Players make decisions about reporting symptoms and watch as the case condition changes. The game's rapid pace underscores the importance of daily temperature taking, IV site assessment, and early symptom reporting. Points are awarded for correctly managing and answering questions. |
| · **Depression Prevention.** Based on Depression Awareness and Management, (National Institutes of Health, U.S. Department of Health and Human Services) depression prevention includes mood self-monitoring with relaxing music as well as video scenes that illustrate mood-elevating activities. This is a safe approach for IV patients who have limited oral intake, such that depression medications are not absorbed. Family members, also have reactive depression. |
| · **Problem-solving IV Care Partnerships.** Video scenarios and specific algorithms for the 48 known frequently recurring IV homecare problems are based on national clinical guidelines and FDA updates and alerts. Prompts for working with healthcare professionals to address problems are sent via the iPad. |
| · **Power Nap.** A short, effective music-guided 10-minute daytime restorative nap addresses fatigue. Short Daytime Restorative Naps [45] is guided by self-selected music (soft rock, slow hip-hop, classical, etc.); each selection initially reduces cadence and is shown to relax into REM sleep. An interactive game measures subjects' fatigue via reaction time (a measure of sleep deprivation). |
| · **Taking Care of You.** Engagement in interactive healthy living activities is tracked using automated prompts. Motivating prompts are used in our KUMC employee healthy living "Stick-to-it-iveness" software. This program has successfully engaged 600 adults in long-term health activities. |
| · **Family Success Stories.** Exemplar stories from experienced families about managing IV home care show challenges met. Each is reviewed by healthcare professionals and de-identified prior to placement online. |
| · **Scripted Guides.** To promote comprehensive communication with their healthcare providers (MD and / or RN) or insurance providers, these guides assist and empower families. |
| · **Oley Foundation News.** FDA alerts and important new IV catheter information is available from the national multidisciplinary family advocacy organization. |
| · **Resource Information** – links to National Caregiving Alliance and other family caregiving resources. |
| · **Multilingual and Other Website Links.** Easy-to-use Internet pages that provide automated language translation. |
| **Social Media and Games for families:** |
| **Hand Hygiene** https://itunes.apple.com/gb/app/hand-hygiene-training |
| • **Breathe2Relax** interactive app from Natl. Center for Telehealth with breathing skills for stressful situations. Includes video demo and charts for mapping use. |
| • *SPARX* is a digital game where the player creates an avatar that fights GNATs (Gloomy Negative Automatic Thoughts). The player participates in confidence-building skills to achieve a positive outlook, avoid depression, and seek help when needed. |
| • **Social Media Safety Guide.** The safety on social media intervention uses the "Be a Good Cyber Citizen Guide" developed by the National Crime Prevention Council. It includes criteria for determining the quality of web information per US Media Common Sense and the Cyber bullies Zombie games. |

**Table 3:** IV Home care intervention website content and IV home care illustrated instructions.

frequently asked questions and answers, decision-making guides, and online health assessment tools using game formats with immediate feedback. The IVhomeCare self-monitoring checklist webpage is also available on the IVhomeCare website. Mobile group sessions reinforce the use of these online resources by families (Table 4).

## IVhome care interventions website

On our intervention website, families have access to the Oley Foundation's (family advocacy association) Lifeline Newsletter for information about IV home-care and a scientifically developed IV catheter complication chart. Other links direct families to caregiving advocacy resources such as the National Caregiving Alliance and the Family Alliance for Caregiving. Our website also provides a guide to help determine the quality of the information found on websites [48].

**Infection-prevention monitoring intervention:** Patients with IV catheters need to adhere to strict aseptic techniques and IV procedures, and their family members need to learn the complex technological treatments that are required to avoid life-threatening IV catheter infections [49]. The infection prevention intervention of IVhomeCare consists of short video scenes: (1) demonstrating the proper technical procedures of hand washing, guided by a catchy jingle; (2) maintaining IVhomeCare equipment; (3) sanitizing the infusion area; and (4) practicing home asepsis for infection prevention.

The lyrics for the music jingle "All You Need for Good Hand washing: Water, Soap and Time" teach about removing jewelry, cleaning areas under nails, and vigorously scrubbing for the duration recommended by the Centers for Disease Control and Prevention (CDC). The music has repeating lyrics, which drastically improves memory and task attention and is similar to the large effect of mnemonic devices [45]. The short video scenes and hand washing song can easily be accessed from the bottom of the iPad home screen under the Video and Music icons. Music interventions have been associated with cognitive improvements and reduced distractibility, all important in IV catheter infection prevention [50].

The infection prevention intervention was extended by including the recommended updates for home care based on scientific guidelines from the CDC for IV device-related infection control in the home setting [51,52]. Thus, follow-up reminders and prompts for safe IV procedures can be readily communicated via the mobile technology (iPad mini).

**Mood and emotion monitoring intervention:** The magnitude of daily care and the patient's chronic illness may negatively impact the whole family and lead to episodes of depressive moods and emotional upset [53]. Situational or reactive depression is common in both IVhomeCare patients and their family members and may result from, for example, worrying about the patient's IV catheter infections or illness exacerbations and the financial strain of costly out-of-pocket healthcare expenses [54]. This may lead to emotional burnout and repeated episodes of depression.

Patient or family member depression can interfere with IV care because depression is characterized by the inability to: (1) concentrate during IV procedure training; (2) maintain IVhomeCare skills; or (3) make problem-solving decisions [55]. Patients and family members who are experiencing depressive symptoms have more difficulty adhering to aseptic techniques, resulting in a higher likelihood of unintentional touch contamination during IV catheter care [56]. Thus, our IVhomeCare intervention includes self-monitoring of mood and emotion. This intervention acknowledges that depressive episodes do occur and asks patients and family members to: (1) monitor their daily mood and common emotional reactions; (2) identify personal techniques that they have used to decrease situational depression; and (3) recognize early warning signs associated with major depressive disorders and share such symptoms with healthcare professionals. Specifically, a monitoring scale is provided for families to rate mood and engagement in mood-elevating activities [57].

Family members can use a page of the daily self-monitoring checklist to write about their emotions, problem-solving techniques,

| IV Care Daily Self-monitoring Checklist and Depression / Mood and Fatigue Monitoring Activities | | | | | | | |
|---|---|---|---|---|---|---|---|
| **INSTRUCTIONS**: Click on the actions below just before connecting or after disconnecting from the IV. Put a checkmark in the box on the days you monitor your own Infection Principles (hand washing, temperature check, safe procedures, and OBSERVING YOUR IV site). Emotions and Sleepiness ratings will lead you to determine your Depression / Mood and Fatigue Levels. We especially want you to do a mood-elevating activity any time you rate your mood at 5 or lower. One important activity for raising mood is writing about your feelings and emotions at least 3 times a week (make one of the days on the weekend). See Table 5. Use Problem-solving with health professionals when needed, and use the short naps to manage your fatigue. | | | | | | | |
| ☺ **HAND WASHING OF IV USER and FAMILY** | Sun | Mon | Tue | Wed | Thu | Fri | Sat |
| 1. Use soap and water. | | | | | | | |
| 2. Scrub vigorously for at least 1 minute. Rinse and dry. | | | | | | | |
| 3. Use a clean paper towel. | | | | | | | |
| 4. Apply alcohol-based hand sanitizer. | | | | | | | |
| ☺ **MONITOR TEMPERATURE OF IV USER** | Sun | Mon | Tue | Wed | Thu | Fri | Sat |
| **EXAMPLE:** | **96.8** | 97.0 | 98.8 | 99.0 | 99.4 | 99.8 | **100.4** |
| 1. Report abnormal temperature (> or < normal). | | | | | | | |
| 2. Report to MD / RN when chills occur. | | | | | | | |
| 3. Monitor when patient doesn't feel well. | | | | | | | |
| ☺ **SAFE WORK PROCEDURES FOR IV CARE** | Sun | Mon | Tue | Wed | Thu | Fri | Sat |
| 1. Check for expiration dates / sterile packages intact. | | | | | | | |
| 2. Dispose of needles / syringes safely. | | | | | | | |
| 3. Use friction with alcohol on hub connections. | | | | | | | |
| ☺ **IV CATHETER INSERTION SITE. Check for:** | Sun | Mon | Tue | Wed | Thu | Fri | Sat |
| 1. Redness. | | | | | | | |
| 2. Pain / tenderness. | | | | | | | |
| 3. Swelling. | | | | | | | |
| 4. Drainage. | | | | | | | |
| ☺ **EMOTIONS and FATIGUE OF IV USER and FAMILY** | Sun | Mon | Tue | Wed | Thu | Fri | Sat |
| 1. Rate your feelings of being sad, unhappy, depressed, gloomy, or discouraged. If rate is ≤ 5, write about your feelings on the back page.  1  2  3  4  5  6  7  8  9  10  *Worst Feelings*          *Best Feelings* | | | | | | | |
| 2. If rate is ≤ 5, did you do an activity that you think is relaxing, enjoyable, and will help with a positive outlook. Circle Yes (Y) or No (N). | Y / N | Y / N | Y / N | Y / N | Y / N | Y / N | Y / N |
| 3. Your dozing off or falling asleep: watching TV, sitting reading, while in a car or stopped for traffic light. **Use short nap music if > 2.**  0      1      2      3  *Never   Slight   Moderate   High* | | | | | | | |
| ☺ **TRY THESE MOOD ELEVATING ACTIVITIES:** | Sun | Mon | Tue | Wed | Thu | Fri | Sat |
| Write the number of times each day you used: | | | | | | | |
| 1. The checklist above. | | | | | | | |
| 2. Soothing music, hugging, back or necks rub. | | | | | | | |
| 3. Enjoyable activity (e.g. hobby, reading, phone calls). | | | | | | | |
| 4. Talking to others (friends, family, peers, etc.). | | | | | | | |
| 5. Short 10-minute nap. | | | | | | | |
| 6. Use problem-solving with health professionals. | | | | | | | |

**Table 4:** The IV home care self-monitoring checklist webpage.

and partnership with peer groups (Table 5). Journal writing about current mood and emotions has been shown to increase a positive outlook and may enhance the body's immune system to help avoid infections [58].

**Problem-solving partnership intervention:** The problem-solving approach is based on the American College of Physicians' Homecare Guide [59] and has been used in other clinical trials [60]. Patients are assisted in identifying their problems and potential solutions by partnering with healthcare professionals. Increased patient longevity and closer physician-patient relationships have been found when problem-solving partnerships were used [61]. The partnership for problem-solving processes can result in increasing patient symptom management and improving physical status, emotional health, and quality of life [62]. Problem-solving skills are practiced by families in each session and reinforced using the pre-loaded videos, the self-monitoring checklist, and the group discussions. Patient concerns

about IVhomeCare were facilitated during group discussions and problem-solving skills were reinforced.

**Fatigue monitoring and restorative nap intervention:** Fatigue management helps improve IVhomeCare because fatigue decreases concentration and increases the risk for poor health [45]. Patients and family members often experience fatigue due to the demands of daily, time-consuming IV infusion procedures. This includes adherence to strict aseptic techniques for hook-up and disconnect from the IV pump. Infusions are often completed over 12 hours, resulting in frequent night-time sleep interruptions, as well as interruptions from IV pump alarms and frequent bathroom trips [63,64]. The National Sleep Foundation (NSF) states that many people are alert in the morning and see a dip in alertness in afternoon. Thus, our IVhomeCare intervention provides a guide to a short da-time nap to help restore levels of energy to last through the evening. The nap music with focused imagery induces relaxation and is accompanied with softly spoken instructions

| INSTRUCTIONS: At least three times a week (use one weekend day), write about your use of Infection Prevention Principles, Emotions Monitoring, and Problem Solving and Partnering with others. Note any actions you took to improve your mood, as well as writing about IV homecare self-management problems / concerns and problem-solving techniques used. | |
|---|---|
| **EXAMPLES:** | |
| 2 / 26 - Every Monday, Wednesday and Saturday morning, I check soap and other IV supplies. We called for new order so we don't run out on the weekend.<br>Also called the MD on Friday as 99.8 was 1 degree higher than usual. I'm worried an infection is coming on that is so frightening. | Date:_____ |
| 2 / 28 - The cat jumped on the counter while I was hooking up to the IV tubing. Fortunately nothing was contaminated.<br>2 / 28 - I followed step-by-step to pace myself when I had too many changes in my supplies and a new infusion pump. First, I focused on having an awareness of these problems and stopped myself from wishing the insurance hadn't changed and decided to solve this. I had to get organized; I took the next step to practice all the new stuff - it worked! | Date:_____ |
| 2 / 30 - I wish I had the energy to do my hobby. So I tried using the nap tape, that little snooze was refreshing.<br>2 / 30 - Tomorrow I am going to exercise. I will start by walking up the block and back. My goal is walk all the way around the block. | Date:_____ |
| 2 / 31 - My emotions rating was a 4 yesterday, so I listened to my favorite CD on the way to the hospital to visit my husband. I was pretty calm by the time I got there. When I got back home, I spent time in my garden. | Date:_____ |

**Table 5:** Monitoring your emotions and reactions through writing.

| Topics During Peer Support | Examples |
|---|---|
| Encouraged activities outside of IV home-care. | Shared experiences on ways to stay engaged, including visiting with friends / family, drawing, taking care of pets, gardening, traveling, etc. |
| Discussed ways for saving money in day-to-day treatment. | Shared experience on keeping a notebook comparing IV equipment costs. |
| Increased understanding of healthcare professionals directions. | Suggested using iPad to record medical appointments and hospital discharge to assist with remembering complex instructions given. |
| Shared common experiences. | Shared similar struggles with fatigue as others described. |
| Gained emotional support. | Praised a young family caregiver for making time for her young children between infusion cares. |
| Suggestions on how to engage with the healthcare team. | Shared how to approach medical team with questions and provide them information. |

**Table 6:** Topics of peer support discussion and examples of each found in mobile sessions.

for deep breathing and muscle relaxation. The gradually decreasing music tempo induces relaxation by synchronizing with a wake-to-sleep heart rate cadence. Our previous clinical trial showed a recuperative effect (0.45) [65], as did another study (0.67) [66]. A Cochrane review also reports that music enhances healthy sleep [67].

To use this sleep promotion intervention, a nap icon is easily accessible with the music choices on the iPad mini home screen. Families choose their preferred music genre such as classical, country, easy-listening, slow hip-hop, or soft rock. The gradually decreasing music tempo initiates sleep. After 20 minutes of napping, a wake-up alarm sounds because sleeping longer than 20 minutes has been shown to cause grogginess and more difficulty falling asleep at bedtime. Families are taught that naps are not a substitute for a full night's sleep.

**Family healthy living activities intervention:** Because the demands of managing complex home care often result in family members neglecting their own health, we engage them in basic healthy living activities, namely healthy eating and monitoring recommended sugar and salt intake and daily walk exercise. Participants were instructed to consult with their primary care provider before engaging in these healthy activities [68]. These behaviors, suggested by the NIH, are evidence-based programs and safe even for older adults [69-71]. The goal is to increase families' awareness of the importance of improving a healthy dietary intake and increasing exercise. These activities are reinforced through follow-up automated prompts with persuasive statements sent via iPad mini email accounts [72-74]. Self-monitoring and prompts have significantly improved healthy eating and are associated with increased exercise [75,76].

**Daily self-monitoring and early symptom reporting using the interactive checklist:** Using a self-monitoring checklist available 24/7 on the IVhomeCare intervention website (Table 4) patients are encouraged to review their IV catheter homecare routines with their healthcare providers and to discuss any daily care concerns during the mobile group sessions. This checklist has been verified to result

in daily self-monitoring and early symptom reporting, which can ward off serious health-related complications such as IV catheter infections [11]. The checklist guides patients to monitor signs and symptoms of infections. It instructs them to adhere to a complex twice daily IV infusion routine including hand washing, recording body temperature, maintaining a clean work area and asepsis during IV care procedures, and assessing the IV catheter site for inflammation and signs of infection. Self-monitoring skills are developed and practiced by patients and reinforced by the healthcare professionals. Problem-solving partnerships with professionals improve daily self-monitoring skills, and peer group discussions provide social support. With practice, patients learn how to weave self-management skills into their daily IVhomeCare routines [77,78].

## Results

### Analyses

There were 126 participants (55 patients and 71 family caregivers) who attended the videoconferencing group sessions (n = 41) via iPad. Participants' average age was 41.87 (SD = 19.95), ranging from 14 to 79 years, and 53% (n = 19) were male. Of the 126 participants, 70 (81.2%) had experience with IV home-care for between 1 to 5 years.

Our mobile-based intervention was evaluated by patient and family ratings of their use of mobile health care on the iPad tablet [79]. On a scale from 1 to 5 (1 = not helpful and 5 = very helpful), patients and family caregivers rated the IVhomeCare intervention program, materials, and technical support as helpful (4) [80]. Subjects reported that they enjoyed meeting with professionals who provided information and facilitated discussion with other families in the iPad group sessions (M = 4.69, SD = 0.68). Sharing their challenges in the iPad group sessions gave them a sense of connection with health professionals and others managing IVhomeCare (M = 4.60, SD = 0.69). The peer support was rated as giving them the sense of not being the only one who gets the blues (M = 4.43, SD = 0.85). Participants gained problem-solving skills by hearing effective ways to manage IVhomeCare from others through iPad group discussion ($M = 4.60$, $SD = 0.60$). In addition, almost all participants (94%) rated themselves as highly satisfied with the overall interventions and would recommend this type of mobile delivery to others ($M = 4.40$, $SD = 1.33$). Further, Table 6 provides examples from the discussed topics that illustrate how peer support was helpful to families.

All the patients and family members received the IVhomeCare intervention via iPad group sessions free of charge. All participants also indicated that they were willing to pay from $25 to $50 per iPad group session with multidisciplinary healthcare professionals and other patients and family caregivers, even if insurance would not cover this cost. However, they also felt that the sessions were worth being covered by insurance companies. Both professionals and patients rated these approaches as saving them time and money. Patients also mentioned the convenience of the intervention and in knowing "their health care professional can be in close contact." One disadvantage that professionals described however was the lack of connection between the iPad data collected and electronic medical records. Professionals wished to have a direct connection for storing the photographs taken of patients during their iPad exams and for typing their notes into a patient's legal medical record. It is likely those technical features will be developed in the near future.

## Discussion

As shown in Table 2, there are major advantages to translation of the IVhomeCare interventions for delivery via iPad mini tablets with 4G access. First, the mobile technology provides easy access to evidence-based online resources with increased ease of use and availability virtually anytime and anyplace that meet patients and family members' needs. This includes access to information and guidance that supports independent self-management for complex IV care. Second, the mobile technology supports group interactions in real-time using secure videoconferencing. This allows patients and families to connect with other patients and families around the country, as well as with healthcare facilitators. There is the added benefit that these iPad based support groups decrease social isolation. And third, rather than gathering at a hospital or clinic, the iPad meetings do not require travel or introduce additional infection risk. Such mobile approaches maximize patient and caregiver confidence in managing complex lifelong daily home IV care.

Families participating in the group discussions receive support and encouragement from other families who have successfully self-managed IVhomeCare issues, and they receive information and guidance from healthcare professionals at a distance. During the peer group discussion, patients and family members shared their experiences and offered support to one another (Table 6). One subject who was developmentally disabled was comfortable using the iPad and engaging with health professionals during the audiovisual sessions and openly participated in the group discussion. Yet family indicated she was reluctant to share during in-person doctor visits. In another case where the subject was deaf, a family member typed what was being discussed during the group iPad session so their parent could follow and participate in the group discussion.

The costs for setting up mobile videoconferencing sessions can be offset by saving out-of-pocket travel costs and travel time [81], as well as providing the opportunity for the health professionals to meet with the whole family [82]. Previous research found the average out-of-pocket health care costs for these patients was $17,000 annually with travel to local and distant specialists [83]. The cost for the iPad support group sessions is considerably less than the cost for travel to meet individual healthcare providers, a single emergency department visit, or one inpatient hospitalization for catheter-related infection. Immediate examination of IV catheter sites and communication with patients on their health status will lead to early diagnosis and treatment initiation, which in turn will reduce morbidity and mortality in the long term. The current emphasis on preventing and reducing morbidity and mortality rates from catheter-related infections supports the translation of our IVhomeCare intervention.

A significant advantage that patients described was the opportunity for group discussion sessions with other patients and their family members. These sessions allowed them to share their day-to-day issues and the solutions they use for their lifelong required IV infusions, managing the fatigue that is common, and the importance of maintaining a positive outlook. Several patients confirmed that the post appointment iPad group discussion was the first time they had met others with the same illness and the same needs, issues, and challenges for the life sustaining IV treatments. However, one common disadvantage in introducing new healthcare delivery systems is merging the new technology with electronic health records as noted by healthcare professionals.

## Conclusion

Due to the extensive daily care demands and financial hardship of complex IV infusion technology, using a mobile distance connection to 24/7 IVhomeCare web-based information and access to healthcare via the mobile platform iPad, for guiding families to use evidence-based interventions, is important to continue testing. Mobile technology can assist participants to avoid life-threatening infections, monitor emotions and fatigue, keep healthy, and decrease costs such as reduced travel for visiting healthcare providers.

Patients who require daily IV infusions at home must adhere to strict technical procedures for survival. Patients and family members need support in mastering the complex technological skills that are required to avoid life-threatening infections. Family education will be effective if the family members perceive that they are well prepared to manage home care, especially when visits to healthcare providers and specialists are both time-consuming and costly. In this study evidence-based interventions were translated to the mobile environment. This includes the asynchronous educational videoconferencing components.

New technologies are constantly arising, even during the course of this intervention. The "lessons learned" in this study may continue to apply in the mobile environment, even as the specific device improves and continues to advance intervention possibilities in the healthcare arena. The early identification of IV catheter infection and potentially fatal IV sepsis may well encourage the use of mobile technologies. This study adds to the knowledge base for assessing technology delivery of health care on health outcomes and translation of the evidenced-based intervention to homecare practice.

## Disclosure

The project was part of a larger study supported by the National Institute of Biomedical Imaging and Bioengineering (NIBIB) R01 EB015911, C. Smith, Principal Investigator. In addition, this study is partially supported from National Institutes of Health (NIH) U54 RR031295 *Trail Blazer Award* #UL1TR000001 to C. Smith *from Frontiers: The Heartland Institute for Clinical and Translational Research*. The content is solely the responsibility of the authors and does not necessarily represent the official views of the National Institute of Biomedical Imaging and Bioengineering or the Frontiers: Heartland Institute.

## Acknowledgement

We are grateful for the recommendations and Telemedicine expertise contributed to this trial by Dr. Ryan Spaulding (Associate Vice Chancellor for Community Engagement, University of Kansas Medical Center). And to Chang-Ming (Jeremy) Ko, MS MA our Technical Specialist. The authors extend their appreciation to Sally Barhydt (Publication Consultant) for her editorial support and to all patients who participated in this study for their time and use of mobile healthcare interventions.

## References

1. Bozzetti F, Staun M, Van Gossum A (2006) Home Parenteral Nutrition. Cambridge: CABI.

2. Winkler MF, Smith CE (2015) The Impact of Long-Term Home Parenteral Nutrition on the Patient and the Family: Achieving Normalcy in Life. J Infus Nurs 38: 290-300.

3. Piamjariyakul U, Yadrich DM, Ross VM, Smith CE, Clements F, et al. (2010) Complex home care: Part 2-family annual income, insurance premium, and out-of-pocket expenses. Nurs Econ 28: 323-329.

4. Smith CE, Piamjariyakul U, Yadrich DM, Ross VM, Gajewski B, et al. (2010) Complex home care: part III-economic impact on family caregiver quality of life and patients' clinical outcomes. Nurs Econ 28: 393-399.

5. Winkler MF, Smith CE (2014) Clinical, social and economic impacts of home parenteral nutrition dependence in short bowel syndrome. J Parenter Enteral Nutr 38: 32S-37S.

6. Winkler MF (2007) American Society of Parenteral and Enteral Nutrition Presidential Address: food for thought: it's more than nutrition. J Parenter Enteral Nutr 31: 334-340.

7. Smith CE, Cha JJ, Kleinbeck SVM, Clements FA, Cook D, et al. (2002) Feasibility of in-home telehealth for conducting nursing research. Clin Nurs Res 11: 220-233.

8. Fitzgerald SA, Yadrich DM, Werkowitch M, Piamjariyakul U, Smith CE (2011) Creating patient and family education websites: Design and content of the home parenteral nutrition family caregivers website. Comput Inform Nurs 29: 637-645.

9. Spaulding R, Smith CE, Nelson EL, Yadrich D, Werkowitch M, et al. (2014) iCare: mHealth Clinic Appointments Using iPad Minis Between Multiple Professionals and Intravenous Dependent Patients in their Homes; ATA Case Study. American Telemedicine Association.

10. Smith CE, Spaulding R, Piamjariyakul U, Werkowitch M, Yadrich DM, et al. (2015) mHealth Clinic Appointment PC Tablet: Implementation, Challenges and Solutions. J Mob Technol Med 4: 21-32.

11. Smith CE, Curtas S, Kleinbeck SVM, Werkowitch M, Mosier M, et al. (2003) Clinical trial of interactive and videotaped educational interventions to reduce catheter-related infection, reactive depression, and rehospitalizations for sepsis in patients receiving home parenteral nutrition. J Parenter Enteral Nutr 27: 135-145.

12. Yadrich DM, Fitzgerald SA, Werkowitch M, Smith CE (2012) Creating patient and family education web sites: assuring accessibility and usability standards. Comput Inform Nurs 30: 46-54.

13. Kelly K, Smith CE, Eskenazi L (2007) Technology and family caregiving: Advantages and challenges for delivering education, training, and support. Gerontologist 47: 272-273.

14. Smith CE (2007) Testing virtual nurse caring by picture phones and internet with home parenteral nutrition patients. Clin Nutr 2: 43-44.

15. Saqui O, Chang A, McGonigle S, Purdy B, Fairholm L, et al. (2007) Telehealth videoconferencing: improving home parenteral nutrition patient care to rural areas of Ontario, Canada. J Parenter Enteral Nutr 31: 234-239.

16. Smith CE (2007) Testing Virtual Nurse Caring of Picture Phones and Internet with Home Parenteral Nutrition Patients. Clinical Nutrition 2: 43-44.

17. National Alliance for Caregiving (2014) Catalyzing Technology to Support Family Caregiving. Key steps to better support family caregiving through the use of mobile, online, and in-home technologies.

18. Schulz DN, Kremers SP, Vandelanotte C, Van Adrichem MJ, Schneider F, et al. (2014) Effects of a web-based tailored multiple-lifestyle intervention for adults: a two-year randomized controlled trial comparing sequential and simultaneous delivery modes. J Med Internet Res 16: e26.

19. Howard L (2006) Home parenteral nutrition: survival, cost, and quality of life. Gastroenterology 130: S52-59.

20. Smith CE, Curtas S, Kleinbeck SV, Werkowitch M, Mosier M, et al. (2003) Clinical trial of interactive and videotaped educational interventions reduce infection, reactive depression, and rehospitalizations for sepsis in patients on home parenteral nutrition. J Parenter Enteral Nutr 27: 137-145.

21. The Oley Foundation (2013) Tools for living better on home IV and tube feedings: Tips for traveling with Home PEN.

22. Winkler MF, Ross VM, Piamjariyakul U, Gajewski B, Smith CE (2006) Technology dependence in home care: impact on patients and their family caregivers. Nutr Clin Pract 21: 544-556.

23. Smith CE, Leenerts MH, Gajewski BJ (2003) A systematically tested intervention for managing reactive depression. Nurs Res 52: 401-409.

24. West DM (2015) How Mobile Devices are Transforming Healthcare.

25. Gaskamp CD (2004) Quality of life and changes in health insurance in long-term home care. Nurs Econ 22: 135-139.

26. National Alliance for Caregiving (2014) Catalyzing Technology to Support Family Caregiving. Key steps to better support family caregiving through the use of mobile, online, and in-home technologies.

27. Holopainen A, Galbiati F, Voutilainen K (2007) Use of Smart Phone Technologies to Offer Easy-to-Use and Cost-Effective Telemedicine Services. First International Conference on the Digital Society, Guadeloupe.

28. Fox S (2014) The social life of health information.

29. Lewis N (2014) Health IT Sees Key Market in Family Caregivers. Healthcare Information Week.

30. Fiordelli M, Diviani N, Schulz PJ (2013) Mapping mHealth research: a decade of evolution. J Med Internet Res 15: e95.

31. Wolbring G, Lashewicz B (2014) Home care technology through an ability expectation lens. J Med Internet Res 16: e155.

32. Spaulding R, Smith CE, Nelson EL, Yadrich D, Werkowitch M, et al. (2014) iCare: Testing iPad Visual Sessions Between Interdisciplinary Professionals and Families Managing Intravenous Dependent Patients in their Homes; ATA Case Study. American Telemedicine Assoication.

33. Lewis N (2014) Health IT Sees Key Market in Family Caregivers. Healthcare Information Week.

34. Metzger LC (2010) Education materials for home nutrition support consumers. Nutr Clin Pract 25: 451-470.

35. Smith C (2008) Technology and Web-based support. Am J Nurs 108: 64-68.

36. Fitzgerald SA, Macan Yadrich D, Werkowitch M, Piamjariyakul U, Smith CE (2011) Creating patient and family education web sites: design and content of the home parenteral nutrition family caregivers web site. Comput Inform Nurs 29: 637-645.

37. Winkler MF, Smith CE (2015) The Impact of Long-Term Home Parenteral Nutrition on the Patient and the Family: Achieving Normalcy in Life. J Infus Nurs 38: 290-300.

38. Smith CE (2005) Website algorithms. Gerotechnology 3: 217.

39. Winkler MF, Smith CE (2015) The Impact of Long-Term Home Parenteral Nutrition on the Patient and the Family: Achieving Normalcy in Life. J Infus Nurs 38: 290-300.

40. Smith CE, Curtas S, Robinson JM (2001) Case study of patients helping patients program. Nutrition 17: 175-176.

41. Fox S (2012) Pew Internet News: Wireless Internet Use, the Internet and the Recession, Mobile technology use. Pew Research Center's Internet and American Life Project.

42. Baker TB, Gustafson DH, Shah D (2014) How can research keep up with eHealth? Ten strategies for increasing the timeliness and usefulness of eHealth research. J Med Internet Res 16: e36.

43. Fitzgerald SA, Macan Yadrich D, Werkowitch M, Piamjariyakul U, Smith CE

(2011) Creating patient and family education web sites: design and content of the home parenteral nutrition family caregivers web site. Comput Inform Nurs 29: 637-645.

44. Smith CE, Spaulding R, Piamjariyakul U, Werkowitch M, Yadrich DM, et al. (2015) mHealth Clinic Appointment PC Tablet: Implementation, Challenges and Solutions. J Mob Technol Med 4: 21-32.

45. Smith CE, Werkowitch M, Piamjariyakul U, Clements F, Schaag H, et al. (2009) Music Strategies to Improve Health and Fatigue. Music: Composition, Interpretation and Effects. Hauppauge, NOVA, Science Publishers, New York.

46. National Institutes of Health (2014) Turning discovery into health. Clear Communication: A NIH health literacy initiative.

47. Yadrich DM, Fitzgerald SA, Werkowitch M, Smith CE (2012) Creating patient and family education web sites: assuring accessibility and usability standards. Comput Inform Nurs 30: 46-54.

48. HON (2009) Confidentiality, proper attribution, justifiability and validity of the health information provided code. Health on the Net Foundation.

49. Infusion Nurses Society (2011) Infusion nursing standards of practice. J Infusion Nurs 34: S1-S109.

50. Smith CE (2009) Empirically Verified Music-Based Sleep Promotion and Infections Control Interventions for ICU. Intense Times 2009: 72-76.

51. CDC (2013) CDC-Guidelines Hand Hygiene.

52. Mermel LA, Allon M, Bouza E, Craven DE, Flynn P, et al. (2009) Clinical practice guidelines for the diagnosis and management of intravascular catheter-related infection: 2009 Update by the Infectious Diseases Society of America. Clin Infect Dis 49: 1-45.

53. Pihl-Lesnovska K, Hjortswang H, Ek AC, Frisman GH (2010) Patients' perspective of factors influencing quality of life while living with Crohn disease. Gastroenterol Nurs 33: 37-44.

54. Smith CE, Piamjariyakul U, Yadrich DM, Ross VM, Gajewski B, et al. (2010) Complex home care: part III-economic impact on family caregiver quality of life and patients' clinical outcomes. Nurs Econ 28: 393-399.

55. Stiles-Shields C, Keefer L (2015) Web-based interventions for ulcerative colitis and Crohn's disease: systematic review and future directions. Clin Exp Gastroenterol 8: 149-157.

56. Huang VW, Reich KM, Fedorak RN (2014) Distance management of inflammatory bowel disease: systematic review and meta-analysis. World J Gastroenterol 20: 829-842.

57. Smith CE, Leenerts MH, Gajewski BJ (2003) A systematically tested intervention for managing reactive depression. Nurs Res 52: 401-409.

58. Eisdorfer C, Czaja SJ, Loewenstein DA, Rubert MP, Arguelles S, et al. (2003) The effect of a family therapy and technology-based intervention on caregiver depression. Gerontologist 43: 521-531.

59. Houts PS, Nezu AM, Nezu CM, Bucher JA, Lipton A (1994) American College of Physician's Home Care Guide for Cancer: How to Care for Family and Friends at Home. Philadelphia: American College of Physicians.

60. Toseland R, Rivas R (2005) An introduction to group work practice. (5th edn) Needham Heights, MA: Allyn and Bacon.

61. Fortinsky KJ, Fournier MR, Benchimol EI (2012) Internet and electronic resources for inflammatory bowel disease: a primer for providers and patients. Inflamm Bowel Dis 18: 1156-1163.

62. Roberts J, Brown GB, Streiner D, Gafni A, Pallister R, et al. (1995) Problem solving counseling or phone-call support for outpatients with chronic illness: Effective for whom? Canadian J of Nurs Res 27: 111-137.

63. Smith CE (2009) Empirically Verified Music-Based Sleep Promotion and Infections Control Interventions for ICU. IntenseTimes.

64. Winkler MF, Hagan E, Wetle T, Smith C, Maillet JO, et al. (2010) An exploration of quality of life and the experience of living with home parenteral nutrition. J Parenter Enteral Nutr 34: 395-407.

65. Smith CE, Dauz E, Clements F, Werkowitch M, Whitman R (2009) Patient education combined in a music and habit-forming intervention for adherence to continuous positive airway (CPAP) prescribed for sleep apnea. Patient Educ Couns 74: 184-190.

66. Twiss E, Seaver J, McCaffrey R (2006) The effect of music listening on older adults undergoing cardiovascular surgery. Nurs Crit Care 11: 224-231.

67. Cepeda MS, Carr DB, Lau J, Alvarez H (2006) Music for pain relief. Cochrane Database Syst Rev- CD004843.

68. CDC (2008) Center for Disease Control & Prevention and the Kimberly-Clark Corporation. Assuring Healthy Caregivers, A Public Health Approach to Translating Research into Practice: The RE-AIM Framework. Neenah, WI: Kimberly-Clark Corporation.

69. Sternfeld B, Block C, Quesenberry CP Jr, Block TJ, Husson G, et al. (2009) Improving diet and physical activity with ALIVE: a worksite randomized trial. Am J Prev Med 36: 475-483.

70. National Institute on Aging (2009) Exercise & Physical Activity and Healthy Eating. National Council on the Aging Evidence-based Health Promotion Series.

71. NIA (2015) New National Institute on Aging Website: Go4Life.

72. Schulz DN, Kremers SP, Vandelanotte C, Van Adrichem MJ, Schneider F, et al. (2014) Effects of a web-based tailored multiple-lifestyle intervention for adults: A two-year randomized controlled trial comparing sequential and simultaneous delivery modes. J Med Internet Res 16: e26.

73. Fogg BJ (2003) Persuasive Technology: Using Computers to Change What We Think and Do. Morgan Kaufmann Publishers, San Francisco.

74. Guerini M, Stock O, Zancanaro M (2007) A taxonomy of strategies for multimodal persuasive message generation. Applied Artificial Intelligence 21: 99-136.

75. Helsel DL, Jakicic JM, Otto AD (2007) Comparison of techniques for self-monitoring eating and exercise behaviors on weight loss in a correspondence-based intervention. J Am Diet Assoc 107: 1807-1810.

76. Normand MP (2008) Increasing physical activity through self-monitoring, goal setting, and feedback. Behavioral Interventions 23: 227-236.

77. American Society for Parenteral and Enteral Nutrition (2004) Safe practices for PN.

78. Green BB, Anderson ML, Ralston JD, Catz S, Fishman PA, et al. (2011) Patient ability and willingness to participate in a web-based intervention to improve hypertension control. J Med Internet Res 13: e1.

79. Smith CE (2010) Internet Nursing Care: Patient Teaching in the Internet Age. NOVA Science. Hauppauge, NOVA Science Publishers, NY.

80. Houck S, Kilo C, Scott J (2013) Family Practice Management. Improving Patient Care Group Visits 101.

81. Piamjariyakul U, Smith CE, Ross VM, Yadrich D, Williams AR, et al. (2010) Part II: HPN Family Income, Insurance Premiums and Put-of-Pocket Expenses. Nurs Econ 28: 255-264.

82. Kim H, Spaulding R, Werkowitch M, Yadrich D, Piamjariyakul U, et al. (2014) Costs of multidisciplinary parenteral nutrition care provided at a distance via mobile tablets. J Parenter Enteral Nutr 38: 50S-7S.

83. Piamjariyakul U, Ross VM, Yadrich DM, Williams AR, Howard L, et al. (2010) Complex home care: Part I-Utilization and costs to families for health care services each year. Nurs Econ 28: 255-263.

# Permissions

The contributors of this book come from diverse backgrounds, making this book a truly international effort. This book will bring forth new frontiers with its revolutionizing research information and detailed analysis of the nascent developments around the world.

We would like to thank all the contributing authors for lending their expertise to make the book truly unique. They have played a crucial role in the development of this book. Without their invaluable contributions this book wouldn't have been possible. They have made vital efforts to compile up to date information on the varied aspects of this subject to make this book a valuable addition to the collection of many professionals and students.

This book was conceptualized with the vision of imparting up-to-date information and advanced data in this field. To ensure the same, a matchless editorial board was set up. Every individual on the board went through rigorous rounds of assessment to prove their worth. After which they invested a large part of their time researching and compiling the most relevant data for our readers.

The editorial board has been involved in producing this book since its inception. They have spent rigorous hours researching and exploring the diverse topics which have resulted in the successful publishing of this book. They have passed on their knowledge of decades through this book. To expedite this challenging task, the publisher supported the team at every step. A small team of assistant editors was also appointed to further simplify the editing procedure and attain best results for the readers.

Apart from the editorial board, the designing team has also invested a significant amount of their time in understanding the subject and creating the most relevant covers. They scrutinized every image to scout for the most suitable representation of the subject and create an appropriate cover for the book.

The publishing team has been an ardent support to the editorial, designing and production team. Their endless efforts to recruit the best for this project, has resulted in the accomplishment of this book. They are a veteran in the field of academics and their pool of knowledge is as vast as their experience in printing. Their expertise and guidance has proved useful at every step. Their uncompromising quality standards have made this book an exceptional effort. Their encouragement from time to time has been an inspiration for everyone.

The publisher and the editorial board hope that this book will prove to be a valuable piece of knowledge for researchers, students, practitioners and scholars across the globe.

# List of Contributors

**Minghui Li**
School of Engineering, University of Glasgow, Glasgow, United Kingdom

**Yilong Lu**
School of Electrical and Electronic Engineering, Nanyang Technological University, Singapore

**Bo He**
School of Information Science and Engineering, Ocean University of China, Qingdao, China

**Balamurali BT Nair, Esam AS Alzqhoul and Bernard J Guillemin**
Forensic and Biometrics Research Group (FaB), The University of Auckland, Auckland, New Zealand
Department of Electrical and Computer Engineering, The University of Auckland, Auckland, New Zealand

**Ambassa Joel Yves and Peng Hao**
School of Electronics and Information, Jiangsu University of Science and Technology, 2 Mengxi Road Jingkou Zhenjiang Jiangsu 212003, PR China

**Dhruv Sharma**
Software Engineer, PayPal, Chennai, India

**Kayiram Kavitha**
Researcher, Pune, India

**R Gururaj**
Department of CS and IS, BITS Pilani, Hyderabad Campus, Hyderabad, India

**Shuncong Zhong**
Laboratory of Optics, Terahertz and Non-destructive Testing & Evaluation, School of Mechanical Engineering and Automation, Fuzhou University, Fuzhou 350108, China
Fujian Key Laboratory of Medical Instrument and Pharmaceutical Technology, Fuzhou 350108, P.R. China

**Li Cao and Qiukun Zhang**
Laboratory of Optics, Terahertz and Non-destructive Testing & Evaluation, School of Mechanical Engineering and Automation, Fuzhou University, Fuzhou 350108, China

**Xinbin Fu**
Xiamen Special Equipment Inspection Institute, Xiamen 361000, P. R. China

**Maryam Banitalebi Dehkordi, Antonio Frisoli, Edoardo Sotgiu and Claudio Loconsole**
Perceptual Robotics laboratory (PERCRO), Scuola Superiore Sant' Anna, Pisa, Italy

**Mingsheng Gao, Jian Li, Wei Li and Ning Xu**
College of IoT Engineering, Hohai University, P.R. China

**G Amouzad Mahdiraji, Peyman Jahanshahi and FR Mahamd Adikan**
Integrated Lightwave Research Group, Faculty of Engineering, University of Malaya, 50603Kuala Lumpur, Malaysia

**Mostafa Ghomeishi**
Integrated Lightwave Research Group, Faculty of Engineering, University of Malaya, 50603Kuala Lumpur, Malaysia
Faculty of Science, Science and Research branch, Islamic Azad University, Tehran, 14778, Iran

**David A Bradley**
Department of Physics, University of Malaya, 50603 Kuala Lumpur, Malaysia
Department of Physics, University of Surrey, Guildford, GU2 7XH, UK

**Mohammed Ezz El Dien, Aliaa AA Youssif and Atef Zaki Ghalwash**
Department of Computer Science, University of Helwan, Helwan, Egypt

**Ching-Kun Chen and Chun-Liang Lin**
Department of Electrical Engineering, National Chung Hsing University, Taichung, Taiwan

**Shyan-Lung Lin**
Department of Automatic Control Engineering, Feng Chia University, Taichung, Taiwan

**Cheng-Tang Chiang**
Boson Technology Co., LTD, Taichung, Taiwan

**Nayak JA, Rambabu CH and Prasad VVKDV**
ECE Department, Gudlavalleru Engineering College, Gudlavalleru, AP, India

**Andrew Richardson, Jordan Rendall and Yongjun Lai**
Mechanical and Materials Engineering, Queens University, Kingston, Ontario, K7L 3N6, Canada

**Farrukh Shahzad**
Information and Computer Science, King Fahd University of Petroleum and Minerals, Dhahran, Saudi Arabia

**Sabzpoushan SH**
Biomedical Engineering Department, School of Electrical Engineering, Iran University of Science and Technology (IUST), Iran

**Maleki A and Miri F**
School of Computer Engineering, Iran University of Science and Technology (IUST), Iran

**Manar Arafat**
Department of Computer Science, An-Najah National University, Nablus, Palestine

**Sameer Bataineh**
Faculty of Computer and Information Technology, Jordan University of Science and Technology, Jordan

**Issa Khalil**
Qatar Foundation, Qatar

**Rajagopal D and Thilakavalli K**
Department of Computer Applications, K.S. Rangasamy College of Arts and Science, Tiruchengode, Namakkal, Dt-637 215, India

**Labrak L and Abouchi N**
Nanotechnology Institute of Lyon, Lyon, France

**Rammouz R**
Nanotechnology Institute of Lyon, Lyon, France
Applied Physics Laboratory, Beirut, Lebanon

**Constantin J, Zaatar Y and Zaouk D**
Applied Physics Laboratory, Beirut, Lebanon

**Ali Ajam, Ridwan Hossain, Nishat Tasnim, Luis Castanuela, Raul Ramos and Yoonsu Choi**
Department of Electrical Engineering, University of Texas Rio Grande Valley, Edinburg, Texas, 78539, USA

**Dongchul Kim**
Department of Computer Science, University of Texas Rio Grande Valley, Edinburg, Texas, 78539, USA

**Darwish IM and Elqafas SM**
Institute of Postgraduate Studies and Research, Selangor, Malaysia

Arab Academy for Science, Technology and Maritime Transport, Giza Governorate, Egypt

**Alkhafaji MK, Sahbudin RKZ, Ismail AB and Hashim SJ**
Department of Computer and Communication System Engineering, Center of Excellence of Wireless and Photonics Network (WIPNET), UPM, 43400-Serdang, Malaysia

**Debraj Basu, Gourab Sen Gupta, Giovanni Moretti and Xiang Gui**
School of Engineering and Advanced Technology, Massey University, New Zealand

**Sinha H and Sinha GR**
Shri Shankaracharya Technical Campus Bhilai, Chhattisgarh, India

**Meshram MR**
Department of Electronics and Telecommunication Engineering, Government Engineering College, Jagdalpur, Chhattisgarh, India

**Nnamani Kelvin N and Alumona TL**
Electronics and Computer Engineering, Nnamdi Azikiwe University, Awka, Anambra State, Nigeria

**Azizi A**
Department of Engineering, German University of Technology, Halban, Oman

**Mobki H**
Department of Mechanical Engineering, University of Tabriz, Iran

**Rezazadeh G**
Department of Mechanical Engineering, Urmia University, Iran

**Farhana Mustafa and Padma Lohiya**
Department of E&TC, D.Y. Patil College of Engineering, Akurdi, Pune, India

**Saritaş S, Kundakci M, Gürbulak B and Yildirim M**
Department of Physics, Ataturk University, 25250, Erzurum, Turkey

**Turgut E**
Department of Electrical and Energy, Aşkale Vocational School of Higher Education, Ataturk University, 25250, Erzurum, Turkey

**Pandesswaran C, Surender S and Karthik KV**
St. Joseph's College of Engineering, Chennai, India

**Ubolrat Piamjariyakul, Marilyn Werkowitch, Donna Macan Yadrich and Noreen Thompson**
School of Nursing, University of Kansas Medical Center, USA

**Carol E Smith**
School of Nursing, University of Kansas Medical Center, USA
School of Preventive Medicine and Public Health, University of Kansas Medical Center, USA

**Dedrick Hooper**
Center for Telemedicine and Telehealth, University of Kansas Medical Center, USA

**Eve-Lynn Nelson**
Center for Telemedicine and Telehealth, University of Kansas Medical Center, USA
Pediatrics and Telemedicine, University of Kansas Medical Center, USA
Institute for Community Engagement, University of Kansas Medical Center, USA

# Index